MORPHOANATOMICAL ATLAS OF GRASS LEAVES, CULMS, AND CARYOPSES

On the Cover:

A	**B**	**C**
D	**E**	**F**

MORPHOANATOMICAL ATLAS OF GRASS LEAVES, CULMS, AND CARYOPSES

Dhara Gandhi, PhD
Susy Albert, PhD

APPLE
ACADEMIC
PRESS

Apple Academic Press Inc.
4164 Lakeshore Road
Burlington ON L7L 1A4
Canada

Apple Academic Press Inc.
1265 Goldenrod Circle NE
Palm Bay, Florida 32905
USA

First issued in paperback 2021

© 2021 by Apple Academic Press, Inc.

Exclusive co-publishing with CRC Press, a Taylor & Francis Group

No claim to original U.S. Government works
ISBN-13: 978-1-77188-848-6 (hbk)
ISBN-13: 978-1-77463-963-4 (pbk)
ISBN-13: 978-0-42932-808-4 (eBook)

Library and Archives Canada Cataloguing in Publication

Title: Morphoanatomical atlas of grass leaves, culms, and caryopses / Dhara Gandhi, PhD, Susy Albert, PhD.

Names: Gandhi, Dhara, author. | Albert, Susy, author.

Description: Includes bibliographical references and index.

Identifiers: Canadiana (print) 20200224514 | Canadiana (ebook) 2020022459X | ISBN 9781771888486 (hardcover) | ISBN 9780429328084 (ebook)

Subjects: LCSH: Grasses—Identification.

Classification: LCC QK495.G74 G36 2020 | DDC 548/.92—dc23

Library of Congress Cataloging-in-Publication Data

Names: Gandhi, Dhara, author. | Albert, Susy, author.

Title: Morphoanatomical atlas of grass leaves, culms, and caryopses / Dhara Gandhi, Susy Albert.

Description: Palm Bay, Florida, USA : Apple Academic Press, 2020. | Includes bibliographical references and index. | Summary: "This new volume, Morphoanatomical Atlas of Grass Leaves, Culms, and Caryopses, features the studied anatomical details of different parts of 100 wild grass species and provides a comprehensive overview of existing knowledge. Each of the three sections of the volume (leaf grass, culm, and caryopses) discusses and illustrates the diagnostic histological features, along with statistical analyses on the quantitative and qualitative data. The descriptions of these grasses, particularly those growing in the grasslands of the Panchmahal and Dahod districts of India, are supplemented with microphotographs and keys for the taxa concentrate upon diagnostic characters above the rank of genus, which will be helpful for the easy identification of the grasses, even in their vegetative stages before flowering. The cluster analysis provides uses the statistical analysis program Minitab for each part on the basis of the diagnostic features. Due to the shortage of diagnostic morphological characteristics of grass leaves, culms, and caryopses, the study of complementary characters, like the anatomical features, has favored the clarification of taxonomic relationships between species and their correct identification. In this volume, readers will be able to easily identify the grass species based on the anatomical features described here. The volume will be of great interest both to grass specialists and as well as to generalists seeking state-of-the-art information on the diversity of grasses, the most ecologically and economically important of the families of flowering plants"-- Provided by publisher.

Identifiers: LCCN 2020016843 (print) | LCCN 2020016844 (ebook) | ISBN 9781771888486 (hardcover) | ISBN 9780429328084 (ebook)

Subjects: LCSH: Grasses--Identification.

Classification: LCC QK495.G74 G356 2020 (print) | LCC QK495.G74 (ebook) | DDC 584/.44--dc23

LC record available at https://lccn.loc.gov/2020016843

LC ebook record available at https://lccn.loc.gov/2020016844

Apple Academic Press also publishes its books in a variety of electronic formats. Some content that appears in print may not be available in electronic format. For information about Apple Academic Press products, visit our website at **www.appleacademicpress.com** and the CRC Press website at **www.crcpress.com**

About the Authors

Dhara Gandhi, PhD, is a researcher working in the Department of Botany of the Faculty of Science at The Maharaja Sayajirao University of Baroda in Vadodara, India. She has several years of experience in research. She worked as a project fellow in a research project funded by the Gujarat Forest Department, India. Her specialization is in plant taxonomy and plant anatomy, with her main area of research concentrated on the diversity and characterization of grasses. She has published several research papers. Dr. Gandhi completed her PhD in grass morphology and anatomy at the Department of Botany, Faculty of Science, The M. S. University of Baroda, India, under the guidance of Prof. Susy Albert.

Susy Albert, PhD, is a Professor in the Department of Botany, The M. S. University of Baroda, India. She has over 26 years of research and teaching experience and has published over 70 research papers in reputable journals. Four students have procured PhDs under her guidance, and she is presently guiding five students for their PhD programs. Her main area of research work is on fungal wood degradation and its biotechnological applications, mainly for biopulping in the paper industry, and bioremediation. In association with a leading industry in Gujarat, India, she is currently working on the degradation of plastic; the main aim of the project is to identify and characterize potential fungal strains that could be used for the degradation of municipal solid waste, mainly the plastic wastes. Other research areas of interests are on morphometrics and developmental anatomy of monocots mainly grasses and some dicots. Dr. Albert completed her PhD in developmental Plant Anatomy from the M S University of Baroda, Vadodara, Gujarat, India, under the guidance of renowned Plant Anatomist Prof. J. J. Shah.

Contents

Abbreviations

a	aleurone layer
BS	barrel shaped
c	cuticle
Cole	coleorhiza
Coleo	coleoptiles
CVB	central vascular strand
d	depletion layer
e	epidermis
e	endosperm
em	embryo
ep	epiblast
ES	elongated shaped
ic	intercellular space
III PVB	tertiary pheripheral vascular bundle
II IVB	secondary intercellular vascular bundle
I IVB	primary intercellular vascular bundle
IVB	internal vascular bundle
kc	kranz cell
m	mestome cells
me	mesocotyl
mx	metaxylem
p	Pericarp
pc	protoxylem cavity (lysigenous cavity)
PEPcase	phosphoenolpyruvate carboxylase
ph	phloem
pi	pth, p-parenchyma
pl	plumule
PVB	peripheral vascular bundles
px	protoxylem
r	radicle
rc	radial chlorenchyma cells
RS	rectangular shaped
S	scutellum
sc	sclerenchyma

si	silica
Sl	scutellum slit
SS	square shaped
st	stomata
St	starch grain

Preface

This book includes studied anatomical details of different parts of 100 grass species. Caryopses and culm anatomy with a wide diversity have not been considered in the other related books and articles. The book provides a comprehensive overview of existing knowledge on the grasses. It will be of great interest both for the grass specialist and also the generalist seeking state-of-the-art information on the diversity of grasses, the most ecologically and economically important of the families of flowering plants. The descriptions, supplemented with photographs and keys for the taxa, concentrate on the diagnostic characters above the rank of genus, which will be helpful in easy identification of the grasses even in their vegetative stages before the flowering. In general, the Poaceae taxonomy is based on the characteristic reproductive features such as the type of inflorescence, Structure of spikelets etc.

The book is unique with descriptions of the anatomy of 100 grass species supplemented with detailed microphotographs of each species, making it very easy to identify to the species level on the basis of diagnostic histological features. Leaf, culms, and caryopses anatomy has been studied in detail, and characteristic features are supplemented with microphotographs. A dichotomous key to the identification of the grass species on the basis of diagnostic anatomical features of each part has been separately prepared. Cluster analysis using Minitab has been prepared for each part on the basis of the diagnostic features.

The book will be great importance to students of taxonomy and can be used for identifying grass species. The book will be beneficial in workshops and training pertaining to grass taxonomy and identification. Other than this, field taxonomists studying monocot taxonomy will be benefitted.

Introduction to the Book

Anatomical studies have provided important diagnostic features that have been proven to be a good phylogenetic tool for systematics. The family Poaceae is one of the largest and most diverse families. Grasses are uniformly distributed on all continents and in all climatic zones. On the basis of morphological and anatomical characters, grasses can be identified by their diagnostic features.

In this book, 100 wild grasses that are growing in the grasslands of Panchmahal and Dahod districts have been studied anatomically and categorized on the basis of the leaf, culm, and caryopses anatomical features.

The book has been divided into three sections. The first section deals with characteristic features of leaf. Anatomical characters of leaf blade, leaf sheath, and ligule have been described in detail and supplemented with microphotographs. A dichotomous key to the identification of all the 100 species has been presented. Separately a key to identify the different species of a particular genera also been recorded and presented.

The second section covers culm anatomy with its detailed anatomical features and microphotographs, and based on the characters a dichotomous key has been prepared. Caryopses anatomy, covered in third section, also describes detailed anatomical features and is supplemented with photographs. Main characters considered are scutellum type, starch grains, pericap, type of embryo, type of epiblast, etc., on the basis of which a dichotomous key is also prepared. A separate genera key is also prepared for both culm and caryopses anatomy. Statistical analysis done for leaf, culm, and caryopses anatomy and cladogram was prepared based on the quantitative and qualitative data.

Leaf, an important part of plant, which structurally varies depending on the environment, has been studied anatomically. Different anatomical features included in the study are outline of lamina, presence or absence of ribs and furrows, keel structure, epidermis, mesophyll structure, position of vascular bundles in leaf blade, shapes of 1st order, 2nd order, 3rd order vascular bundles, relationship of vascular elements, structure of vascular bundles, characterization of vascular bundle sheath, sclerenchyma structure at abaxial, adaxial and at margin of leaf lamina, types of bulliform cells, colorless cells, presence and absence of mestome. Leaf sheath and ligule anatomy of the 100 species have been described. To describe anatomical features of leaf sheath of different species, the features have been adapted as per that of the leaf lamina. Features described by Ellis (1976) have been referred to for the description with the aid of the anatomical feature and a key and cladogram have been prepared.

Culm anatomy of the grass species has been studied and described with photographs. Both qualitative and quantitative features like shape in cross section, epidermis, hypodermis, ground tissue, type of sclerenchyma, I and II IVBs, III PVB, kranz anatomy, shape of kranz arc, shape of kranz cell, chloroplast type, mesotme and features like cross sectional area, number of vascular bundles, size of vascular bundles, size of metaxylem, number of kranz cells have been considered and depicted. Using quantitative and qualitative features, a cladogram has been prepared.

To describe Caryopses anatomy, features like course of vascularization, presence or absence of epiblast, presence or absence of cleft and arrangements of embryonic leaves in cross section, pericarp, angle of embryo with respect to anterior-posterior position of caryopses, coleoptiles, plumule, mesocotyl, coleorhiza, types of epiblast, scutellum, angle of vascularization, endosperm, type of starch grains, types of aleurone layer have been taken into consideration the study. Quantitative features considered are cross-section size of caryopses, seed coat thickness, aleurone cell size, occupied % of endosperm, thickness of endosperm, occupied % of embryo, thickness of embryo, number of starch grains per endospermic cell, etc. As for the other grass parts, a key on the basis of Caryopsis anatomical features has been prepared along with a separate identification key of different species belonging to the same genera.

The taxonomy of grasses is very complex due to the great morphological similarity between species and the high degree of overlap in the ranges of variation. Due to the shortage of diagnostic morphological characters, the study of complementary characters like the anatomical features has favored the clarification of taxonomic relationships between species and their correct identification. Readers, especially researchers, academicians, and students in the field of plant systematics, will be able to very easily identify the grass species based on the anatomical features of the vegetative parts and the caryopses described in this book.

Introduction

Grasses exists with thousands of species. They are important as ornamental, edible crops, lawn, etc. Flowers of grasses are showy, and the component parts are often much reduced. It is well established that foliar anatomy and epidermal micromorpology are important in the characterization of broad groups within the members of Poaceae. Anatomical studies of grasses have provided some important diagnostic features in coastal and inter-coastal parts (Metcalfe, 1960; Ogundipe and Olatunji, 1992; Keshavarzi et al., 2002). Metcalfe (1960) in his book, *Anatomy of the Monocotyledons, I. Gramineae*, and several papers dealing with the anatomy of grasses and anatomical peculiarities such as root, culm, and leaf anatomy and epidermal characteristics, etc., have been used in grass systematics at generic and species level. Duval-Jouve (1875) was the first to attempt the use of grass leaf anatomy revealed by trasverse sections in systematics. Subsequently many authors have worked on leaf anatomy of grasses, especially leaf lamina. Other parts like leaf sheath, ligule, culm, and caryopsis have been studied by only a few authors. Leaf micromorphological characterization has been considered to a great extent by many authors. Watson and Dallwitz (1992) published a detailed description of the leaf structure in the genera *Sehima* and *Thelepogon*, and described the diagnostic generic characters. Data on the anatomical characters have proved to be an important tool in grass identification and classification. Caryopses, the unique type of fruit in Poaceae, is also an important part of great taxonomic interest. Characteristic morphological and morphometric features of different tribes and genera have been studied and described by many authors. Reeder (1957) investigated the taxonomic significance of grass embryo and indicated that embryological characters have proven to be of great importance in grass systematics, particularly at the tribe and subfamily level. Wleaherwas (1930) studied the endosperms of *Zea* and *Coix*, and he summarized that the embryo of *Zea* and *Coix* is present only on one side by the endosperm but later on it was surrounded expect at the base. Endosperm shows at maturity a higher degree of differentiation than is ordinarily attributed to it. Rost (1973) studied caryopsis coat in mature caryopses of *Setaria lutescens*. Rost et al. (1990) studied caryopses anatomy of the *Briza maxima*. They observed that seed coat cuticle extends all around the caryopses, except in the placental pad region. Rost, also in his study, indicated the anatomy of the pericarp to be remarkably similar in all the grasses studied. However, detailed anatomical studied of caryopsis revealing the detailed description of the different parts of the caryopses are very meager. In the present studies, we attempt to describe the anatomical features of culm, leaf parts and caryopses, and keys to identification of different species on the basis of their characteristic features have been prepared.

List of Grass Species Studied

Sr. No.	Name of Plant
Tribe: Maydeae	
1.	*Chionachne koenigii*(Spr.) Thw.
2.	*Coix lachryma-jobi* L.
Tribe: Andropogoneae	
3.	*Andropogon pumilus* Roxb.
4.	*Apluda mutica*L.
5.	*Arthraxon lanceolatus* (Roxb.) Hochst.
6.	*Bothriochloa pertusa* (L.) A. Camus
7.	*Capillipedium huegelii* (Hack.) A. Camus
8.	*Chrysopogon fulvus* (Spreng.) Chiov.
9.	*Cymbopogon martini* (Roxb.) W. Waston
10.	*Dichanthium annulatum* (Forsk.) Stapf
11.	*Dichanthium caricosum* (L.) A.Camus
12.	*Hackelochloa granularis* (L.) O.
13.	*Heteropogon contortus var. genuinus sub var. typicus* Blatt. & McCann
14.	*Heteropogon contortus var. genuinus sub var. hispidissimus* Blatt. & McCann
15.	*Heteropogon ritchiei* (Hook.f.) Blatt. & McCann
16.	*Heteropogon triticeus* (R.Br.) Stapf ex Craib.
17.	*Imperata cylindrica* (L.) Raeusch.
18.	*Ischaemum indicus* (Houtt.) Merr.
19.	*Ischaemum molle* Hook.f.
20.	*Ischaemum pilosum* (Klein ex Wild.) Wt.
21.	*Ischaemum rugosum* Salib.
22.	*Iseilema laxum* Hack.
23.	*Ophiuros exaltatus* (L.) Kuntze
24.	*Rottboellia exaltata* L.
25.	*Saccharum spontaneum* L.
26.	*Sehima ischaemoides* Forsk.
27.	*Sehima nervosum* (Rottler) Stapf
28.	*Sehima sulcatum* (Hack.) A. Camus

Sr. No.	Name of Plant
29.	*Sorghum halepense* (L.) Pers.
30.	*Thelepogn elegans*Roth ex R. & S.
31.	*Themeda cymbaria* (Roxb.) Hack.
32.	*Themeda laxa* (Anderess.) A. Camus
33.	*Themeda triandra* Forssk.
34.	*Themeda quadrivalvis* (L.) Kuntze
35.	*Triplopogon ramosissimus* (Hack.) Bor
36.	*Vetivaria zinzanoides* (L.) Nash

Tribe: Paniceae

37.	*Alloteropsis cimicina* (L.) Stapf
38.	*Brachiaria distachya* (L.) Stapf
39.	*Brachiaria eruciformis* (J. E. Smith) Griseb
40.	*Brachiaria ramosa* (L.) Stapf
41.	*Brachiaria reptans* (L.) C.A.Gardner & C. E. Hubb
42.	*Cenchrus biflorus* Roxb.
43.	*Cenchrus ciliaris* L.
44.	*Cenchrus setigerus* Vahl
45.	*Digitaria ciliaris* (Retz.) Koeler
46.	*Digitaria granularis* (Trin. Ex Spr.) Henr.
47.	*Digitaria longiflora* (Retz.) Pers.
48.	*Digitaria stricta* Roth
49.	*Echinochloa colona* (L.) Link
50.	*Echinochloa crusgalli* (L.) Beauv.
51.	*Echinochloa stagnina* (Retz.) Beauv.
52.	*Eremopogon foveolatus* (Del.) Stapf
53.	*Eriochloa procera* (Retz.) C.E. Hubb.
54.	*Oplismenus burmannii* (Retz.) Beauv.
55.	*Oplismenus compositus* (L.) P.Beauv.
56.	*Panicum antidotale* Retz.
57.	*Panicum maximum*Jacq.
58.	*Panicum miliaceum*L.
59.	*Panicum trypheron* Schult
60.	*Paspalidium flavidum* (Retz.) A. Camus
61.	*Paspalidium geminatum* (Forssk.) Stapf
62.	*Paspalum scrobiculatum* L.

Sr. No.	Name of Plant
63.	*Pennisetum setosum* (Sw.) L. C. Rich
64.	*Setaria glauca* (L.) P. Beauv.
65.	*Setaria tomentosa* (Roxb.) Kunth
66.	*Setaria verticillata* (L.) P. Beauv.

Tribe: Isachneae

67.	*Isachne globosa* (Thumb.) O. Kuntze

Tribe: Aristideae

68.	*Aristida adscensionis* L.
69.	*Aristida funiculata* Trin. & Rupr.

Tribe: Perotideae

70.	*Perotis indica* (L.) O. Kuntze

Tribe: Chlorideae

71.	*Chloris barbata* Sw.
72.	*Choris montana* Roxb.
73.	*Chloris virgata* Sw.
74.	*Cynodon dactylon* (L.) Pers.
75.	*Melanocenchris jaequemontii* Jaub. & Spach
76.	*Oropetium villosulum* Stapf ex Bor.
77.	*Schoenefeldia gracilis* Kunth
78.	*Tetrapogon tenellus* (Roxb.) Chiov.
79.	*Tetrapogon villosus* Desf.

Tribe: Eragrosteae

80.	*Acrachne racemosa* (Heyne ex Roth) Ohwi
81.	*Dactyloctenium aegyptium* (L.) Beauv.
82.	*Dactyloctenium scindicus* Boiss.
83.	*Desmostachya bipinnata* (L.) Stapf
84.	*Dinebra retroflexa* (Vahl.) Panz.
85.	*Eleusine indica* (L.) Gaerth
86.	*Eragrostiella bachyphylla* (Stapf) Bor
87.	*Eragrostiella bifaria* (Vahl.) Bor
88.	*Eragrostis cilianensis* (All.) Link.
89.	*Eragrostis ciliaris* (L.) R. Br.
90.	*Eragrostis japonica* (Thumb.) Trin.
91.	*Eragrostis nutans* (Retz.) Nees & Steud
92.	*Eragrostis pilosa* (L.) P.Beauv.

Sr. No.	Name of Plant
93.	*Eragrostis tenella* (L.) Beauv.
94.	*Eragrostis tremula* (Lam.) Hochst. ex Steud.
95.	*Eragrostis unioloides* (Retz.) Nees.
96.	*Eragrostis viscosa* (Retz.) Trin.
Tribe: Sporoboleae	
97.	*Sporobolus coromandelianus* (Retz.) Kunth
98.	*Sporobolus diander* (Retz.) P. Beauv.
99.	*Sporobolus indicus*
Tribe: Zoysieae	
100.	*Tragrus biflorus* (Roxb.) Schult.

Materials and Methods

LEAF AND CULM ANATOMY

For anatomical study, grass species collected from the field were thoroughly washed, and the different parts were separated and fixed in FAA (Formalin: Acetic acid:Alcohol) for further processing.

A leaf in grass can be distinctly differentiated into blade, sheath, and ligule. Different parts of the leaf, viz. leaf blade, leaf sheath, and ligule (the joint of the leaf blade and leaf sheath) were prepared for the study of internal features. About 3–4 mm small pieces of samples were put in FAA for 48 h. Fixed samples were then washed 2–3 times with regular water and transferred to 3% commercial hydrofluoric acid for 24 hours or more, depending upon the nature of the material for removing silica. Excess amount of hydrofluoric acid was washed in running water for 12–18 h and transferred to 70% alcohol. The washed samples were dehydrated in tertiary butyl alcohol series (TBA series), embedded in paraffin wax (56°C–58°C). Paraffin blocks were cut and fixed on the wooden blocks and trimmed. 12–15 μm sections were obtained on rotary microtome, and slides were prepared for staining. Prepared slides were stained with Safranin and Fast Green and mounted in DPX (Di thalate xylene).

CARYOPSES ANATOMY

Mature spikelets were collected and stored in polythene bags and brought to the laboratory. Caryopses were carefully separated out from the glumes with the help of dissecting microscope (especially the smaller ones). The collected grass caryopses were immersed in distilled water for 15–20 days with a few drops of formalin to prevent microbial attack, depending upon the nature of the caryopses for the softening. Soaked caryopses were then fixed in FAA (Formalin: Acetic acid: Alcohol) for 48 h, washed 2–3 times with tap water and then dehydrated in tertiary butyl alcohol series (TBA series). Embedding was done with paraffin wax (56°C–58°C). Wax blocks were cut and fixed on the wooden blocks and trimmed. The trimmed paraffin blocks were exposed and immersed with exposed surface of sample in 50% glycerine for 20–25 days to soften the tissue. 12–15 μm sections were obtained on rotary microtome and stained with Toludene blue and mounted in DPX (Di phthalate xylene).

DICHOTOMUS KEY

Generally a dichotomus key is prepared to identify the different genera or species within particular genera. The key to and description of each of the 100 grass species of the Panchmahal and Dahod districts of Gujarat are treated in the present study, are artificial, and do not reflect phylogenetic relations. The keys do not necessarily emphasize features of evolutionary importance, nor do they have closely related plants appearing next to one another. All the keys to genera and species are dichotomous and numbered. It consists of a series of couplets, paired and contrasting statements (leads) that describe one or more featured of the plant. The couplet lead begins with a characteristic feature and then alternative states to guide users efficiently to identification. The paired statements are numbered. At each step in the key is a choice to be made as to which one of the statements best fits the unknown sample that is being attempted to identify. Each statement in the dichotomous key ends in the name of a genera and its species or in a number directing to a subsequent dichotomy. The present study represents the state of knowledge of Poaceae in the Panchmahal and Dahod districts and is an important addition to grass literature. A linked type of key was prepared. In dichotomous keys the statement of characters and the taxa to be identified are inextricably intermingled.

At the end of the species a description of a single genera key to identify them is also provided. A dichotomous key differentiating the 100 species on the basis of leaf anatomical characters has been represented at the end of the discussion.

CLUSTER ANALYSIS

The qualitative and quantitative anatomical data were used to perform a cluster analysis using Minitab software (Version 16) and a dendro-gram in order to explore the correspondence between anatomical and taxonomic features.

CHAPTER 1

Leaf Anatomy

ABSTRACT

The leaf is an important part of the plant, which structurally varies depending on the environment. A major characteristic feature of leaf lamina has been characterized according to Ellis (1976). These characters include outline of lamina, presence or absence of ribs and furrows, keel structure, epidermis, mesophyll structure, position of vascular bundles in leaf blade, shapes of 1st, 2nd, and 3rd order vascular bundles, relationship of vascular elements, structure of vascular bundles, characterization of vascular bundle sheath, sclerenchyma structure at abaxial, adaxial and at margin of leaf lamina, types of bulliform cells, colorless cells, and presence and absence of mestome.

In this chapter characteristic leaf anatomical features of 100 grass species have been described. Four different types of leaf blade outline are noted. Leaf lamina showed presence or absence of ribs and furrows. Vascular bundles are centrally located and arranged in the vertical plane of the blade. The arrangement of the chlorenchyma cells appears to be of fundamental taxonomic significance. Chlorenchyma cells are arranged radially, irregular or incomplete radially. Based on this feature the grass species can be divided into two major categories: panicoids and festucoids, with chlorenchyma cells arranged radially belonging to panicoids and chlorenchyma cells arranged irregularly then it belongs to festucoids. Vascular bundles show either regular arrangement or irregular arrangement. If they are regularly arranged then they are placed centrally or abaxial to the leaf blade. Other than this, few species showed double bundle sheath surrounding the vascular bundle, i.e., presence of mestome or inner bundle sheath.

Most of the leaf sheath anatomical characters resemble the leaf lamina. For assessing anatomical features of leaf sheath characteristic features described by Ellis (1976) has been are considered for in the present study. Based on the outline shape, leaf sheath can be categorized into three: "V" or "U" or round shaped. Ligule anatomy shows variations in different species. It showed either "V" shaped or round shaped. Anatomically, ligule has very simple structure. It has adaxial and abaxial epidermis; between them one or two or many layered mesophyll cells are present. Based on all above characteristic features, unique features of individual species have been identified and dichotomous key has been prepared and dendrogram prepared by using cluster analysis.

1.1   INTRODUCTION

Anatomical studies of grasses have provided some important diagnostic features in coastal and intercoastal parts (Metcalfe, 1960; Ogundipe and Olatunji, 1992; Keshavarzi et al., 2002).

Leaf anatomy has proved to be a good phylogenetic tool for grass systematic. Typically grass leaf consists of mainly two parts, a base, that is, leaf sheath and a blade, that is, leaf lamina, and the joint portion of these two parts is called ligule. Leaf sheath is generally amplexicaul which may be open to the base or closed and tubular for all of its length. The lamina is bifacial and possesses a series of vascular bundles.

Metcalfe's (1960) work on "Anatomy of the monocotyledons, I. Gramineae" and several papers have appeared, those dealing with the anatomy of grasses and anatomical peculiarities, such as root, culm, leaf anatomy, epidermal characteristics, and so on have been used in grass systematic at generic and specific level. He described the anatomy of 206 genera and 413 species based on his own observations. After that, in addition, the literature has been compiled with the author's own results thus bringing the total number of genera that have been treated up to 345. That means out of 898 genera, 345 genera were worked out by Metcalfe (1960).

Kaufman et al. (1956), Stebbins and Jain (1960), De Wet (1960), Stebbins and Shah (1960), Stebbins and Khush (1961), Brown (1958, 1960, 1974a, 1974b, 1975), Deshpande and Sarkar (1962), Chakravarty and Verma (1966), Picket-Heaps and Northcote (1966), Jauhar and Joshi (1967), Hitch and Sharman (1968, 1971), Blackman (1969), Paliwal (1969), Sangster and Parry (1969), Kok (1972), Crookston and Moss (1973), and Kuo et al. (1974) have contributed to the anatomy of the Gramineae.

Patel (1976) studied anatomy of root, culm, leaf, mature node, and leaf epidermal surfaces of 51 species. In the present study, 101 species have been studied and out of that 30 species have been worked out by Patel. The work includes few microphotographs and camera lucida drawings; detailed photographs for anatomical characterization were not given.

Kesar et al. (2003) studied morphology and anatomy of *Aristida stricta* and concluded that a reassessment of the disjunction in the species *A. stricta* and *Aristida begrichiana* is conspecific. Keshavarzi et al. (2007) studied the anatomy of five species of *Eremopyrum* and they concluded that on the basis of anatomical observations, these species can be differentiated/identified easily and classify *Eremopyrum confusum* as a separate subgroup of *Eremopyrum bonaepartis*. Abbasi et al. (2010) studied anatomical features of *Puccinellia dolicholepis* and compared with *Puccinellia bulbosa*.

The parts of the grass leaf are variously interpreted by morphologists. Bugnon (1921) concluded that the blade is equivalent to the leaf base of the dicotyledon and that the sheath of the grass leaf is a new structure, having no equivalence with dicotyledonous plants. Arber (1918, 1923) uphold the phyllode theory which states that the blade and sheath of the grass leaf correspond to the petiole and base of the dicotyledonous leaf. According to Hitchcock (1922), the sheath, petiole, ligule, and blade of the grass leaf are homologous to the leaf base, petiole, stipules, and blades, respectively, of the dicotyledonous leaf.

Grass leaf anatomy, as revealed by transverse sections, has been emphasized as a very fundamental character (Avdulov, 1931; Brown and Emery, 1957; Brown et al., 1957; Hubbard, 1934; Yakovlev, 1950; Tateoka, 1956, 1957; Stebbins, 1956; Reeder, 1957; Row and Reeder, 1957). Duval-Joyve (1857) was the first to attempt to use it for systematic studies. The character used was the position of the bands of bulliform cells in relation to the nerves. In Paniceae and Andropogoneae, bulliform cells are present over the tertiary nerves and these bulliform cells are present on both the epidermis of Paniceae. Schwendener (1890) discussed the nature of the two-bundle sheaths which surround the vascular bundle. The inner part, the mestome sheath which has characters of the endodermis, has been reported to be present in all grasses (Duval-Joyve, 1857), or present in some grass groups but absent from others (Schwendener, 1896). Apart from that, chlorenchymatous tissue shows various cellular arrangements, which are present external to parenchyma sheath.

Finally, Avdulov (1931) recognized two basic types of leaf anatomy in the grass family.

Type 1, which has a thick-walled mestome sheath, is connected by sclerenchyma to the upper and lower epidermises, a poorly developed parenchyma sheath, and irregularly arranged chlorenchyma. This type is mostly found in Festuceae, Agrostideae, Hordeae, Aveneae, and Phalarideae (Prat, 1936). This type of anatomy is called "festucoid" type of leaf anatomy.

Type 2: This is characterized by the large size of the parenchyma sheath cells which separate the xylem from the sclerenchyma next to the upper epidermis and by the radial arrangement of the chlorenchyma cells. This type of anatomy is mostly found in Paniceae, Andropogoneae, Maydeae, Chlorideae, and Zoysieae. Prat (1936) represented this type of anatomy as "panicoid" type of leaf anatomy.

These two basic types of anatomy have been accepted by almost all (Vickery, 1935; Prat, 1937; Burbridge, 1946; Tateoka, 1956, 1957), but Stebbins (1956) illustrated four types of anatomy. He added "bambusoid" and "chloridoid" to the "panicoid" and "festucoid" of Avdulov and Prat.

Some other anatomical characters have been used in phylogenetic studies are as follows: kranz sheath, patterns in vascular bundles (Columbus, 1996; Cerros-Tlatilpa, 1999), sclerenchyma patterns (Siqueiros Delgado and Herrera Arrieta, 1996), position and form of the chloroplasts (Columbus, 1996), and shape and position of bulliform cells (Columbus, 1996; Cerros-Tlatipla, 1999).

Based on the characteristics of the parenchyma sheath, Brown (1958) proposed six types of leaves, whereas Nikolaevskic (1972) shows that only three types of structure are present in Poaceae. But Clayton and Renvoize (1986) give more realistic subtypes of leaves structure and also introduced a new (kranz) crown type of structure.

Hsu (1965) in his extensive survey of the anthecial epidermis of the Paniceae recognized four basic patterns, two of which are included in *Brachiaria* (sensu Stapf). His findings led him to question spikelet orientation as a primary generic delimator in *Brachiaria*. However, his survey of *Brachiaria* was restricted to only six species. Brown (1977) conducted a comparative study of foliar vascular anatomy and photosynthetic pathway in the Paniceae. He recognized a group, including Brachiaria, with a kranz sheath derived from a parenchyma sheath and exhibiting the C_4, PEP-carboxykinase photosynthetic pathway.

Anatomy is related to biochemistry and C_4-type grasses usually have kranz anatomy. Radiate mesophyll is a characteristic of C_4 grasses, but C_4 photosynthesis has multiple independent origins in the Poaceae (Kellogg, 2000; Giussani et al., 2001).

Grass ligules are small outgrowths at the junction of the leaf sheath and the blade. Their morphology and anatomy may be important for identification of some critical grass species (Zuloaga et al., 1989; Judziewicz and Clark, 1993; Fuente and Ortunez, 2001). Study of the ligule anatomy provides information on the evolutionary relationships among certain species (Chaffey, 1984), contributing to a more natural taxonomic system. Ligule anatomy is a relatively constant feature; it is also important for identifying individuals living in nonoptimal environments (Neumann, 1938). Chaffey (1994) demonstrated the functional anatomy of ligules of 49 grass species from 10 tribes.

1.2   ANATOMICAL FEATURES

General Monocot Leaf Anatomy

Monocotyledonous leaves are commonly isobilateral with adaxial and abaxial surface of leaf having similar anatomical arrangements of leaf tissue and mesophyll being hardly differentiated into palisade and spongy parenchyma cells.

Epidermis

The epidermis is generally uniseriate or sometimes multiseriate and composed of more or less oval cells with cuticle on outer walls. The upper epidermis generally possesses motor cells or bulliform cells. These cells help in rolling down of leaf to avoid excessive loss of water when the wind velocity is high. A major characteristic feature in grass leaves are silica deposition in various morphological forms. Stomata are evenly distributed on both the surfaces of epidermal layer.

Mesophyll

Isobilateral leaf mesophyll cells do not show differentiation into palisade and spongy parenchyma and thus the tissues are composed of mainly isodiametric cells with intercellular spaces (mostly spongy type). Chloroplasts are abundantly present in these cells.

Conducting Strand

Vascular bundles are collateral and closed type as found in dorsiventral leaf, but they are arranged in parallel series. These vascular bundles are of monomorphic or dimorphic (smaller and larger size) in nature. When dimorphic, the larger ones are fewer in number. Generally, the vascular bundle contains xylem on the upper side and phloem on the lower side surrounded by parenchymatous cells known as bundle sheath cells. The bundle sheath cells contain plastids with starch grains. These plastids present in the bundle sheath cells lack grana or poor in grana but possess the machinery to fix the carbon dioxide with the help of enzyme ribulose-1,5-bisphosphate carboxylase/oxygenase and thus consequently forms starch. Absence of grana makes plastid nongreen and eventually the bundle sheath colorless. On the other hand, the mesophyll cells also have plastids with well-developed grana that generates high amount of ATP and NADPH (assimilatory power) and have special enzyme phosphoenolpyruvate carboxylase (PEPcase) that can capture the CO_2 very efficiently from the atmosphere and send trapped CO_2 and assimilatory power to bundle sheath cells for fixation.

Grass Leaf Anatomy

As the leaves have a parallel venation, grass leaves can be divided into two major parts, that is, midrib and leaf blade. The leaf-blade region which comprises two epidermis with mesophyll tissue in between them within which vascular bundles are arranged parallely. In midrib region, two epidermis and vascular bundles are present in parenchymatous and chlorenchymatous cells. Terminologies in the present study used to describe the leaves of different grass species are in accordance to Ellis (1976). Only the types observed in the present study have been included from the description key proposed by Ellis (1976).

Outline of the Lamina in Transverse Section

The lamina of most grass leaves in transverse section is flattened, and it may or may not have a conspicuous keel or midrib. Leaves of some grasses have well-defined midribs. When a midrib is present, it generally projects more prominently from the abaxial than from the adaxial surface and then constitutes a keel. In an individual leaf, the midrib or keel usually becomes progressively larger toward the base of the lamina. When a midrib or keel is described as being conspicuous, moderately conspicuous, or not conspicuous, it refers to sections taken in the halfway position that was adopted as a standard. The portion of the lamina on either side of the midrib is relatively thin, and the lamina frequently becomes still thinner toward the margins or may again be thicker and they are frequently well supported by sclerenchyma.

The outline of open leaf lamina can be mainly categorized into four types, based on their shape: (I) expanded, (II) V shaped, (III) U shaped, and (IV) inrolled. The three categories have been further categorized by taking into consideration features like angle of arms, shape of arms, symmetry of two halves, nature of folding, degree of folding, and so on.

Outline of Lamina for Open Leaves

Categorizations of the outline are as follows:

I. Expanded: Lamina is flattened, a line with connecting both margins and the median bundle straight:

The nature of leaf blade varies and it may be

1. Flat, even, or straight (Fig. 1.1A and B)
2. Undulating gently (Fig. 1.1C)
3. Distinctly wavy or undulating
   (a) Corrugated leaf, that is, waves rounded (Fig. 1.1D)
   (b) Pleated leaf, that is, waves pointed or angled (Fig. 1.1E)

II. V-shaped: Two halves of the lamina folded toward each other on either side of the midrib, that is, a line connecting both margins with the median bundle angled at the median bundle:

Further, it has been categorized on the basis of the following three characteristics:

A. Angel formed by the two arms of the lamina at the midrib
   1. Narrow; closed V, that is, less than 45° to each other (Fig. 1.1F)
   2. Standard V, that is, between 45° and 90° to each other (Fig. 1.1G)
   3. Broad, open V, that is, more than 90° to each other (Fig. 1.1H)
   4. Wide, very open V, that is, almost 180°; presence of projecting keel gives appearance of V-shaped (Fig. 1.1I)

B. Shape of arms of the lamina:
   1. Straight (Fig. 1.1J)
   2. Outwardly bowed; concave (Fig. 1.1)
   3. Outwardly curving; convex (Fig. 1.1K)
   4. Outline heart shaped (Fig. 1.1L)
   5. Distinctly wavy or undulating with rounded wave, that is, corrugated leaf (Fig. 1.1M)
   6. Irregularly wavy or bent (Fig. 1.1N)

C. Symmetry of two halves of the lamina
   1. Symmetrical on either side of the median vascular bundle (Fig. 1.1J)
   2. Two halves asymmetrical on either side of the median vascular bundle (Fig. 1.1N)

III. U shaped: Two halves of the lamina curved upwards on either side of the midrib or with a prominent rounded keel, that is, no definite angle formed with the midrib.

Shape of the U formed is tall, narrow, that is, vertically elongated (Fig. 1.1O).

IV. Inrolled: Lamina rolled inwards the adaxial surface.
Nature of rolling may be

1. Convolute: Inrolled from one margin only; margins wrapped around each other (Fig. 1.2A)
2. Involute: Inrolled from both margins (Fig. 1.2B)

Ribs and Furrows on the Leaf Surface

The adaxial surface of the leaf may be either flat or longitudinally ribbed, the height of ribs varying in different species. The ribs are separated from one another by wide or narrow grooves. Ribs are generally characteristic of and more fully developed on the adaxial than the abaxial surface, but abaxial ribs also occur, and in some grasses, the adaxial and abaxial grooves opposite each other, the leaf in transverse section has the appearance of a number of almost circular areas each enclosed by a pair of opposed ribs and having a vascular bundle at the center, adjacent circles being connected to one another by much narrower portions of the lamina where the grooves are situated.

The characterizations of ribs and furrows present on the leaf surface have been categorized into the different types. Characteristic features included are the presence and absence of furrows and further the depth of the ribs and furrows. Accordingly, the studied species could be categorized into the following types:

Two major categories based on the presence or absences are

I. No ribs and furrows or ribs and furrows are absent
II. Ribs and furrows present. This is further categorized on the basis of its presence on adaxial/abaxial surface or on both surfaces. Also, further, it is divided on the basis of the depth of the furrows. Depth of the furrows has been considered in comparison to the leaf thickness, that is, the depth of the larger ribs in the central region of the lamina between the margin and the median vascular bundle regarded as leaf thickness.

The shape of the ribs showed variation so the shape of the ribs as seen in transverse section has been included. Variations have been observed in the distribution of the different ribs in association with the vascular bundles as given below:

I.   Ribs are distributed over all the vascular bundles

II. Ribs are not being associated necessarily with vascular bundles. The elevation of the ribs may be similar over all the vascular bundles or they may be of varying size.

Taking into consideration the abovementioned features, three major categories are further divided as follows:

I. No ribs or furrows present on either surface, that is, lamina surface straight or only slightly undulating with no regular pattern associated with the vascular bundles (Fig. 1.2C)

II. Adaxial ribs and furrows present
Further categorized on basis of depth, shape, and distribution of furrows:

A: Depth of adaxial furrows in comparison to the leaf thickness, that is, the depth of the larger ribs in the central region of the lamina between the margin and the median vascular bundle regarded as leaf thickness:

1.   Slight, shallow furrows, that is, less than a quarter of the leaf thickness (Fig. 1.2D)
2.   Medium furrows, that is, a quarter to one-half of the leaf thicknesses (Fig. 1.2E)
3.   Deep furrow, that is, more than one-half of the leaf thickness (Fig. 1.2F)

B: Shape of adaxial furrows

1.   Wide, open furrow, that is, obtuse angle formed by furrow sides at base (Fig. 1.2G)
2.   Narrow furrow, that is, sides of furrow almost vertical:
     (a)   In form of cleft (Fig. 1.2H)
     (b)   Base fairly broad but sides steep (Fig. 1.2I)

C: Distribution of adaxial furrows (important for acicular leaves)
1.   Furrows between all vascular bundles (Fig. 1.2H)
2.   Furrows between first and second-order vascular bundles, that is, present over third-order vascular bundles (Fig. 1.2I)
3.   Deepest furrow on either side of the median vascular bundle (Fig. 1.2J)

D: Shape of adaxial ribs as seen in transection

1. Rounded with obtuse ribs and apex rounded:
    (a)   Situated over the vascular bundles (Fig. 1.2K)
2. Flat-topped with square ribs and apex flattened
    (a)   Sides rounded with flat top (Fig. 1.2I)
3. Triangular ribs with apex pointed:
    (a)   One vascular bundle in each rib (Fig. 1.2L)
    (b)   Grooved at apex (Fig. 1.2M)
4. Massive ribs, that is, very large usually with bases narrow in relation
    to upper parts of rib: rounded apices (Fig. 1.2J)

E: Distribution of different ribs in association with the vascular bundles:
1. Ribs present over all vascular bundles
    (a)   Similar ribs over all vascular bundles (Fig. 1.2H)
    (b)   Ribs over first and second-order vascular bundles larger than
    those over the third-order vascular bundles (Fig. 1.2F)
2. Ribs not associated with all vascular bundles
    (a)   Ribs present only over first and second-order vascular bundles
    (Fig. 1.2E)

III. Abaxial ribs and furrows present or absent

A: Abaxial rib absent:
1. Surface smooth or with few undulations not regularly associated
    with the vascular bundles (Fig. 1.2C)
2. Grooves present on abaxial surface
    (a)   Opposite larger vascular bundles (Fig. 1.2N)
    (b)   Between vascular bundles; often shallow (Fig. 1.2O)
    (c)   Groove on either side of midrib (Fig. 1.2R)

B: Ribs present on abaxial surface
1. Size of abaxial ribs
    (a)   Taller than the adaxial ribs (Fig. 1.2P)
    (b)   Same size as adaxial ribs, that is, section resembles a string of
    beads; moniliform (Fig. 1.2Q)
    (c)   Smaller than adaxial ribs (Fig. 1.2H)
    (d)   Slight undulation associated with vascular bundles (Fig. 1.2G)
2. Distribution of abaxial ribs

    (a)   Present opposite all vascular bundles (Fig. 1.2G and S)
3. Composition of abaxial ribs

    (a)   Composed of girder or strand of sclerenchyma in contact with
    the epidermis (Fig. 1.2S and T)

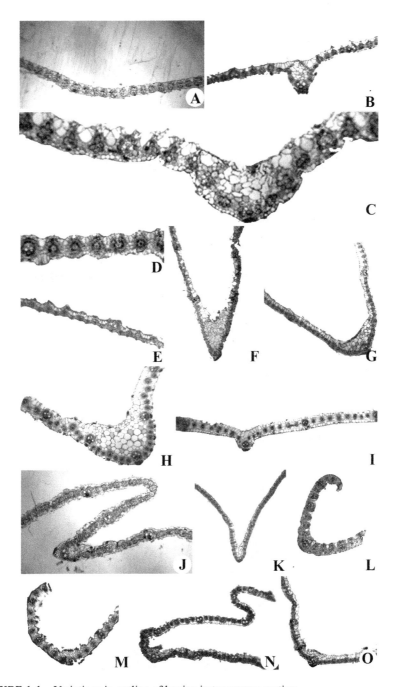

FIGURE 1.1   Variations in outline of lamina in transverse section.

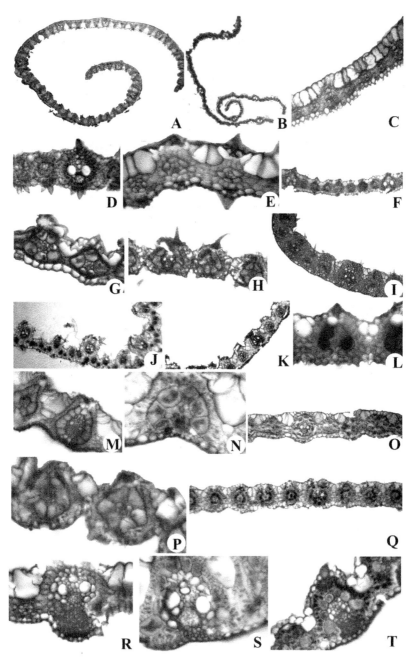

FIGURE 1.2   Characteristic features of ribs and furrows of lamina.

Median Vascular Bundle, Midrib, and Keel Structure

Grass leaves show difference in their midrib shape and arrangements of first-order vascular bundles/median vascular bundles. Numbers of vascular bundles are also different. Keel is longitudinal rib present on leaf. When parenchyma or bulliform cells are developed in association with the median bundle or bundles, the whole structure, often incorporating many bundles, is termed a keel. Keel and midrib are distinguished by the presence or absence of parenchyma associated with the median vascular bundle. It is termed a "midrib" if the median bundle is solitary, structurally distinct and without associated parenchyma and a "keel" if parenchyma or bulliform cells are associated with the median bundles. Thus, one or many vascular bundles may be incorporated in a keel unless there is a marked thickening in relation to the rest of the lamina. The number of vascular bundles in a keel is in general, roughly proportional to the size of the keel. The number of vascular bundles present in the keel shows difference from one to many, so this feature has been included for observation and categorization. Shape of keel in transverse section, sclerenchyma, and air lacunae associated with the keel is some of the other features considered. Accordingly, the classification is as follows:

I. Median bundle only present. The entire or more than half the blade width sectioned and no structurally distinct midrib is distinguishable (Fig. 1.3A)

II. Midrib bundle only present and is distinguishable from other first-order vascular bundles; no associated parenchyma developed, that is, keel inconspicuous:

    A. Projection of midrib varies
        1. Projects abaxially
            a. Sclerenchyma causes projection (Fig. 1.3B)
            b. Projection due to position or size of bundle (Fig. 1.3C)
        2. Projection inconspicuous or marked; not projecting abaxially only:
            a. Blade expanded or slightly inrolled (Fig. 1.3D)
            b. Lamina V shaped but no parenchyma at angle (Fig. 1.3E)
            c. Acicular or terete leaves; parenchyma may be developed but no specifically in association with the midrib (Fig. 1.3F)

III. Keel developed; colorless parenchyma and/or bulliform cells associated with the median bundle; sclerenchyma development may also occur

A: Number of vascular bundles comprising keel; in V-shaped leaves, if there is no thickening in the keel area, the median bundle is considered to constitute the keel

1. One vascular bundle comprising the keel
    a. Parenchyma of small, rounded cells surrounding or immediately adaxial to the median bundle (Fig. 1.3G)
    b. Bulliform cells in adaxial epidermis above-median bundle which makes the
        •   Leaf expanded, flat or slightly inrolled (Fig. 1.3D)
        •   Leaf V shaped with
            o   Bulliform cells penetrating into leaf on either side of median bundle forming an inverted "V" (Fig. 1.3D)
            o   Bulliform cells elongated and arranged in U above median bundle (Fig. 1.3H)
            o   Elongated bulliform cells in "U" as well as layers of elongated parenchyma cells (Fig. 1.3I)
2. Three vascular bundles present in keel
    a. Adaxial groove present (Fig. 1.3J)
    b. Adaxial groove absent (Fig. 1.3K)
3. Many vascular bundles present in keel with all vascular bundles abaxially arranged. The median bundle may or may not be prominent.
    •   Median bundle indistinguishable from other first-order bundles:
        o   One first-order bundle and smaller bundles comprise keel (Fig. 1.3L)
        o   Three first-order bundles and other smaller bundles comprise keel (Fig. 1.3O)
    •   Median bundle structurally distinct (size, bundle sheath) from other first-order bundles
        o   Large median bundle and other third and second-order bundles only in keel (Fig. 1.3G)
        o   Normal first-order bundles present in addition to the median bundle
B. Shape of keel in transverse section. Three categories:
1. Not really distinct from leaf outline (Fig. 1.3A)
2. V-shaped, pointed, or inverted triangular keel:

        a.   Leaf V-shaped with V-shaped keel (Fig. 1.3E)

        b.   Leaf V-shaped with inverted triangle-shaped keel (Fig. 1.3L)

        c.   Leaf expanded with inverted triangular or pointed keel (Fig. 1.3I)

   3.   Rounded or semicircular keel; leaf not necessarily U shaped:

        a.   U-shaped keel; slightly thicker than rest of lamina (Fig. 1.3H)

        b.   U-shaped keel; much thicker than rest of lamina (Fig. 1.3J)

        c.   Rounded or semicircular keel, that is, adaxial side of the keel flat (Fig. 1.112O)

        d.   Rounded keel with raised, flattened adaxial side (Fig. 1.3M)

        e.   Rounded keel with single, central adaxial groove (Fig. 1.3I)

C. Sclerenchyma associated with the keel on adaxial and abaxial surface:

   1.   Adaxial sclerenchyma:

        a.   Strands in subepidermal layers (Fig. 1.3O)

        b.   Strands fused forming a hypodermal band (Fig. 1.3G)

   2.   Abaxial sclerenchyma:

        a.   Most bundles with abaxial strands (Fig. 1.3J)

        b.   Most bundles with abaxial girders (Fig. 1.3N)

        c.   Central or solitary bundle with anchor-shaped girder (Fig. 1.3P)

D: Air spaces or lacunae incorporated in the keel; definite lacunae present or enlarged parenchyma cells of keel breaking down: one air space in keel (Fig. 1.3Q)

Vascular Bundles

The vascular bundles, apart from those in the midrib, are generally arranged in a single row, and they are embedded more or less in the middle of the mesophyll, although they may be closer to the abaxial or adaxial surface in different species. The vascular bundles are of different sizes, and it is usual to find bundles of different sizes or orders not only in different species but also within a single leaf. The bundles are arranged in such a way that those of different orders alternate with one another. Vascular bundles are of first, second, and third-order arrangement varying.

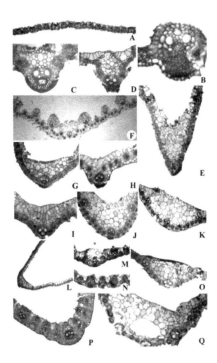

FIGURE 1.3 Variations in structure of keel.

Arrangement of Vascular Bundles

I. Number: Number of first-order vascular bundles varies. Accordingly, total number of first-order bundles in half the width of the lamina; includes the median vascular bundle and the variation can be categorized as follows:

 A. One first-order vascular bundle in blade section, that is, median bundle
 B. Two first-order vascular bundles in half lamina, that is, 3 in entire blade
 C. Three first-order vascular bundles in half lamina, that is, 5 in entire blade
 D. Four first-order vascular bundles in half lamina, that is, 7 in entire blade
 E. Five first-order vascular bundles in half lamina, that is, 9 in entire blade
 F. Six first-order vascular bundles in half lamina, that is, 11 in entire blade
 G. Seven first-order vascular bundles in half lamina, that is, 13 in entire blade
 H. Eight first-order vascular bundles in half lamina, that is, 15 in entire blade
 I. Nine first-order vascular bundles in half lamina, that is, 17 in entire blade
 J. Ten or more first-order vascular bundles in half lamina

II. Arrangement and alternation of different orders of vascular bundles along the width of blade vary. The part of lamina between the margin and the midrib has been observed to determine the arrangement and alternation of the different order bundles.

    A.   Variable from median bundle to margin; no regular pattern discernible, that is, there was no regular pattern in which different-order vascular bundles were arranged.

        1.   Progressively fewer first and more third-order bundles toward the margin

        2.   Progressively more first and fewer second and third-order bundles toward the margin

        3.   Arrangement near margin differs from remainder

    B.   Regular arrangement from median bundle to margin, that is, the different-order vascular bundles were arranged in a specific pattern as described below:

        1.   Number of third-order bundles between consecutive second and first-order bundles at a position halfway to margin.

            a.   Third-order bundles between consecutive larger bundles absent

            b.   One third-order bundle between consecutive larger bundles

            c.   Two third-order bundles between consecutive larger bundles

            d.   Three third-order bundles between consecutive larger bundles

            e.   Four third-order bundles between consecutive larger bundles

            f.   Five third-order bundles between consecutive larger bundles

            g.   Six third-order bundles between consecutive larger bundles

            h.   Seven third-order bundles between consecutive larger bundles

            i.   Eight third-order bundles between consecutive larger bundles

            j.   Nine third-order bundles between consecutive larger bundles

            k.   Ten or more third-order bundles between consecutive larger bundles

        2.   Number of second-order bundles between consecutive first-order bundles

            a.   Second-order bundles between consecutive first-order bundles absent

            b.   One second-order bundle between consecutive first-order bundles

    c.   Two second-order bundles between consecutive first-order bundles

    d.   Three second-order bundles between consecutive first-order bundles

    e.   Four second-order bundles between consecutive first-order bundles

    f.   Five or more second-order bundles between consecutive first-order bundles

III. Position of vascular bundles within the leaf blade. The position of the vascular bundles may be oriented in the center, adaxial or abaxial side of the leaf blade in transaction.

    A. Same level of positioning for all bundles of different orders:

       1.   All bundles situated in the center of the blade (Fig. 1.4A)

       2.   All bundles situated closer to the abaxial surface (Fig. 1.4B)

       3.   All bundles situated closer to the adaxial surface (Fig. 1.4C)

    B. Different levels of positioning for bundles of different orders:

       1.   First-order bundles central and third-order bundles abaxial (Fig. 1.4D)

       2.   Third-order bundles central and first-order bundles displaced adaxially in ribs (Fig. 1.4E)

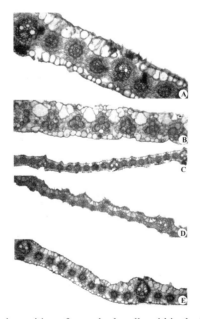

FIGURE 1.4   Variations in position of vascular bundle within the leaf blade.

All the above features were mainly based on the morphological appearance of the tissue system. Further characterization of the leaf blade has been done on the basis of the anatomical features of the different layers of tissue.

Epidermis

Epidermis is the outermost layer which is covered by cuticle. Adaxial and abaxial show different shaped epidermal cells. In between epidermal cells, silica and stomata are also seen. Grass leaves also have specialized cells which are known as bulliform cells. They are large, bubble-shaped cells occurring in groups. These cells are present on the adaxial surface of the leaf. They are generally present near the midvein. These cells are large, empty, and colorless.

Description of the epidermal layer in the present study includes all the different types of cells, that is, epidermis, bulliform, and prickles present in this layer. Characteristic features of bulliform cells have been considered followed by the epidermal cells. Adaxial and abaxial epidermal layers and their associated cells have been categorized separately. Abaxial surface of the epidermis is homogenous and made of the same type of cells.

Adaxial and abaxial epidermis of the leaf; cells of the upper and lower epidermis considered separately and the adaxial epidermis is fully described before commencing with the abaxial epidermis:

  I.   Cells of adaxial epidermis
  II. Cells of abaxial epidermis

Bulliform Cells

Colorless cells forming part of the epidermis but differing from the remaining epidermal cells by being markedly larger and inflated. Characteristically, they are large and are restricted to groups often intimately associated with colorless cells but all gradations are found. Thus, small solitary bulliform cells may be present or the epidermis may be comprised primarily of large, rounded, and inflated cells.

Mainly bulliform cells present on the adaxial surface of leaf blade, but few species have bulliform cells on abaxial surface also. Nine major categories could be identified and are as follows:

I. Bulliform cells are absent in epidermis: Consists of small cells; often thickened epidermal cells (Fig. 1.5A)

II. Most part of epidermis made up of bulliform cells: Not arranged in regular groups; epidermal cells rounded and inflated the same size or slightly larger than the cells of the bundle sheath:

    A.   Almost all the epidermal cells are inflated (Fig. 1.5B)

    B.   Isolated large and inflated cells interrupted with irregular small groups of cells. The irregular group of cells may be located.

        1.   On sides of deep furrows (Fig. 1.5C)

        2.   Not in furrows but on the flat surface. This feature is noted usually in leaves with flat surfaces (Fig. 1.5D)

III. Groups of bulliform cells present in epidermis which are well defined and regular; distinct from normal epidermal cells. They may be

    A.   Small bulliform cells; not conspicuously larger than the normal epidermal cells which is

        1.   Not associated with colorless cells

        2.   Closely associated with colorless cells; indistinct (Fig. 1.5E)

    B.   Extensive groups of large, inflated bulliform cells extending over one or many vascular bundles. It may or may not be associated with colorless parenchyma.

        1.   Not associated with colorless parenchyma. Distribution and size of the cells vary.

            a.   Distribution of extensive bulliform groups:

               •   Present throughout the epidermis; may be slightly reduced opposite the larger bundles (Fig. 1.5F)

               •   Present in most of the epidermis but not present opposite the first-order bundles and usually reduced over the second-order bundles (Fig. 1.5G)

            b.   Size of constituent cells; taken in areas opposite the third-order bundles

               •   Occupy less than one-fourth of the leaf thickness (Fig. 1.5H)

               •   Occupy more than one-fourth of the leaf thickness

        2.   Associated with colorless parenchyma cells together with which it forms a zone of colorless cells above or below the chlorenchyma; may be disrupted over the first-order bundles:

            a.   Colorless zone comprises more than half of the leaf thickness

            b.   Colorless zone comprises less than half of the leaf thickness

IV. Restricted groups of large, inflated bulliform cells (*Zea* type) having parallel-sided cells, that is, inner tangential wall same length or only slightly shorter than the outer tangential wall:

    A.   Bulliform cells projecting above the level of the epidermis. Few bulliform cells form cushion cells of macrohairs (Fig. 1.5I)

    B.   Bulliform cells in level with the general epidermal surface (Fig. 1.5J)

V. Restricted groups of tall and narrow bulliform cells which are not inflated and having parallel-sided cells or with outer tangential wall slightly shorter than the inner wall. The lateral walls are long and straight (Fig. 1.5K)

VI. Fan-shaped groups bulliform cells each cell of the group is with outer tangential wall shorter than inner tangential wall; median cell of group appreciably larger than the remaining cells of the group:

    A.   Bulliform cells situated at bases of furrows (Fig. 1.5L)

    B.   Not found at bases of furrows as it is present in leaves without furrows; some cells parallel-sided (Fig. 1.5M)

VII. Resemble fan-shaped due to central cell of the bulliform group being larger and of different shapes than the rest which may be small by comparison; not necessarily with shorter inner tangential walls, it has been designated (*Sporobolus* type). The shape and size of the central cell varies.

    A.   Shape of central cell of the group which is always narrow at epidermis and often recurved or straight at area of contact with lateral bulliform cells of group:

        1.   Inflated, rounded; short area of contact with lateral cells (Fig. 1.5N)

        2.   Inflated, fan-shaped (Fig. 1.5O)

        3.   Narrower than deep; shield-shaped (Fig. 1.5P)

        4.   Elongated with parallel sides (Fig. 1.5Q)

        5.   Elongated rather pointed base; diamond-shaped (Fig. 1.5R)

    B.   Size of central cell:

        1.   Relatively small; not much larger than bundle sheath parenchyma (Fig. 1.5S)

        2.   Occupy one-fourth to half of the leaf thickness (Fig. 1.5T)

        3.   Occupy more than half of the leaf thickness (Fig. 1.5U)

VIII. Narrow groups of bulliform cells and intimately associated large colorless parenchyma cells penetrating deep into the mesophyll; the contact with the adaxial epidermis made by the bulliform cell superficial (**Arundo type**) (Fig. 1.6A)

IX. Bulliform cells and closely associated colorless cells forming an extensive column or girder extending from the base of an adaxial furrow deep into leaf:

- A. Nature of column or girder:
    1. Column uniseriate (Fig. 1.6B)
    2. Column biseriate (Fig. 1.6C)
    3. Column multiseriate
- B. Position of column:
    1. Extends from base of furrow to abaxial epidermis (Fig. 1.6B)
    2. Extends from base of furrow toward a vascular bundle

Typical Epidermal Cells and Appendages

Includes all epidermal structures as seen in transverse section excluding the bulliform cells which has been described in the above section. Characteristic features of the cuticle, macrohairs, prickles, and papillae as seen in transverse section have also been incorporated.

I. Cuticle and thickening of epidermal cell walls

- A. Outer tangential wall of the epidermal cells:
    1. Outer walls not thickened and with a very thin cuticle (Fig. 1.6D)
    2. Outer walls slightly thickened or with a thin cuticle (Fig. 1.6E)
    3. Outer walls thickened and covered by a distinct, thick cuticle continuous over the epidermal cells. The thickness of cuticle varies:
        a. Cuticle and cell wall equal to or greater than the depth of the average epidermal cells (Fig. 1.6F)
        b. Cuticle and cell wall less than the depth of the average epidermal cells (Fig. 1.6G)
    4. Outer tangential wall of each epidermal cell thickened individually:
        a. Outer wall occupies half or more of the depth of the cells (Fig. 1.6H)
        b. Outer wall occupies less than half of the depth of the cells (Fig. 1.6I)
        c. Irregular; cells of different sizes comprise the epidermis (Fig. 1.6J)
- B. Thickening of radial and inner tangential walls of epidermal cells:
    1. All walls of epidermal cells thickened
    2. Only outer wall thickened
    3. No walls thickened

II. Macrohairs, as seen in transverse section and nature of hair bases, are categorized as

- A. No macrohairs visible
- B. Superficial bases, that is, not sunken into the leaf or embedded between inflated bulliform cells: epidermal cells not modified to form cushion cells associated with the base of the hair
  - a. Swollen base not much larger than and situated between normal epidermal cells; included here are probably some elongated prickle hairs:
    - Hair short and thick (Fig. 1.6K)
  - b. Constriction above bulbous base embedded in normal epidermal cells
    - Hair thickened and stiff (Fig. 1.6L)

III. Sunken bases, that is, sunken in leaf or embedded between large, inflated epidermal cells or bulliform cells

- A. Constriction above bulbous base embedded between large epidermal cells
  1. Size and shape of hairs
     - a. Hairs slender but relatively short (Fig. 1.6M)
     - b. Hairs thick and short (Fig. 1.6N)
     - c. Hairs long and very slender (Fig. 1.6O)
  2. Nature of bulliform or epidermal cells surrounding base
     - a. Unraised bulliform cell groups (Fig. 1.6P)
- B. Base bulbous and not constricted; embedded between bulliform cells

IV. Prickles and hooks as seen in transverse section:

- A. No hooks or prickles visible
- B. Small hooks present in the epidermis; usually located in intercostal areas:
  1. With short, curved barbs (Fig. 1.6P)
  2. With straight, pointed, and slender barbs:
     - a. Present between typical epidermal cells (Fig. 1.6Q)
     - b. Present between inflated bulliform-like cells (Fig. 1.6R)
- C. Thickened prickles present in epidermis; usually located opposite the vascular bundles
  1. Pointed broad prickle; base not bulbous
  2. Small with short barb (Fig. 1.7A)

V. Papillae on the epidermal cells as seen in transverse section:

- A. Shape of the outer tangential wall of the epidermal cells:

1. Outer walls of epidermal cells not arched to any marked degree; flattened (Fig. 1.7B)
2. Outer walls of epidermal cells arched but not papillose (Fig. 1.7C)
3. Arching of outer wall exaggerated; papillose; inflated papillae as wide as epidermal cells:
   a. Thin-walled wide papillae scattered throughout the epidermis (Fig. 1.7D)
   b. Entire or major part of epidermis composed of thin-walled wide papillae (Fig. 1.7E)
4. Papillae wide; as wide as or slightly narrower than epidermal cells; not inflated and thin-walled: Distal, outer wall markedly thickened (Fig. 1.7F)
5. Papillae narrower than the epidermal cells
   a. Size of papillae
      • Relatively broad but not much more than half the width of the epidermal cells (Fig. 1.7G)
      • Much less than half the width of the cell (Fig. 1.7H)

The following two types of papillae have not been described by Ellis but were observed in the present study.

      • Long, as wide as cell (Fig. 1.7J)
      • Long, balloon or similar to Ω shaped (Fig. 1.7K)
   b. Number per cell as seen in transverse section:
      • One per cell (Fig. 1.7I)
      • Bifurcate; two per cell (Fig. 1.7G)

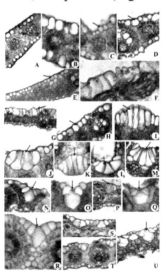

FIGURE 1.5   Characteristic features of bulliform cells in the epidermal layer in leaf blade.

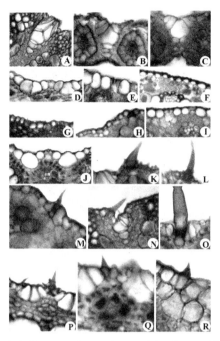

FIGURE 1.6   Characteristic features of appendages arising from epidermal cells of leaf blade.

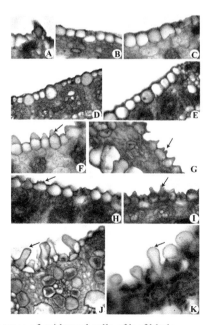

FIGURE 1.7   Different types of epidermal cells of leaf blade.

Mesophyll

Mesophyll is present between the adaxial and the abaxial epidermis occupying all the space in the leaf not occupied by the vascular bundles, bundle sheath, and sclerenchyma. It consists of chlorenchyma, but in some, it is often partly composed of translucent cells, which are arranged in different patterns which appear to depend on the affinities of the grasses concerned. The chlorenchyma and translucent cells occupy all of the space in the leaf that is not taken up by the sclerenchyma or the vascular bundles and their surrounding sheaths. There is no differentiation of mesophyll into palisade and spongy was observed among studied species. Based on the arrangement of the chlorenchyma cells, mesophyll can be classified into two major categories: (1) panicoid, that is, radiate chlorenchyma (2) festucoid, that is, nonradiate chlorenchyma.

Chlorenchyma

All the tissue containing chloroplasts excepting the parenchyma bundle sheath which may or may not contain chloroplasts is described under bundle sheaths. It does not include the ground tissue between the adaxial and abaxial epidermises especially if any translucent cells without chloroplasts are present. Chlorenchyma may be

I. Radiate, that is, chlorenchyma radially arranged around the vascular bundles which are often close together (panicoid). It may be interrupted by sclerenchyma girders and thus divided into two groups on either side of the vascular bundle:
- A. Number of layers of chlorenchyma cells arranged around the bundles may be
    - 1. Single layer of cells which may be
        - a. Tabular (Fig. 1.8A)
        - b. Isodiametric (Fig. 1.8B)
    - 2. Two layers of radiating cells (Fig. 1.8C)
    - 3. Numerous layers of radiating cells; cells long and narrow\ isachne type (Fig. 1.8D)
- B. Extent of radiating cells around bundles varies and may be
    - 1. Radiating cells completely surrounding bundles (Fig. 1.8A)
    - 2. Interrupted above bundles by sclerenchyma or colorless parenchyma girders (Fig. 1.8E)
    - 3. Interrupted below bundles by sclerenchyma or colorless parenchyma girders (Fig. 1.8F)

4. Reduced to two strips of chlorenchyma by large girders or colorless parenchyma (Fig. 1.8G)
C. Relationships of successive radiating mesophyll groups to each other could be categorized as
    1. Separated by bulliform and colorless cell groups
        a. Completely divided; colorless cells continuous to abaxial epidermis (Fig. 1.8G)
        b. Adaxially divided; colorless cells in adaxial half of leaf (Fig. 1.8A)
    2. Separated by irregular chlorenchyma and intercellular air-spaces (Fig. 1.8B)
    3. Radiating cells of successive bundles adjoin one another

II. Indistinctly or incompletely radiating chlorenchyma, that is, intermediate between radiate type and that not arranged in a definite pattern:
A. Continuous between bundles (Fig. 1.8C)
B. Chlorenchyma divided by groups of colorless cells; tending to be radiate (Fig. 1.8D)

III. Irregular chlorenchyma, that is, chlorenchyma does not radiate or arranged in a definite pattern; vascular bundles are usually widely spaced (festucoid)
A. Vertical arrangement of chlorenchyma in the mesophyll between successive vascular bundles
    1. Occupying the major or entire area between the adaxial and abaxial epidermises
        a. All cells of the chlorenchyma similar; homogenous
            • Regular small cells; isodiametric; tightly packed (Fig. 1.8E)
            • Irregular; cells of different size and shape; often with intercellular air-spaces
        b. Cells palisade-like in the adaxial chlorenchyma (Fig. 1.8F)
        c. Chlorenchyma cells adjoining the bundles larger
    2. Occupying the lower abaxial half of the leaf thickness
    3. Occupying the lower abaxial third of the leaf thickness
    4. Confined to small patches surrounding abaxial and adaxial grooves which contain stomata
B. Horizontal arrangement of chlorenchyma in the mesophyll between successive vascular bundles
    1. Continuous between bundles

     a. Strap-shaped; horizontally elongated; bundles widely spaced (Fig. 1.8G)

     b. Tall and narrow groups; bundles close together (Fig. 1.8I)

     c. U-shaped; occupying the sides and bases of furrows or around bulliform cell groups (Fig. 1.8H)

  2. Mesophyll groups divided by colorless cell groups

IV. Arm cells present or comprise chlorenchyma with walls invaginated

  A. Invaginations extend almost to opposite wall; cells divided into compartments

  B. Invaginations relatively short

V. Lacunae or air spaces present in mesophyll

  A. Extent of lacunae

  1. Between all or most adjacent vascular bundles

  2. Continuous air spaces above third-order bundles (Fig. 1.8J)

  B. Aerenchyma associated with the lacunae

    1. Distinct lacunae bounded by compact mesophyll with no aerenchyma traversing the lacunae

    2. Lacunae distinct but traversed by colorless aerenchyma cell chains

    3. Irregularly defined air spaces in diffuse mesophyll with many intercellular spaces and chlorophyll-bearing aerenchyma; often subtending the stomata

    4. Stellate cells representing a diaphragm interspersed by sclerotic strands present

Colorless Cells

Translucent cells constitute the remaining cells in the zones between successive vascular bundles, excluding the chlorenchyma, the sclerenchyma, and the bulliform cells. They vary in size and may be larger or smaller than the bulliform cells but are always without chlorenchyma and never part of the epidermis, that is, not in contact with the surface.

I. Absent, no colorless cells present

II. Well-defined groups of colorless cells present

  A. Closely associated with inflated bulliform cells or groups:

    1. Size of colorless cells:

     a. Smaller than bulliform cells; uninflated; often same size as chlorenchyma cells

      b.   Similar in size or shape to bulliform cells; inflated (Fig. 1.8G)

   2.   Width of extension of bulliform cell group; composed of colorless cells:

      a.   Narrower than bulliform cell group (Fig. 1.8B)

      b.   Same width as bulliform cell group

   3.   Number of extensions from each bulliform cell group:

      a.   One extension from each group (Fig. 1.8D)

      b.   Two extensions from each group; one on either side of the vascular bundle (Fig. 1.8M)

   4.   Extent of intrusion into the mesophyll by colorless cells:

      a.   Girders extend to the opposite epidermis (Fig. 1.8H)

      b.   Strands not extending to the opposite epidermis (Fig. 1.8I)

      c.   Band occupying the adaxial half of the leaf composed of colorless cells

B.   Not closely associated with large bulliform cells:

   1.   Origin at the base of furrows:

      a.   Girder-like extension to the opposite epidermis between the bundles

      b.   Girder-like extensions to the bundle sheath or toward the bundle sheath

   2.   Origin on flat epidermis or with slight furrows:

      a.   Girder-like extension extends to the opposite epidermis

         •   Column of girder only uni or biseriate

         •   Girder multiseriate

      b.   Girder of colorless cells does not extend to the opposite epidermis

III. Well-defined groups of smallish colorless cells: Not associated with large or well-defined bulliform groups; usually adaxial:

A.   Adaxial half of leaf consisting entirely of colorless cells

B.   Abaxial half of leaf consisting mainly of colorless cells

C.   Present on adaxial side of small vascular bundles only

D.   Form arches over third-order bundles and extend to the abaxial epidermis

E.   Abaxial irregular groups; may be associated with sclerenchyma strands

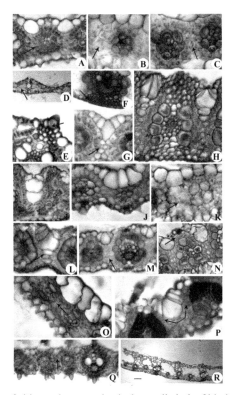

FIGURE 1.8   Nature of chlorenchyma and colorless cells in leaf blade.

Description of Individual Cells

I. Structure of colorless cells
    A.   Inflated, large parenchyma cells; resemble bulliform cells in size and shape

    B.   Smaller, thin-walled parenchyma cells; rounded in shape
        1.   Cells regular in size and shape; tissue uniform in appearance
        2.   Cells irregular in size and shape; tissue irregular in appearance

II. Thick-walled parenchyma or collenchyma cells

Vascular Bundles

Vascular bundles are commonly more or less circular or elliptical in outline, but sometimes it may be angular. Present study shows the grass species comprising third, second, and first-order vascular bundles. When the shapes of the vascular bundles are described it includes the shape formed by the cells within the bundle sheath.

Description of Vascular Bundle

I. Third-order vascular bundles: They are usually very small bundles often with xylem and phloem indistinguishable and consisting of only few lignified cells and few phloem elements; when the third-order vascular bundles are not obviously smaller than basic type bundles then they are often distinguishable by the absence of sclerenchyma strands and/or the presence of bulliform cell groups adaxially:

    A.   Absence of third-order vascular bundles

    B.   Shape of third-order bundles in transection

        1.   Round in outline; usually with many small parenchyma sheath cells:

            a.   Circular in outline (Fig. 1.9A)

            b.   Elliptical; vertically elongated (Fig. 1.9B)

        2.   Angular in outline:

            a.   Square-shape or pentagonal, that is, surrounded by a sheath of 4 or 5 large parenchyma cells (Fig. 1.9C)

            b.   Hexagonal or octagonal, that is, surrounded by 6 or more relatively small parenchyma cells (Fig. 1.9D)

            c.   More or less triangular in outline with 9–10 small bundle sheath cells (Fig. 1.9E)

            d.   Vertically elongated; tall and narrow (Fig. 1.9F)

    C.   Nature of vascular tissue of third-order bundles:

        1.   Xylem and phloem groups distinguishable (Fig. 9B)

        2.   Vascular tissue consists of only a few indistinguishable vascular elements (Fig. 1.9C)

II. Second-order vascular bundles: Xylem and phloem easily distinguishable; bundles usually fairly large, often of similar size to the first-order bundles; no conspicuously large metaxylem vessels or lysigenous cavities present; sclerenchyma arrangement usually the same as for first-order bundles:

    A.   Second-order bundles absent

    B.   Shape of second-order bundles present:

        1.   Rounded in outline; usually with many small parenchyma sheath cells:

            a.   Circular in outline (Fig. 1.10R)

            b.   Elliptical; vertically elongated (Fig. 1.10S)

    2.   Angular in outline; parenchyma sheath of relatively few large cells; inner mestome sheath may also be present:

            a.   Hexagonal or octagonal (Fig. 1.9I)

    b.   Triangular in shape; often rather large (Fig. 1.9J)

III. First-order or basic-type vascular bundles: The vascular bundle has a large metaxylem vessel present on either side of protoxylem elements, lysigenous cavity commonly present and is associated with sclerenchymatous girders or strands. Based on the characteristic features of the different cells of the vascular bundles, the following categories were observed:

   A.   Shape of first-order bundles in section:
      1.   Circular or round in outline (Fig. 1.9K)
      2.   Elliptical; vertically elongated (Fig. 1.9L)
      3.   Egg-shaped; broadest side adaxial (Fig. 1.9M)
   B.   Relationship of phloem cells and the vascular fibers:
      1.   Phloem adjoins the inner or parenchyma sheath (Fig. 1.9L)
      2.   Phloem completely surrounded by thick-walled fibers (Fig. 1.9K)
      3.   Phloem divided by intrusion of small fibers resulting in sclerosed phloem (Fig. 1.9N)
   C.   Nature of lysigenous cavity and protoxylem:
      1.   Lysigenous cavity and enlarged protoxylem vessel present (Fig. 1.9O)
      2.   Lysigenous cavity but no protoxylem vessel presents (Fig. 1.9L)
      3.   Enlarged protoxylem vessel present but no lysigenous cavity (Fig. 1.9M)
      4.   No lysigenous cavity or protoxylem vessel presents (Fig. 1.9P)
   D.   Size of metaxylem vessels in relation to parenchyma sheath cells in transection:
      1.   Narrow vessels; parenchyma sheath cells wider than vessels (Fig. 1.9Q)
      2.   Wide vessels, that is, vessels with width equal to or slightly greater than parenchyma sheath cells (Fig. 1.9L)
      3.   Very wide vessels, that is, width of vessels very much more than that of parenchyma sheath cells (Fig. 1.9O)
   E.   Shape of metaxylem vessels as seen in transection:
      1.   Angular in transection (Fig. 1.9L and M)
      2.   Circular in transection (Fig. 1.9N and O)

IV. Median/midrib vascular bundles: Based on the characteristic features of the cells of the band comprising the median bundle and their shape. The further categories have been presented as follows:

   A.   Shape:

1. Obovate (Fig. 1.10A)
2. Inverted triangular (Fig. 1.10B)
3. Egg (Fig. 1.10C)
4. Oval (Fig. 1.10D–F)
5. Round (Fig. 1.10G–K)
B. Phloem cells relationship with vascular fibers:
1. Phloem adjoins the inner or parenchyma sheath (Fig. 1.10D)
2. Phloem complete surrounded by thick-walled fibers (Fig. 1.10A, G)
C. Nature of lysigenous cavity and protoxylem:
1. Lysigenous cavity and enlarged protoxylem vessel present (Fig. 1.10D)
2. Lysigenous cavity but no protoxylem vessel presents (Fig. 1.10G)
3. Enlarged protoxylem vessel present but no lysigenous cavity (Fig. 1.10J)
D. Size of metaxylem vessels in relation to parenchyma sheath cells in cross-section:
1. Narrow vessels, that is, parenchyma sheath cells wider than vessels (Fig. 1.10E, F)
2. Wide vessel, that is, vessels with width equal to or slightly greater than parenchyma sheath cells (Fig. 1.10A, C)
3. Very wide vessels, that is, width of vessels very much more than that of parenchyma sheath cells (Fig. 1.10J)

Vascular Bundle Sheath

Bundle sheath may or may not be present around vascular bundles. Almost all vascular bundles surrounded, either completely or partly, by one or more bundle sheath layers, each sheath consisting of a single layer of cells. The outer sheath, when two are present, or the single sheath where there is but one, generally consists of parenchymatous cells with thin, or not more than slightly thickened walls. The cells of inner sheath are nearly always smaller in diameter and with thicker walls than those of the outer sheaths. In general, single sheaths are characteristic of the panicoid and double sheaths of the festucoid grasses. Structure of the sheaths surrounding the different orders of vascular bundles shows variations. Characteristics included are shape of bundle sheath, extent of bundle sheath, composition of extension, size of cells comprising sheath, and shape of cells. It is categorized as given below.

I. Bundle sheaths surrounding the vascular bundles absent

II. Outer or single parenchymatous bundle sheath. It includes vascular bundles with single and those with double sheaths:

    A.   Shape of bundle sheath
        1.   Sheath round, circular (Fig. 1.9A, G, and K)
        2.   Sheath elliptical, vertically elongated (Fig. 1.9N)
        3.   Sheath triangular with adaxial apex (Fig. 1.9E and J)
        4.   Sheath horseshoe shaped (Fig. 1.9P)
    B.   Extent of bundle sheath around the vascular bundle:
        1.   Sheath complete; completely surrounding the bundle (Fig. 1.9C)
        2.   Sheath incomplete interrupted by sclerenchymatous girders of various sizes:
            a.   Adaxial interruption:
                •   Slight interruption caused by a narrow girder of one to three fibers wide (Fig. 1.10A)
                •   Wide interruption caused by broad girder of more than three fibers wide or by colorless parenchyma (Fig. 1.10A)
            b.   Abaxial interruption:
                •   Slight interruption caused by a narrow girder of one to three fibers wide (Figs. 1.9Q and 1.11G)
                •   Wide interruption caused by broad girder of more than three fibers wide or by colorless parenchyma (Fig. 1.9M and P)

III. Extensions of the bundle sheath, comprising parenchyma cells associated with the sheaths and not part of the bulliform groups. The parenchyma cells extend to adjacent sclerenchyma girders or strands or to the epidermis:

    A.   Extensions of the bundle sheath absent (Fig. 1.10M)
    B.   Nature of the extensions:
        1.   Adaxial extension of the bundle sheath:
            a.   Narrow extension:
                •   Uniseriate; consisting of one column of cells
                •   Biseriate; consisting of two columns of cells
            b.   Broad extension; tri or multiseriate (Fig. 1.10N)
        2.   Abaxial extension of the bundle sheath: broad extension; tri or multiseriate (Fig. 1.10O)
    C.   Composition of the extensions
        1.   Consist of large, thin-walled colorless cells; the same size or bigger than sheath cells (Fig. 1.10T)

2. Consist of relatively small, thin-walled colorless cells; smaller than the sheath cells (Fig. 1.10P)
3. Consist of thickened parenchyma cells (Fig. 1.10Q)
4. Consist of colorless cells gradually decreasing in size as walls increase in thickness and eventually merge into the sclerenchyma strand (Fig. 1.10R)

D. Extent of the extensions of the bundle sheath:
1. Extend to and in contact with the epidermis or an inconspicuous sclerenchyma strand (1–4 fibers) (Fig. 1.10R)
2. Extend to conspicuous sclerenchyma strand (Fig. 1.10S)
3. Extend to sclerenchyma strand almost extending to the bundle sheath (Fig. 1.6C)
4. Extend to colorless subepidermal parenchyma (Fig. 1.10T)

IV. Number of cells comprising the bundle sheath

A. Four parenchyma cells comprise the sheath
B. Five parenchyma cells comprise the sheath
C. Six parenchyma cells comprise the sheath
D. Seven parenchyma cells comprise the sheath
E. Eight parenchyma cells comprise the sheath
F. Nine parenchyma cells comprise the sheath
G. Ten parenchyma cells comprise the sheath
H. Eleven parenchyma cells comprise the sheath
I. Twelve parenchyma cells comprise the sheath
J. Thirteen parenchyma cells comprise the sheath
K. Fourteen parenchyma cells comprise the sheath
L. Fifteen or more parenchyma cells comprise the sheath

V. Structure of the parenchyma sheath cells: Cells of each order of vascular bundle considered separately and all cell types included:

A. Not well differentiated from the chlorenchyma cells
B. Distinct from chlorenchyma cells:
1. Shape of cells comprising sheath; includes all shapes relevant for each order of vascular bundle:
   a. Inflated, round, circular (Fig. 1.9G)
   b. Radial walls straight; tangential walls inflated (Fig. 1.9C)
   c. Radial and outer tangential walls straight; inner tangential wall inflated (Fig. 1.9D)
   d. Radial and inner tangential walls straight; outer tangential wall inflated; fan shaped (Fig. 1.9E)
   e. Cells elliptical; elongated; often rather irregular (Fig. 1.10P)

2. Arrangement of cells of different shapes comprising the bundle sheath:
   a. All cells similar in shape (Fig. 1.9H)
   b. Two enlarged elongated cells abaxially situated:
      • Adjoining girder of sclerenchyma (Fig. 1.9P)
      • Normal parenchyma sheath cells between the elongated cells (Fig. 1.9L)
   c. Cells at adaxial side of sheath elongated vertically; elliptical (Fig. 1.10D)
3. Size of cells comprising sheath:
   a. Cells of sheath similar in size:
      • Large and inflated; generally larger than mesophyll cells; conspicuous (Fig. 1.10E)
      • Not markedly larger than the mesophyll cells; conspicuous (Fig. 1.10D)
      • Cells smaller than the mesophyll cells; inconspicuous (Fig. 1.10C)
      • Cells smaller than the inner bundle sheath cells
      • Cells small; resemble cells of a mestome sheath, that is, thickened walls often especially the inner tangential walls; parenchyma sheath apparently absent
   b. Cells of sheath of various sizes
      • Cells of different sizes comprise sheath; irregular (Fig. 1.10O)
      • Gradation in size with largest cells adaxially situated (Fig. 1.10H)
      • Gradation in size with largest cells abaxially situated (Fig. 1.10P)
      • Gradation with smallest cells in the center; abaxial and adaxial cells larger (Fig. 1.10I)
      • Gradation with largest cells in the center on each side (Fig. 1.10K)
4. Chloroplast structure of the bundle sheath cells:
   a. Translucent; sheath cells without chloroplasts
   b. With small chloroplasts; not distinct from chloroplasts of the chlorenchyma
   c. With large, specialized chloroplasts
      • Chloroplasts fill entire cell lumen (Fig. 1.10E)

- Chloroplasts concentrated near the outer tangential wall (Fig. 1.9L)
- Chloroplasts concentrated near the inner tangential wall (Fig. 1.9N)
- Chloroplasts centrally situated (Fig. 1.9O)

Inner or Mestome Bundle Sheath or Endodermis

Bundle sheath in many grass species are made up of two layers in which case the inner layer is designated as "mestome," it is only applicable when two sheaths are present; often difficult to distinguish from fibrous ground tissue of the vascular bundle; in contact with the metaxylem, that is, if metaxylem vessels in contact with parenchyma sheath then there is no inner bundle sheath:

I. Extent of sheath
  A. No inner sheath presents (Fig. 1.10H)
  B. Sheath complete; completely surrounding the xylem and phloem (Fig. 1.10F)
  C. Sheath incomplete due to interruptions of sclerenchyma girders:
    1. Adaxial interruption (Fig. 1.10C)
    2. Abaxial interruption (Fig. 1.10E)
  D. Sheath reduced; intermediate type; interruptions not due to sclerenchyma girders: adaxial; opposite xylem only (Fig. 1.10T)

II. Structure of cells of inner bundle sheath in region adjacent to phloem; cells of each order of vascular bundle considered separately:

  A. Relatively large with inner tangential and radial walls thickened (Fig. 1.10D)
  B. Small, with uniformly thickened walls:
    1. Walls heavily thickened; lumen small (Fig. 1.10E)
    2. Walls not conspicuously thickened (Fig. 1.10F)
  C. Larger than outer bundle sheath; parenchymatous; with chloroplasts
  D. Adaxially situated cells larger than lateral cells of the sheath (Fig. 1.10D)

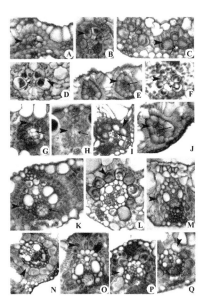

FIGURE 1.9   Characteristic features of First, Second and Third order vascular bundles and bundle sheath cells of leaf blade.

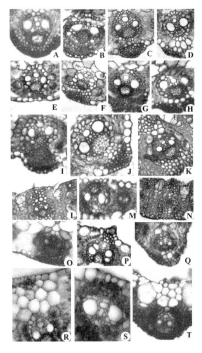

FIGURE 1.10   Characteristic features of median/midrib vascular bundles and bundle sheath cells of leaf blade.

Sclerenchyma

In grass leaf, support is provided by fibrous and other types of thick-walled cells that collectively constitute the sclerenchyma. This is in the form of subepidermal longitudinal strands or girders which follow the course of the vascular bundles. In some species, sclerenchyma appears as islands of lignified tissue lying above and below or on one side only, of each vascular bundle, it is known as strands. Sometimes, it appears as girders extending from either or each epidermis to the bundle sheath and is known as girdered.

Adaxial sclerenchyma: Sclerenchyma located and associated above the vascular bundles:

I. Adaxial sclerenchyma absent (Fig. 1.11A)

II. Sclerenchymatous strand present: It is not in contact with the bundle sheath and is separated by parenchyma or mesophyll from sheath

- A. Minute strand consisting of only a few subepidermal fibers (Fig. 1.11B)
- B. Small strand; inconspicuous; epidermal cells unaltered over the strand
    1. Shallow, forming a strip; only two to four fibers deep; subepidermal (Fig. 1.11C)
    2. Narrow; only two to four fibers wide (Fig. 1.11D)
- C. Well-developed conspicuous strand. Epidermal cells over strand usually small and thick-walled
    1. As deep as wide as seen in transection
    2. Wider than deep as seen in transection; in form of band
        a. Arched; follows the shape of the adaxial rib (Fig. 1.11E)
        b. Straight, horizontal band

III. Sclerenchymatous girder present. The girder is in contact with or interrupts the bundle sheath:

- A. Very small girder; inconspicuous; epidermal cells usually unaltered over girder (Fig. 1.11F)
- B. Small girder; epidermal cells over the girder usually small and thick walled
    1. Small thin subepidermal strip (Fig. 1.11H)
    2. Narrow; deeper than wide
    3. As deep as wide as seen in transection; equidimensional (Fig. 1.11G)

    C. Well-developed girder; conspicuous; epidermal cells over girder small and thickened
        1. The girder is relatively wide and deep band; as wide or wider than the vascular bundle:
           a. Arched; follows the shape of the adaxial rib (Fig. 1.11I)
           b. Straight; horizontal band (Fig. 1.11J)
           c. Narrowing toward bundle (Fig. 1.11K)
      2. Relatively narrow and deep girder; narrower than the vascular bundle (Fig. 1.11A)
      3. The girder is T-shaped with the horizontal band of "T" connected to bundle or bundle sheath by a vertical sclerenchymatous band of the stem:
          a. Stem is short, that is, it is shorter than the horizontal band of "T"
          b. Stem is relatively long; as long as or longer than horizontal band of "T"
      4. Inversely anchor-shaped; arched band follows shape of adaxial rib:
          a. Stem short and sturdy; more than triseriate
          b. Stem long or tall and sturdy; more than triseriate
          c. Stem short and thin or narrow; one to three seriates
          d. Stem long or tall and thin or narrow; one to three seriates

IV. Girder extends from bundle sheath to bulliform or colorless cells: Not to epidermis:
    A. Fibers interrupt the cells of the single or outer bundle sheath (Fig. 1.11K)
    B. Fibers in contact with the cells of the single or outer bundle sheath (Fig. 1.11J)

Abaxial sclerenchyma: Sclerenchyma found below the vascular bundles

I. Abaxial sclerenchyma absent (Fig. 1.11A)

II. Sclerenchymatous strand present: Not in contact with the bundle sheath; separated by parenchyma or mesophyll from sheath:
    A. Minute strand consisting of only a few subepidermal fibers (Fig. 1.11C)
    B. Small strand; inconspicuous; epidermal cells unaltered under the strand:
        1. Shallow, forming a strip; only two to four fibers deep; subepidermal (Fig. 1.11B)
        2. Narrow but fairly deep; only two to four fibers wide
        3. As deep as wide as seen in transection (Fig. 1.12C)

C. Well-developed strand; conspicuous; epidermal cells under strand usually small and thick walled:
1. As deep as wide as seen in transection
2. Wider than deep as seen in transection; in form of a band:
    a. Arched, follows the shape of the abaxial rib
    b. Straight, horizontal band

III. Sclerenchymatous girder present. The girder is in contact with or interrupts the bundle sheath

A. Very small girder; inconspicuous; epidermal cells usually unaltered under girder (Fig. 1.5N)
B. Small girder; epidermal cells under the girder usually small and thick walled:
1. Small, thin subepidermal strip (Fig. 1.12D)
2. Narrow, deeper than wide
3. As deep as wide as seen in transection; equidimensional
C. Well-developed girder; conspicuous; epidermal cells under the girder usually small and thickened:
1. Relatively wide and deep band; as wide or wider than the vascular bundle:
    a. Arched, follow the shape of the abaxial rib (Fig. 1.12E)
    b. Straight, horizontal band (Fig. 1.11K)
    c. Narrowing toward bundle; triangular or trapezoidal (Fig. 1.11G)
2. Relatively narrow and deep girder; narrower than the vascular bundle
3. Inverted T-shaped; horizontal band connected to bundle or bundle sheath by stem:
    a. Stem is short, that is, it is shorter than the horizontal band of "T" (Fig. 1.12F)
    b. Stem is relatively long; as long as or longer than horizontal band of "T"
4. Anchor-shaped; arched band follows shape of abaxial rib:
    a. Stem short and sturdy; more than triseriate (Fig. 1.12A)
    b. Stem long or tall and sturdy; more than triseriate (Fig. 1.12E)
    c. Stem short and thin or narrow; one to three seriates
    d. Stem long or tall and thin or narrow; one to three seriates
5. Continuous or almost continuous abaxial hypodermal band
D. Girder extends from bundle sheath to bulliform or colorless cells; not to the epidermis

1. Fibers interrupt the cells of the single or outer bundle sheath (Fig. 1.12A)
2. Fibers in contact with the cells of the single or outer bundle sheath

Sclerenchyma between bundles

These are supernumerary strands not associated with the vascular bundles but situated intermediately between them usually beneath the abaxial epidermis opposite the bulliform cell groups or furrows. The supernumerary strands exclude sclerenchyma at the leaf margin and of the keel or midrib.

I. Sclerenchyma is absent between the vascular bundles

II. Abaxial strands of sclerenchymatous fibers

    A. Opposite adaxial furrows with or without bulliform cells

    B. Opposite bulliform cell groups not associated with furrows

III. Hypodermal band: Sclerenchyma is arranged in the form of more or less continuous abaxial band:

    A. Extensions of girder or strand groups

    B. Hypodermal layer continuous

Sclerenchyma in leaf margin

This includes all mechanical tissue situated in or directly associated with the margin of the lamina. It may incorporate or fuse with the ultimate, the penultimate or other lateral bundles. The margin of the leaf may remain straight or is sometimes curved. Also, the marginal tip in cross-section when viewed has different shapes in different species. Also, many of the species have leaf margin capped with sclerenchyma.

I. Sclerenchyma absent in the margin

II. Cap of sclerenchyma at the margin which is not in contact with the lateral bundle. It can be further categorized into

    A. Size of the sclerenchymatous cap:
1. Very small; consists of a couple fibers (Fig. 1.12H)
2. Relatively small; less than the width of a third-order vascular bundle (Fig. 1.12J)
3. Well-developed; wider than third-order vascular bundles (Fig. 1.12K)

    B. Shape of sclerenchymatous cap:
1. Rounded cap (Fig. 1.12L)
2. Pointed cap (Fig. 1.12K)

3. Narrow, very pointed projection (Fig. 1.12H)
4. Crescent-shaped cap; sclerenchyma extends shortly along both abaxial and adaxial side of leaf
5. Curved in shape with sclerenchyma extending along adaxial side of the leaf (Fig. 1.12M)
6. Curved in shape with sclerenchyma extending along abaxial side of the leaf

C. Epidermal cells at margin:
1. Not fibrous; thin-walled or outer walls thickened; distinct from fibers (Fig. 1.12N)
2. Small, fibrous; thickened on all walls; similar to fibers in transection (Fig. 1.12O)

D. Angular prickles of margin:
1. Enlarged bases present in margin
2. No prickle bases visible in margin

E. Nature of junction with mesophyll and remainder of lamina:
1. Adjoins normal mesophyll cells
2. Specialized enlarged parenchyma cells near cap or ultimate bundle
3. Lateral intercellular canal near margin; lysigenous duct not associated with the bundle

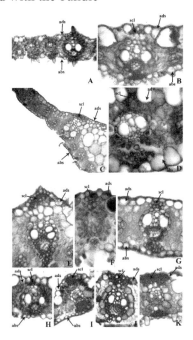

FIGURE 1.11 Variations of sclerenchyma cells in leaf blade.

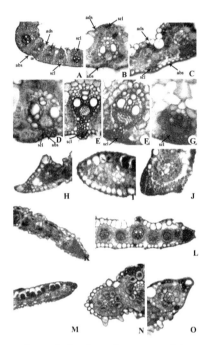

FIGURE 1.12 Variations of marginal sclerenchymatous girder and sclerenchymatous cells in leaf blade.

Leaf-Sheath Anatomy

Characteristic features of leaf sheath cells are similar to the leaf lamina cells up to some extent. It showed adaxial and abaxial epidermis, in between them ground tissue is present in which the vascular bundles were arranged. Vascular bundles were distributed toward the abaxial surface. Most of the characters like outline of leaf sheath, vascular bundle arrangement, vascular bundle structure, vascular bundle sheath structure are similar with the structure of leaf lamina. But leaf lamina shows bulliform cells, papillae structure on epidermal cell, macrohairs, prickles, chlorenchymatous cells, colorless cells which are not present in leaf sheath transection.

Outline of Leaf Sheath

Outline of leaf sheath showed two types of shapes, either V shaped or round shaped. In both the types, margin of the leaf sheath is overlapping. The degree of overlapping is variable. It may show single or double inrolling. Categorization is as follows:

I. V shaped: Where two halves of the leaf sheath folded toward inner side:
   A: Angle formed by the two arms of the lamina at the midrib is

1. Narrow, that is, closed "V," that is, less than 45° to each other (Fig. 1.13A)
2. Standard "V," that is, between 45° and 90° to each other (Fig. 1.13B)
3. Broad, open "V," that is, more than 90° to each other (Fig. 1.13C)
4. Wide, very open "V," that is, almost 180°; presence of projecting keel gives appearance of V-shaped (Fig. 1.13D)

B: Shape of arms of the lamina:
1. Straight (Fig. 1.13C)
2. Outwardly bowed; concave (Fig. 1.13E)
3. Outwardly curving; convex (Fig. 1.13F)
4. Outline heart shaped (Fig. 1.13G)
5. Distinctly wavy or undulating

II. U shaped (Fig. 1.13H)

III. Round shaped wherein leaf sheath is circular in outline and two halves of that overlapping each other:

A. margins are inrolled, not overlapping (Fig. 1.13I)
B. margins are inrolled, single overlapping (Fig. 1.13J, K)
C. margins are inrolled, double overlapping (Fig. 1.13L)

Longitudinal Ribs and Furrows

The longitudinal ribs and furrows may or may not be present. If they are present, then it was observed only on abaxial surface. Adaxial surface never showed presence of ribs and furrows.

I. Ribs and furrows are absent

II. Abaxial ribs and furrows present: Distribution of abaxial ribs varied

1. Present opposite first-order vascular bundles only
2. Present opposite all vascular bundles

Vascular Bundle Arrangement

Vascular system is comprised of first-order, second-order, and third-order bundles similar to the leaf lamina. They showed different arrangements:

1. First, second, and third-order vascular bundle arranged sequentially one after other (Fig. 1.13N)
2. Two consecutive third-order vascular bundle between first and second-order vascular bundle (Fig. 1.13M)

3. Three consecutive third-order vascular bundle between first and second-order vascular bundle (Fig. 1.13O)
4. One consecutive third-order vascular bundle between first and second-order vascular bundle (Fig. 1.13P)

Epidermis

Single layered epidermal cells are present. Adaxial epidermal cells are always flat, rectangular, or barrel shaped and cuticle is absent. Abaxial epidermal cells are small, elongated, or barrel shaped with thick cuticle. Other than this, there was no distinct difference observed. Papillae, bulliform cells, and prickles are absent in leaf sheath.

Ground Tissue

Parenchymatous with air spaces; four types of air spaces were observed:
1. Vertically oval-shaped air cavities (Fig. 1.13Q)
2. Horizontally oval-shaped air cavities (Fig. 1.13R)
3. Oval shaped, big, air cavity, very few cavities present (Fig. 1.13T)
4. Round-shaped air cavity (Fig. 1.13S)

Vascular Bundle

Vascular bundles are covered with bundle sheath and they contain phloem, metaxylem, and protoxylem. Based on their shape and type of bundle sheath cells, they could be differentiated into the following categories:

Description of Vascular Bundles

I. Third-order vascular bundles were usually very small bundles often with xylem and phloem indistinguishable and consisting of only a few lignified cells and a few phloem elements. When vascular bundle is not obviously smaller than basic type bundles it is often distinguishable by the absence of sclerenchyma strands and/or the presence of bulliform cell groups adaxially:
A. Third-order bundles absent in section
B. Shape of third-order bundles in section:
1. Round in outline; usually with many small parenchyma sheath cells
a. Circular to elliptical in outline (Fig. 1.14A)
b. Elliptical; vertically elongated (Fig. 1.14B)
2. Angular in outline:

        a. Square-shape or pentagonal, that is, surrounded by a sheath of four or five large parenchyma cells
        b. Hexagonal or octagonal, that is, surrounded by 6 or more relatively small parenchyma cells (Fig. 1.14C, E)
        c. Triangular in outline (Fig. 1.14D)
        d. Vertically elongated; tall and narrow
  C. Nature of vascular tissue of third-order bundles:
     1. Xylem and phloem groups distinguishable (Fig. 1.14D)
     2. Vascular tissue consists of only a few indistinguishable vascular elements (Fig. 1.14A)

II. Second-order vascular bundles: Xylem and phloem easily distinguishable; bundles usually fairly large, often of similar size to the first-order bundles; no conspicuously large metaxylem vessels or lysigenous cavities present; sclerenchyma arrangement usually the same as for first-order bundles:

  A. Second-order bundles absent in section
  B. Shape of second-order bundles in section:
     1. Rounded in outline; usually with many small parenchyma sheath cells:
       a. Circular in outline (Fig. 1.14F)
       b. Elliptical; vertically elongated (Fig. 1.14G, H)
     2. Angular in outline; parenchyma sheath of relatively few large cells; inner mestome sheath may also be present
       a. Hexagonal or octagonal (Fig. 1.14I–K)
       b. Triangular in shape; often rather large

III. First-order or basic-type vascular bundles: Large metaxylem vessel present on either side of protoxylem elements; lysigenous cavity commonly present; associated with sclerenchyma girders or strands:

  A. Shape of first-order bundles in section:
     1. Circular or round in outline (Fig. 1.14L)
     2. Elliptical; vertically elongated (Fig. 1.14N)
     3. Egg-shaped; broadest side adaxial (Fig. 1.14R)
  B. Relationship of phloem to vascular fibers:
     1. Phloem adjoins the inner or parenchyma sheath (Fig. 1.14O)
     2. Phloem is completely surrounded by thick-walled fibers (Fig. 1.14L)
  C. Nature of lysigenous cavity and protoxylem:
     1. Lysigenous cavity and enlarged protoxylem vessel present (Fig. 1.14P)

      2. Lysigenous cavity but no protoxylem vessel presents (Fig. 1.14M)
      3. No lysigenous cavity or protoxylem vessel presents (Fig. 1.14R)
  D. Size of metaxylem vessels in relation to parenchyma sheath cells in transection:
      1. Narrow vessels, that is, parenchyma sheath cells wider than vessels (Fig. 1.14R)
      2. Wide vessels, that is, vessels with width equal to or slightly greater than parenchyma sheath cells (Fig. 1.14N)
      3. Very wide vessels, that is, width of vessels very much more than that of parenchyma sheath cells (Fig. 1.14P)
  E. Shape of metaxylem vessels as seen in T/S
      1. Angular in transection (Fig. 1.14N)
      2. Circular in transection (Fig. 1.14L)

Vascular Bundle Sheath

Structure of the sheaths surrounding the different orders of vascular bundles shows variations. Characters are included are shape of bundle sheath, extent of bundle sheath, composition of extension, size of cells comprising sheath, and shape of cells. It is categorized as given below:

I. No bundle sheaths surrounding the vascular bundles

II. Outer or single bundle sheath: Parenchyma sheath; includes vascular bundles with single and those with double sheaths:
  A. Shape of bundle sheath:
      1. Sheath round, circular (Fig. 1.14L)
      2. Sheath elliptical, vertically elongated (Fig. 1.14N)
      3. Sheath triangular with adaxial apex
      4. Sheath horseshoe shaped (Fig. 1.14J)
  B. Extent of bundle sheath:
      1. Sheath complete; completely surrounding the bundle (Fig. 1.14M)
      2. Sheath incomplete due to interruption of sclerenchyma girders of various sizes:
         a. Abaxial interruption:
            • Slight interruption caused by a narrow girder of one to three fibers wide (Fig. 1.14I)
            • Wide interruption caused by broad girder of more than three

III. Number of cells comprising the bundle sheath

   A.   Four parenchyma cells comprise the sheath
   B.   Five parenchyma cells comprise the sheath
   C.   Six parenchyma cells comprise the sheath
   D.   Seven parenchyma cells comprise the sheath
   E.   Eight parenchyma cells comprise the sheath
   F.   Nine parenchyma cells comprise the sheath
   G.   Ten parenchyma cells comprise the sheath
   H.   Eleven parenchyma cells comprise the sheath
   I.   Twelve parenchyma cells comprise the sheath
   J.   Thirteen parenchyma cells comprise the sheath
   K.   Fourteen parenchyma cells comprise the sheath
   L.   Fifteen or more parenchyma cells comprise the sheath

IV. Structure of the parenchyma sheath cells: Cells of each order of vascular bundle considered separately and all cell types included

   A.   Not well differentiated from the ground tissue
   B.   Distinct from ground tissue:
      1.   Shape of cells comprising sheath; include all shapes relevant for each order of vascular bundle:
         a.   Inflated, round, circular (Fig. 1.14P)
         b.   Radial walls straight; tangential walls inflated (Fig. 1.14N)
         c.   Radial and outer tangential walls straight; inner tangential wall inflated
         d.   Radial and inner tangential walls straight; outer tangential wall inflated; fan shaped (Fig. 1.14I)
         e.   Cells elliptical; elongated; often rather irregular
      2.   Arrangement of cells of different shapes comprising the bundle sheath:
         a.   All cells similar in size (Fig. 1.9N)
         b.   Two enlarged elongated cells abaxially situated:
            •   Adjoining girder of sclerenchyma
            •   Normal parenchyma sheath cells between the elongated cells
      3.   Size of cells comprising sheath:
         a.   Cells of sheath similar in size:
            •   Large and inflated; generally larger than ground tissue; conspicuous (Fig. 1.14J)
            •   Not markedly larger than the ground tissue; conspicuous (Fig. 1.14N)

- Cells smaller than the ground tissue; inconspicuous (Fig. 1.14P)
  b. Cells of sheath of various sizes:
  - Cells of different sizes comprise sheath; irregular
  - Gradation in size with largest cells adaxially situated (Fig. 1.14N)
  - Gradation in size with largest cells abaxially situated
  - Gradation with smallest cells in the center; abaxial and adaxial cells larger (Fig. 1.14I)

Inner or Mestome Bundle Sheath or Endodermis

Bundle sheath in many grass species is made up of two layers in which case the inner layer is designated as "mestome." It is only applicable when two sheaths are present and is often difficult to distinguish from fibrous ground tissue of the vascular bundle.

I. Extent of sheath

  A. Inner sheath absent (Fig. 1.14N)
  B. Sheath complete; completely surrounding the xylem and phloem (Fig. 1.14I)
  C. Sheath incomplete due to interruptions of sclerenchyma girders:
  1. Abaxial interruption (Fig. 1.14K)

Sclerenchyma

In grass leaf sheath, sclerenchyma cells present only on abaxial surface. Adaxial surface and margin do not show sclerenchymatous cells.

Abaxial sclerenchyma; sclerenchyma found below the vascular bundles:

I. Abaxial sclerenchyma absent (Fig. 1.11A)

II. Strand present; sclerenchymatous not in contact with the bundle sheath; separated by parenchyma or mesophyll from sheath:

  A. Minute strand consisting of only a few subepidermal fibers
  B. Small strand; inconspicuous; epidermal cells unaltered under the strand:
  1. Shallow, forming a strip; only two to four fibers deep; subepidermal (Fig. 1.13T)
  C. Well-developed strand; conspicuous; epidermal cells under strand usually small and thick-walled:
  1. Wider than deep as seen in transection; in form of a band:

> a. Arched, follows the shape of the abaxial rib (Fig. 1.14M)
> b. Straight, horizontal band (Fig. 1.14N)

III. Girder present: Sclerenchymatous in contact with or interrupts the bundle sheath:

A. Very small girder; inconspicuous; epidermal cells usually unaltered under girder (Fig. 1.14L)

B. Small girder; epidermal cells under the girder usually small and thick-walled:
   1. Small, thin subepidermal strip (Fig. 1.14I)
   2. Narrow, deeper than wide (Fig. 1.14J)

C. Well-developed girder; conspicuous; epidermal cells under the girder usually small and thickened:
   1. Relatively wide and deep band; as wide or wider than the vascular bundle:
      a. Arched, follow the shape of the abaxial rib (Fig. 1.14R)
      b. Straight, horizontal band (Fig. 1.14K)

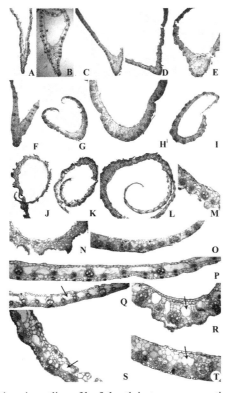

FIGURE 1.13   Variations in outline of leaf sheath in transverse section.

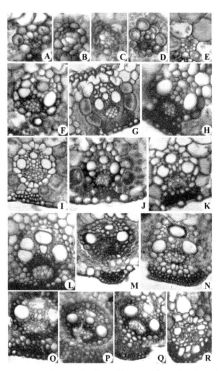

FIGURE 1.14   Different types of vascular bundles and bundle sheath present in leaf sheath.

Ligule Anatomy

Chaffey (1994) studied functional anatomy of ligules in different grass species. According to him, most of the ligules consisted of three layers: the mesophyll and two epidermal cell layers. Mesophyll in those ligules contained chloroplasts. Mesophyll layers may be one/two/up to eight-cell layers in thickness.

Connection of the ligule between leaf blade and leaf sheath is at right angles to the longitudinal axis of leaf sheath. At the cut, only the median part of the ligule connects to the leaf sheath tissue. The lateral parts are free and converge at the cut. Grasses either show vascularized ligule or nonvascularized ligule.

From the studied species different connection of ligule between leaf blade and leaf sheath were observed, which are the following:

Based on the shape of leaf blade or leaf sheath, broadly it can be categorized into two:

I.   "V" shaped
II. Round shaped

I. *"V" shaped*

Type 1: Ligule attached to the lamina up to the half-length of it and is free and sharp "V" shaped at the point of connection (Fig. 1.15A)

Type 2: Ligule attached to the lamina up to one-third length of it and free, blunt, narrow, "V" shaped at the point of connection (Fig. 1.15B)

Type 3: Ligule attached to the lamina more than half-length of it and blunt "V" shaped at the connection point and narrow (Fig. 1.15C)

Type 4: Ligule attached to the lamina till the three-fourth than is free and blunt, broad "V" shaped at the connection (Fig. 1.15D)

Type 5. Ligule attached less than half-length of the lamina and at the connection point it is very broad, blunt "V" shaped (Fig. 1.15E)

Type 6: Ligule attached to the lamina only one-fourth portions and at connection point it was very broad almost like straight "V" shaped with small sharp angled (Fig. 1.15F).

II. *Round shaped*

Type 9: Ligule attached to the midrib region and little with lamina region, and at the connection, it is broad round shaped (Fig. 1.15I)

Type 8: Ligule attached to the half width of the leaf sheath region and it is round shaped at the connection point (Fig. 1.15H)

Type 7: Ligule attached to the leaf sheath more than the half width and very broad round shaped at the connections point (Fig. 1.15G)

Type 10: Ligule attached only to the midrib region and broad round shaped at connections point (Fig. 1.15J)

Type 11: Ligule is attached to the midrib and few parts of lamina region because it is hairy ligule (Fig. 15K).

Generally, anatomically ligule shows adaxial and abaxial surface and between them mesophyll cells may be have chloroplast or not. Within these simple anatomical structures, few variations are observed in studied species.

Type 1: Single layered mesophyll between adaxial and abaxial surfaces (Fig. 1.15L)

Type 2: Two layered mesophyll cells between adaxial and abaxial (Fig. 1.15M)

Type 3: Three layered mesophyll cells between adaxial and abaxial (Fig. 1.15O)

Type 4: More than three layered mesophyll cells between adaxial and abaxial (Fig. 1.15N)

Type 5: Two layered mesophyll cells between adaxial and abaxial surface but on abaxial surface few sclerenchyma cells present at regular interval (Fig. 1.15P).

Type 6: Two layered mesophyll cells between adaxial and abaxial surfaces abaxial surface almost covered with sclerenchyma cells (Fig. 1.15Q)

Type 7: Single layered mesophyll cells between adaxial and abaxial surfaces and at some places sclerenchyma cells present (Fig. 1.15R)

Chaffey (1994) observed few cells which are round to oblong shaped with thickened wall and thickening of lignin, so it is considered as sclerenchyma cells of ligule.

Few species show presences of prickles on the surfaces, especially those have hairy ligule. Among the studied species two types of prickle were observed:

I. Thickened prickle, bulbous base, barbed (Fig. 1.15S, arrow)
II: Thickened prickle, bulbous base, sharp, pointed (Fig. 1.15T, arrow)

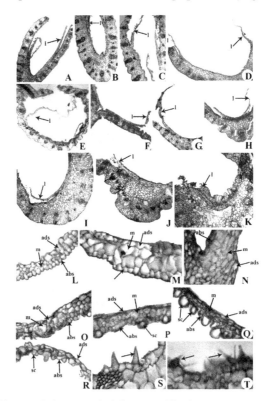

FIGURE 1.15   Characteristic anatomical features of ligule.

Characteristic Features in Leaves of the Different Grass Species

The characteristic features noted in following different leaf parts, that is, leaf lamina, leaf sheath, and ligule of the grass species have been described. Microphotographs depicting the features have been put up after the description of each species, individually. Description of genera with more than one species has been considered together. Variations in the characteristic features of the different species have been depicted in the given descriptions at the appropriate points. Microphotographs of the species have been given separately. At the end of the species description of a single genus, key to identify them is also provided. A dichotomous key differentiating the 100 species on the basis of leaf anatomical characters has been represented at the end of the discussion.

1. *Chionachne koenigii*

Leaf lamina (Fig. 1.16A–G)

Outline of the lamina: Expanded with flat, even leaf blade (Fig. 1.16A); ribs and furrows absent.

Keel is well developed, semicircular with an indistinguishable median vascular bundle, three first-order vascular bundles and a small third-order vascular bundle arranged abaxially. Small, rounded parenchyma cells are present surrounding the median vascular bundle adaxially. Sclerenchyma present on abaxial surface of the keel is associated with vascular bundles in form of girders.

Epidermis: Adaxial epidermal surface has restricted group of large inflated bulliform cells of two different types (1) in level with epidermal surface (Fig. 1.16F, arrow) and (2) projecting above the epidermal surface (Fig. 1.16G, arrow). Epidermal cells have thick cuticle. Prickles, macrohiars, and hooks are absent.

Mesophyll: Indistinctly or incompletely radiating, small, isodiametric, compactly homogenous chlorenchymatous cells occupying the major area between the adaxial and abaxial epidermises; chlorenchyma cells are continuous strap shaped and elongated between bundles; colorless cells absent.

Vascular bundles: Five first-order vascular bundles regularly arranged from median bundle to the margin of half lamina. Six third-order bundles between consecutive larger bundles and six second-order bundles between consecutive first-order bundles present. All the vascular bundles present at same level and toward the abaxial surface.

Third-order vascular bundles: Square with xylem and phloem groups distinguishable (Fig. 1.16F)

Second-order vascular bundles: Circular shaped (Fig. 1.16E)

First-order vascular bundles (Fig. 1.16G)

Shape: Circular

Relationship of phloem cells and the vascular fibers: Phloem adjoins the inner or parenchyma sheath

Nature of lysigenous cavity and protoxylem: Enlarged protoxylem vessel present but no lysigenous cavity

Size of metaxylem vessels in relation to parenchyma sheath cells in cross-section: Very wide vessels

Shape of metaxylem vessels: Circular

Median vascular bundles (Fig. 1.16D)

Shape: Circular

Relationship of phloem cells and the vascular fibers: Phloem adjoins the inner or parenchyma sheath

Nature of lysigenous cavity and protoxylem: Enlarged protoxylem vessel present but no lysigenous cavity

Size of metaxylem vessels in relation to parenchyma sheath cells in cross-section: Very wide vessels

Vascular bundle sheath

Third-order vascular bundle's bundle sheath (Fig. 1.16F, arrowhead)

Shape: Round

Shape of bundle sheath cell: Cells elliptical and elongated; often rather irregular

Second-order vascular bundle's bundle sheath (Fig. 1.16E, arrowhead)

Shape: Round

Shape of bundle sheath cell: Cells elliptical and elongated; often rather irregular

Number of bundle sheath cell: 10

Nature of bundle sheath (either complete or incomplete): Incomplete, abaxially interrupted by sclerenchyma cells

Chloroplast position in bundle sheath cell: Chloroplast concentrated near the inner tangential wall

Mestome/Inner sheath: Absent

First-order vascular bundle's bundle sheath (Fig. 1.16G)

Shape: Round, circular

Extent of bundle sheath around the vascular bundle: Sheath incomplete, abaxially interrupted by one to three layers of sclerenchymatous fibers

Extension of bundle sheath: Abaxial extension of the bundle sheath: broad extension; tri or multiseriate

Number of cells comprising the bundle sheath: 15–16 parenchyma cells

Bundle sheath cell shape: All cells are elliptical, elongated in shape; chloroplast concentrated near the outer tangential wall of cell

Mestome/Inner sheath: Absent

Sclerenchyma

Adaxial sclerenchyma: Absent

Abaxial sclerenchyma: Well-developed girder, wider than deep (Fig. 1.16C–E, arrow)

Sclerenchyma between bundles: Absent

Sclerenchyma in leaf margin: Forming a well-developed (Fig. 1.16B, arrow) rounded cap. Epidermal cells of margin are thin with outer thick tangential wall (Fig. 1.16B)

Leaf-sheath anatomy (Fig. 1.16H–K)

Outline of lamina: "V" shaped, broad, outwardly bowed, margins are inrolled. Ribs and furrows absent (Fig. 1.16H)

Epidermis

Adaxial epidermal cells: Flat, rectangular

Abaxial epidermal cells: Barrel shaped with thick cuticle. Restricted groups of tall and narrow bulliform cells which are not inflated and having parallel-sided cells or with outer tangential wall slightly shorter than the inner wall (Fig. 1.16K)

Ground tissue: Parenchymatous ground tissue, vertically oval-shaped air cavity present (Fig. 1.16H, arrow)

Vascular bundle

Arrangement of vascular bundle: One consecutive third-order vascular bundle between first and second-order vascular bundle

Third-order vascular bundle: Circular with vascular tissue consisting of only few indistinguishable vascular elements

Second-order vascular bundle: Circular shaped

First-order vascular bundle: Circular with phloem adjoining the inner or parenchyma sheath, lysigenous cavity absent but protoxylem vessel present; circular very wide metaxylem vessel elements (Fig. 1.16J)

Vascular bundle sheath (Fig. 1.16J)

Shape: Round, circular

Extent of bundle sheath around the vascular bundle: Sheath complete

Number of cells comprising the bundle sheath: 14–15 parenchyma cells

Bundle sheath cell shape: Different sized round, cells smaller than the ground tissue

Mestome/Inner sheath: Absent

Sclerenchyma: In form of vertical well-developed strand on abaxial surface (Fig. 1.16I, arrow)

Ligule anatomy (Fig. 1.16L, M)

Shape: Ligule attached to the lamina more than half-length of it, blunt narrow "V" shaped at the connection point (Fig. 1.16L, arrow)

Anatomy: Two layered mesophyll cells between adaxial and abaxial epidermis (Fig. 1.16M)

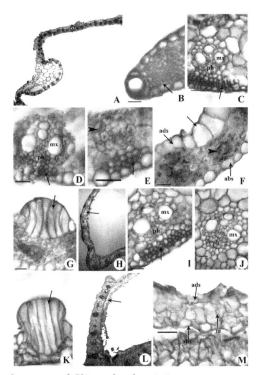

FIGURE 1.16   Leaf anatomy of *Chionachne koenigii.*

2. *Coix lachryma-jobi*

Leaf lamina (Fig. 1.17A–H)

Outline of the lamina: Expanded with flat, even leaf blade (Fig. 1.17A); ribs and furrows absent.

Keel is well developed, semicircular with an indistinguishable median vascular bundle, four first-order vascular bundles, and small second and third-order vascular bundle arranged abaxially and keel breaking down by one enlarged air space. Small, rounded parenchyma cells are present surrounding the vascular bundle adaxially.

Epidermis: Adaxial epidermal surface is majorly made up of inflated, slightly thick-walled bulliform cells and also few restricted group of bulliform cells (Fig. 1.17B, arrow) present which are slightly above the level of epidermal surface. Prickles, macrohiars, hooks are absent.

Mesophyll: Indistinctly or incompletely radiating, small, isodiametric, compactly arranged homogenous chlorenchyma cells occupying the major area between the adaxial and abaxial epidermises; chlorenchyma cells are continuous strap shaped and elongated between bundles; colorless cells absent.

Vascular bundles: Nine first-order vascular bundles regularly arranged from median bundle to marginal vascular bundle in half lamina. Six third-order bundles between consecutive larger bundles and two second-order bundles between consecutive first-order bundles present. All the vascular bundles present at same level and toward the abaxial surface.

Third-order vascular bundles: Circular (Fig. 1.17D, arrow) and elliptical (Fig. 1.17E, arrow), vascular tissue consisting of only few indistinguishable vascular elements

Second-order vascular bundles: Circular (Fig. 1.17F, arrow)

First-order vascular bundles (Fig. 1.17G)

Shape: Circular

Relationship of phloem cells and the vascular fibers: Phloem adjoins the inner or parenchyma sheath

Nature of lysigenous cavity and protoxylem: Lysigenous cavity and enlarged protoxylem vessel present

Size of metaxylem vessels in relation to parenchyma sheath cells in cross-section: Very wide vessels

Shape of metaxylem vessels: Circular

Median vascular bundles (Fig. 1.17H, arrow)

Shape: Circular

Relationship of phloem cells and the vascular fibers: Phloem adjoins the inner or parenchyma sheath

Nature of lysigenous cavity and protoxylem: Lysigenous cavity and enlarged protoxylem vessel present

Size of metaxylem vessels in relation to parenchyma sheath cells in cross-section: Very wide vessels

Vascular bundle sheath

Third-order vascular bundle's bundle sheath (Fig. 1.17D, E)

Shape: Round

Shape of bundle sheath cell: Cells elliptical and elongated; often rather irregular

Second-order vascular bundle's bundle sheath (Fig. 1.17F)

Shape: Round

Shape of bundle sheath cell: Cells elliptical and elongated; often rather irregular

Number of bundle sheath cell: 12–14

Nature of bundle sheath (either complete or incomplete): Complete

Chloroplast position in bundle sheath cell: Chloroplasts centrally situated

Mestome/Inner sheath: Absent

First-order vascular bundle's bundle sheath (Fig. 1.17G)

Shape: Round, circular

Extent of bundle sheath around the vascular bundle: Sheath complete

Extension of bundle sheath: Broad abaxial and narrow adaxial extension of the bundle sheath

Number of cells comprising the bundle sheath: 17–18 parenchyma cells

Bundle sheath cell shape: Elliptical, elongated cells; chloroplasts centrally situated

Mestome/Inner sheath: Absent

Sclerenchyma

Adaxial sclerenchyma: Minute strand consisting of only a few subepidermal fibers

Abaxial sclerenchyma: Well-developed strand, wider than deep

Sclerenchyma between bundles: Absent

Sclerenchyma in leaf margin: Forming a well-developed pointed fibrous cap. Epidermal cells of margin are small, thick walled (Fig. 1.17C)

Leaf-sheath anatomy (Fig. 1.17I–L)

Outline of lamina: "U" shaped. Minute ribs and furrows on abaxial surface present opposite to all vascular bundles (Fig. 1.17I)

Epidermis

Adaxial epidermal cells: Flat, rectangular

Abaxial epidermal cells: Square to barrel shaped with thick cuticle. Stomata present (Fig. 1.17K, arrow)

Ground tissue: Parenchymatous ground tissue, horizontally oval to square-shaped air cavity present (Fig. 1.17J, arrow)

Vascular bundle

Arrangement of vascular bundle: One consecutive third-order vascular bundle between first and second-order vascular bundle

Third-order vascular bundle: Circular with vascular tissue consisting of only few indistinguishable vascular elements (Fig. 1.17K)

Second-order vascular bundle: Circular

First-order vascular bundle: Circular with phloem adjoins the inner or parenchyma sheath, lysigenous cavity and enlarged protoxylem vessel present; circular wide metaxylem vessel element (Fig. 1.17L)

Vascular bundle sheath (Fig. 1.17K, L, arrow)

Shape: Round, circular

Extent of bundle sheath around the vascular bundle: Sheath complete

Number of cells comprising the bundle sheath: 19–20 parenchyma cells

Bundle sheath cell shape: Different sized elliptical, elongated cells smaller than the ground tissue

Mestome/Inner sheath: Absent

Sclerenchyma: In form of straight horizontal well-developed girder on abaxial surface

Ligule anatomy (Fig. 1.17M, N)

Shape: Ligule attached to the half width of the leaf sheath region and it is round shaped at the connection point (Fig. 1.17M, arrow)

Anatomy: Two layered mesophyll cells between adaxial and abaxial epidermis (Fig. 1.17N)

3. *Andropogon pumilus*

Leaf lamina (Fig. 1.18A–E)

Outline of the lamina: Expanded with flat, even leaf blade (Fig. 1.18A); ribs and furrows absent.

Keel is well developed, semicircular with single median bundle, two third-order vascular bundles arranged abaxially; small, rounded parenchyma cells are present surrounding the median vascular bundle adaxially. Sclerenchyma present on abaxial surface of keel is associated with central bundle in the form of small, anchor-shaped girder.

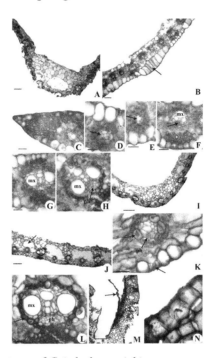

FIGURE 1.17 Leaf anatomy of *Coix lachryma-jobi.*

Epidermis: Adaxial epidermal surface with restricted fan-shaped bulliform cells with larger, shield-shaped central cell (Fig. 1.18C, arrow). Bulliform cells occupying one-fourth to half of the leaf thickness. Short, slender hairs are present (Fig. 1.18D, arrow). Prickles, hooks are absent.

Mesophyll: Radiate, single layered isodiametric chlorenchyma, that is, panicoid type, completely surrounding bundles and radiating cells of successive bundles adjoin one another; colorless cells absent.

Vascular bundles: Two first-order vascular bundles regularly arranged from median bundle to marginal bundle in half lamina. Four third-order bundles between consecutive larger bundles and four second-order bundles between consecutive first-order bundles present. All the vascular bundles present at same level and placed centrally in leaf blade.

Third-order vascular bundles: Hexagonal with vascular tissue consisting of only few indistinguishable vascular elements (Fig. 1.18C).

Second-order vascular bundles: Hexagonal

First-order vascular bundles (Fig. 1.18E)

Shape: Circular

Relationship of phloem cells and the vascular fibers: Completely surrounded by thick-walled fibers

Nature of lysigenous cavity and protoxylem: Lysigenous cavity and enlarged protoxylem vessel present

Size of metaxylem vessels in relation to parenchyma sheath cells in cross-section: Wide vessels

Shape of metaxylem vessels: Circular

Median vascular bundles

Shape: Round

Relationship of phloem cells and the vascular fibers: Completely surrounded by thick-walled fibers

Nature of lysigenous cavity and protoxylem: Lysigenous cavity and enlarged protoxylem vessel present

Size of metaxylem vessels in relation to parenchyma sheath cells in cross-section: Wide vessels

Vascular bundle sheath

Third-order vascular bundle's bundle sheath (Fig. 1.18C)

Shape: Elliptical

Shape of bundle sheath cell: Radial and outer tangential walls straight; inner tangential wall inflated

Second-order vascular bundle's bundle sheath

Shape: Elliptical

Shape of bundle sheath cell: Radial and outer tangential walls straight; inner tangential wall inflated

Number of bundle sheath cell: 8–9

Nature of bundle sheath (either complete or incomplete): Complete

Chloroplast position in bundle sheath cell: chloroplasts fill entire cell lumen

Mestome/Inner sheath: Absent

First-order vascular bundle's bundle sheath (Fig. 1.18E, arrowhead)

Shape: Round, circular

Extent of bundle sheath around the vascular bundle: Absent

Number of cells comprising the bundle sheath: 11–12 parenchyma cells

Bundle sheath cell shape: Radial and outer tangential walls straight; inner tangential wall inflated; all cells are similar in shape and similar sized as ground tissue; chloroplasts fill entire cell lumen

Mestome/Inner sheath: Absent

Sclerenchyma

Adaxial sclerenchyma: Absent

Abaxial sclerenchyma: Small, narrow strand but fairly deep; only two to four fibers wide

Sclerenchyma between bundles: Absent

Sclerenchyma in leaf margin: Forming very small pointed fibrous cap. Epidermal cells thin with outer thick tangential wall (Fig. 1.18B)

Leaf-sheath anatomy (Fig. 1.18F–I)

Outline of lamina: Broad "V" shaped with inrolled margins and arms forms heart shape (Fig. 1.18F). Ribs and furrows absent

Epidermis

Adaxial epidermal cells: Flat, rectangular

Abaxial epidermal cells: Barrel shaped with thick cuticle

Ground tissue: Parenchymatous with absence of air cavities

Vascular bundle

Arrangement of vascular bundle: One consecutive third-order vascular bundle between first and second-order vascular bundle

Third-order vascular bundle: Pentagonal shaped, hexagonal, vascular tissue consists of only a few indistinguishable elements vascular (Fig. 1.18I)

Second-order vascular bundle: Circular shaped (Fig. 1.18H)

First-order vascular bundle: Circular with phloem adjoins the inner or parenchyma sheath, lysigenous cavity and enlarged protoxylem vessel present, circular wide metaxylem vessel elements (Fig. 1.18G)

Vascular bundle sheath (Fig. 1.18G, H, arrowhead)

Shape: Round, circular

Extent of bundle sheath around the vascular bundle: Sheath incomplete, abaxially interrupted by sclerenchyma cells

Number of cells comprising the bundle sheath: 19–20 parenchyma cells

Bundle sheath cell shape: Elliptical, elongated cells often interrupted by normal parenchyma sheath cells; cells are smaller than the ground tissue

Mestome/Inner sheath: Absent

Sclerenchyma: In form of straight horizontal well-developed girder on abaxial surface

Ligule anatomy (Fig. 1.18J, K)

Shape: "V" shaped, Ligule attached to the lamina till three-fourth portion then is free and blunt, broad "V" shaped at the connection (Fig. 1.18J, arrow)

Anatomy: Single layered mesophyll cells between adaxial and abaxial epidermis (Fig. 1.18K)

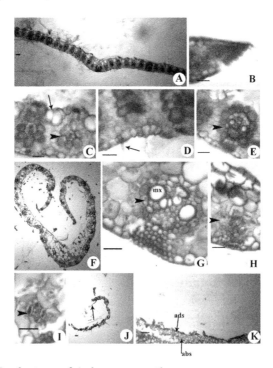

FIGURE 1.18   Leaf antomy of *Andropogon pumilus.*

4. *Apluda mutica*

Leaf lamina (Fig. 1.19A–E)

Outline of the lamina: "V" shaped (Fig. 1.19A) with wide angle having straight arms of lamina and two halves are symmetrical on either side of the median vascular bundle; furrows and ribs present between all vascular bundles (Fig. 1.19B); shallow and narrow furrows present on both the surface. Triangular-shaped ribs present (Fig. 1.19B).

Keel is well developed, U shaped (Fig. 1.19C), slightly thicker than rest of lamina, and projecting abaxially due to position and size of single median vascular bundle. Small, rounded parenchyma cells are present surrounding the median vascular bundle adaxially. Sclerenchyma present on adaxial surface in form of strands and on abaxial surface associated with vascular bundles in form of strands (Fig. 1.19C).

Epidermis: Adaxial epidermal has restricted group of large isolated and inflated bulliform cells present on sides of deep furrows interrupted with irregular small groups of epidermal cells which covered with thick cuticle (Fig. 1.19D); hooks present (Fig. 1.19D, arrow); short macrohairs are present (Fig. 1.19D).

Mesophyll: Radiate, single layered isodiametric chlorenchyma completely surrounding bundles; colorless cells absent (Fig. 1.19D).

Vascular bundles: Two first-order vascular bundles regularly arrangement from median to marginal bundle in half lamina. Eight to nine third-order bundles between consecutive larger bundles and eight to nine second-order bundles between consecutive first-order bundles present. All the vascular bundles present at same level and placed centrally in leaf blade.

Third-order vascular bundles: Elliptical with vascular tissue consisting of only few indistinguishable vascular elements (Fig. 1.19E, arrow)

Second-order vascular bundles: Circular shaped

First-order vascular bundles

Shape: Elliptical (Fig. 1.19D)

Relationship of phloem cells and the vascular fibers: Phloem adjoins the inner or parenchyma sheath

Nature of lysigenous cavity and protoxylem: Lysigenous cavity and enlarged protoxylem vessel present

Size of metaxylem vessels in relation to parenchyma sheath cells in cross-section: Wide vessels

Shape of metaxylem vessels: Circular

Median vascular bundles (Fig. 1.19C)

Shape: Obovate

Relationship of phloem cells and the vascular fibers: Phloem completely surrounded by thick-walled fibers

Nature of lysigenous cavity and protoxylem: Lysigenous cavity present, protoxylem vessel absent

Size of metaxylem vessels in relation to parenchyma sheath cells in cross-section: Wide vessels

Vascular bundle sheath

Third-order vascular bundle's bundle sheath

Shape: Elliptical

Shape of bundle sheath cell: Cells elliptical and elongated

Second-order vascular bundle's bundle sheath

Shape: Round

Shape of bundle sheath cell: Cells elliptical and elongated

Number of bundle sheath cell: 7–8

Nature of bundle sheath (either complete or incomplete): Complete

Chloroplast position in bundle sheath cell: Chloroplast concentrated near the outer tangential wall

Mestome/Inner sheath: Absent

First-order vascular bundle's bundle sheath

Shape: Elliptical

Extent of bundle sheath around the vascular bundle: Complete

Extension of bundle sheath: Absent

Number of cells comprising the bundle sheath: 14–15 parenchyma cells

Bundle sheath cell shape: All cells are elliptical, elongated in shape; chloroplast concentrated near the outer tangential wall of cell

Mestome/Inner sheath: Absent

Sclerenchyma

Adaxial sclerenchyma: Minute strand consisting of only a few subepidermal fibers

Abaxial sclerenchyma: Small, narrow strand but fairly deep; only two to four fibers wide

Sclerenchyma between bundles: Absent

Sclerenchyma in leaf margin: Forming very small rounded fibrous cap. Epidermal cells of margin are thin with outer thick tangential wall

Leaf-sheath anatomy (Fig. 1.19F–H)

Outline of lamina: Round shaped; ribs and furrows present opposite to first-order vascular bundle on abaxial surface

Epidermis

Adaxial epidermal cells: Flat, rectangular

Abaxial epidermal cells: Rectangular shaped with thick cuticle

Ground tissue: Parenchymatous with vertically oval/oblong shaped air cavity present (Fig. 1.19F)

Vascular bundle

Arrangement of vascular bundle: One consecutive third-order vascular bundle between first and second-order vascular bundle

Third-order vascular bundle: Circular shaped, vascular tissue consists of only a few indistinguishable vascular elements

Second-order vascular bundle: Circular shaped (Fig. 1.19G)

First-order vascular bundle: Circular with phloem completely surrounded by thick-walled fibers, lysigenous cavity, and enlarged protoxylem vessel present; circular-shaped wide metaxylem vessel elements (Fig. 1.19H)

Vascular bundle sheath

Shape: Round, circular (Fig. 1.19H)

Extent of bundle sheath around the vascular bundle: Sheath complete

Number of cells comprising the bundle sheath: 13–14 parenchyma cells

Bundle sheath cell shape: Elliptical, elongated cells smaller than the ground tissue

Mestome/Inner sheath: Absent

Sclerenchyma: In form of straight horizontal well-developed strand on abaxial surface

Ligule anatomy (Fig. 1.19I, J)

Shape: Ligule attached to the midrib region and little with lamina region, and at the connection, it is broad round shaped (Fig. 1.19I, arrow)

Anatomy: More than three layered mesophyll cells between adaxial and abaxial surfaces (Fig. 1.19J)

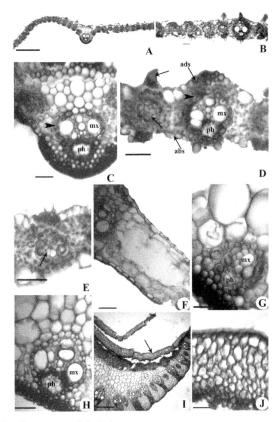

FIGURE 1.19   Leaf anatomy of *Apluda mutica.*

5. *Arthraxon lanceolatus*

Leaf lamina (Fig. 1.20A–E)

Outline of the lamina: Expanded with pleated, distinctly undulating leaf blade; ribs and furrows present on both surfaces; wide, deep furrows present between all vascular bundles; all ribs similar, triangular shape present over all vascular bundles. Keel is not well developed.

Epidermis: Epidermal surface has barrel to rectangular-shaped cells with thin cuticle. Bulliform cells are absent; pointed broad prickle present (Fig. 1.20E, arrow)

Mesophyll: Indistinctly or incompletely radiating, small, isodiametric; cells are compactly homogenous chlorenchymatous cells occupying the major area between the adaxial and abaxial epidermises; chlorenchyma cells are continuous between bundles; colorless cells absent.

Vascular bundles: Three first-order vascular bundles regularly arrangement from median to marginal bundle in half lamina. Three third-order bundles between consecutive larger bundles and three to four second-order bundles between consecutive first-order bundles present. All the vascular bundles present at same level and placed centrally in leaf blade.

Third-order vascular bundles: Circular with vascular tissue consisting of only few indistinguishable vascular elements (Fig. 1.20D)

Second-order vascular bundles: Elliptical shaped (Fig. 1.20C)

First-order vascular bundles (Fig. 1.20B)

Shape: Obovate

Relationship of phloem cells and the vascular fibers: Phloem adjoins the inner or parenchyma sheath

Nature of lysigenous cavity and protoxylem: Enlarged protoxylem vessel present, lysigenous cavity absent

Size of metaxylem vessels in relation to parenchyma sheath cells in cross-section: Wide vessels

Shape of metaxylem vessels: Circular

Vascular bundle sheath

Third-order vascular bundle's bundle sheath (Fig. 1.20D, arrow)

Shape: Round

Shape of bundle sheath cell: Cells elliptical and elongated

Second-order vascular bundle's bundle sheath

Shape: Elliptical

Shape of bundle sheath cell: Cells elliptical and elongated

Number of bundle sheath cell: 17

Nature of bundle sheath (either complete or incomplete): Complete

Chloroplast position in bundle sheath cell: Translucent

Mestome/Inner sheath: Absent

First-order vascular bundle's bundle sheath

Shape: Elliptical

Extent of bundle sheath around the vascular bundle: Complete

Extension of bundle sheath: Absent

Number of cells comprising the bundle sheath: 23–24 parenchyma cells comprise the sheath

Bundle sheath cell shape: All cells are elliptical, elongated in shape; chloroplast translucent

Mestome/Inner sheath: Absent

Sclerenchyma

Adaxial and abaxial sclerenchyma: Minute strand consisting of only a few subepidermal fibers

Sclerenchyma between bundles: Absent

Sclerenchyma in leaf margin: Forming a very small pointed cap. Epidermal cells of margin are thin-walled

Leaf-sheath anatomy (Fig. 1.20F, G)

Outline of lamina: Round shaped within rolled margins. Shallow ribs and furrows present over all vascular bundles (Fig. 1.20F)

Epidermis

Adaxial epidermal cells and abaxial epidermal cells: Small, flat, rectangular shaped with thick cuticle

Ground tissue: Parenchymatous with oval-shaped air cavity present (Fig. 1.20F, arrow)

Vascular bundle (Fig. 1.20G)

Arrangement of vascular bundle: Two consecutive third-order vascular bundles between first and second-order vascular bundle

Third-order vascular bundle: Circular with vascular tissue consisting of only few indistinguishable vascular elements

Second-order vascular bundle: Elliptical shaped

First-order vascular bundle: Elliptical with phloem completely surrounded by thick-walled fibers, lysigenous cavity, and enlarged protoxylem vessel present; circular wide metaxylem vessel element

Vascular bundle sheath

Shape: Elliptical

Extent of bundle sheath around the vascular bundle: Sheath incomplete, interrupted by sclerenchyma cells

Number of cells comprising the bundle sheath: 14–15 parenchyma cells

Bundle sheath cell shape: All cells are elliptical, elongated in shape

Mestome/Inner sheath: Absent

Sclerenchyma: In form of well-developed wide girder present on abaxial surface

Ligule anatomy (Fig. 1.20H–J)

Shape: Ligule attached along half width of the leaf sheath region and it is round shaped at connection point (Fig. 1.20H, arrow)

Anatomy: Two layered mesophyll cells between adaxial and abaxial surfaces; sclerenchyma cells (Fig. 1.20I, arrow) and prickles (Fig. 1.20J, arrow) present on abaxial surface

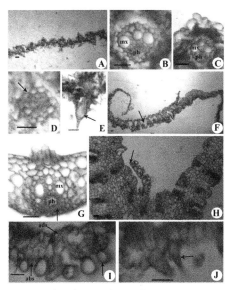

FIGURE 1.20 Leaf anatomy of *Arthraxon lanceolatus*.

6. *Bothriochloa pertusa*

Leaf lamina (Fig. 1.21A–F)

Outline of the lamina: "V" shaped with wide angle having convex-shaped outwardly curving of the arms. Two halves are symmetrical on either side of the median vascular bundle (Fig. 1.21A). Ribs and furrows are absent on the leaf surface.

Keel is well developed, V shaped with an indistinguishable median vascular bundle, one first and second-order vascular bundle and two third-order vascular bundles arranged abaxially. Small, rounded parenchyma cells are present surrounding the median vascular bundle adaxially. Sclerenchyma present on adaxial surface of keel is associated with vascular bundles in form of straight horizontal well-developed strands.

Epidermis: Adaxial epidermal surface has isolated, large, inflated bulliform cells (Fig. 1.21D, arrow) interrupted by small groups of inflated and slightly outer thickened walled epidermal cells which are covered with thin cuticle. Abaxial surface has barrel-shaped epidermal cells with cuticle. Few epidermal cells have cell wall equal to or greater than the depth of the average epidermal cells (Fig. 1.21D, E). Long, slender macrohairs present (Fig. 1.21C).

Mesophyll: Radiate, single layered isodiametric chlorenchyma which completely surrounds vascular bundles; compact homogenous chlorenchymatous cells occupy the major area between the adaxial and abaxial epidermises; chlorenchyma cells continuous between bundles; colorless cells absent.

Vascular bundles: Two first-order vascular bundles regularly arranged from median to marginal bundle in half lamina. Two to three third-order bundles between consecutive larger bundles and two to three second-order bundles between consecutive first-order bundles present. All the vascular bundles are present at the same level and centrally placed in leaf blade.

Third-order vascular bundles: Hexagonal shaped and vascular tissue consisting of only a few indistinguishable vascular elements (Fig. 1.21D)

Second-order vascular bundles: Elliptical shaped (Fig. 1.21E)

First-order vascular bundles (Fig. 1.21F)

Shape: Circular

Relationship of phloem cells and the vascular fibers: Phloem adjoins the inner or parenchyma sheath

Nature of lysigenous cavity and protoxylem: Lysigenous cavity and enlarged protoxylem vessel present

Size of metaxylem vessels in relation to parenchyma sheath cells in cross-section: Narrow vessels

Shape of metaxylem vessels: Circular

Vascular bundle sheath

Third-order vascular bundle's bundle sheath (Fig. 1.21D, arrowhead)

Shape: Round

Shape of bundle sheath cell: Cells elliptical and elongated

Second-order vascular bundle's bundle sheath (Fig. 1.21E, arrowhead)

Shape: Elliptical

Shape of bundle sheath cell: Cells elliptical and elongated

Number of bundle sheath cell: 10–11

Nature of bundle sheath (either complete or incomplete): Incomplete, abaxially interrupted by subepidermal sclerenchyma strands

Chloroplast position in bundle sheath cell: Chloroplast centrally situated

Mestome/Inner sheath: Absent

First-order vascular bundle's bundle sheath (Fig. 1.21E, arrowhead)

Shape: Round, circular

Extent of bundle sheath around the vascular bundle: Sheath incomplete, abaxially interrupted by parenchyma cells

Extension of bundle sheath: Broad adaxial extension of the bundle sheath

Number of cells comprising the bundle sheath: 16–17 parenchyma cells

Bundle sheath cell shape: All cells are elliptical, elongated in shape and shows gradation in size toward abaxial side; chloroplast centrally in the cell

Mestome/Inner sheath: Absent

Sclerenchyma

Adaxial sclerenchyma: Minute strand consisting of only a few subepidermal fibers

Abaxial sclerenchyma: Minute strand consisting of only a few subepidermal fibers (Fig. 1.21F, arrow)

Sclerenchyma between bundles: Absent

Sclerenchyma in leaf margin: Forming well-developed fibrous pointed cap (Fig. 1.21B). Epidermal cells of margin are thick walled and covered with cuticle

Leaf-sheath anatomy (Fig. 1.21G–J)

Outline of lamina: V shaped with standard angle and inrolled margins; ribs and furrows absent (Fig. 1.21G)

Epidermis

Adaxial and abaxial epidermal cells: Small, flat, rectangular shaped with thick cuticle (Fig. 1.21H)

Ground tissue: Parenchymatous with absence of air cavity

Vascular bundle

Arrangement of vascular bundle: One consecutive third-order vascular bundle between first and second-order vascular bundle

Third-order vascular bundle: Circular with vascular tissue consists of only a few indistinguishable vascular elements (Fig. 1.21J)

Second-order vascular bundle: Oval shaped (Fig. 1.21H)

First-order vascular bundle: Circular with phloem completely surrounded by thick-walled fibers, lysigenous cavity, and enlarged protoxylem vessel present; circular wide metaxylem vessel elements (Fig. 1.21I)

Vascular bundle sheath (Fig. 1.21H, J, arrowhead)

Shape: Round, circular

Extent of bundle sheath around the vascular bundle: Sheath incomplete, interrupted by parenchyma cells adaxially and sclerenchyma cells abaxially

Number of cells comprising the bundle sheath: 14–15 parenchyma cells comprise the sheath

Bundle sheath cell shape: Elliptical, elongated in shape; cells smaller than the ground tissue

Mestome/Inner sheath: Absent

Sclerenchyma: In form of straight horizontal well-developed broad strand on abaxial surface (Fig. 1.21F, arrow)

Ligule anatomy (Fig. 1.21K, L)

Shape: Ligule attached to the lamina more than half-length of it and blunt "V" shaped and narrow at the connection point (Fig. 1.21K, arrow)

Anatomy: Single layered mesophyll cells between adaxial and abaxial surfaces (Fig. 1.21L)

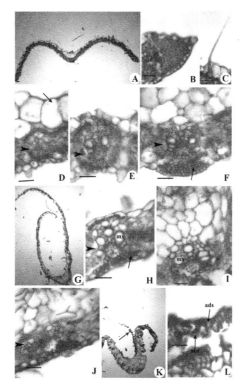

FIGURE 1.21   Leaf anatomy of *Bothriochloa pertusa.*

7. *Capillipedium huegelii*

Leaf lamina (Fig. 1.22A–F)

Outline of the lamina: "V" shaped with wide angle (Fig. 1.22A), having straight shape of the arms. Two halves are symmetrical on either side of the median vascular bundle; ribs and furrows present on leaf surface; medium, wide furrow present between first and second-order vascular bundles; round tall ribs on abaxial surface and triangular shorter ribs on adaxial surface present over first and second-order vascular bundles.

Keel is well developed, V shaped with single median vascular bundle (Fig. 1.22A). Small, round parenchyma present surround the median vascular bundle adaxially. Sclerenchyma on abaxial surface of keel associated with vascular bundles in form of well-developed, wide girder (Fig. 1.22B, arrow).

Epidermis: Adaxial epidermal surface has isolated large, inflated bulliform cells interrupted with small barrel to rectangular shaped epidermal cells which are covered with thick cuticle; short prickles present (Fig. 1.22D, arrow); abaxial surface has barrel to rectangular shaped epidermal cells covered with cuticle and single papillae per cell (Fig. 1.22E, arrow); short, slender macrohairs present (Fig. 1.22F, arrow)

Mesophyll: Radiate, single layered, tubular cells completely surrounding vascular bundles; compactly arranged homogenous chlorenchymatous cells occupy the major area between the adaxial and abaxial epidermises; chlorenchyma cells continuous between bundles; colorless cells absent.

Vascular bundles: Eight first-order vascular bundles regularly arranged from median to marginal bundle in half lamina. Six to seven third-order bundles between consecutive larger bundles and four to five second-order bundles between consecutive first-order bundles present. All the vascular bundles present at same level and centrally placed in leaf blade.

Third-order vascular bundles: Hexagonal with vascular tissue consisting of only a few indistinguishable vascular elements (Fig. 1.22D).

Second-order vascular bundles: Circular shaped (Fig. 1.22C)

First-order vascular bundles (Fig. 1.22E)

Shape: Triangular

Relationship of phloem cells and the vascular fibers: Completely surrounded by thick-walled fibers

Nature of lysigenous cavity and protoxylem: Lysigenous cavity and enlarged protoxylem vessel present

Size of metaxylem vessels in relation to parenchyma sheath cells in cross-section: Very wide vessels

Shape of metaxylem vessels: Circular

Median vascular bundles (Fig. 1.22B)

Shape: Obovate

Relationship of phloem cells and the vascular fibers: Completely surrounded by thick-walled fibers

Nature of lysigenous cavity and protoxylem: Lysigenous cavity and enlarged protoxylem vessel present

Size of metaxylem vessels in relation to parenchyma sheath cells in cross-section: Very wide vessels

Vascular bundle sheath

Third-order vascular bundle's bundle sheath (Fig. 1.22D, arrowhead)

Shape: Round

Shape of bundle sheath cell: Cells elliptical and elongated

Second-order vascular bundle's bundle sheath (Fig. 1.22C, F)

Shape: Circular

Shape of bundle sheath cell: Cells elliptical and elongated

Number of bundle sheath cell: 11–12

Nature of bundle sheath (either complete or incomplete): Complete

Chloroplast position in bundle sheath cell: Chloroplast centrally placed

Mestome/Inner sheath: Absent

First-order vascular bundle's bundle sheath

Shape: Triangular

Extent of bundle sheath around the vascular bundle: Complete

Extension of bundle sheath: Absent

Number of cells comprising the bundle sheath: 13–14 parenchyma cells comprise the sheath

Bundle sheath cell shape: All cells are elliptical, elongated in shape; chloroplast placed centrally

Mestome/Inner sheath: Absent

Sclerenchyma

Adaxial sclerenchyma: Minute strand consisting of only a few subepidermal fibers

Abaxial sclerenchyma: Narrow strand, few fiber cells follow the ribs and furrows

Sclerenchyma between bundles: Absent

Sclerenchyma in leaf margin: Forming a small pointed fibrous cap. Epidermal cells of margin are thin with outer thick tangential wall

Leaf-sheath anatomy (Fig. 1.22G–I)

Outline of lamina: "V" shaped with standard angle and overlapping inrolled margins (Fig. 1.22G). Minute ribs and furrows are present over the first-order vascular bundle

Epidermis

Adaxial epidermal cells: Flat, rectangular

Abaxial epidermal cells: Small, rectangular to barrel shaped with thick cuticle

Ground tissue: Parenchymatous with absence of air cavity

Vascular bundle

Arrangement of vascular bundle: Two consecutive third-order vascular bundles between first and second-order vascular bundle

Third-order vascular bundle: Circular with vascular tissue consisting of only few indistinguishable vascular elements (Fig. 1.22H)

Second-order vascular bundle: Circular shaped

First-order vascular bundle: Circular with phloem completely surrounded by thick-walled fibers, lysigenous cavity, and enlarged protoxylem vessel present; circular very wide metaxylem vessel elements

Vascular bundle sheath (Fig. 1.22H, I, arrowhead)

Shape: Round, circular

Extent of bundle sheath around the vascular bundle: Sheath incomplete, interrupted by few sclerenchyma cells

Number of cells comprising the bundle sheath: 15–16 parenchyma cells

Bundle sheath cell shape: Elliptical, elongated, and smaller than the ground tissue

Mestome/Inner sheath: Absent

Sclerenchyma: In form of straight horizontal well-developed narrow strand on abaxial surface

Ligule anatomy (Fig. 1.22J, K)

Shape: Ligule attached to the lamina till three-fourth portions then is free and blunt, round shaped at the connection (Fig. 1.22J, arrow)

Anatomy: Two layered mesophyll cells between adaxial and abaxial surfaces (Fig. 1.22K)

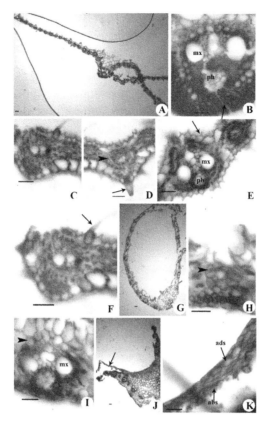

FIGURE 1.22   Leaf anatomy of *Capillipedium huegelii.*

8. *Chrysopogon fulvus*

Leaf lamina (Fig. 1.23A–F)

Outline of the lamina: "V" shaped with broad angle having straight arms (Fig. 1.23A). Two halves are symmetrical on either side of the median vascular bundle. Ribs and furrows are absent on the leaf surface.

Keel is well developed, V shaped with three first-order vascular bundles, six to seven second and third-order vascular bundles. Median and first-order vascular bundles appear to be of same size. Small, rounded parenchyma cells and chlorenchyma cells are present surrounding adaxial to the median vascular bundle.

Epidermis: Most part of adaxial epidermis made up of rounded, inflated bulliform cells (Fig. 1.23E, arrow); abaxial epidermis with round epidermal cells with thick cuticle; few cells have papillae (Fig. 1.23D, arrow)

Mesophyll: Radiate, single layered, isodiametric cells completely surrounding bundles; compactly arranged homogenous chlorenchymatous cells occupy the major area between the adaxial and abaxial epidermises and are continuous between bundles; colorless cells absent.

Vascular bundles: Six first-order vascular bundles regularly arrange from median to marginal bundle in half lamina; two to three third-order bundles between consecutive larger bundles and two to three second-order bundles between consecutive first-order bundles present; all the vascular bundles are present at same level and placed centrally

Third-order vascular bundles: Pentagonal with vascular tissue consisting of only few indistinguishable vascular elements (Fig. 1.23B, arrow)

Second-order vascular bundles: Circular shaped (Fig. 1.23F)

First-order vascular bundles

Shape: Oval

Relationship of phloem cells and the vascular fibers: Completely surrounded by thick-walled fibers

Nature of lysigenous cavity and protoxylem: Lysigenous cavity present, protoxylem vessel absent

Size of metaxylem vessels in relation to parenchyma sheath cells in cross-section: Narrow vessels

Shape of metaxylem vessels: Circular

Vascular bundle sheath

Third-order vascular bundle's bundle sheath (Fig. 1.23B, arrowhead)

Shape: Round

Shape of bundle sheath cell: Cells elliptical and elongated

Second-order vascular bundle's bundle sheath (Fig. 1.23F, arrowhead)

Shape: Round

Shape of bundle sheath cell: Cells elliptical and elongated

Number of bundle sheath cell: 8–9

Nature of bundle sheath (either complete or incomplete): Incomplete, abaxially interrupted by sclerenchyma

Chloroplast position in bundle sheath cell: Chloroplast concentrated outer tangential wall

Mestome/Inner sheath: Absent

First-order vascular bundle's bundle sheath

Shape: Round, circular

Extent of bundle sheath around the vascular bundle: Sheath incomplete, abaxially interrupted by sclerenchyma girder

Extension of bundle sheath: Broad adaxial extension of the bundle sheath

Number of cells comprising the bundle sheath: 13–14 parenchyma cells

Bundle sheath cell shape: All cells are elliptical, elongated in shape; chloroplast concentrated outer tangentially wall

Mestome/Inner sheath: Absent

Sclerenchyma

Adaxial sclerenchyma: Absent

Abaxial sclerenchyma: Well-developed narrow girder present (Fig. 1.23F, arrow)

Sclerenchyma between bundles: Absent

Sclerenchyma in leaf margin: Forming very small pointed fibrous cap (Fig. 1.23C, arrow); epidermal cells at margin are thin with outer thick tangential wall

Leaf-sheath anatomy (Fig. 1.23G–J)

Outline of lamina: V shaped with standard angle, concave inrolled margins (Fig. 1.23G); ribs and furrows absent

Epidermis

Adaxial and abaxial epidermal cells: Small, flat, rectangular with thick cuticle

Ground tissue: Parenchymatous with oval-shaped air cavity present (Fig. 1.23H, arrow)

Vascular bundle

Arrangement of vascular bundle: One consecutive third-order vascular bundle between first and second-order vascular bundle

Third-order vascular bundle: Circular with vascular tissue consisting of only few indistinguishable vascular elements (Fig. 1.23I, arrow)

Second-order vascular bundle: Oval shaped

First-order vascular bundle: Circular with phloem completely surrounded by thick-walled fibers; enlarged protoxylem vessel present, lysigenous cavity absent; circular wide metaxylem vessel elements

Vascular bundle sheath (Fig. 1.23I, arrowhead)

Shape: Round, circular

Extent of bundle sheath around the vascular bundle: Sheath incomplete, interrupted by parenchyma cells

Number of cells comprising the bundle sheath: 14–15 parenchyma cells

Bundle sheath cell shape: Elliptical, elongated, and smaller than the ground tissue

Mestome/Inner sheath: Absent

Sclerenchyma: In form of straight horizontal well-developed broad strand present on abaxial surface (Fig. 1.23J, arrow)

Ligule anatomy (Fig. 1.23K, L)

Shape: Ligule attached to the lamina more than half-length of it and blunt "V" shaped and narrow at the connection point (Fig. 1.23K)

Anatomy: Many layered mesophyll cells between adaxial and abaxial surfaces (Fig. 1.23L, arrow)

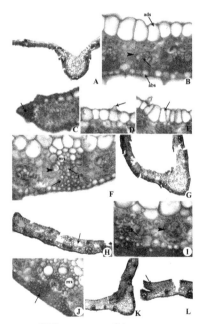

FIGURE 1.23 Leaf anatomy of *Chrysopogon fulvus*.

9. *Cymbopogon martinii*

Leaf lamina (Fig. 1.24A–F)

Outline of the lamina: Expanded with flat, even leaf blade, ribs and furrows absent (Fig. 1.24A, B)

Keel is well developed, semicircular shaped with single median vascular bundle and two first-order and two second-order vascular bundles. Round parenchyma cells surrounding median vascular bundle adaxially. Sclerenchyma present on adaxial surface in form of strands and on abaxial surface associated with vascular bundles in form of girder.

Epidermis: Adaxial epidermal surface has extensive group of large, inflated bulliform cells present throughout epidermis (Fig. 1.24D), size reducing opposite larger bundles and extending over many vascular bundles (Fig. 1.24E); abaxial epidermal cells barrel to round shaped covered with cuticle

Mesophyll: Radiate, single layered, isodiametric cells completely surrounding bundles; chlorenchyma cells continuous between bundles and occupy the major area between the adaxial and abaxial epidermises; colorless cells absent.

Vascular bundles: Seven to eight first-order vascular bundles regularly arranged from median to marginal bundle in half lamina. Eight third-order bundles and two second-order bundles between two consecutive first-order bundles present. All the vascular bundles present at same level and placed abaxially.

Third-order vascular bundles: Hexagonal with vascular tissue consisting of only few indistinguishable vascular elements (Fig. 1.24D)

Second-order vascular bundles: Circular shaped (Fig. 1.24E)

First-order vascular bundles (Fig. 1.24C)

Shape: Elliptical

Relationship of phloem cells and the vascular fibers: Completely surrounded by thick-walled fibers

Nature of lysigenous cavity and protoxylem: Protoxylem vessel present, lysigenous cavity absent

Size of metaxylem vessels in relation to parenchyma sheath cells in cross-section: Wide vessels

Shape of metaxylem vessels: Circular

Median vascular bundles

Shape: Elliptical

Relationship of phloem cells and the vascular fibers: Completely surrounded by thick-walled fibers

Nature of lysigenous cavity and protoxylem: Lysigenous cavity and enlarged protoxylem vessel present

Size of metaxylem vessels in relation to parenchyma sheath cells in cross-section: Wide vessels

Vascular bundle sheath

Third-order vascular bundle's bundle sheath (Fig. 1.24D, arrowhead)

Shape: Round

Shape of bundle sheath cell: Cells elliptical and elongated

Second-order vascular bundle's bundle sheath (Fig. 1.24E, arrowhead)

Shape: Round

Shape of bundle sheath cell: Cells elliptical and elongated

Number of bundle sheath cell: 7–8

Nature of bundle sheath (either complete or incomplete): Complete

Chloroplast position in bundle sheath cell: Chloroplast concentrated near outer tangential wall

Mestome/Inner sheath: Absent

First-order vascular bundle's bundle sheath (Fig. 1.24C, arrowhead)

Shape: Elliptical

Extent of bundle sheath around the vascular bundle: Complete

Extension of bundle sheath: Narrow, subepidermal strip of girder present on abaxial surface of the bundle sheath

Number of cells comprising the bundle sheath: 15–16 parenchyma cells

Bundle sheath cell shape: All cells are elliptical, elongated in shape; chloroplast concentrated near outer tangential wall

Mestome/Inner sheath: Absent

Sclerenchyma

Adaxial sclerenchyma: Minute strand consisting of only a few subepidermal fibers

Abaxial sclerenchyma: Narrow, subepidermal strip of girder

Sclerenchyma between bundles: Absent

Sclerenchyma in leaf margin: Forming a well-developed pointed fibrous cap. Epidermal cells thin with thick tangential wall

Leaf-sheath anatomy (Fig. 1.24G–I)

Outline of lamina: Round shaped with single overlapping inrolled margins; minute, shallow ribs and furrows present (Fig. 1.24G)

Epidermis

Adaxial epidermal cells: Flat, rectangular

Abaxial epidermal cells: Small, flat, rectangular shaped with thick cuticle

Ground tissue: Parenchymatous ground tissue, air cavity absent

Vascular bundle

Arrangement of vascular bundle: One consecutive third-order vascular bundle between first and second-order vascular bundle

Third-order vascular bundle: Circular shaped, vascular tissue consists of only a few indistinguishable vascular elements (Fig. 1.24I)

Second-order vascular bundle: Oval shaped

First-order vascular bundle: Circular shaped, phloem completely surrounded by thick-walled fibers, lysigenous cavity and enlarged protoxylem vessel present; circular very wide metaxylem vessel elements (Fig. 1.24H)

Vascular bundle sheath (Fig. 1.24H, I, arrowhead)

Shape: Round, circular

Extent of bundle sheath around the vascular bundle: Sheath incomplete, interrupted by sclerenchyma cells

Number of cells comprising the bundle sheath: 22–24 parenchyma cells

Bundle sheath cell shape: Elliptical, elongated, and smaller than the ground tissue

Mestome/Inner sheath: Absent

Sclerenchyma: In form of narrow well-developed girder present on abaxial surface

Ligule anatomy (Fig. 1.24J, K)

Shape: Ligule attached to the half width of the leaf sheath region and it is round shaped at the connection point (Fig. 1.24J, arrow)

Anatomy: More than three layered mesophyll cells between adaxial and abaxial surfaces (Fig. 1.24K)

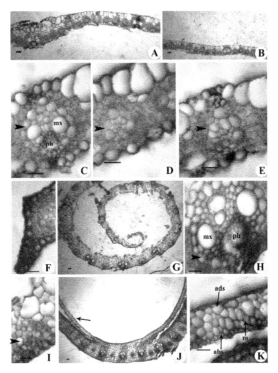

FIGURE 1.24   Leaf anatomy of *Cymbopogon martinii.*

Dicanthium

10. *Dicanthium annulatum* (Fig. 1.25)

11. *Dicanthium caricosum* (Fig. 1.26)

Leaf lamina (Figs. 1.25, 1.26A–E)

Outline of the lamina: "V" shaped with narrow angle and irregularly wavy (Fig. 1.25A, 1.26A)

Keel is well developed, inverted "T" shaped with single median vascular bundle and two second-order and four third-order vascular bundles. Rounded parenchymas present surrounding the median vascular bundle abaxially. Sclerenchyma present on abaxial surface in form of horizontal well-developed strands (Figs. 1.25C, 1.26D, arrow).

Epidermis: Adaxial epidermal surface has barrel to square-shaped epidermal cells interrupted by restricted groups of large, inflated bullifom cells in level with epidermal cells (Fig. 1.25D, arrow; Fig. 1.26B, arrow).

Abaxial epidermal cells have barrel to rectangular shaped epidermal cells covered with cuticle.

Mesophyll: Radiate, single layered, tubular cells completely surrounding bundles; chlorenchyma cells continuous between bundles and occupying the major area between the adaxial and abaxial epidermises; colorless cells absent.

Vascular bundles: Five to six first-order vascular bundles in *D. annulatum*, seven to eight first-order vascular bundles in *D. caricosum* regularly arranged from median to marginal bundle in half lamina.

Four third-order bundles and three second-order bundles between two consecutive first-order bundles present; all the vascular bundles present at same level and placed abaxially

Third-order vascular bundles (Figs. 1.25B, 1.26E, 1.26B)

Pentagonal shaped in *D. annulatum*; pentagonal (Fig. 1.26E) and elliptical shaped both in *D. caricosum* (Fig. 1.26B), xylem and phloem groups distinguishable

Second-order vascular bundles: Circular shaped

First-order vascular bundles (Figs. 1.25, 1.26C)

Shape: Obovate

Relationship of phloem cells and the vascular fibers: Completely surrounded by thick-walled fibers

Nature of lysigenous cavity and protoxylem: Protoxylem vessel present, lysigenous cavity absent

Size of metaxylem vessels in relation to parenchyma sheath cells in cross-section: Very wide vessels

Shape of metaxylem vessels: Circular

Median vascular bundles (Fig. 1.26D)

Shape: Obovate

Relationship of phloem cells and the vascular fibers: Completely surrounded by thick-walled fibers

Nature of lysigenous cavity and protoxylem: Protoxylem vessel present, lysigenous cavity absent

Size of metaxylem vessels in relation to parenchyma sheath cells in cross-section: Very wide vessels

Vascular bundle sheath

Third-order vascular bundle's bundle sheath

Shape: Round in *D. annulatum* (Fig. 1.25B, D, arrowhead); round (Fig. 1.26E, arrowhead) and elliptical (Fig. 1.26B, arrowhead) both in *D. caricosum*

Shape of bundle sheath cell: Cells elliptical and elongated

Second-order vascular bundle's bundle sheath

Shape: Round

Shape of bundle sheath cell: Cells elliptical and elongated

Number of bundle sheath cell: 8–9

Nature of bundle sheath (either complete or incomplete): Complete

Chloroplast position in bundle sheath cell: Chloroplast centrally placed

Mestome/Inner sheath: Absent

First-order vascular bundle's bundle sheath

Shape: Elliptical

Extent of bundle sheath around the vascular bundle: Incomplete, adaxially interrupted by parenchyma and abaxially interrupted by sclerenchyma cells

Extension of bundle sheath: Narrow, subepidermal strip of girder present on abaxial surface of the bundle sheath

Number of cells comprising the bundle sheath: 13–14 parenchyma cells

Bundle sheath cell shape: All cells are elliptical, elongated with different sizes; chloroplast centrally placed

Mestome/Inner sheath: Absent

Sclerenchyma

Adaxial sclerenchyma: Absent

Abaxial sclerenchyma: Minute strand consisting of only a few subepidermal fibers

Sclerenchyma between bundles: Absent

Sclerenchyma in leaf margin: Forming a well-developed fibrous pointed cap; epidermal cells of margin are thin with thick tangential wall (Figs. 1.25F, 1.26C)

Leaf-sheath anatomy (Figs. 1.25, 1.26F–H)

Outline of lamina: Round shaped (Figs. 1.25, 1.26F); ribs and furrows absent

Epidermis

Adaxial epidermal and abaxial epidermal cells: Small, flat, rectangular shaped with thick cuticle

Ground tissue: Parenchymatous with absence of air cavity

Vascular bundle

Arrangement of vascular bundle: One consecutive third-order vascular bundle between first and second-order vascular bundle

Third-order vascular bundle: Circular shaped, xylem and phloem groups distinguishable (Figs. 1.25H and 1.26G)

Second-order vascular bundle: Circular shaped

First-order vascular bundle: Circular shaped, phloem completely surrounded by thick-walled fibers, protoxylem vessel present but no lysigenous cavity; circular very wide metaxylem vessel elements (Figs. 1.25G and 1.26H)

Vascular bundle sheath (Fig. 1.25G, H, arrowhead)

Shape: Round, circular

Extent of bundle sheath around the vascular bundle: Sheath incomplete, interrupted by sclerenchyma cells

Number of cells comprising the bundle sheath: 12–14 parenchyma cells comprise the sheath

Bundle sheath cell shape: Elliptical, elongated, and smaller than the ground tissue

Mestome/Inner sheath: Absent

Sclerenchyma: In form of wide, horizontal band on abaxial surface

Ligule anatomy (Figs. 1.25 and 1.26I, J)

Shape: Ligule attached along half width of the leaf sheath region and it is round shaped at the connection point (Figs. 1.25 and 1.26I, arrow)

Anatomy: Three layered mesophyll cells between adaxial and abaxial surfaces (Figs. 1.25 and 1.26J), *D. annulatum* shows absence of sclerenchyma while *D. caricosum* shows presence of sclerenchyma on abaxial surface of ligule

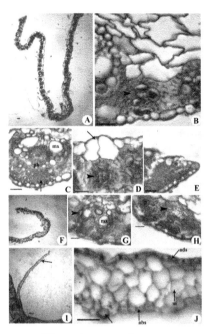

FIGURE 1.25   Leaf anatomy of *Dicanthium annulatum.*

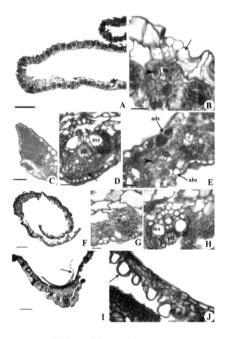

FIGURE 1.26   Leaf anatomy of *Dicanthium caricosum.*

DIFFERENTIATING FEATURES

Sclerenchyma absent on abaxial surface of ligule ... *D. annulatum*

Sclerenchyma present on abaxial surface of ligule ... *D. caricosum*

12. *Hackelochloa granularis*

Leaf lamina (Fig. 1.27A–D)

Outline of the lamina: Expanded and distinctly wavy (Fig. 1.27A); ribs and furrows present on leaf surface; adaxial surface with shallow, wide furrows between all vascular bundles; abaxial surface with ribs taller than the adaxial ribs and groove present between all vascular bundles.

Keel is not well developed; sclerenchyma is present on abaxial surface in form of narrow subepidermal strands.

Epidermis: Adaxial epidermal surface has barrel to round-shaped epidermal cells covered with thick cuticle. Abaxial epidermal cells have barrel to rect-angular shaped epidermal cells covered with cuticle.

Mesophyll: Radiate, single layered, isodiametric cells completely surrounding bundles; compact-arranged chlorenchyma cells continuous between bundles, occupying the major area between the adaxial and abaxial epidermises; colorless cells absent.

Vascular bundles: Four to five first-order vascular bundles in half lamina regularly arranged from median to marginal bundle. Two third-order bundles and one second-order bundles between two consecutive first-order bundles present. All the vascular bundles present at same level and placed abaxially

Third-order vascular bundles: Circular with xylem and phloem groups distinguishable (Fig. 1.27C)

Second-order vascular bundles: Circular shaped (Fig. 1.27D)

First-order vascular bundles

Shape: Elliptical

Relationship of phloem cells and the vascular fibers: Phloem adjoins the inner or parenchyma sheath

Nature of lysigenous cavity and protoxylem: Lysigenous cavity and enlarged protoxylem vessel present

Size of metaxylem vessels in relation to parenchyma sheath cells in cross-section: Wide vessels

Shape of metaxylem vessels: Circular

Vascular bundle sheath

Third-order vascular bundle's bundle sheath (Fig. 1.27C, arrowhead)
Shape: Round
Shape of bundle sheath cell: Cells elliptical and elongated
Second-order vascular bundle's bundle sheath (Fig. 1.27D, arrowhead)
Shape: Round
Shape of bundle sheath cell: Cells elliptical and elongated
Number of bundle sheath cell: 7–8
Nature of bundle sheath (either complete or incomplete): Complete
Chloroplast position in bundle sheath cell: Chloroplast small and indistinct
Mestome/Inner sheath: Absent
First-order vascular bundle's bundle sheath
Shape: Elliptical
Extent of bundle sheath around the vascular bundle: Complete
Extension of bundle sheath: Absent
Number of cells comprising the bundle sheath: 19–20 parenchyma cells
Bundle sheath cell shape: All cells are elliptical, elongated with different sizes; chloroplast small and indistinct
Mestome/Inner sheath: Absent

Sclerenchyma

Adaxial sclerenchyma: Absent
Abaxial sclerenchyma: Well-developed broad girder present
Sclerenchyma between bundles: Absent
Sclerenchyma in leaf margin: Forming a very small rounded cap; epidermal cells of margin are thin with thick tangential wall; prickle present at margin (Fig. 1.27B, arrow)
Leaf-sheath anatomy (Fig. 1.27E–G)
Outline of lamina: Round shaped (Fig. 1.27E); ribs and furrows absent
Epidermis
Adaxial epidermal cells and abaxial epidermal cells: Small, flat, rectangular shaped with thick cuticle

Ground tissue: Parenchymatous with horizontally oval-shaped air cavity present (Fig. 1.27E)

Vascular bundle

Arrangement of vascular bundle: One consecutive third-order vascular bundle between first and second-order vascular bundle

Third-order vascular bundle: Circular shaped, xylem and phloem groups distinguishable (Fig. 1.27G)

Second-order vascular bundle: Circular shaped

First-order vascular bundle: Circular shaped, Phloem adjoins the inner or parenchyma sheath, cavity and enlarged protoxylem vessel present; circular wide metaxylem vessel elements (Fig. 1.27F)

Vascular bundle sheath (Fig. 1.27F, G, arrowhead)

Shape: Round, circular

Extent of bundle sheath around the vascular bundle: Sheath incomplete, interrupted by sclerenchyma cells abaxially and parenchyma cells adaxially

Number of cells comprising the bundle sheath: 12–14 parenchyma cells

Bundle sheath cell shape: Elliptical, elongated, and smaller than the ground tissue

Mestome/Inner sheath: Absent

Sclerenchyma: In form of wide, horizontal band well-developed girder present on abaxial surface

Ligule anatomy (Fig. 1.27H, I)

Shape: Ligule attached to the leaf sheath more than the half width and very broad round shaped at the connection point (Fig. 1.27H, arrow)

Anatomy: Two layered mesophyll cells between adaxial and abaxial (Fig. 1.27I)

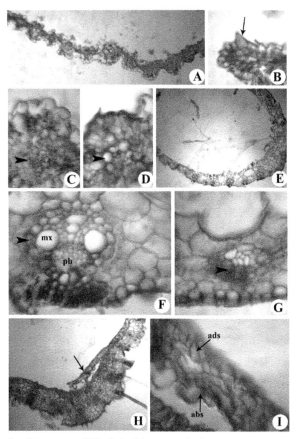

FIGURE 1.27   Leaf anatomy of *Hackelochloa granularis*.

Heteropogon

13. *Heteropogon contortus* var. *genuinus* subvar. *typicus* (Fig. 1.28)

14. *Heteropogon contortus* var. *genuinus* subvar. *hispidissimus* (Fig. 1.29)

15. *Heteropogon ritcheii* (Fig. 1.30)

16. *Heteropogon triticeus* (Fig. 1.31)

Leaf lamina (Figs. 1.28A–E, 1.29A–E, 1.30A–D, 1.31A–E)

Outline of the lamina: "V" shaped with wide angle having straight arms (Figs. 1.28A, 1.29A, B, 1.30A, 1.31A). Two halves are symmetrical on either side of the median vascular bundle.

Ribs and furrows are present on the leaf surface. Adaxial surface has wide, shallow furrow in all the three species except *H. triticeus* in which the furrows were median and wide. Triangular ribs present with pointed apex in *H. contortus* var. *genuinus* subvar. *typicus*, *H. contortus* var. *genuinus* subvar. *hispidissimus*, and *H. ritcheii* and round ribs with round apex in *H. triticeus* present only over first and second-order vascular bundles. Abaxial surface has shallow groove in *H. contortus* var. *genuinus* subvar. *typicus*, *H. contortus* var. *genuinus* subvar. *hispidissimus*, and *H. ritcheii* and smooth or few undulations in *H. triticeus* between vascular bundles. Abaxial ribs smaller than adaxial ribs.

Keel is well-developed, bulliform cells penetrating into leaf on either side of median bundle forming inverted "V" shape with single median vascular bundle in *H. contortus* var. *genuinus* subvar. *typicus* and *H. contortus* var. *genuinus* subvar. *hispidissimus*, "V"-shaped keel with single median vascular bundle in *H. ritcheii* and "V"-shaped keel without single median vascular bundle in *H. triticeus*.

Two second-order and 3 third-order vascular bundles in *H. contortus* var. *genuinus* subvar. *typicus*, *H. contortus* var. *genuinus* subvar. *hispidissimus*, and *H. ritcheii* and 9 first-order, 9 second-order, and 22–24 third-order vascular bundles in *H. triticeus*. Small, round parenchyma cells present surrounding the median vascular bundle adaxially.

Sclerenchyma on adaxial surface in form of subepidermal strands and on abaxial surface associated with vascular bundles in form of well-developed girder (Fig. 1.28D).

Epidermis: Adaxial epidermal surface with isolated large, inflated bulliform cells (Figs. 1.28B, arrow; 1.29F, 1.30D), interrupted by irregular, different sized small groups of epidermal cells (Fig. 1.29D); *H. triticeus* has bulliform cells smaller than barrel-shaped epidermal cells (Fig. 1.31G, arrow); stomata present (Fig. 1.29D, arrow).

Abaxial epidermal cells are large, round in *H. contortus* var. *genuinus* subvar. *typicus*, *H. contortus* var. *genuinus* subvar. *hispidissimus* and *H. ritcheii* while barrel shaped in *H. triticeus*; all four species showed presence of single papillae on one cell (Figs. 1.28C, arrow, 1.29D, arrow, 1.30D, 1.31D, arrow); prickles also present (Fig. 1.28B).

Mesophyll: Radiate, single layered, tubular chlorenchyma cells completely surrounding bundles; compact homogenous chlorenchymatous cells occupy the major area between the adaxial and abaxial epidermises; chlorenchyma cells continuous between bundles; colorless cells absent.

Vascular bundles: Five first-order vascular bundles in *H. contortus* var. *genuinus* subvar. *typicus*, *H. contortus* var. *genuinus* subvar. *hispidissimus*, and *H. ritcheii*, 15–17 first-order vascular bundles in *H. triticeus* regularly arranged from median to marginal bundle in half lamina.

Three third and second-order bundles in *H. contortus* var. *genuinus* subvar. *typicus*, *H. contortus* var. *genuinus* subvar. *hispidissimus*, and *H. ritcheii*, 65–70 third-order and 15–17 second-order vascular bundles in *H. triticeus* between consecutive first-order bundles present.

All the vascular bundles are present at the same level and placed centrally in *H. contortus* var. *genuinus* subvar. *typicus*, *H. contortus* var. *genuinus* subvar. *hispidissimus*, and *H. ritcheii* while in *H. triticeus* it is abaxially.

Third-order vascular bundles (Figs. 1.28B; 1.29D, 1.30D, 1.31D, F)

Pentagonal shaped in *H. contortus* var. *genuinus* subvar. *typicus*, *H. contortus* var. *genuinus* subvar. *hispidissimus*, and *H. ritcheii*, circular shaped (Fig. 1.31D) and elliptical (Fig. 1.31F) in *H. triticeus*; vascular tissue consisting of only few indistinguishable vascular elements

Second-order vascular bundles: Circular shaped in *H. contortus* var. *genuinus* subvar. *typicus*, *H. contortus* var. *genuinus* subvar. *hispidissimus*, and *H. ritcheii*, elliptical shaped in *H. triticeus* (Fig. 1.31C)

First-order vascular bundles (Figs. 1.28C, 1.29C, 1.30C, E, 1.31E)

Shape: Circular

Relationship of phloem cells and the vascular fibers: Completely surrounded by thick-walled fibers

Nature of lysigenous cavity and protoxylem: Protoxylem vessel present with absence of lysigenous cavity in *H. contortus* var. *genuinus* subvar. *typicus*, *H. contortus* var. *genuinus* subvar. *hispidissimus* while presence of lysigenous cavity in *H. ritcheii* and *H. triticeus*

Size of metaxylem vessels in relation to parenchyma sheath cells in cross-section: Wide vessels in all species

Shape of metaxylem vessels: Circular

Median vascular bundles

Shape: Round

Relationship of phloem cells and the vascular fibers: Completely surrounded by thick-walled fibers

Nature of lysigenous cavity and protoxylem: Lysigenous cavity and enlarged protoxylem vessel present

Size of metaxylem vessels in relation to parenchyma sheath cells in cross-section: Wide vessels

Vascular bundle sheath

Third-order vascular bundle's bundle sheath (Figs. 1.28B, 1.29D, 1.30D, 1.31D and F, arrowhead)

Shape: Round in *H. contortus* var. *genuinus* subvar. *typicus*, *H. contortus* var. *genuinus* subvar. *hispidissimus*, and *H. ritcheii* while round (Fig. 1.31D) and elliptical (Fig. 1.31F) in *H. triticeus.*

Shape of bundle sheath cell: Cells elliptical and elongated

Second-order vascular bundle's bundle sheath

Shape: Round in *H. contortus* var. *genuinus* subvar. *typicus*, *H. contortus* var. *genuinus* subvar. *hispidissimus*, and *H. ritcheii* while elliptical in *H. triticeus*

Shape of bundle sheath cell: Cells elliptical and elongated

Number of bundle sheath cell: 9–10 in *H. contortus* var. *genuinus* subvar. *typicus*, *H. contortus* var. *genuinus* subvar. *hispidissimus*, 11 in *H. ritcheii*, and 12–13 in *H. triticeus*

Nature of bundle sheath (either complete or incomplete): Complete in *H. contortus* var. *genuinus* subvar. *typicus*, *H. contortus* var. *genuinus* subvar. *hispidissimus*, and *H. ritcheii* and incomplete sheath in *H. triticeus* in which it abaxially interrupted by sclerenchyma cells

Chloroplast position in bundle sheath cell: Chloroplast centrally placed

Mestome/Inner sheath: Absent

First-order vascular bundle's bundle sheath (Figs. 1.28C, 1.29C, 1.30C, 1.31C, arrowhead)

Shape: Round, circular

Extent of bundle sheath around the vascular bundle: Sheath incomplete, abaxially interrupted by sclerenchyma

Extension of bundle sheath: Broad abaxial extension of the bundle sheath

Number of cells comprising the bundle sheath: 13–14 parenchyma cells

Bundle sheath cell shape: All cells are elliptical, elongated in shape; chloroplast centrally placed

Mestome/Inner sheath: Absent

Sclerenchyma

Adaxial sclerenchyma: Small strand present above the first-order vascular bundle

Abaxial sclerenchyma: Well-developed strand, wider than deep and it follows shape of abaxial rib in *H. contortus* var. *genuinus* subvar. *typicus*, *H. contortus* var. *genuinus* subvar. *hispidissimus*, and *H. ritcheii* and narrow strands present in *H. triticeus*

Sclerenchyma between bundles: Absent

Sclerenchyma in leaf margin: Forming very small round fibrous cap; epidermal cells of margin are thin with outer thick tangential wall (Figs. 1.28E, 1.29B, 1.31H)

Leaf-sheath anatomy (Figs. 1.28F–H, 1.29G–I, 1.30E–H, 1.31G–I)

Outline of lamina: "V" shaped with standard angle with inrolled margins (Figs. 1.28F, 1.29G, 1.30E, 1.31I)

Epidermis

Adaxial and abaxial epidermal cells: Small, flat, rectangular shaped with thick cuticle

Ground tissue: Parenchymatous with absence of air cavity in *H. contortus* var. *genuinus* subvar. *typicus*, *H. contortus* var. *genuinus* subvar. *hispidissimus*, and *H. ritcheii* and oval-shaped air cavity present in *H. triticeus*

Vascular bundle

Arrangement of vascular bundle: Two consecutive third-order in *H. contortus* var. *genuinus* subvar. *typicus*, *H. contortus* var. *genuinus* subvar. *hispidissimus*, and *H. ritcheii* and six to seven consecutive third-order vascular bundles in *H. triticeus* between first and second-order vascular bundle.

Third-order vascular bundle: Pentagonal shaped in *H. contortus* var. *genuinus* subvar. *typicus* and *H. contortus* var. *genuinus* subvar. *hispidissimus* (Figs. 1.28, 1.29H), circular in *H. ritcheii* and *H. triticeus*; vascular tissue consisting of only a few indistinguishable vascular elements

Second-order vascular bundle: Elliptical in *H. contortus* var. *genuinus* subvar. *typicus*, *H. contortus* var. *genuinus* subvar. *hispidissimus*, and *H. ritcheii* and circular in *H. triticeus*.

First-order vascular bundle (Figs. 1.28G, 1.29H, I, 1.30F, 1.31M): Circular shaped, phloem completely surrounded by thick-walled fibers, lysigenous cavity, and enlarged protoxylem vessel present; circular wide in *H. contortus* var. *genuinus* subvar. *typicus*, *H. contortus* var. *genuinus* subvar. *hispidissimus*, and *H. ritcheii* and very wide in *H. triticeus* metaxylem vessel elements

Vascular bundle sheath (Figs. 1.28G, 1.29H, I, arrowhead, 1.30F, 1.31M)

Shape: Obovate in *H. contortus* var. *genuinus* subvar. *typicus*, *H. contortus* var. *genuinus* subvar. *hispidissimus*, and *H. ritcheii*, circular in *H. triticeus*

Extent of bundle sheath around the vascular bundle: Sheath incomplete, interrupted by parenchyma cells

Number of cells comprising the bundle sheath: 14–15 in *H. contortus* var. *genuinus* subvar. *typicus* and *H. contortus* var. *genuinus* subvar. *hispidissimus*, 12–13 in *H. ritcheii* and 20–22 in *H. triticeus* parenchyma cells

Bundle sheath cell shape: Elliptical, elongated, and smaller than the ground tissue

Mestome/Inner sheath: Absent

Sclerenchyma: In form of straight horizontal well-developed girder in *H. contortus* var. *genuinus* subvar. *typicus*, *H. contortus* var. *genuinus* subvar. *hispidissimus*, and *H. ritcheii*; wide, deep horizontal well-developed girder in *H. triticeus* present on abaxial surface (Figs. 1.28G, H, arrow, 1.30F, arrow)

Ligule anatomy (Figs. 1.28I, 1.29J, K, 1.30I, 1.31O)

Shape: Ligule attached to the lamina only one-fourth portions, and at connection point, it was very broad almost like straight "V" shaped with small sharp angled in *H. contortus* var. *genuinus* subvar. *typicys* and *H. contortus* var. *genuinus* subvar. *hispidissimus*, ligule attached to the lamina more than half-length of it and blunt "V" shaped at the connection in *H. ritcheii* and ligule attached less than half-length of the lamina and at the connection point it is very broad, blunt "V" shaped in *H. triticeus*.

Anatomy: Many layered in *H. contortus* var. *genuinus* subvar. *typicus* and *H. contortus* var. *genuinus* subvar. *hispidissimus* and two layered in *H. ritcheii* and *H. triticeus* mesophyll cells between adaxial and abaxial surfaces (Figs. 1.28I, 1.29J, 1.30I, arrow); prickles present on abaxial surface only in *H. contortus* var. *genuinus* subvar. *hispidissimus* (Fig. 1.29K, arrow)

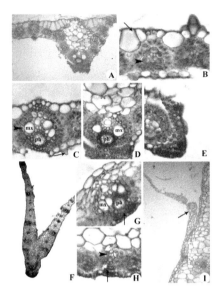

FIGURE 1.28   Leaf anatomy of *Heteropogon contortus var. genuinus sub var. typicus.*

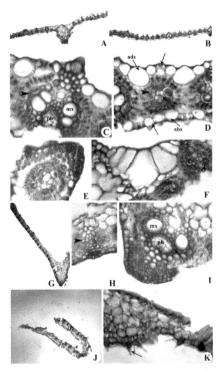

FIGURE 1.29   Leaf anatomy of *Heteropogon contortus var. genuinus sub var. hispidissimus.*

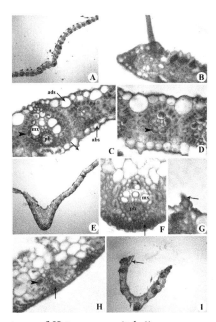

FIGURE 1.30 Leaf anatomy of *Heteropogon ritcheii.*

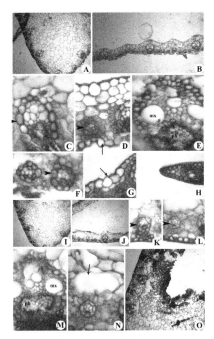

FIGURE 1.31 Leaf anatomy of *Heteropogon triticeus.*

DIFFERENTIATING FEATURES

Vascular bundles arranged toward abaxial side of leaf blade … *H. triticeous*
Vascular bundles arranged in the center of the leaf blade
Two layered mesophyll present in ligule … *H. ritchiei*
More than three layered of mesophyll present in ligule
Presence of prickles on abaxial surface of ligule … *H. contortus* var. *genuinus* subvar. *hispidissimus*
Absence of prickles on abaxial surface of ligule … *H. contortus* var. *genuinus* subvar. *typicus*

17. *Imperata cylindrica*

Leaf lamina (Fig. 1.32A–E)

Outline of the lamina: Expanded and gently undulating shape of the arms (Fig. 1.32A). Two straight halves are symmetrical on either side of the median vascular bundle; ribs and furrows are present on the leaf surface; very shallow and wide furrows present between first-order vascular bundles.

Keel is well developed, round with single first-order bundle, two second and third-order bundles. Small, rounded parenchyma cells are present surrounding the median vascular bundle adaxially. Sclerenchyma present on abaxial surface of keel associated with vascular bundles in form anchor-shaped girder.

Epidermis: Adaxial epidermal surface with fan-shaped bulliform cells (Fig. 1.32E) without furrows or with a very shallow furrow; round abaxial epidermal cells with cuticle and cell wall equal to or greater than the depth of the average epidermal cells; long, slender macrohairs present.

Mesophyll: Indistinctly or incompletely radiating chlorenchyma cells which were divided by group of colorless cells; smaller colorless cells closely associated with bulliform cells with two extensions (Fig. 1.32E, arrow)

Vascular bundles: Four first-order vascular bundles in half and progressively fewer first and more third-order bundles toward margin. Three third-order bundles between consecutive larger bundles and 1 second-order bundles between consecutive first-order bundles present. All the vascular bundles present at same level and placed centrally.

Third-order vascular bundles: Elongated shaped and vascular tissue consisting of only few indistinguishable vascular elements (Fig. 1.32C)

Second-order vascular bundles: Oval shaped

First-order vascular bundles (Fig. 1.32D)

Shape: Elliptical

Relationship of phloem cells and the vascular fibers: Completely surrounded by thick-walled fibers

Nature of lysigenous cavity and protoxylem: Lysigenous cavity and enlarged protoxylem vessel present

Size of metaxylem vessels in relation to parenchyma sheath cells in cross-section: Narrow vessels

Shape of metaxylem vessels: Circular

Vascular bundle sheath

Third-order vascular bundle's bundle sheath (Fig. 1.32C, arrowhead)

Shape: Round

Shape of bundle sheath cell: Cells with straight radial walls and tangential inflated walls

Second-order vascular bundle's bundle sheath

Shape: Round

Shape of bundle sheath cell: Cells with straight radial walls and tangential inflated walls

Number of bundle sheath cell: 9–10

Nature of bundle sheath (either complete or incomplete): Incomplete, abaxially interrupted by sclerenchyma

Chloroplast position in bundle sheath cell: Translucent sheath cell

Mestome/Inner sheath: Absent

First-order vascular bundle's bundle sheath (Fig. 1.32D, arrowhead)

Shape: Elliptical

Extent of bundle sheath around the vascular bundle: Sheath incomplete, abaxially interrupted by sclerenchyma

Extension of bundle sheath: Broad abaxial extension of the bundle sheath

Number of cells comprising the bundle sheath: 12–13 parenchyma cells

Bundle sheath cell shape: Cells with straight radial walls and tangential inflated walls; translucent chloroplast in sheath cell

Mestome/Inner sheath: Absent

Sclerenchyma

Adaxial sclerenchyma: Well-developed strand present

Abaxial sclerenchyma: Well-developed narrow and deep girder

Sclerenchyma between bundles: Absent

Sclerenchyma in leaf margin: Forming a well-developed rounded cap; consists of a couple of fibers; epidermal cells of margin are thin with thick radial wall

Leaf-sheath anatomy (Fig. 1.32F–I)

Outline of lamina: "U" shaped (Fig. 1.32F)

Epidermis

Adaxial epidermal cells: Flat, rectangular

Abaxial epidermal cells: Small, flat, square shaped with thick cuticle

Ground tissue: Parenchymatous with round-shaped air cavity (Fig. 1.32F)

Vascular bundle

Arrangement of vascular bundle: One consecutive third-order vascular bundle between first and second-order vascular bundle

Third-order vascular bundle: Hexagonal with vascular tissue consisting of only few indistinguishable vascular elements (Fig. 1.32G)

Second-order vascular bundle: Circular shaped (Fig. 1.32H)

First-order vascular bundle: Elliptical shaped, phloem completely surrounded by thick-walled fibers (Fig. 1.32I, arrow), lysigenous cavity and enlarged proto-xylem vessel present; circular narrow metaxylem vessel elements (Fig. 1.32I)

Vascular bundle sheath (Fig. 1.32G, H, arrowhead)

Shape: Elliptical

Extent of bundle sheath around the vascular bundle: Sheath incomplete, interrupted by sclerenchyma cells

Number of cells comprising the bundle sheath: 13–14 parenchyma cells

Bundle sheath cell shape: Cells with straight radial walls and tangential inflated walls; cells similar in size of the ground tissue

Mestome/Inner sheath: Absent

Sclerenchyma: In form of narrow, deep well-developed girder present on abaxial surface

Ligule anatomy (Fig. 1.32J, K)

Shape: Ligule attached to the lamina more than half-length of it and blunt "V" shaped at the connection point and narrow (Fig. 1.32J, arrow)

Anatomy: Two layered mesophyll cells between adaxial and abaxial surfaces (Fig. 1.32K, arrow)

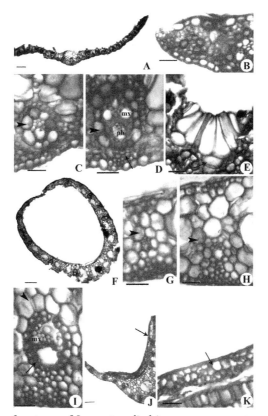

FIGURE 1.32 Leaf anatomy of *Imperata cylindrica*.

Ischaemum

18. *Ischaemum indicus* (Fig. 1.33)

19. *Ischaemum molle* (Fig. 1.34)

20. *Ischaemum pilosum* (Fig. 1.35)

21. *Ischaemum rugosum* (Fig. 1.36)

Leaf lamina (Figs. 1.33A–E, 1.34A–E, 1.35A–I, 1.36A–G)

Outline of the lamina: "V" shaped with standard angled having straight shape of the arms in *I. indicus* and *I. rugosum* (Figs. 1.33A, 1.36A), with wide angled having straight arms *I. pilosum* (Fig. 1.35A) or with outwardly curving arms in *I. molle* (Fig. 34A)

Two halves are symmetrical on either side of the median vascular bundle. Ribs and furrows are absent on the leaf surface in all the species except *I. pilosum*. Ribs are present on first and second-order vascular bundles and the furrows are shallow, round, wide, and with a sharp apex (Fig. 1.35I).

Keel is well developed, "V" shaped with single median vascular bundle, presence of two second and four third-order vascular bundles in *I. indicus*; five to six third-order vascular bundles *I. pilosum*. Keel is semicircular in *I. molle* or inverted triangular shaped in *I. rugosum* (Fig. 1.36B)

Small, round parenchyma cells are present surrounding the median vascular bundle adaxially. Sclerenchyma present on adaxial surface in form of strand in the subepidermal layers and on abaxial surface associated with vascular bundles in form of well-developed strands in all species.

Epidermis: Adaxial epidermal surface has most part of the epidermis made up of inflated bulliform cells in *I. indicus* and *I. pilosum* (Figs. 1.33B, 1.35H). Abaxial epidermal cells are barrel shaped with one in (Figs. 1.34D, 1.36D) or two papillae per cell in *I. rugosum* (Fig. 1.36E, arrow). Stomata could be observed interrupting the epidermal layer in *I. pilosum* and *I. rugosum* (Figs. 1.35G, 1.36D, arrow). Prickles, macrohiars, hooks are absent.

Mesophyll: Chlorenchyma is not arranged in any specific pattern. Compactly arranged homogenous layer of chlorenchyma cells is present between the adaxial and abaxial epidermises. Chlorenchyma cells continuous between bundles and arranged in form of a horizontally elongated strap; colorless cells absent.

Vascular bundles: Two to three first-order vascular bundles present in all the species except *I. pilosum* which has five to six first-order vascular bundles arranged in a regular manner from median to marginal bundles in half lamina.

I. indicus and *I. rugosum* have three third-order and two second-order bundles, *I. molle* has seven to eight third-order and two to three second-order bundles and *I. pilosum* has four third-order and three second-order bundles between consecutive larger bundles.

All the vascular bundles present at same level and placed abaxially in *I. indicus* and *I. rugosum* or centrally in *I. molle*. Vascular bundles placed at different levels in *I. pilosum*, that is, first and second-order bundles present in center and third-order bundles present on abaxial surface of leaf blade.

Third-order vascular bundles (Figs. 1.33E, 1.34C, 1.35D, 1.36E)

Pentagonal and circular in *I. indicus*, circular in *I. molle*, only pentagonal in *I. pilosum* and *I. rugosum*; xylem and phloem groups distinguishable in all except *I. molle* has vascular tissue consists of only a few indistinguishable vascular elements

Second-order vascular bundles: Circular in all species except in *I. pilosum* elliptical

First-order vascular bundles (Figs. 1.33D, 1.35F, 1.36G)

Shape: Circular in all species except in *I. pilosum* elliptical

Relationship of phloem cells and the vascular fibers: Completely surrounded by thick-walled fibers

Nature of lysigenous cavity and protoxylem: Enlarged protoxylem present in all species and lysigenous cavity absent only in *I. indicus*

Size of metaxylem vessels in relation to parenchyma sheath cells in cross-section: Wide vessels

Shape of metaxylem vessels: Angular

Median vascular bundles (Figs. 1.33C, 1.34B, 1.35E, 1.36F)

Shape: Round

Relationship of phloem cells and the vascular fibers: Complete surrounded by thick-walled fibers

Nature of lysigenous cavity and protoxylem: Enlarged protoxylem present in all species and lysigenous cavity absent only in *I. indicus*

Size of metaxylem vessels in relation to parenchyma sheath cells in cross-section: Wide vessels

Vascular bundle sheath

Third-order vascular bundle's bundle sheath (Figs. 1.33E, 1.34C, arrowhead, 1.36E)

Shape: Round

Shape of bundle sheath cell: Fan shaped in *I. indicus* and *I. rugosum*; elliptical in *I. molle* and *I. pilosum*

Second-order vascular bundle's bundle sheath

Shape: Round

Shape of bundle sheath cell: Radial wall straight, tangential walls inflated in *I. indicus* and *I. rugosum* while round to elliptical in *I. molle* and *I. pilosum*

Number of bundle sheath cell: 10–11 (*I. indicus*), 7–8 (*I. molle*), 14 (*I. rugosum*), 12–13 (*I. pilosum*)

Nature of bundle sheath (either complete or incomplete): Complete in all species except *I. indicus* in which it is incomplete, interrupted by sclerenchyma on abaxial side

Chloroplast position in bundle sheath cell: Chloroplast concentrated near the inner (*I. indicus*) or outer tangential wall (*I. rugosum*, *I. pilosum*) or centrally placed (*I. molle*)

Mestome/Inner sheath: Absent

First-order vascular bundle's bundle sheath

Shape: Round, circular

Extent of bundle sheath around the vascular bundle: Incomplete interrupted by sclerenchyma on abaxial side in *I. indicus* and *I. molle*, complete in *I. rugosum* and *I. pilosum*

Extension of bundle sheath: Absent in *I. rugosum* and *I. molle*. Wide abaxial extension of the bundle sheath in *I. indicus* and abaxial and/or adaxial extension in *I. pilosum*.

Number of cells comprising the bundle sheath: 14–15 in *I. indicus*, *I. molle*, 19–20 in *I. rugosum* and *I. pilosum* parenchyma cells

Bundle sheath cell shape: All cells are elliptical and elongated; chloroplast concentrated near the inner in *I. indicus* or outer tangential wall in *I. rugosum* and *I. pilosum* or centrally placed in *I. molle*

Mestome/Inner sheath: Absent

Sclerenchyma

Adaxial sclerenchyma: Absent in all species except *I. pilosum* with well-developed horizontal strand

Abaxial sclerenchyma: Well-developed narrow girder in all species except *I. rugosum* with wide girder

Sclerenchyma between bundles: Absent

Sclerenchyma in leaf margin: Forming very small pointed fibrous cap. Epidermal cells of margin are thin with outer thick tangential wall

Leaf-sheath anatomy (Figs. 1.33F–I, 1.34F–I, 1.35J–L, 1.36H–J)

Outline of lamina: Round shaped with double overlapping inrolled margins in *I. indicus*, *I. rugosum* (Figs. 1.33F, 1.36H) while with inrolled margins in *I. molle*, *I. pilosum* (Figs. 1.34F, 1.35J)

Minute ribs and furrows present opposite to first-order vascular bundle on abaxial surface in *I. indicus*, *I. rugosum* while ribs and furrows absent in *I. molle*, *I. pilosum*

Epidermis

Adaxial epidermal cells: Flat, rectangular

Abaxial epidermal cells: Barrel shaped with thick cuticle

Bulliform cells inflated, projecting above the level of epidermal cells only in *I. rugosum* (Fig. 1.36I, arrow)

Ground tissue: Parenchymatous with vertically oval-shaped air cavities in *I. indicus*, *I. molle*, and *I. rugosum* (Figs. 1.33G, 1.34G); air cavities absent in *I. pilosum*

Vascular bundle

Arrangement of vascular bundle: One consecutive third-order vascular bundle between first and second-order vascular bundle

Third-order vascular bundle (Figs. 1.33I, 1.34H, 1.35K)

Circular in *I. indicus*, *I. rugosum* and *I. molle*, hexagonal in *I. rugosum*. Xylem and phloem groups distinguishable in all except *I. molle* with vascular tissue consisting of only few indistinguishable vascular elements

Second-order vascular bundle: Circular shaped

First-order vascular bundle (Figs. 1.33H, 1.34I, 1.35L, 1.36J)

Circular with phloem completely surrounded by thick-walled fibers. Enlarged protoxylem present in all species and lysigenous cavity absent only in *I. indicus*; circular wide metaxylem vessel elements (Fig. 1.32H)

Vascular bundle sheath (Fig. 1.33I, arrowhead)

Shape: Round, circular

Extent of bundle sheath around the vascular bundle: Complete in *I. rugosum* while incomplete in all other species abaxially being interrupted by sclerenchyma and adaxially by parenchyma cells in *I. indiucm* and only abaxially interrupted by sclerenchyma in *I. molle* and *I. pilosum*

Number of cells comprising the bundle sheath: 12–13 in *I. indiucm*, 14–15 in *I. molle*, 17–18 in *I. pilosum* and 19–20 in *I. rugosum* parenchyma cells

Bundle sheath cell shape: Different sized elliptical, elongated cells; cells smaller than the ground tissue

Mestome/Inner sheath: Absent

Sclerenchyma: In form of wide well-developed grider present on abaxial surface (Fig. 1.33H, I, arrow) in *I. indiucm*, *I. pilosum* while well-developed broad girder present on abaxial surface (Fig. 1.34H, arrow) in *I. molle*, *I. rugosum*

Ligule anatomy (Figs. 1.33J, K, 1.34J, K, 1.35M, N, 1.36K, L)

Shape: Ligule attached only to the midrib region and broad round shaped at connection point in *I. indiucm*, *I. pilosum*, *I. rugosum* (Figs. 1.33J, arrow, 1.35M, 1.36K), ligule attached to the midrib region and slightly with lamina region, and at the connection, it is broad round shaped (Fig. 1.34J) in *I. molle*.

Anatomy: Three layered mesophyll cells between adaxial and abaxial surfaces (Figs. 1.33K, 1.34K, 1.35N) in *I. indiucm*, *I. pilosum*, *I. molle* while more than three layered in *I. rugosum* (Fig. 1.36L)

DIFFERENTIATING FEATURES

Ribs and furrows absent on leaf lamina … *I. pilosum*
Ribs and furrows present on leaf lamina
Third-order vascular bundle sheath cells are elliptical …*I. molle*
Third-order vascular bundle sheath cells are fan shaped

Chloroplast concentrated near inner tangential wall … *I. indicus*

Chloroplast concentrated near outer tangential wall … *I. rugosum*

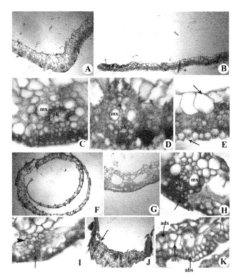

FIGURE 1.33 Leaf anatomy of *Ischaemum indicus.*

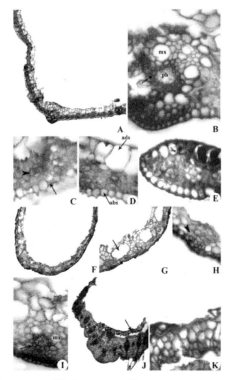

FIGURE 1.34 Leaf anatomy of *Ischaemum molle.*

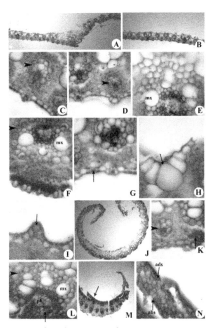

FIGURE 1.35   Leaf anatomy of *Ischaemum pilosum.*

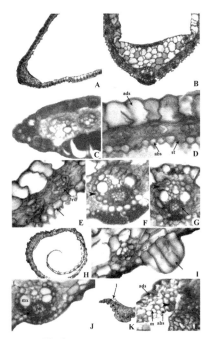

FIGURE 1.36   Leaf anatomy of *Ischaemum rugosum.*

22. *Iseilema laxum*

Leaf lamina (Fig. 1.37A–D)

Outline of the lamina: Expanded with flat, straight leaf blade. Ribs and furrows are present on the leaf surface. Adaxial surface with medium furrow between all vascular bundles and triangular ribs present above all vascular bundles. Abaxial surface has shallow ribs and furrows. Keel is inconspicuous.

Epidermis: Adaxial epidermal surface with fan-shaped bulliform cells and small round epidermal cells. Abaxial epidermal cells are barrel shaped with thick cuticle which occupies less than half of the depth of the cells.

Mesophyll: Radiate, single layered, isodiametric chlorenchyma cells completely surrounding bundles; compact homogenous chlorenchymatous cells occupying the major area between the adaxial and abaxial epidermises; chlorenchyma cells continuous between bundles and forming a horizontally elongated strap; colorless cells absent.

Vascular bundles: Four first-order vascular bundles in half lamina regularly arranged from median to marginal bundle. Six third-order bundles between consecutive larger bundles present. All the vascular bundles are at same level and placed centrally.

Third-order vascular bundles: Pentagonal shaped and vascular tissue consisting of only few indistinguishable vascular elements (Fig. 1.37B)

Second-order vascular bundles: Circular shaped (Fig. 1.37C)

First-order vascular bundles

Shape: Circular

Relationship of phloem cells and the vascular fibers: Phloem adjoins the inner or parenchyma sheath

Nature of lysigenous cavity and protoxylem: Lysigenous cavity and enlarged protoxylem vessel present

Size of metaxylem vessels in relation to parenchyma sheath cells in cross-section: Narrow vessels

Shape of metaxylem vessels: Circular

Vascular bundle sheath

Third-order vascular bundle's bundle sheath (Fig. 1.37B, arrowhead)

Shape: Triangular

Shape of bundle sheath cell: Fan shaped

Second-order vascular bundle's bundle sheath (Fig. 1.37C, arrowhead)

Shape: Round

Shape of bundle sheath cell: Elliptical and fan shaped

Number of bundle sheath cell: 8–9

Nature of bundle sheath (either complete or incomplete): Complete

Chloroplast position in bundle sheath cell: Chloroplast placed centrally

Mestome/Inner sheath: Absent

First-order vascular bundle's bundle sheath

Shape: Circular

Extent of bundle sheath around the vascular bundle: Complete

Extension of bundle sheath: Absent

Number of cells comprising the bundle sheath: 8–9 parenchyma cells

Bundle sheath cell shape: Few cells are fan shaped and few elliptical

Mestome/Inner sheath: Absent

Sclerenchyma

Adaxial sclerenchyma: Absent

Abaxial sclerenchyma: Minute strand consisting of only a few subepidermal fibers

Sclerenchyma between bundles: Absent

Sclerenchyma in leaf margin: Forming fibrous pointed cap; epidermal cells of margin are thin with thick tangential wall

Leaf-sheath anatomy (Fig. 1.37E–H)

Outline of lamina: Round-shaped inrolled margins; minute ribs and furrows present

Epidermis

Adaxial epidermal cells: Flat, rectangular

Abaxial epidermal cells: Broad, barrel shaped with thick cuticle

Ground tissue: Parenchymatous with oval-shaped air cavity (Fig. 1.37E, arrow)

Vascular bundle

Arrangement of vascular bundle: One consecutive third-order vascular bundle between first and second-order vascular bundle

Third-order vascular bundle: Circular shaped with vascular tissue consisting of only few indistinguishable vascular elements (Fig. 1.37G)

Second-order vascular bundle: Circular shaped

First-order vascular bundle: Circular shaped with phloem adjoins the inner or parenchyma sheath; lysigenous cavity and enlarged protoxylem vessel present; circular wide metaxylem vessel elements (Fig. 1.37H)

Vascular bundle sheath (Fig. 1.37G, arrowhead)

Shape: Half circle

Extent of bundle sheath around the vascular bundle: Sheath incomplete, interrupted by sclerenchyma cells

Number of cells comprising the bundle sheath: 9–10 parenchyma cells

Bundle sheath cell shape: Fan shaped; cells smaller than the ground tissue

Mestome/Inner sheath: Absent

Sclerenchyma: In form of well-developed broad girder present on abaxial surface (Fig. 1.37H, arrow)

Ligule anatomy (Fig. 1.37I, J)

Shape: Ligule attached to the leaf sheath more than the half width and very broad round shaped at the connections point (Fig. 1.37I, arrow)

Anatomy: Two layered mesophyll cells between adaxial and abaxial surface, abaxial few sclerenchyma cells present on abaxial surface (Fig. 1.37J, arrow)

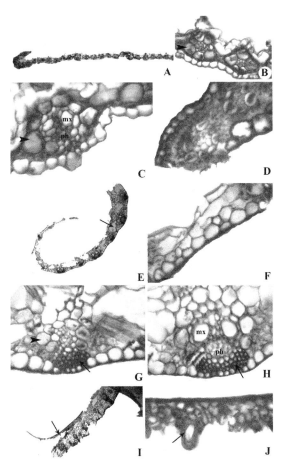

FIGURE 1.37   Leaf anatomy of *Iseilema laxum.*

23. *Ophiuros exaltatus*

Leaf lamina (Fig. 1.38A–G)

Outline of the lamina: "V" shaped with wide angle having straight arms (Fig. 1.38A). Two halves are symmetrical on either side of the median vascular bundle. Ribs and furrows are absent on the leaf surface.

Keel is well developed, inverted triangular shaped with 3 first-order vascular bundle, 6–7 second-order and 9–10 third-order vascular bundles (Fig. 1.38C). Small, rounded parenchyma cells are present surrounding the median vascular bundle adaxially. Sclerenchyma present on abaxial surface associated with vascular bundles in form of well-developed wide-anchored girder.

Epidermis: Adaxial epidermal surface has groups of bulliform cells which are similar in size with epidermal cells. Abaxial epidermis has round to square-shaped epidermal cells with cuticle. Stomata (Fig. 1.38E, arrow) and silica bodies (Fig. 1.38D, arrow) present on abaxial surface

Mesophyll: Radiate, single layered, tubular chlorenchyma cells completely surrounding bundles; compact homogenous chlorenchymatous cells occupying the major area between the adaxial and abaxial epidermises; chlorenchyma cells arranged in form of horizontal elongated strap; colorless cells absent.

Vascular bundles: Nine to 10 first-order vascular bundles in half lamina regularly arranged from median to marginal bundle. Two to three third-order bundles between consecutive larger bundles and five to six second-order bundles between consecutive first-order bundles present. All the vascular bundles present at same level and placed centrally

Third-order vascular bundles: Circular shaped and vascular tissue consisting of only few indistinguishable vascular elements (Fig. 1.38G)

Second-order vascular bundles: Circular shaped

First-order vascular bundles (Fig. 1.38F)

Shape: Elliptical

Relationship of phloem cells and the vascular fibers: Completely surrounded by thick-walled fibers

Nature of lysigenous cavity and protoxylem: Lysigenous cavity and enlarged protoxylem vessel present

Size of metaxylem vessels in relation to parenchyma sheath cells in cross-section: Wide vessels

Shape of metaxylem vessels: Circular

Vascular bundle sheath

Third-order vascular bundle's bundle sheath (Fig. 1.38G, arrowhead)

Shape: Round

Shape of bundle sheath cell: Cells elliptical and elongated

Second-order vascular bundle's bundle sheath

Shape: Round

Shape of bundle sheath cell: Cells elliptical and elongated

Number of bundle sheath cell: 7–8

Nature of bundle sheath (either complete or incomplete): Complete

Chloroplast position in bundle sheath cell: Chloroplast toward outer tangential wall

Mestome/Inner sheath: Absent

First-order vascular bundle's bundle sheath (Fig. 1.38F, arrowhead)

Shape: Elliptical

Extent of bundle sheath around the vascular bundle: Complete

Extension of bundle sheath: First-order vascular bundles showed adaxial and abaxial broad extension, extend to bundle sheath

Number of cells comprising the bundle sheath: 20–22 parenchyma cells comprise the sheath

Bundle sheath cell shape: Different sized elliptical, elongated cells; chloroplast concentrated toward outer tangential wall

Mestome/Inner sheath: Absent

Sclerenchyma

Adaxial sclerenchyma: Well-developed broad girder present above first-order vascular bundle

Abaxial sclerenchyma: Well-developed girder below the first-order vascular bundle

Sclerenchyma between bundles: Absent

Sclerenchyma in leaf margin: Forming well-developed rounded cap; epidermal cells of margin are thin with outer thick tangential wall (Fig. 1.38B)

Leaf-sheath anatomy (Fig. 1.38H–J)

Outline of lamina: Round-shaped inrolled margins; minute ribs and furrows absent (Fig. 1.38H)

Epidermis

Adaxial and abaxial epidermal cells: Small, flat, rectangular shaped with thick cuticle

Ground tissue: Parenchymatous with round-shaped air cavity (Fig. 1.38I, arrow)

Vascular bundle

Arrangement of vascular bundle: Six consecutive third-order vascular bundles between first and second-order vascular bundle

Third-order vascular bundle: Pentagonal shaped with vascular tissue consisting of only few indistinguishable vascular elements (Fig. 1.38J)

Second-order vascular bundle: Circular shaped

First-order vascular bundle: Elliptical shaped with phloem completely surrounded by thick-walled fibers; lysigenous cavity and enlarged proto-xylem vessel present; circular wide metaxylem vessel elements

Vascular bundle sheath (Fig. 1.38J, arrowhead)

Shape: Elliptical

Extent of bundle sheath around the vascular bundle: Incomplete, inter-rupted by parenchyma cells

Number of cells comprising the bundle sheath: 12–13 parenchyma cells

Bundle sheath cell shape: Different sized elliptical, elongated cells; chloroplast concentrated toward outer tangential wall; cells are smaller than the ground tissue

Mestome/Inner sheath: Absent

Sclerenchyma: In form of straight horizontal well-developed anchored girder present on abaxial surface

Ligule anatomy (Fig. 1.38K, L)

Shape: Ligule attached to the lamina more than half width and very broad round shaped at the connection point (Fig. 1.38K, arrow)

Anatomy: Two layered mesophyll cells between adaxial and abaxial surfaces (Fig. 1.38L)

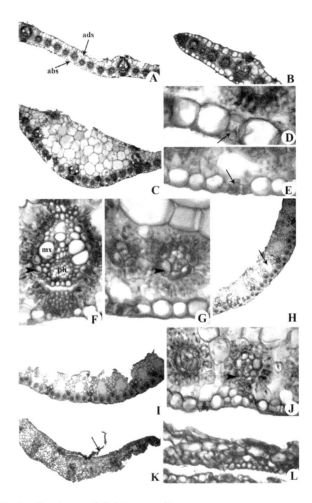

FIGURE 1.38   Leaf anatomy of *Ophiuros exaltatus.*

24. *Rottboellia exaltata*

Leaf lamina (Fig. 1.39A–G)

Outline of the lamina: "V" shaped with wide angle and having straight arms (Fig. 1.39A). Two halves are symmetrical on either side of the median vascular bundle. Ribs and furrows are absent on the leaf surface.

Keel is well developed, V shaped with six first-order vascular bundles, four second, and eight to nine third-order vascular bundles; median vascular bundle surrounded by small round parenchyma cells (Fig. 1.39B);

sclerenchyma in form of strands adaxially and on abaxially surface associated with vascular bundles.

Epidermis: Adaxial epidermal surface shows two types of bulliform cells: (1) most of epidermal cell which is inflated and big bulliform cells (Fig. 1.39G, arrow) and (2) restricted group of inflated bulliform cells projecting above the epidermal surface (Fig. 1.39E) and at the base of bulliform cell, hair is present (Fig. 1.39E, arrow). Abaxial epidermal cells rectangular to barrel shaped with cuticle.

Mesophyll: Radiate, single layered, isodiametric chlorenchyma cells completely surrounding bundles; compact homogenous chlorenchymatous cells occupy the major area between the adaxial and abaxial epidermises; chlorenchyma cells form an elongated strap continuous between bundles; colorless cells absent.

Vascular bundles: Six to seven first-order vascular bundles in half lamina regularly arranged from median to marginal bundles. Eight to nine third-order bundles between consecutive larger bundles and three to four second-order bundles between consecutive first-order bundles present; all the vascular bundles present at same level and centrally placed

Third-order vascular bundles: Hexagonal with vascular tissue consisting of only few indistinguishable vascular elements (Fig. 1.39D)

Second-order vascular bundles: Elliptical shaped

First-order vascular bundles (Fig. 1.39F)

Shape: Elliptical

Relationship of phloem cells and the vascular fibers: Phloem adjoins the inner or parenchyma sheath

Nature of lysigenous cavity and protoxylem: Lysigenous cavity and enlarged protoxylem vessel present

Size of metaxylem vessels in relation to parenchyma sheath cells in cross-section: Wide vessels

Shape of metaxylem vessels: Circular

Vascular bundle sheath

Third-order vascular bundle's bundle sheath (Fig. 1.39D)

Shape: Round

Shape of bundle sheath cell: Cells elliptical and elongated

Second-order vascular bundle's bundle sheath

Shape: Round

Shape of bundle sheath cell: Cells elliptical and elongated

Number of bundle sheath cell: 9–10

Nature of bundle sheath (either complete or incomplete): Complete

Chloroplast position in bundle sheath cell: Chloroplast concentrated toward outer tangential wall

Mestome/Inner sheath: Absent

First-order vascular bundle's bundle sheath (Fig. 1.39F)

Shape: Elliptical

Extent of bundle sheath around the vascular bundle: Sheath incomplete, abaxially interrupted by sclerenchyma

Extension of bundle sheath: Narrow abaxial extension of the bundle sheath

Number of cells comprising the bundle sheath: 23–25 parenchyma cells

Bundle sheath cell shape: Different sized, elliptical, elongated cells; chloroplast concentrated toward outer tangential wall

Mestome/Inner sheath: Absent

Sclerenchyma

Adaxial sclerenchyma: Absent

Abaxial sclerenchyma: Well-developed strand, wider than deep and it follows shape of abaxial rib

Sclerenchyma between bundles: Absent

Sclerenchyma in leaf margin: Forming very small fibrous pointed cap; epidermal cells of margin are with thick tangential wall

Leaf-sheath anatomy (Fig. 1.39H–L)

Outline of lamina: Round-shaped inrolled margins (Fig. 1.39H); ribs and furrows absent

Epidermis

Adaxial epidermal cells: Flat, rectangular

Abaxial epidermal cells: Small, flat, rectangular shaped with outer tangential wall covered with very thick cuticle. It also shows short trichomes (Fig. 1.39L, arrow)

Ground tissue: Parenchymatous with an absence of air cavity

Vascular bundle

Arrangement of vascular bundle: One consecutive third-order vascular bundle between first and second-order vascular bundle

Third-order vascular bundle: Hexagonal with vascular tissue consisting of only few indistinguishable vascular elements

Second-order vascular bundle: Elliptical shaped

First-order vascular bundle: Elliptical shaped with phloem adjoins the inner or parenchyma sheath. Lysigenous cavity and enlarged protoxylem vessel present; circular wide metaxylem vessel elements (Fig. 1.39I)

Vascular bundle sheath (Fig. 1.39I, J, arrowhead)

Shape: Elliptical

Extent of bundle sheath around the vascular bundle: Sheath incomplete, interrupted by sclerenchyma cells

Number of cells comprising the bundle sheath: 13–14 parenchyma cells

Bundle sheath cell shape: Round similar in size than ground tissue

Mestome/Inner sheath: Absent

Sclerenchyma: In form of well-developed arched girder on abaxial surface (Fig. 1.39J, arrow)

Ligule anatomy (Fig. 1.39M, N)

Shape: Ligule attached to the leaf sheath more than the half width and very broad round shaped at the connection point (Fig. 1.39M, arrow)

Anatomy: Two layered mesophyll cells between adaxial and abaxial surfaces (Fig. 1.39N)

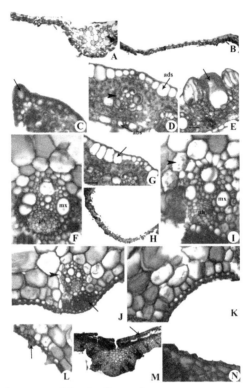

FIGURE 1.39   Leaf anatomy of *Rottboellia exaltata.*

25. *Saccharum spontaneum*

Leaf lamina (Fig. 1.40A–H)

Outline of the lamina: "V" shaped with wide angle having outwardly curving arms (Fig. 1.40A, B). Two halves are symmetrical on either side of the median vascular bundle; ribs and furrows are absent on the leaf surface.

Keel is well developed, V shaped with 15–17 first-order, 16–17 second-order, and 28–29 third-order vascular bundles. Small, round parenchyma cells are present surrounding the median vascular bundle adaxially; sclerenchyma present on adaxial surface in form of subepidermal strands and on abaxial surface associated with vascular bundles in form of girders.

Epidermis: Adaxial epidermal surface has two types of bulliform cells: (1) restricted group of inflated bulliform cells in level with the epidermal surface (Fig. 1.40H, arrow) and (2) fan-shaped bulliform cells (Fig. 1.40E, arrow); abaxial epidermal cells round to square shaped (Fig. 1.40D, arrow) with thick cuticle.

Mesophyll: Incomplete radiate chlorenchyma cells completely surrounding bundles; compact homogenous chlorenchymatous cells occupying the major area between the adaxial and abaxial epidermises; chlorenchyma cells continuous between bundles; colorless cells absent.

Vascular bundles: Seven first-order vascular bundles in half lamina regularly arranged from median to marginal bundles. Thirty-three to 34 third-order bundles between consecutive larger bundles and 14–16 second-order bundles between consecutive first-order bundles. All the vascular bundles present at same level and placed abaxially.

Third-order vascular bundles: Hexagonal (Fig. 1.40C) or elliptical (Fig. 1.40D) with vascular tissue consisting of only few indistinguishable vascular elements.

Second-order vascular bundles: Elliptical shaped

First-order vascular bundles (Fig. 1.40G)

Shape: Elliptical

Relationship of phloem cells and the vascular fibers: Phloem completely surrounded by sclerenchyma cells

Nature of lysigenous cavity and protoxylem: Lysigenous cavity and enlarged protoxylem vessel present

Size of metaxylem vessels in relation to parenchyma sheath cells in cross-section: Wide vessels

Shape of metaxylem vessels: Circular

Vascular bundle sheath

Third-order vascular bundle's bundle sheath

Shape: Round (Fig. 1.40C, arrowhead) and elliptical (Fig. 1.40D, arrowhead)

Shape of bundle sheath cell: Cells elliptical and elongated

Second-order vascular bundle's bundle sheath

Shape: Round

Shape of bundle sheath cell: Cells elliptical to round

Number of bundle sheath cell: 11–12

Nature of bundle sheath (either complete or incomplete): Complete

Chloroplast position in bundle sheath cell: Chloroplast placed centrally

Mestome/Inner sheath: Absent

First-order vascular bundle's bundle sheath (Fig. 1.40G, arrowhead)

Shape: Elliptical

Extent of bundle sheath around the vascular bundle: Sheath incomplete, abaxially interrupted by sclerenchyma

Extension of bundle sheath: Narrow abaxial extension of the bundle sheath

Number of cells comprising the bundle sheath: 17–19 parenchyma cells comprise the sheath

Bundle sheath cell shape: Different sized, elliptical, elongated cells; chloroplast placed centrally

Mestome/Inner sheath: Absent

Sclerenchyma

Adaxial sclerenchyma: Absent

Abaxial sclerenchyma: Well-developed girder, narrower deeper than wide (Fig. 1.40C, arrow)

Sclerenchyma between bundles: Absent

Sclerenchyma in leaf margin: Forming very small fibrous pointed cap; epidermal cells of margin are with thick tangential wall

Leaf-sheath anatomy (Fig. 1.40I–N)

Outline of lamina: Round-shaped inrolled margins (Fig. 1.40K); ribs and furrows absent

Epidermis

Adaxial epidermal cells: Flat, rectangular

Abaxial epidermal cells: Small, flat, rectangular and with outer tangential wall have double thick cuticle

Ground tissue: Parenchymatous with round-shaped air cavity (Fig. 1.40K)

Vascular bundle

Arrangement of vascular bundle: One consecutive third-order vascular bundle between first and second-order vascular bundle

Third-order vascular bundle: Elliptical shaped with vascular tissue consisting of only few indistinguishable vascular elements (Fig. 1.40N)

Second-order vascular bundle: Elliptical shaped (Fig. 1.40L)

First-order vascular bundle: Elliptical shaped with phloem completely surrounded by sclerenchyma; lysigenous cavity and enlarged protoxylem vessel present; circular very wide metaxylem vessel elements (Fig. 1.40J)

Vascular bundle sheath (Fig. 1.40J, L, arrowhead)

Shape: Elliptical

Extent of bundle sheath around the vascular bundle: Complete

Number of cells comprising the bundle sheath: 16–17 parenchyma cells

Bundle sheath cell shape: All cells are round to elliptical in shape; cells smaller than the ground tissue

Mestome/Inner sheath: Absent

Sclerenchyma: In form of well-developed narrow girder present on abaxial surface

Ligule anatomy (Fig. 1.40O, P)

Shape: Ligule attached to the midrib region and partially with lamina region and at the connection, it is broad round shaped (Fig. 1.40O, arrow)

Anatomy: Three layered mesophyll cells between adaxial and abaxial surfaces (Fig. 1.40P)

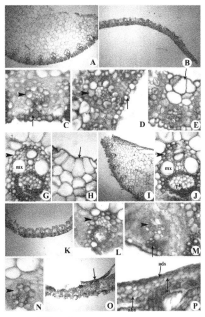

FIGURE 1.40 Leaf anatomy of *Saccharum spontaneum.*

Sehima

26. *Sehima ischaemoides* (Fig. 1.41)

27. *Sehima nervosum* (Fig. 1.42)

28. *Sehima sulcatum* (Fig. 1.43)

Leaf lamina (Figs. 1.41A–E, 1.42A–F, 1.43A–E)

Outline of the lamina: Expanded with undulating surface in *S. ischaemoides* (Fig. 1.41A, B), straight leaf blade in *S. nervosum* and *S. sulcatum* (Figs. 1.42A, 1.43A).

Ribs and furrows are present on the leaf surface in all except *S. nervosum*. Medium in *S. ischaemoides* or shallow furrows in *S. sulcatum* on adaxial and abaxial surface present between first and second-order vascular bundles and triangular ribs with pointed apex present only over first and second-order vascular bundles.

Keel formed due to sclerenchyma projecting outward abaxially is well developed, semicircular with single median vascular bundle. Sclerenchyma is present on abaxial surface associated with vascular bundles in form of girders.

Epidermis: Adaxial epidermal layer with fan-shaped bulliform cells present at bases of furrows in *S. ischaemoides* (Fig. 1.41D) or large, inflated restricted bulliform cells on adaxial surface of *S. nervosum* and *S. sulcatum* (Figs. 1.42E, F, arrow, 1.43E, arrow). Long hairs present on abaxial surface (Fig. 1.42B). Abaxial epidermal cells have barrel-shaped epidermal cells with thick cuticle. Prickles are present in *S. ischaemoides* (Fig. 1.41D, arrow).

Mesophyll: Radiate, single layered, isodiametric chlorenchyma cells which completely surround bundles; compactly arranged homogenous chlorenchyma cells occupy the major area between the adaxial and abaxial epidermises and are continuous between bundles; colorless cells absent.

Vascular bundles: Three to four first-order vascular bundles in *S. ischaemoides*, two to three first-order vascular bundles in *S. nervosum* and *S. sulcatum* arranged regularly from median to marginal bundle in half lamina.

Five to six third-order bundles between consecutive larger bundles and three to four second-order bundles between consecutive first-order bundles. First and second-order vascular bundles centrally placed while third-order vascular bundles present abaxially to leaf blade.

Third-order vascular bundles (Figs. 1.41E, 1.42E, 1.43E)

Circular in *S. ischaemoides*), pentagonal in *S. nervosum* and *S. sulcatum* with xylem and phloem distinguishable

Second-order vascular bundles: Circular shaped

First-order vascular bundles

Shape: Circular

Relationship of phloem cells and the vascular fibers: Completely surrounded by thick-walled fibers

Nature of lysigenous cavity and protoxylem: Enlarged protoxylem vessel present and lysigenous cavity absent

Size of metaxylem vessels in relation to parenchyma sheath cells in cross-section: Wide vessels

Shape of metaxylem vessels: Circular

Median vascular bundles (Figs. 1.41C, 1.42D, 1.43C)

Shape: Circular

Relationship of phloem cells and the vascular fibers: Completely surrounded by thick-walled fibers

Nature of lysigenous cavity and protoxylem: Enlarged protoxylem vessel present and lysigenous cavity absent

Size of metaxylem vessels in relation to parenchyma sheath cells in cross-section: Wide vessels

Shape of metaxylem vessels: Circular

Vascular bundle sheath

Third-order vascular bundle's bundle sheath

Shape: Round

Shape of bundle sheath cell: Fan shaped

Second-order vascular bundle's bundle sheath

Shape: Round

Shape of bundle sheath cell: Cells elliptical and elongated

Number of bundle sheath cell: 7–8

Nature of bundle sheath (either complete or incomplete): Incomplete, abaxially interrupted by sclerenchyma

Chloroplast position in bundle sheath cell: Translucent in *S. ischaemoides* or placed centrally as in *S. nervosum* and *S. sulcatum*

Mestome/Inner sheath: Absent

First-order vascular bundle's bundle sheath

Shape: Round

Extent of bundle sheath around the vascular bundle: Sheath incomplete, abaxially interrupted by sclerenchyma

Extension of bundle sheath: Broad abaxial extension of the bundle sheath

Number of cells comprising the bundle sheath: 13–14 in *S. ischaemoides*, 15–16 in *S. nervosum* and 12–13 in *S. sulcatum* parenchyma cells

Bundle sheath cell shape: All cells are elliptical, elongated in shape; chloroplast translucent in *S. ischaemoides* or placed centrally as in *S. nervosum* and *S. sulcatum*

Mestome/Inner sheath: Absent

Sclerenchyma

Adaxial sclerenchyma: Absent

Abaxial sclerenchyma: Subepidermal strands of two to four fibers in *S. ischaemoides* and *S. nervosum*, well-developed small girder in *S. sulcatum*

Sclerenchyma between bundles: Absent

Sclerenchyma in leaf margin: Forming very small pointed fibrous cap. Epidermal cells thick tangential wall.

Leaf-sheath anatomy (Figs. 1.41F–H, 1.42G, H, 1.43F–H)

Outline of lamina: Round in all species (Figs. 1.41F, 1.42G, 1.43F). Ribs and furrows present in all except *S. sulcatum*. Abaxial surface has furrows which are medium and wide in *S. ischaemoides*, shallow and wide with rounded ribs in *S. nervosum*.

Epidermis

Adaxial epidermal cells: Flat, rectangular

Abaxial epidermal cells: Small, flat, rectangular shaped with thick cuticle

Ground tissue: Parenchymatous with absence of air cavity

Vascular bundle

Arrangement of vascular bundle: One consecutive third-order vascular bundle between first and second-order vascular bundle

Third-order vascular bundle (Figs. 1.41H, 1.42H, 1.43H): Circular with distinguishable xylem and phloem

Second-order vascular bundle: Circular shaped

First-order vascular bundle (Fig. 1.43G): Circular with phloem completely surrounded by thick-walled fibers. Lysigenous cavity present and enlarged protoxylem vessel absent; circular wide metaxylem vessels elements

Vascular bundle sheath

Shape: Round

Extent of bundle sheath around the vascular bundle: Complete in all except *S. ischaemoides* in which bundle sheath is incomplete because it is interrupted adaxially by parenchyma cells and abaxially by sclerenchyma cells.

Number of cells comprising the bundle sheath: 11–12 in *S. nervosum* and *S. sulcatum*, 12–13 in *S. ischaemoides* parenchyma cells

Bundle sheath cell shape: All cells are elliptical, elongated in shape and smaller than the ground tissue cells

Mestome/Inner sheath: Absent

Sclerenchyma: In form of straight horizontal well-developed broad abaxial strand present

Ligule anatomy (Figs. 1.41I, J, 1.42I, 1.43I)

Shape: Ligule attached to half width of leaf sheath and it is round shaped at the connection point (Figs. 1.41I, 1.42I, 1.43I)

Anatomy: More than three layered mesophyll cells between adaxial and abaxial surfaces (Fig. 41J)

DIFFERENTIATING FEATURES

Circular-shaped third-order vascular bundles … *S. ischaemoides*
Pentagonal-shaped third-order vascular bundles
Shallow furrows present on abaxial leaf surface, subepidermal strands of two to four fibers on abaxial leaf surface … *S. nervosum*
Deep furrows present on abaxial leaf surface, well-developed sclerenchymatous girder present on abaxial leaf surface … *S. sulcatum*

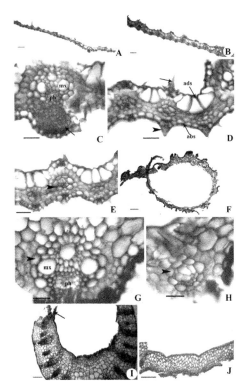

FIGURE 1.41   Leaf anatomy of *Sehima ischaemoides*.

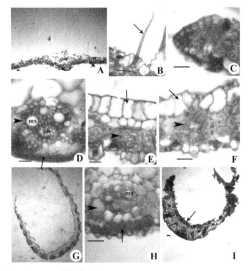

FIGURE 1.42   Leaf anatomy of *Sehima nervosum*.

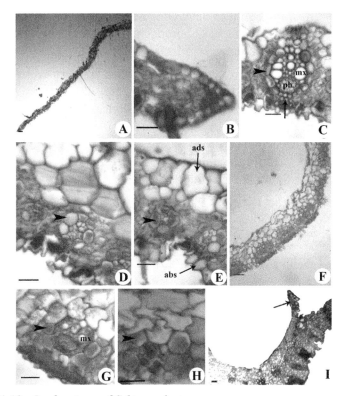

FIGURE 1.43   Leaf anatomy of *Sehima sulcatum.*

29. *Sorghum halepense*

Leaf lamina (Fig. 1.44A–G)

Outline of the lamina: "V" shaped with wide angles having straight arms. Two halves are symmetrical on either side of the median vascular bundle. Ribs and furrows are absent on the leaf surface.

Keel is a well-developed, inverted triangle shaped with three first-order vascular bundles, four to five second-order vascular bundles, six to seven third-order vascular bundles. Small, round parenchyma cells surrounds the median vascular bundle adaxially. Sclerenchyma is present on adaxial surface in form of well-developed subepidermal strands and on abaxial surface associated with vascular bundles in form of girder.

Epidermis: Adaxial epidermis major part made up of inflated, large bulliform cells; prickles present (Fig. 1.44G, arrow); abaxial epidermal cells barrel shaped with thick cuticle; stomata present (Fig. 1.44F, arrow)

Mesophyll: Radiate, single layered, isodiametric cells completely surrounding bundles; compact homogenous chlorenchymatous cells occupying the major area between adaxial and abaxial epidermises; Chlorenchyma cells continuous forming a horizontally elongated strap between bundles; colorless cells absent.

Vascular bundles: Nine to 10 first-order vascular bundles in half lamina regularly arranged from median to marginal bundles. Seven to eight third-order bundles between consecutive larger bundles and two to three second-order bundles between consecutive first-order bundles. All the vascular bundles are centrally placed at same level.

Third-order vascular bundles: Circular with vascular tissue consisting of only few indistinguishable vascular elements (Fig. 1.44F)

Second-order vascular bundles: Circular shaped

First-order vascular bundles (Fig. 1.44E)

Shape: Elliptical

Relationship of phloem cells and the vascular fibers: Completely surrounded by thick-walled fibers

Nature of lysigenous cavity and protoxylem: Enlarged protoxylem vessel present and lysigenous cavity absent

Size of metaxylem vessels in relation to parenchyma sheath cells in cross-section: Very wide vessels

Shape of metaxylem vessels: Circular

Median vascular bundles (Fig. 1.44D)

Shape: Round

Relationship of phloem cells and the vascular fibers: Complete surrounded by thick-walled fibers

Nature of lysigenous cavity and protoxylem: Enlarged protoxylem vessel present and lysigenous cavity absent

Size of metaxylem vessels in relation to parenchyma sheath cells in cross-section: Very wide vessels

Vascular bundle sheath

Third-order vascular bundle's bundle sheath (Fig. 1.44F, arrowhead)

Shape: Round

Shape of bundle sheath cell: Cells elliptical and elongated

Second-order vascular bundle's bundle sheath

Shape: Round

Shape of bundle sheath cell: Cells elliptical and elongated

Number of bundle sheath cell: 12–14

Nature of bundle sheath (either complete or incomplete): Complete

Chloroplast position in bundle sheath cell: Chloroplast positioned toward outer tangential wall

Mestome/Inner sheath: Absent

First-order vascular bundle's bundle sheath (Fig. 1.44E)

Shape: Circular

Extent of bundle sheath around the vascular bundle: Incomplete, abaxially interrupted by sclerenchyma cells

Extension of bundle sheath: Narrow extension toward both epidermises present

Number of cells comprising the bundle sheath: 21–22 parenchyma cells

Bundle sheath cell shape: All cells are elliptical and elongated; chloroplast positioned toward outer tangential wall

Mestome/Inner sheath: Absent

Sclerenchyma

Adaxial sclerenchyma: Minute strand consisting of only a few subepidermal fibers

Abaxial sclerenchyma: Well-developed girder below the first-order vascular bundle

Sclerenchyma between bundles: Absent

Sclerenchyma in leaf margin: Forming well-developed (Fig. 1.44C, arrow) round cap; epidermal cells at margin are fibrous, thick walled

Leaf-sheath anatomy (Fig. 1.44H–K)

Outline of lamina: Round shaped (Fig. 1.44H, I); ribs and furrows absent

Epidermis

Adaxial epidermal cells: Flat, rectangular

Abaxial epidermal cells: Small, flat, rectangular shaped with thick cuticle

Ground tissue: Parenchymatous with horizontal oval-shaped air cavity (Fig. 1.44I, arrow)

Vascular bundle

Arrangement of vascular bundle: One consecutive third-order vascular bundle between first and second-order vascular bundle

Third-order vascular bundle: Elliptical with vascular tissue consisting of only a few indistinguishable vascular strands (Fig. 1.44K)

Second-order vascular bundle: Elliptical shaped

First-order vascular bundle: Elliptical shaped with phloem completely surrounded by thick-walled fibers (Fig. 1.44J, arrow); lysigenous cavity and enlarged protoxylem vessel present; circular wide metaxylem vessel elements (Fig. 1.44J)

Vascular bundle sheath (Fig. 1.44J, K, arrowhead)

Shape: Round, circular

Extent of bundle sheath around the vascular bundle: Sheath incomplete and interrupted by sclerenchyma cells

Number of cells comprising the bundle sheath: 19–20 parenchyma cells comprise the sheath

Bundle sheath cell shape: All cells are elliptical, elongated in shape; cells smaller than the ground tissue

Mestome/Inner sheath: Absent

Sclerenchyma: In form of straight horizontal well-developed abaxial strands present

Ligule anatomy (Fig. 1.44L, M)

Shape: Ligule attached to the midrib region and broad round shaped at connection point (Fig. 1.44M, arrow)

Anatomy: More than three layered mesophyll cells between adaxial and abaxial surfaces (Fig. 1.44L)

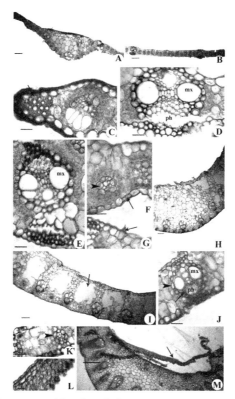

FIGURE 1.44   Leaf anatomy of *Sorghum halepense.*

30. *Thelepogon elegans*

Leaf lamina (Fig. 1.45A–G)

Outline of the lamina: Expanded with flat, straight leaf blade (Fig. 1.45A, B). Ribs and furrows are absent on the leaf surface.

Keel is well developed, semicircular with two first-order vascular bundle, three to four third-order vascular bundles. Sclerenchyma present on abaxial surface associated with vascular bundles in form of narrow strands.

Epidermis: Adaxial epidermis major part of adaxial epidermis made up of inflated bulliform cells interrupted by small group of epidermal cells (Fig. 1.45D, arrow); long, slender constriction of hair above bulbose base, embedded in large, inflated bulliform cells (Fig. 1.45G); barrel-shaped abaxial epidermal cells with thick cuticle.

Mesophyll: Radiate, two layered, isodiametric chlorenchyma cells which completely surround the bundles; compact homogenous chlorenchymatous cells occupy the major area between the adaxial and abaxial epidermises and continuous between bundles; colorless cells absent.

Vascular bundles: Four to five first-order vascular bundles in half lamina regularly arranged from median to marginal bundles. Three third-order bundles between consecutive larger bundles and three second-order bundles between consecutive first-order bundles present. All bundles are at same level and placed centrally.

Third-order vascular bundles: Hexagonal with xylem and phloem distinguishable (Fig. 1.45D)

Second-order vascular bundles: Circular shaped (Fig. 1.45E)

First-order vascular bundles (Fig. 1.45F)

Shape: Circular

Relationship of phloem cells and the vascular fibers: Phloem adjoins the inner or parenchyma sheath

Nature of lysigenous cavity and protoxylem: Enlarged protoxylem vessel and lysigenous cavity absent

Size of metaxylem vessels in relation to parenchyma sheath cells in cross-section: Wide vessels

Shape of metaxylem vessels: Angular

Vascular bundle sheath

Third-order vascular bundle's bundle sheath (Fig. 1.45D, arrowhead)

Shape: Round

Shape of bundle sheath cell: Round shaped

Second-order vascular bundle's bundle sheath (Fig. 1.45E, arrowhead)

Shape: Round

Shape of bundle sheath cell: Cells round in shape

Number of bundle sheath cell: 9–10

Nature of bundle sheath (either complete or incomplete): Complete

Chloroplast position in bundle sheath cell: Chloroplast small and indistinct

Mestome/Inner sheath: Absent

First-order vascular bundle's bundle sheath (Fig. 1.45F, arrowhead)

Shape: Round, circular

Extent of bundle sheath around the vascular bundle: Complete

Extension of bundle sheath: Absent

Number of cells comprising the bundle sheath: 12–13

Bundle sheath cell shape: All cells are elliptical to round in shape; chloroplast small and indistinct

Mestome/Inner sheath: Absent

Sclerenchyma

Adaxial sclerenchyma: Absent

Abaxial sclerenchyma: Small, two to four fibers in form of strands

Sclerenchyma between bundles: Absent

Sclerenchyma in leaf margin: Forming a well-developed round cap; epidermal cells at margin are with thick outer tangential wall (Fig. 1.45C)

Leaf-sheath anatomy (Fig. 1.45H–J)

Outline of lamina: "V" shaped with inrolled margins; ribs and furrows absent (Fig. 1.45H)

Epidermis

Adaxial epidermal cells: Flat, rectangular

Abaxial epidermal cells: Small, flat, rectangular shaped with thick cuticle

Ground tissue: Parenchymatous with absence of air cavity

Vascular bundle

Arrangement of vascular bundle: One consecutive third-order vascular bundle between first and second-order vascular bundle

Third-order vascular bundle: Circular shaped with xylem and phloem distinguishable (Fig. 1.45J)

Second-order vascular bundle: Circular shaped

First-order vascular bundle: Elliptical shaped with phloem adjoins the inner or parenchyma sheath; lysigenous cavity absent and enlarged protoxylem vessel present; circular wide metaxylem vessels elements (Fig. 1.45I)

Vascular bundle sheath (Fig. 1.45I, J, arrowhead)

Shape: Elliptical

Extent of bundle sheath around the vascular bundle: Complete

Number of cells comprising the bundle sheath: 13–14 parenchyma cells

Bundle sheath cell shape: All cells are elliptical, elongated in shape; there is gradation in size of cells with large cells toward abaxial and adaxial surfaces, cells smaller than the ground tissue

Mestome/Inner sheath: Absent

Sclerenchyma: In form of straight horizontal well-developed broad strand present on abaxial surface (Fig. 1.45I, arrow)

Ligule anatomy (Fig. 1.45K)

Shape: Ligule attached to the half width of the leaf sheath region and it is round shaped at the connection point (Fig. 1.45K, arrow)

Anatomy: Three layered mesophyll cells between adaxial and abaxial surfaces

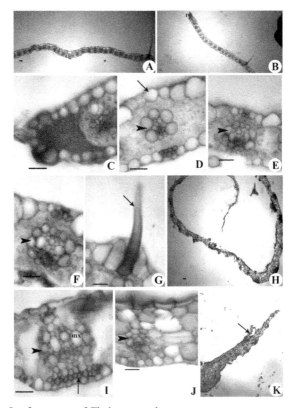

FIGURE 1.45   Leaf anatomy of *Thelepogon elegans*.

Themeda

31. *Themeda cymbaria* (Fig. 1.46)

32. *Themeda laxa* (Fig. 1.47)

33. *Themeda triandra* (Fig. 1.48)

34. *Themeda quadrivalvis* (Fig. 1.49)

Leaf lamina (Figs. 1.46A–G, 1.47A–D, 1.48A–F, 1.49A–E)

Outline of the lamina: "V" shaped with standard angled in *T. cymbaria* and *T. laxa* (Figs. 1.46A, B, 1.47A), narrow angled in *T. triandra* (Fig. 1.48A) having outwardly curving convex-shaped arms or broad angled with straight arms in *T. quadrivalvis* (Fig. 1.49A).

Two halves are symmetrical on either side of the median vascular bundle. Ribs and furrows are absent in all species except in *T. cymbaria* in which it has medium, wide furrows between first and second-order vascular bundles and triangular-shaped ribs with pointed apex only on adaxial surface.

Keel is well developed, V shaped with single median vascular bundle, 15–16 third-order and three to four second-order bundles in *T. cymbaria*, two large, eight to nine third and three to four second order in *T. laxa*, 10–11 third order and two to three second-order bundles in *T. triandra* and two to three third-order bundles in *T. quadrivalvis*. Small, round parenchyma cells present surrounding the median vascular bundle adaxially. Sclerenchyma present on abaxial surface associated with vascular bundles in form of strands.

Epidermis: Adaxial epidermal surface has fan-shaped bulliform cells (Fig. 1.46C, arrow) which occupy less than half of the leaf thickness in *T. cymbaria*, most of the adaxial epidermis is made up of large, inflated bulliform cells in *T. laxa* (Fig. 1.47D) while adaxial epidermal surface has restricted group of bulliform cells in *T. triandra* and *T. quadrivalvis* (Figs. 1.48D, 1.49D) which are not larger than the epidermal cells. There are not associated with colorless cells in *T. triandra* and *T. quadrivalvis*.

Abaxial epidermal cells have barrel-shaped cells in *T. cymbaria* and *T. triandra* or elongated, rectangular-shaped cells in *T. laxa* or small, rectangular to barrel-shaped cells in *T. quadrivalvis* with thick cuticle. Papillae present on few epidermal cells one papillae per cell in *T. cymbaria* and *T. triandra* (Figs. 1.46G, 1.48D). Pointed trichomes present on abaxial surface in *T. cymbaria* (Fig. 1.46E). Restricted group of large, inflated bulliform cells projecting above epidermis are present on abaxial surface in *T. laxa*, *T. triandra* (Figs. 1.47D, 1.48F, arrow).

Mesophyll: Radiate, single layered, isodiametric cells completely surrounding bundles; compactly arranged homogenous chlorenchyma cells occupy the major area between the adaxial and abaxial epidermises; chlorenchyma cells form horizontal elongated strap continuous between bundles; colorless cells absent.

Vascular bundles: Vascular bundles regularly arranged from median to marginal bundles in half lamina. Three first orders in *T. cymbaria* and *T. triandra*, four to five first order in *T. laxa*, and five first order in *T. quadrivalvis*. Four to five third order and two to three second order in *T. cymbaria*, three to four third order, and one to two second order in *T. laxa*, seven to eight third order and two to three second order in *T. triandra*, three to four third order and two to three second order in *T. quadrivalvis* bundles between consecutive first-order bundles present. First and second-order vascular bundles placed centrally and third-order vascular bundles placed abaxially of leaf blade in *T. cymbaria*, *T. laxa* while in *T. triandra*, *T. quadrivalvis*, all vascular bundles are placed abaxially at same level.

Third-order vascular bundles (Figs. 1.47D, 1.48E, 1.49D): Hexagonal with vascular tissue consisting of only few indistinguishable vascular elements

Second-order vascular bundles (Figs. 1.46F, 1.47F): Elliptical in *T. cymbaria* and *T. laxa*; circular in *T. triandra* and *T. quadrivalvis*

First-order vascular bundles (Figs. 1.48C, 1.49C)

Shape: Elliptical in all except circular in *T. quadrivalvis*

Relationship of phloem cells and the vascular fibers: Complete surrounded by thick-walled fibers

Nature of lysigenous cavity and protoxylem: Enlarged protoxylem vessel present in all and lysigenous cavity absent only in *T. quadrivalvis*

Size of metaxylem vessels in relation to parenchyma sheath cells in cross-section: Wide vessels

Shape of metaxylem vessels: Circular

Median vascular bundles (Fig. 1.47C)

Shape: Round in *T. cymbaria* and *T. quadrivalvis*, Elliptical in *T. laxa* and *T. triandra*

Relationship of phloem cells and the vascular fibers: Complete surrounded by thick-walled fibers

Nature of lysigenous cavity and protoxylem: Enlarged protoxylem vessel present in all and lysigenous cavity absent only in *T. quadrivalvis*

Size of metaxylem vessels in relation to parenchyma sheath cells in cross-section: Wide vessels

Vascular bundle sheath (Fig. 1.49D, arrow)

Third-order vascular bundle's bundle sheath

Shape: Round

Shape of bundle sheath cell: Cells elliptical and elongated

Second-order vascular bundle's bundle sheath

Shape: Round

Shape of bundle sheath cell: Cells elliptical and elongated

Number of bundle sheath cell: 7–8 cells in all except 8–9 in *T. triandra*

Nature of bundle sheath (either complete or incomplete): Complete in *T. cymbaria* and *T. laxa*, Incomplete and abaxially interrupted by sclerenchyma in *T. triandra* and *T. quadrivalvis*

Chloroplast position in bundle sheath cell: Chloroplast positioned toward outer tangential wall in all species except in *T. laxa* where it is centrally located.

Mestome/Inner sheath: Absent

First-order vascular bundle's bundle sheath

Shape: Elliptical in all species except in *T. cymbaria* which is circular

Extent of bundle sheath around the vascular bundle: Sheath incomplete and abaxially interrupted by sclerenchyma

Extension of bundle sheath: Absent in *T. cymbaria* and *T. quadrivalvis*. In *T. laxa* and *T. triandra*, small abaxial extension of the bundle sheath

Number of cells comprising the bundle sheath: 16–17 in *T. cymbaria*, 12–14 in *T. laxa*, 13–14 in *T. triandra*, 11–12 in *T. quadrivalvis* parenchyma cells

Bundle sheath cell shape: All cells are elliptical and elongated in shape; chloroplast concentrated toward outer tangential wall in all species except in *T. laxa* where it is centrally placed

Mestome/Inner sheath: Absent

Sclerenchyma

Adaxial sclerenchyma: Absent in *T. triandra* and *T. laxa*, minute strand consisting of two to four fibers present above first and second-order vascular bundles in *T. cymbaria* or well-developed strands in *T. quadrivalvis*

Abaxial sclerenchyma: Narrow strand present below first and second-order vascular bundles in *T. laxa* and *T. cymbaria* but well-developed narrow, small girder present in *T. triandra* and *T. quadrivalvis*

Sclerenchyma between bundles: Absent

Sclerenchyma in leaf margin: Forming very small pointed fibrous cap. In *T. quadrivalvis* sclerenchymatous cells curving toward adaxial surface (Fig. 1.49E). Marginal epidermal cells thin with outer thick tangential wall

Leaf-sheath anatomy (Figs. 1.46H–K, 1.47E–G, 1.48G–I, 1.49F–I)

Outline of lamina: V shaped standard angled in *T. laxa* and *T. cymbaria*, narrow angled in *T. triandra* and *T. quadrivalvis* with inrolled margins (Figs. 1.46H, 1.47E, 1.48G, 1.49F).

Ribs and furrows absent in *T. triandra*, *T. quadrivalvis* while shallow furrows with pointed triangular ribs in *T. cymbaria* and round abaxial ribs in *T. laxa* (Fig. 1.46K).

Epidermis

Adaxial epidermal cells: Flat, rectangular

Abaxial epidermal cells: Small, flat, rectangular to barrel shaped with thick cuticle

Ground tissue: Parenchymatous with absence of air cavity

Vascular bundle

Arrangement of vascular bundle: Two in *T. cymbaria*, one in *T. laxa* or three in *T. triandra* and *T. quadrivalvis* consecutive third-order vascular bundle between first and second-order vascular bundle

Third-order vascular bundle (Figs. 1.46J, 1.47F, 1.48I, 1.49H): Pentagonal in all species except in *T. cymbaria* where it is circular with vascular tissue consisting of only few indistinguishable vascular elements

Second-order vascular bundle (Fig. 1.47G): Circular in all species except elliptical in *T. cymbaria*

First-order vascular bundle (Figs. 1.46I, 1.49I): Circular with phloem completely surrounded by thick-walled fibers; enlarged protoxylem vessel and lysigenous cavity present in all species except in *T. quadrivalvis* which showed absence of lysigenous cavity; circular metaxylem wide vessel element

Vascular bundle sheath

Shape: Round, circular

Extent of bundle sheath around the vascular bundle: Sheath incomplete and interrupted by sclerenchyma cells

Number of cells comprising the bundle sheath: 15–16 in *T. cymbaria* and *T. laxa*, 14–15 in *T. triandra*, 13–15 in *T. quadrivalvis* parenchyma cells

Bundle sheath cell shape: All cells are elliptical and elongated in shape; cells smaller than the ground tissue

Mestome/Inner sheath: Absent

Sclerenchyma: In form of narrow strands present on abaxial surface

Ligule anatomy (Figs. 1.46L, M, 1.47H, I, 1.48J, K, 1.49J, K)

Shape: Ligule is attached less than half-length of the lamina and the connection point is very broad, blunt "V" shaped in *T. cymbaria* and *T. laxa* (Figs. 1.46L, 1.47H). Ligule is attached to the lamina more than half-length of it and blunt narrow "V" shaped at the connection point in *T. triandra* and *T. quadrivalvis* (Figs. 1.48J, 1.49J)

Anatomy: Mesophyll cells between adaxial and abaxial surfaces is two layered in all species except three layered in *T. triandra*; few abaxial cells have sclerenchyma cells in *T. cymbaria* and *T. laxa* (Figs. 1.46M, arrow, 1.47I); prickles present on abaxial surface in *T. quadrivalvis* (Fig. 1.49K, arrowhead)

DIFFERENTIATING FEATURES

Vascular bundles placed toward the abaxial surface of leaf blade
Circular-shaped first-order vascular bundle … *T. qurqdrivalvis*
Elliptical-shaped first-order vascular bundle … *T. triandra*
First and second-order vascular bundles placed centrally and third-order vascular bundles present toward abaxial surface of leaf blade.
Chloroplast concentrated near outer tangential wall of mesophyll cells … *T. cymbaria*
Chloroplast placed centrally of mesophyll cells … *T. laxa*

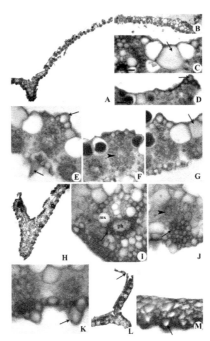

FIGURE 1.46 Leaf anatomy of *Themeda cymbaria.*

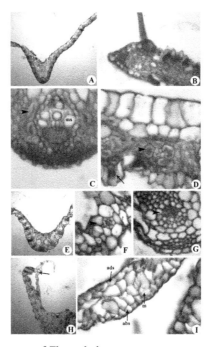

FIGURE 1.47 Leaf anatomy of *Themeda laxa.*

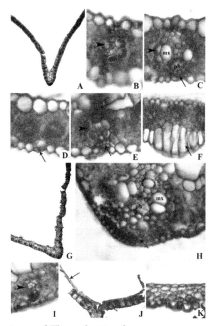

FIGURE 1.48   Leaf anatomy of *Themeda triandra.*

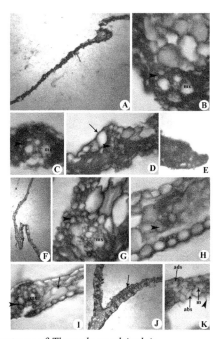

FIGURE 1.49   Leaf anatomy of *Themeda quadrivalvis.*

35. *Triplopogon ramosissimus*

Leaf lamina (Fig. 1.50A–F)

Outline of the lamina: Inrolled from both margin, involute (Fig. 1.50A, B). Ribs and furrows are present on the leaf surface. Shallow furrows with rounded ribs present.

Keel is well developed, semicircular with three large vascular bundles, three to four third-order vascular bundles, and two to three second-order vascular bundles. Small, round parenchyma cells are present surrounding the median vascular bundle adaxially. Sclerenchyma is present on abaxial surface associated with vascular bundles in form of strands.

Epidermis: Adaxial epidermal surface has restricted group of large, inflated bulliform cells which are projecting above epidermis. Short hairs are present with bulbous base sunken in bulliform cells (Fig. 1.50F, arrow). Abaxial epidermal cells are square to rectangular shaped.

Mesophyll: Incomplete radiate chlorenchyma cells; compact homogenous occupy the major area between the adaxial and abaxial epidermises; chlorenchyma cells are arranged in form of horizontal elongated strap continuous between bundles; colorless cells absent.

Vascular bundles: Six first-order vascular bundles in half lamina regularly arranged from median to marginal bundles; four to five third-order bundles between consecutive larger bundles and three to four second-order bundles between consecutive first-order bundles present; all the vascular bundles centrally placed at same level.

Third-order vascular bundles: Hexagonal with vascular tissue consisting of only a few indistinguishable vascular strands

Second-order vascular bundles: Circular shaped

First-order vascular bundles (Fig. 1.50E)

Shape: Circular

Relationship of phloem cells and the vascular fibers: Completely surrounded by thick-walled fibers

Nature of lysigenous cavity and protoxylem: Lysigenous cavity and enlarged protoxylem vessel present

Size of metaxylem vessels in relation to parenchyma sheath cells in cross-section: Wide vessels

Shape of metaxylem vessels: Circular

Median vascular bundles

Shape: Round

Relationship of phloem cells and the vascular fibers: Completely surrounded by thick-walled fibers

Nature of lysigenous cavity and protoxylem: Lysigenous cavity and enlarged protoxylem vessel present

Size of metaxylem vessels in relation to parenchyma sheath cells in cross-section: Wide vessels

Vascular bundle sheath

Third-order vascular bundle's bundle sheath

Shape: Round

Shape of bundle sheath cell: Cells elliptical and elongated

Second-order vascular bundle's bundle sheath

Shape: Round

Shape of bundle sheath cell: Cells elliptical and elongated

Number of bundle sheath cell: 12–13

Nature of bundle sheath (either complete or incomplete): Complete

Chloroplast position in bundle sheath cell: Chloroplast placed centrally

Mestome/Inner sheath: Absent

First-order vascular bundle's bundle sheath

Shape: Round, circular

Extent of bundle sheath around the vascular bundle: Complete

Extension of bundle sheath: Absent

Number of cells comprising the bundle sheath: 15–16 parenchyma cells

Bundle sheath cell shape: All cells are elliptical and elongated in shape; chloroplast positioned centrally in the cell

Mestome/Inner sheath: Absent

Sclerenchyma

Adaxial sclerenchyma: Minute strand consisting of only a few subepidermal fibers

Abaxial sclerenchyma: Well-developed strand, wider than deep and it follows shape of abaxial rib

Sclerenchyma between bundles: Absent

Sclerenchyma in leaf margin: Forming very small fibrous pointed cap; epidermal cells at margin of thin with outer thick tangential wall (Fig. 1.50C, arrow)

Leaf-sheath anatomy (Fig. 1.50G, H)

Outline of lamina: Round shaped with inrolled margins; minute ribs and furrows present

Epidermis

Adaxial epidermal cells: Flat, rectangular

Abaxial epidermal cells: Small, flat, rectangular shaped with thick cuticle

Ground tissue: Parenchymatous with oval-shaped air cavity (Fig. 1.50G, arrow)

Vascular bundle

Arrangement of vascular bundle: Three to four consecutive third-order vascular bundles between first and second-order vascular bundle

Third-order vascular bundle: Circular with vascular tissue consisting of only few indistinguishable vascular elements

Second-order vascular bundle: Circular shaped

First-order vascular bundle: Circular shaped with phloem completely surrounded by thick-walled fibers; lysigenous cavity and enlarged proto-xylem vessel present; circular wide metaxylem vessel elements (Fig. 1.50H)

Vascular bundle sheath

Shape: Round, circular

Extent of bundle sheath around the vascular bundle: Sheath incomplete and interrupted by sclerenchyma cells

Number of cells comprising the bundle sheath: 15–16 parenchyma cells comprise the sheath

Bundle sheath cell shape: All cells are elliptical and elongated in shape; cells smaller than the ground tissue

Mestome/Inner sheath: Absent

Sclerenchyma: In form of well-developed broad girder present on abaxial surface (Fig. 1.50H, arrow)

Ligule anatomy (Fig. 1.50I, J)

Shape: Ligule attached to the midrib region and partially with lamina region and the connection point it is broad round shaped (Fig. 1.50I, arrow)

Anatomy: More than three layered mesophyll cells between adaxial and abaxial surfaces (Fig. 1.50J)

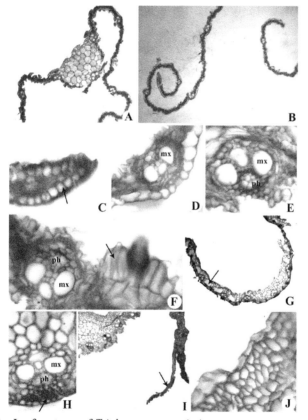

FIGURE 1.50   Leaf anatomy of *Triplopogon ramosissimus.*

36. *Vetivaria zinzanoides*

Leaf lamina (Fig. 1.51A–F)

Outline of the lamina: "V" shaped with wide angle and having outwardly curving arms (Fig. 1.51A). Two halves are symmetrical on either side of the median vascular bundle. Ribs and furrows are absent on the leaf surface. Air cavities present in leaf lamina.

Keel is well developed, V shaped with single median vascular bundle, and two to three third-order vascular bundles. Small, rounded parenchyma cells surround the median vascular bundle adaxially. Sclerenchyma is present on adaxial surface in form of strands and on abaxial surface associated with vascular bundles in form of strands. Elongated bulliform cells form "U" as well as layers of elongated parenchyma cells.

Epidermis: Adaxial epidermal surface has group of bulliform cells which are not associated with colorless cells and occupy less than one-fourth of the leaf thickness. Abaxial epidermal cells barrel shaped and covered with thick cuticle. Stomata is present (Fig. 1.51C and E, arrow).

Mesophyll: Radiate, single layered, isodiametric chlorenchyma cells completely surrounding bundles; compact homogenous chlorenchymatous cells occupy the major area between the adaxial and abaxial epidermises; colorless cells absent.

Vascular bundles: Nine to 10 first-order vascular bundles in half lamina regularly arranged from median to marginal bundles. Three to four third-order bundles between consecutive larger bundles and two to three second-order bundles between consecutive first-order bundles present. All the vascular bundles abaxially placed at same level.

Third-order vascular bundles: Pentagonal (Fig. 1.51B) and round shaped (Fig. 1.51C) with distinguishable xylem and phloem elements

Second-order vascular bundles: Elliptical shaped (Fig. 1.51F)

First-order vascular bundles

Shape: Circular

Relationship of phloem cells and the vascular fibers: Completely surrounded by thick-walled fibers

Nature of lysigenous cavity and protoxylem: Lysigenous cavity and enlarged protoxylem vessel present

Size of metaxylem vessels in relation to parenchyma sheath cells in cross-section: Very wide vessels

Shape of metaxylem vessels: Circular

Median vascular bundles

Shape: Circular

Relationship of phloem cells and the vascular fibers: Completely surrounded by thick-walled fibers

Nature of lysigenous cavity and protoxylem: Lysigenous cavity and enlarged protoxylem vessel present

Size of metaxylem vessels in relation to parenchyma sheath cells in cross-section: Very wide vessels

Shape of metaxylem vessels: Circular

Vascular bundle sheath

Third-order vascular bundle's bundle sheath (Fig. 1.51B, C, arrowhead)

Shape: Round

Shape of bundle sheath cell: Cells elliptical and elongated

Second-order vascular bundle's bundle sheath (Fig. 1.51F, arrowhead)

Shape: Elliptical

Shape of bundle sheath cell: Cells elliptical and elongated

Number of bundle sheath cell: 9–10

Nature of bundle sheath (either complete or incomplete): Complete

Chloroplast position in bundle sheath cell: Chloroplast positioned near outer tangential wall

Mestome/Inner sheath: Absent

First-order vascular bundle's bundle sheath (Fig. 1.51D, arrowhead)

Shape: Round, circular

Extent of bundle sheath around the vascular bundle: Sheath incomplete, abaxially interrupted by sclerenchyma

Extension of bundle sheath: Narrow adaxial extension and well-developed triangular girder present on abaxial side

Number of cells comprising the bundle sheath: 13–14 parenchyma cells

Bundle sheath cell shape: All cells are elliptical, elongated in shape; chloroplast positioned near outer tangential wall

Mestome/Inner sheath: Absent

Sclerenchyma

Adaxial sclerenchyma: Narrow, small, four to six fibers, girder present above first-order vascular bundle

Abaxial sclerenchyma: Well-developed triangular shaped girder present below first-order vascular bundle

Sclerenchyma between bundles: Absent

Sclerenchyma in leaf margin: Forming very small fibrous pointed cap; epidermal cells at margin of thin with thick outer tangential wall

Leaf-sheath anatomy (Fig. 1.51G–J)

Outline of lamina: V shaped narrow angle with inrolled margins; ribs and furrows absent (Fig. 1.51G, H)

Epidermis

Adaxial epidermal cells: Flat, rectangular

Abaxial epidermal cells: Small, flat, rectangular shaped with thick cuticle; stomata present (Fig. 1.51J)

Ground tissue: Parenchymatous with oval-shaped air cavity (Fig. 1.51G, arrow)

Vascular bundle

Arrangement of vascular bundle: Two consecutive third-order vascular bundles between first and second-order vascular bundle

Third-order vascular bundle: Pentagonal with xylem and phloem distinguishable

Second-order vascular bundle: Elliptical shaped

First-order vascular bundle: Elliptical shaped with phloem completely surrounded by thick-walled fibers; lysigenous cavity and enlarged proto-xylem vessel present; circular wide metaxylem vessel elements (Fig. 1.51I)

Vascular bundle sheath

Shape: Round, circular

Extent of bundle sheath around the vascular bundle: Sheath incomplete, interrupted by sclerenchyma cells

Number of cells comprising the bundle sheath: 18–19 parenchyma cells

Bundle sheath cell shape: All cells are elliptical and elongated smaller than the ground tissue

Mestome/Inner sheath: Absent

Sclerenchyma: In form of well-developed broad anchor-shaped girder present on abaxial surface (Fig. 1.51I, arrow)

Ligule anatomy (Fig. 1.51K, L)

Shape: Ligule attached to the lamina more than half-length of it and blunt "V" shaped and narrow at the connection point (Fig. 1.51K, arrow)

Anatomy: Single layered mesophyll cells between adaxial and abaxial surfaces (Fig. 1.51L)

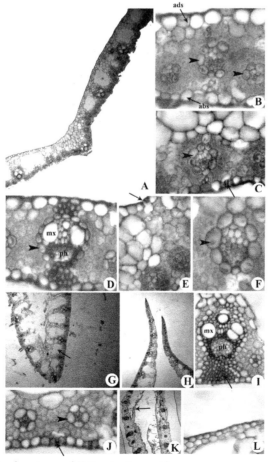

FIGURE 1.51   Leaf anatomy of *Vetivaria zinzanoides*.

37. *Alloteropis cimiciana*

Leaf lamina (Fig. 1.52A–F)

Outline of the lamina: "V" shaped with standard angle with straight arms (Fig. 1.52A, B); two halves symmetrical on either side of the median vascular bundle; ribs and furrows present on the leaf surface; abaxial surface has narrow, medium furrows with rounded ribs.

Keel is well-developed, inverted triangular shaped (Fig. 1.52A) with three first-order, two to three second-order and six to seven third-order vascular bundles. Small, round parenchyma cells surround the median vascular bundle adaxially. Adaxial sclerenchyma present in form of strands in subepidermal layers and abaxial sclerenchyma associated with vascular bundles in form of strands.

Epidermis: Adaxial epidermal surface is majorly made up of inflated bulliform cells (Fig. 1.52C) with outer walls slightly thickened and covered with thin cuticle; abaxial epidermal cells barrel shaped; prickles present on abaxial surface (Fig. 1.52E, arrow)

Mesophyll: Irregular chlorenchyma with small isodiametric cells; compact homogenous chlorenchymatous cells occupy the major area between the adaxial and abaxial epidermises; chlorenchyma cells continuous and horizontally elongated strap shaped between bundles; colorless cells absent.

Vascular bundles: Four to five first-order vascular bundles in half lamina regularly arranged from median to marginal bundles; 9–10 third-order bundles between consecutive larger bundles and three to four second-order bundles between consecutive first-order bundles present; all the vascular bundles present at same level and toward the abaxial surface

Third-order vascular bundles: Circular with xylem and phloem distinguished (Fig. 1.52C)

Second-order vascular bundles: Circular shaped

First-order vascular bundles (Fig. 1.52D)

Shape: Circular

Relationship of phloem cells and the vascular fibers: Completely surrounded by thick-walled fibers

Nature of lysigenous cavity and protoxylem: Protoxylem present and lysigenous cavity is absent

Size of metaxylem vessels in relation to parenchyma sheath cells in cross-section: Very wide vessels

Shape of metaxylem vessels: Circular

Median vascular bundles

Shape: Circular

Relationship of phloem cells and the vascular fibers: Completely surrounded by thick-walled fibers

Nature of lysigenous cavity and protoxylem: Protoxylem present and lysigenous cavity is absent

Size of metaxylem vessels in relation to parenchyma sheath cells in cross-section: Very wide vessels
Shape of metaxylem vessels: Circular

Vascular bundle sheath

Third-order vascular bundle's bundle sheath (Fig. 1.52C, arrowhead)

Shape: Round

Shape of bundle sheath cell: Elliptical in shape

Second-order vascular bundle's bundle sheath

Shape: Round

Shape of bundle sheath cell: Elliptical

Number of bundle sheath cell: 10–11

Nature of bundle sheath (either complete or incomplete): Complete

Chloroplast position in bundle sheath cell: Chloroplast positioned near the outer tangential wall

Mestome/Inner sheath: Absent

First-order vascular bundle's bundle sheath (Fig. 1.52D, arrowhead)

Shape: Round, circular

Extent of bundle sheath around the vascular bundle: Complete

Extension of bundle sheath: Absent

Number of cells comprising the bundle sheath: 17–18 parenchyma cells

Bundle sheath cell shape: All cells are elliptical and elongated in shape; chloroplast positioned near the outer tangential wall

Mestome/Inner sheath: Absent

Sclerenchyma

Adaxial sclerenchyma: Absent

Abaxial sclerenchyma: Well-developed wider strands present below first-order vascular bundles (Fig. 1.52D, arrow)

Sclerenchyma between bundles: Absent

Sclerenchyma in leaf margin: Forming well-developed rounded fibrous cap; epidermal cells at margin are thin with outer thick tangential wall

Leaf-sheath anatomy (Fig. 1.52G–I)

Outline of lamina: "V" shaped with inrolled margins (Fig. 1.52G); ribs and furrows absent

Epidermis

Adaxial epidermal cells: Flat, rectangular

Abaxial epidermal cells: Barrel shaped with thick cuticle (Fig. 1.52I)

Ground tissue: Parenchymatous with vertical oval-shaped air cavity (Fig. 1.52G, arrow)

Vascular bundle

Arrangement of vascular bundle: One consecutive third-order vascular bundle between first and second-order vascular bundle

Third-order vascular bundle: Circular with xylem and phloem distinguishable

Second-order vascular bundle: Circular shaped

First-order vascular bundle: Circular shaped with phloem completely surrounded by thick-walled fibers; protoxylem and lysigenous cavity present; circular wide metaxylem vessels elements (Fig. 1.52H)

Vascular bundle sheath

Shape: Round, circular

Extent of bundle sheath around the vascular bundle: Sheath incomplete and interrupted by sclerenchyma abaxially

Number of cells comprising the bundle sheath: 13–14 parenchyma cells

Bundle sheath cell shape: All cells are elliptical and elongated, smaller than the ground tissue

Mestome/Inner sheath: Absent

Sclerenchyma: In form of straight horizontal well-developed abaxial strand (Fig. 1.52H, arrow)

Ligule anatomy (Fig. 1.52J, K)

Shape: Ligule attached to the midrib and few parts of lamina region, hairy ligule (Fig. 1.52J, arrow)

Anatomy: More than three layered mesophyll cells between adaxial and abaxial surfaces (Fig. 1.52K)

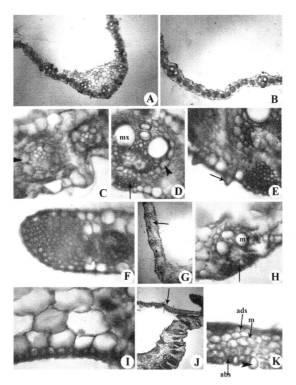

FIGURE 1.52 Leaf anatomy of *Alloteropis cimiciana*.

Brachiaria

38. *Brachiaria distachya* (Fig. 1.53)

39. *Brachiaria eruciformis* (Fig. 1.54)

40. *Brachiaria ramosa* (Fig. 1.55)

41. *Brachiaria reptans* (Fig. 1.56)

Leaf lamina (Figs. 1.53A–G, 1.54A–H, 1.55A–E, 1.56A–H)

Outline of the lamina: Expanded with gently undulating leaf blade in *B. eruciformis*, *B. ramosa*, *B. distachya*, and *B. reptans* (Figs. 1.53A, 1.54A, 1.55A, B, 1.56A). Ribs and furrows are present on the leaf surface. Shallow, wide furrows present between all vascular bundles. Rounded ribs present over all vascular bundles.

Keel inconspicuous in all species except well developed in *B. reptans*, semicircular keel with one first-order vascular bundles and three third-order vascular bundles. Sclerenchyma is associated with vascular bundles in form of girder.

Epidermis: Adaxial epidermal surface has fan-shaped bulliform cells in all the species (Figs. 1.53B, 1.54F, G, arrow, 1.55C, arrow, 1.56G, arrow) and it occupies one-fourth to half portion of the leaf thickness in *B. eruciformis*, *B. ramosa*, *B. distachya* while less than one-fourth portion of the leaf thickness in *B. reptans*. Abaxial epidermis surface has round epidermal cells which have thick cuticle. Long, slender macrohairs with bulbous base sunken in bulliform cells are present in *B. distachya* (Fig. 1.53E, F), prickles present in *B. distachya* (Fig. 1.53G). Trichomes present in *B. reptans* (Fig. 1.56, F, arrow).

Mesophyll: Radiate, single layered, isodiametric chlorenchyma cells completely surround the bundles; compactly arranged homogenous chlorenchymatous cells occupy the major area between the adaxial and abaxial epidermises; colorless cells present only in *B. eruciformis* which are closely related to bulliform cells (Fig. 1.54E, arrow).

Vascular bundles: Four to five first-order vascular bundles in *B. distachya* and *B. eruciformis*, three to four first order in *B. ramosa* and *B. reptans*.

Vascular bundles regularly arranged from median to marginal bundles in half lamina. Three to four third-order and two to three second-order in *B. distachya*, *B. eruciformis*, and *B. ramosa*, four to five third-order and second-order bundles between consecutive first-order bundles present. All the vascular bundles present at same level and placed centrally

Third-order vascular bundles (Figs. 1.53B, 1.54C, E, 1.55C, 1.56E): Pentagonal in *B. distachya* and *B. reptans*, hexagonal in *B. eruciformis* and *B. ramosa* with vascular tissue consisting of only few indistinguishable vascular elements

Second-order vascular bundles: Elliptical in *B. distachya* and *B. ramosa*, circular in *B. eruciformis* and *B. reptans*

First-order vascular bundles (Figs. 1.53C, 1.54D, 1.55D, 1.56C)

Shape: Elliptical in *B. distachya* and *B. ramosa*, circular in *B. eruciformis* and *B. reptans*

Relationship of phloem cells and the vascular fibers: Completely surrounded by thick-walled fibers

Nature of lysigenous cavity and protoxylem: Lysigenous cavity absent and enlarged protoxylem vessel present

Size of metaxylem vessels in relation to parenchyma sheath cells in cross-section: Narrow vessels

Shape of metaxylem vessels: Circular

Median vascular bundles (Figs. 1.53D, 1.55F, 1.56D)

Shape: Elliptical in *B. distachya* and *B. ramosa*, circular in *B. eruciformis* and *B. reptans*

Relationship of phloem cells and the vascular fibers: Completely surrounded by thick-walled fibers

Nature of lysigenous cavity and protoxylem: Lysigenous cavity and enlarged protoxylem vessel present in all species but lysigenous cavity absent in *B. eruciformis* and *B. reptans*

Size of metaxylem vessels in relation to parenchyma sheath cells in cross-section: Narrow vessels

Vascular bundle sheath

Third-order vascular bundle's bundle sheath (Figs. 1.54C, E, 1.55C, arrowhead, 1.56E)

Shape: Round

Shape of bundle sheath cell: Fan shaped

Second-order vascular bundle's bundle sheath

Shape: Elliptical

Shape of bundle sheath cell: Fan shaped

Number of bundle sheath cell: 8–9 in all species except 10–11 in *B. ramosa*

Nature of bundle sheath (either complete or incomplete): Complete in all species except *B. distachya* in which bundle sheath is abaxially interrupted by sclerenchyma

Chloroplast position in bundle sheath cell: Chloroplast centrally placed

Mestome/Inner sheath: Absent

First-order vascular bundle's bundle sheath

Shape: Horseshoe shaped

Extent of bundle sheath around the vascular bundle: Complete in *B. eruciformis* and *B. reptans*, incomplete and abaxially interrupted by sclerenchyma in *B. distachya* and *B. ramosa*

Extension of bundle sheath: Absent in all species except *B. distachya* has broad abaxial extension

Number of cells comprising the bundle sheath: 10–11 parenchyma cells in all species but *B. ramosa* has 12–13 cells

Bundle sheath cell shape: Fan shaped; chloroplast placed centrally in *B. distachya* and *B. ramosa* and near outer tangential wall in *B. eruciformis* and *B. reptans*

Mestome/Inner sheath: Complete sheath present with cells smaller than the outer bundle sheath

Sclerenchyma

Adaxial sclerenchyma: Absent

Abaxial sclerenchyma: Well developed in form of small, narrow girder in *B. distachya* and *B. ramosa* and minute subepidermal strands in *B. eruciformis* and *B. reptans*

Sclerenchyma between bundles: Absent

Sclerenchyma in leaf margin: Forming very small marginal area in *B. distachya* and *B. ramosa* while well developed in *B. eruciformis* and *B. reptans* with pointed fibrous cap; epidermal cells at margin of thin with thick outer tangential wall

Leaf-sheath anatomy (Figs. 1.53H–K, 1.54I–M, 1.55F–H, 1.56I–L)

Outline of lamina: Round with overlapping inrolled margins in *B. distachya*, *B. ramosa*, and *B. reptans* (Figs. 1.53H, 1.54F, 1.55I) while V-shaped inrolled margins of arms in *B. eruciformis* (Fig. 1.56I); ribs and furrows present; shallow, wide furrows present between all vascular bundles on abaxial surface

Epidermis

Adaxial epidermal cells: Flat, rectangular

Abaxial epidermal cells: Small, flat, rectangular shaped with thick cuticle; prickles present (Fig. 1.54L, arrow) in *B. eruciformis*

Ground tissue: Parenchymatous with vertically oval-shaped air cavity in *B. distachya* and *B. eruciformis* (Figs. 1.53K, arrow, 1.54M, arrow), oval shaped in *B. ramosa* (Fig. 1.55F) while absent in *B. reptans*

Vascular bundle

Arrangement of vascular bundle: One consecutive third-order vascular bundle between first and second-order vascular bundle

Third-order vascular bundle (Figs. 1.53J, 1.54M): Circular in *B. distachya*, hexagonal in *B. eruciformis* and *B. reptans*, elliptical in *B. ramosa* with vascular tissue consisting of only few indistinguishable vascular elements

Second-order vascular bundle: Elliptical shaped

First-order vascular bundle (Figs. 1.53I, 1.54J, 1.55H, 1.56J, K): Elliptical with phloem completely surrounded by thick-walled fibers; lysigenous cavity absent and enlarged protoxylem vessel present in all species but lysigenous cavity present in *B. ramosa*; circular wide metaxylem vessel element

Vascular bundle sheath (Fig. 1.56K, arrowhead)

Shape: Horseshoe shaped

Extent of bundle sheath around the vascular bundle: Sheath incomplete, interrupted by sclerenchyma cells on abaxial surface

Number of cells comprising the bundle sheath: 8–9 in *B. distachya* and *B. reptans*, 10–11 in *B. eruciformis* and *B. ramosa* parenchyma cells

Bundle sheath cell shape: Fan-shaped cells; all cells similar in *B. eruciformis* and *B. ramosa* while adaxial cells larger than the other cells of sheath cell in *B. distachya* and *B. reptans*

Mestome/Inner sheath: Complete sheath present with cells are smaller than the outer bundle sheath

Sclerenchyma: In form of well-developed straight horizontal broad strand present on abaxial surface in *B. distachya*, *B. reptans*, and *B. eruciformis* while anchored shaped girder in *B. ramosa*

Ligule anatomy (Figs. 1.53L, M, 1.54N, 1.55I, 1.56M, N)

Shape: Ligule attached to the midrib and few parts of lamina region because it is hairy ligule (Figs. 1.53L, 1.54N, 1.55I, 1.56M)

Anatomy: Three layered in *B. distachya*, *B. ramose*, and *B. eruciformis* while single layered in *B. reptans* mesophyll cells between adaxial and abaxial surfaces (Figs. 1.53M, 1.56N)

DIFFERENTIATING FEATURES

Colorless cells present … *B. eruciformis*
Colorless cells absent
Extension of vascular bundle sheath is present … *B. distachya*
Extension of vascular bundle sheath is absent
Chloroplast placed centrally of mesophyll cells … *B. ramosa*
Chloroplast concentrated near outer tangential walls of mesophyll cells …
B. reptans

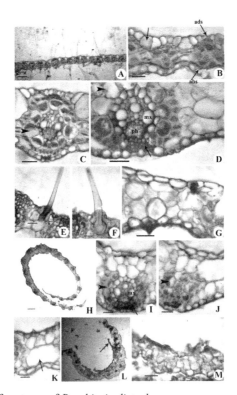

FIGURE 1.53   Leaf anatomy of *Brachiaria distachya*.

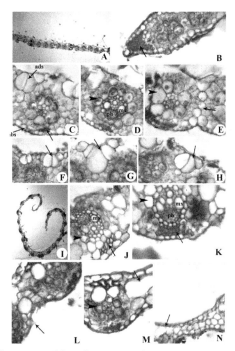

FIGURE 1.54   Leaf anatomy of *Brachiaria eruciformis.*

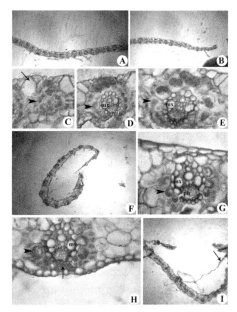

FIGURE 1.55   Leaf anatomy of *Brachiaria ramosa.*

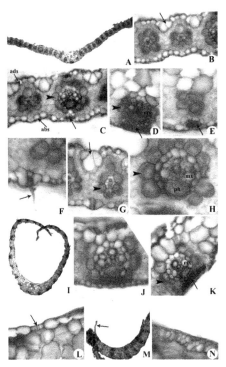

FIGURE 1.56   Leaf anatomy of *Brachiaria reptans*.

Cenchrus

42. *Cenchrus biflorus* (Fig. 1.57)

43. *Cenchrus ciliaris* (Fig. 1.58)

44. *Cenchrus setigerus* (Fig. 1.59)

Leaf lamina (Figs. 1.57A–G, 1.58A–E, 1.59A–K)

Outline of the lamina: "V" shaped, broad angled with irregularly wavy arms in *C. biflorus* and *C. ciliaris* (Figs. 1.57A, 1.58A), wide angled in *C. setigerus* outwardly curving convex arms (Fig. 1.59A); two halves are symmetrical on either side of the median vascular bundle.

Leaf surfaces show ribs on the vascular bundle and furrows in between vascular bundle in *C. biflorus* and *C. ciliaris* and absent in *C. setigerus*. Furrows present between the vascular bundles on the adaxial surface are shallow, wide and the ribs on the vascular bundles are flat in *C. biflorus* while medium, wide furrows and the ribs on the vascular bundles are triangular in *C. ciliaris*. On the abaxial surface furrows are narrow, deeper, medium, and ribs are rounded in *C. biflorus* while shallow furrows and rounded ribs present in *C. ciliaris*.

Keel is semicircular in *C. biflorus* and *C. ciliaris*, "V" shaped in *C. setigerus* with three first-order, six to seven third-order in *C. biflorus*, single medium vascular bundle, three to four third-order and in *C. ciliaris*, single median vascular bundle, 12–14 third-order in *C. setigerus* and two to three second-order vascular bundles. Sclerenchyma is associated with vascular bundles in form of well-developed strands.

Epidermis: Adaxial epidermal surface has extensive groups of large, inflated bulliform cells extending over many vascular bundles except first-order vascular bundle in *C. biflorus* (Fig. 1.57E), restricted group of large, inflated bulliform cells present in level with the general epidermal surface in *C. ciliaris* and *C. setigerus* (Figs. 1.58C, arrow, 1.59J, arrow). Adaxial epidermal cells are rectangular to square shaped in *C. biflorus*, barrel shaped in *C. ciliaris* or large, rectangular shaped in *C. setigerus* covered with thin cuticle; thickened prickle present (Fig. 1.57F, arrow); abaxial epidermal cells are round with outer tangential wall inflated in *C. biflorus*, barrel shaped with thick cuticle in *C. ciliaris*, *C. setigerus*; prickles present on abaxial surface (Fig. 1.57D, arrow) of *C. biflorus*; abaxial epidermis has restricted groups of large, inflated bulliform cells present above the level of general epidermis surface in *C. setigerus* (Fig. 1.59K, arrow) and prickles in *C. setigerus* (Fig. 1.59I, arrow) and stomata in *C. setigerus* (Fig. 1.59H, arrow) present on abaxial.

Mesophyll: Irregular, vertically arranged single layered, isodiametric chlorenchyma cells. Compactly arranged homogenous chlorenchymatous cells occupy the major area between the adaxial and abaxial epidermises, chlorenchyma cells continuous between bundles in *C. biflorus* while incompletely radiating chlorenchyma cells continuous between bundles in *C. ciliaris*, *C. setigerus*; colorless cells absent.

Vascular bundles: Three to four first-order vascular bundles regularly arranged from median to marginal bundles in half lamina.

Five to six third and three to four second-order bundles in *C. biflorus*, four to five third, and two to three second-order bundles in *C. ciliaris* and *C. setigerus* between consecutive first-order bundles present; first-order vascular bundles arranged centrally while third-order vascular bundles placed abaxially

Third-order vascular bundles (Figs. 1.57D, 1.58E, 1.59D): Hexagonal with vascular tissue consisting of only a few indistinguishable vascular strands in *C. biflorus* and *C. ciliaris* while xylem and phloem distinguishable strands in *C. setigerus*

Second-order vascular bundles: Circular (Figs. 1.57E, 1.59E)

First-order vascular bundles (Figs. 1.57C, 1.58D, 1.59G)

Shape: Circular

Relationship of phloem cells and the vascular fibers: Phloem adjoins the inner or parenchyma sheath in *C. biflorus* while completely surrounded by thick-walled fibers in *C. ciliaris* and *C. setigerus*

Nature of lysigenous cavity and protoxylem: Enlarged protoxylem vessel present in all species but lysigenous cavity present only in *C. ciliaris*

Size of metaxylem vessels in relation to parenchyma sheath cells in cross-section: Wide vessels

Shape of metaxylem vessels: Circular

Median vascular bundles (Figs. 1.57G, 1.59F)

Shape: Circular in *C. biflorus* and *C. ciliaris*, egg shaped in *C. setigerus*

Relationship of phloem cells and the vascular fibers: Phloem adjoins the inner or parenchyma sheath in *C. biflorus* while completely surrounded by thick-walled fibers in *C. ciliaris* and *C. setigerus*

Nature of lysigenous cavity and protoxylem: Enlarged protoxylem vessel present in all species but lysigenous cavity present only in *C. ciliaris*

Size of metaxylem vessels in relation to parenchyma sheath cells in cross-section: Wide vessels

Vascular bundle sheath

Third-order vascular bundle's bundle sheath (Figs. 1.57D, 1.58E, arrowhead, 1.59D, arrowhead)

Shape: Round

Shape of bundle sheath cell: Elliptical shaped

Second-order vascular bundle's bundle sheath

Shape: Elliptical in *C. biflorus*, circular in *C. ciliaris* and *C. setigerus*

Shape of bundle sheath cell: Elliptical shaped

Number of bundle sheath cell: 10–11 in *C. biflorus*, 11–12 in *C. ciliaris*, 17–18 in *C. setigerus*

Nature of bundle sheath (either complete or incomplete): Complete

Chloroplast position in bundle sheath cell: Chloroplast positioned toward inner in *C. biflorus* and *C. ciliaris* or outer in *C. setigerus* tangential wall

Mestome/Inner sheath: Absent

First-order vascular bundle's bundle sheath (Figs. 1.57C, 1.58D, 1.59G)

Shape: Circular

Extent of bundle sheath around the vascular bundle: Complete

Extension of bundle sheath: Absent

Number of cells comprising the bundle sheath: 15–16 in *C. biflorus*, 20–22 in *C. ciliaris* and *C. setigerus* parenchyma cells comprise the sheath

Bundle sheath cell shape: Elliptical shaped; chloroplast positioned toward inner in *C. biflorus* and *C. ciliaris* or outer in *C. setigerus* tangential wall

Mestome/Inner sheath: Absent

Sclerenchyma

Adaxial sclerenchyma: Minute subepidermal strands present

Abaxial sclerenchyma: Narrow well-developed tall strands

Sclerenchyma between bundles: Absent

Sclerenchyma in leaf margin: Forming well-developed round fibrous cap (Figs. 1.165B, 1.166B, 1.167B); epidermal cells at margin are thin with thick outer tangential wall

Leaf-sheath anatomy (Figs. 1.57H–K, 1.58F–K, 1.59L–O)

Outline of lamina: "V" shaped, standard angle in *C. biflorus* (Figs. 1.57H, 1.58F, 1.59L) with outwardly bowed arms, outwardly curving in *C. ciliaris* and *C. setigerus*; ribs and furrows present; shallow, wide furrows present between all vascular bundles abaxially

Epidermis

Adaxial epidermal cells: Flat, rectangular with thin cuticle

Abaxial epidermal cells: Flat, rectangular shaped with thick cuticle; trichomes (Fig. 1.58G, arrow), hooks (Fig. 1.58H, arrow), and stomata (Fig. 1.58I, arrow) present in *C. ciliaris*; restricted groups of large, inflated bulliform cells present above the level of general epidermal surface (Fig. 1.59O, arrow) and trichomes (Fig. 1.59N, arrow) present in *C. setigerus*

Ground tissue: Parenchymatous with absence of air cavity

Vascular bundle

Arrangement of vascular bundle: Two to three in *C. biflorus*, one in *C. ciliaris*, three to four in *C. setigerus* consecutive third-order vascular bundles between first and second-order vascular bundle

Third-order vascular bundle (Figs. 1.57I, J, 1.58I, J, 1.59I, J): Pentagonal in *C. biflorus*, hexagonal in *C. ciliaris*, and circular in *C. setigerus* with vascular tissue consisting of only few indistinguishable vascular elements in *C. biflorus* and *C. ciliaris* while xylem and phloem distinguishable strands in *C. setigerus*

Second-order vascular bundle: Elliptical in *C. biflorus* and *C. ciliaris*, circular in *C. setigerus*

First-order vascular bundle: Circular in *C. ciliaris*, *C. biflorus* and *C. setigerus* (Figs. 165K, 166K, 167O) with phloem completely surrounded by thick-walled fibers; enlarged protoxylem vessel present in all species but lysigenous cavity present only in *C. ciliaris*; circular wide metaxylem vessel element

Vascular bundle sheath (Figs. 1.57K, 1.58K, 1.59O)

Shape: Circular

Extent of bundle sheath around the vascular bundle: Complete

Number of cells comprising the bundle sheath: 19–20 in *C. biflorus*, 17–18 in *C. ciliaris*, 22–23 in *C. setigerus* parenchyma cells

Bundle sheath cell shape: Elliptical

Mestome/Inner sheath: Absent

Sclerenchyma: In form of narrow, wide strand on abaxial surface in *C. biflorus* and *C. ciliaris* (Fig. 1.58J, arrow), wider strand on abaxial surface above ribs in *C. setigerus*

Ligule anatomy (Figs. 1.57L, 1.58L, 1.59P, Q)

Shape: Ligule attached to the lamina more than half-length of it and blunt narrow "V" shaped at the connection point (Figs. 1.57L, 1.58L, 1.59P)

Anatomy: Two layered (*C. biflorus*, *C. ciliaris*) and more than three layered (*C. setigerus*) mesophyll cells between adaxial and abaxial surfaces

DIFFERENTIATING FEATURES

More than three layered of mesophyll present in ligule ... *C. setigerus*
Two layered of mesophyll present in ligule
Phloem adjoins the inner or parenchyma sheath in first-order vascular bundle ... *C. biflorus*
Phloem completely surrounded by thick-walled fibers ... *C. ciliaris*

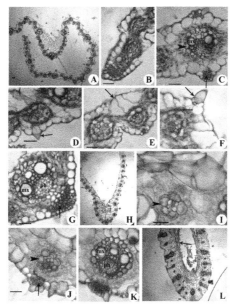

FIGURE 1.57 Leaf anatomy of *Cenchrus biflorus*.

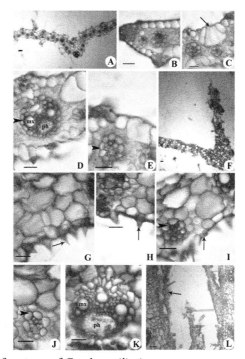

FIGURE 1.58 Leaf anatomy of *Cenchrus ciliaris*.

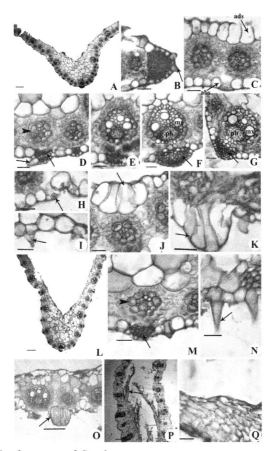

FIGURE 1.59 Leaf anatomy of *Cenchrus setigerus.*

Digitaria

45. *Digitaria ciliaris* (Fig. 1.60)

46. *Digitaria granularis* (Fig. 1.61)

47. *Digitaria longiflora* (Fig. 1.62)

48. *Digitaria stircta* (Fig. 1.63)

Leaf lamina (Figs. 1.60A–F, 1.61A–E, 1.62A–G, 1.63A–G)

Outline of the lamina: Expanded with flat, straight leaf blade in *D. ciliaris*, *D. granularis*, *D. longiflora*, and *D. stricta* (Figs. 1.60A, 1.61A, 1.62A, 1.63A). Two halves are symmetrical on either side of the median vascular bundle. Ribs and furrows are absent on the leaf surface.

Keel is well developed in all species except inconspicuous in *D. ciliaris*. Semicircular keel with single median vascular bundle, three to four third-order vascular bundles in *D. granularis*, two to three second-order vascular bundles and five to six third-order vascular bundles in *D. longiflora* and one to two third-order vascular bundles in *D. stricta*.

Epidermis: Adaxial epidermal surface has most part of it made up of large, inflated by bulliform cells in *D. ciliaris*, *D. longiflora* and *D. stricta* (Figs. 1.60F, 1.62G, arrow, 1.63D) while restricted groups of large, inflated bulliform cells present above the level of general epidermis in *D. granularis* (Fig. 1.61F). Abaxial epidermal cells are barrel shaped in *D. granularis* and *D. ciliaris*, rectangular in *D. longiflora*, barrel to rectangular in *D. stricta* covered with cuticle; small prickles (Fig. 1.60B, arrow) and stomata (Fig. 1.60C, arrow) in *D. ciliaris*.

Mesophyll: Irregular, vertically arranged chlorenchyma cells occupy the major area between the adaxial and abaxial epidermises; chlorenchyma cells continuous between bundles; colorless cells absent.

Vascular bundles: Four to five first-order vascular bundles in *D. ciliaris*, three to four first-order vascular bundles in *D. granularis*, *D. longiflora*, and *D. stricta* regularly arranged from median to marginal bundles in half lamina. Three to four third-order and two to three second-order in *D. ciliaris*, five to six third-order and three to four second-order in *D. granularis*, eight to nine third-order and three to four second-order in *D. longiflora* and seven to eight third-order and two to three second-order in *D. stricta* bundles between consecutive first-order bundles present. All the vascular bundles present at same level and placed centrally.

Third-order vascular bundles (Figs. 1.60F, 1.61F, 1.62F, 1.63D, E): Square in *D. ciliaris*, only pentagonal in *D. granularis*, hexagonal in *D. longiflora*, and pentagonal (Fig. 1.63D) and square (Fig. 1.63E) in *D. stricta* with vascular tissue consisting of only a few indistinguishable vascular element

Second-order vascular bundles (Figs. 1.61D, 1.62D): Elliptical in all species except circular in *D. ciliaris*

First-order vascular bundles (Figs. 1.60E, 1.61E, 1.63F)

Shape: Circular in all species except elliptical in *D. granularis*

Relationship of phloem cells and the vascular fibers: Completely surrounded by thick-walled fibers

Nature of lysigenous cavity and protoxylem: Lysigenous cavity present in all species and protoxylem vessel absent in *D. longiflora*

Size of metaxylem vessels in relation to parenchyma sheath cells in cross-section: Wide vessels

Shape of metaxylem vessels: Circular

Midrib vascular bundles (Figs. 1.60D, 1.63G)

Shape: Round in *D. ciliaris* and *D. longiflora*, obovate in *D. granularis*, egg shaped in *D. stricta*

Relationship of phloem cells and the vascular fibers: Completely surrounded by thick-walled fibers

Nature of lysigenous cavity and protoxylem: Lysigenous cavity present in all species and protoxylem vessel absent in *D. longiflora*

Size of metaxylem vessels in relation to parenchyma sheath cells in cross-section: Wide vessels

Vascular bundle sheath

Third-order vascular bundle's bundle sheath (Figs. 1.60F, arrowhead, 1.61E, arrowhead, 1.62F, arrowhead, 1.63D, E, arrowhead)

Shape: Round

Shape of bundle sheath cell: Cells elliptical and elongated

Second-order vascular bundle's bundle sheath

Shape: Round in *D. ciliaris* and *D. stricta*, elliptical in *D. granularis* and *D. longiflora*

Shape of bundle sheath cell: Cells elliptical and elongated

Number of bundle sheath cell: 8–9 in *D. ciliaris* and *D. stricta*, 11–12 in *D. granularis*, 14–15 in *D. longiflora*

Nature of bundle sheath (either complete or incomplete): Complete

Chloroplast position in bundle sheath cell: Chloroplast positioned toward near outer tangential wall

Mestome/Inner sheath: Absent

First-order vascular bundle's bundle sheath (Fig. 1.60D)

Shape: Circular in all species except in *D. granularis* elliptical

Extent of bundle sheath around the vascular bundle: Complete

Extension of bundle sheath: Absent

Number of cells comprising the bundle sheath: 15–16 in *D. ciliaris*, 17–18 in *D. granularis*, 13–14 in *D. longiflora*, and 16–17 in *D. stricta* parenchyma cells

Bundle sheath cell shape: All cells are elliptical and elongated in shape; chloroplast positioned toward outer tangential wall

Mestome/Inner sheath: Absent

Sclerenchyma

Adaxial sclerenchyma: Absent

Abaxial sclerenchyma: Minute subepidermal strands present

Sclerenchyma between bundles: Absent

Sclerenchyma in leaf margin: Forming very small pointed fibrous cap (Fig. 1.63C, arrow); epidermal cells at margin are thin with thick outer tangential wall

Leaf-sheath anatomy (Figs. 1.60G–J, 1.61F–I, 1.62H–K, 1.63H–L)

Outline of lamina: V shaped, standard angle with inrolled margins in *D. ciliaris*, *D. granularis*, *D. longiflora*, and *D. stricta* (Figs. 1.60G, 1.61F, 1.62H, 1.63H); shallow, wide furrows present on abaxial surface with flat ribs

Epidermis: Flat, rectangular adaxial epidermal cells covered with thin cuticle

Abaxial epidermal cells: Small, flat, rectangular shaped with thick cuticle; restricted groups of large, inflated bulliform cells (Fig. 1.60J, arrow) present above the level of general epidermis surface in *D. ciliaris*

Ground tissue: Parenchymatous ground tissue; all species showed absence of air cavity except *D. granularis* with few big, oval-shaped air cavities (Fig. 1.61H, arrow)

Vascular bundle

Arrangement of vascular bundle: One in *D. ciliaris*, three to four in *D. granularis*, two to three in *D. longiflora* and one to two in *D. stricta* consecutive third-order vascular bundles between first and second-order vascular bundle

Third-order vascular bundle (Figs. 1.60H, 1.61I, 1.62I, 1.63L): Pentagonal in *D. ciliaris*, *D. longiflora*, and *D. stricta*, circular (*D. granularis*) with vascular tissue consisting of only few indistinguishable vascular elements

Second-order vascular bundle (Figs. 1.61G, 1.62I, 1.63L): Circular in all species except in *D. longiflora* elliptical

First-order vascular bundle (Figs. 1.60I, 1.61H, 1.62J, 1.63L, J): Circular in *D. ciliaris* and *D. granularis*, elliptical in *D. longiflora*, circular (Fig. 1.63L)

and elliptical (Fig. 1.63J) in *D. stricta* with phloem completely surrounded by thick-walled fibers; lysigenous cavity and enlarged protoxylem vessel present in *D. ciliaris*, lysigenous cavity absent and protoxylem present in *D. granularis* while lysigenous cavity present and protoxylem present in *D. longiflora* and *D. stricta*; circular shaped wide metaxylem vessels elements

Vascular bundle sheath (Figs. 1.60I, 1.61H, arrowhead, 1.62J, arrowhead, 1.63J, arrowhead)

Shape: Round, circular in all species except in *D. stricta* elliptical

Extent of bundle sheath around the vascular bundle: Complete in all species except in *D. granularis*, incomplete and interrupted by sclerenchyma cells abaxially

Number of cells comprising the bundle sheath: 12–13 in *D. ciliaris*, 11–12 in *D. granularis*, 15–16 in *D. longiflora*, and 19–20 in *D. stricta* parenchyma cells

Bundle sheath cell shape: All cells are elliptical and elongated, smaller than the ground tissue cell

Mestome/Inner sheath: Absent

Sclerenchyma: In form of subepidermal strand in *D. ciliaris* and *D. granularis*, narrow, taller than wider strand in *D. longiflora* and narrow, straight horizontal strand in *D. stricta* present on abaxial surface

Ligule anatomy (Figs. 1.60K, L, 1.61J, K, 1.62L, M, 1.63M, N)

Shape: Ligule attached to the leaf sheath more than the half width and very broad round shaped at the connection point (Figs. 1.60K, 1.61J, 1.62L, 1.63M)

Anatomy: Two layered in *D. ciliaris*, *D. granularis*, and *D. longiflora*, three layered in *D. stricta* mesophyll cells between adaxial and abaxial surfaces (Figs. 1.60L, 1.61L, 1.62M, 1.63N)

DIFFERENTIATING FEATURES

Circular-shaped third-order vascular bundles in leaf sheath ... *D. granularis*
Pentagonal-shaped third-order vascular bundles in leaf sheath
Egg-shaped midrib vascular bundle ... *D. stricta*
Round-shaped midrib vascular bundle
Square-shaped third-order vascular bundles in leaf lamina ... *D. ciliaris*
Angular (penta and hexagonal)-shaped third-order vascular bundles in leaf lamina ... *D. longiflora*

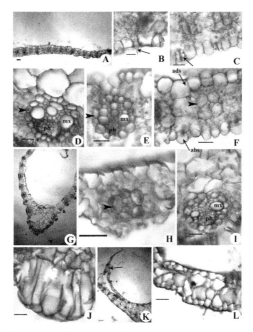

FIGURE 1.60   Leaf anatomy of *Digitaria ciliaris*.

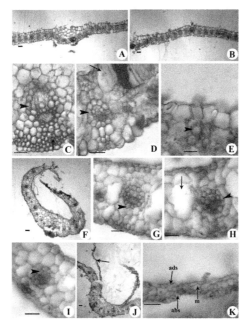

FIGURE 1.61   Leaf anatomy of *Digitaria granularis*.

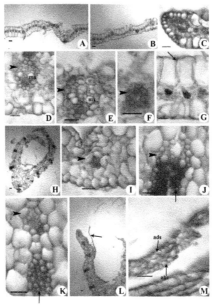

FIGURE 1.62   Leaf anatomy of *Digitaria longiflora.*

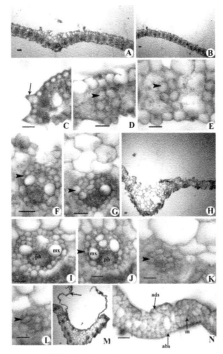

FIGURE 1.63   Leaf anatomy of *Digitaria stircta.*

Echinochloa

49. *Echinochloa colona* (Fig. 1.64)

50. *Echinochloa crusgalli* (Fig. 1.65)

51. *Echinochloa stagnina* (Fig. 1.66)

Leaf lamina (Figs. 1.64A–F, 1.65A–E, 1.66A–E)

Outline of the lamina: "V" shaped in all species (Figs. 1.64A, B, 1.65A, B, 1.66A, B) with broad angled inrolled margin in *E. colona*, broad angled irregularly wavy arms in *E. crusgalli* and wide angled with straight arms in *E. stagnina*. Two halves are symmetrical on either side of the median vascular bundle. Ribs and furrows are absent in all species except present in *E. colona*. Shallow, wide furrows with round ribs present on all vascular bundles.

Keel is well developed, "U" shaped in *E. colona*, "V" shaped in *E. crusgalli* and *E. stagnina* with single median vascular bundles, two second-order and five to six third-order in *E. colona*, four to five third-order in *E. crusgalli* and three to four third-order in *E. stagnina* vascular bundles; small, round parenchyma cells present surrounding the median vascular bundle adaxially. Sclerenchyma is present on abaxial surface associated with vascular bundles in form of strands.

Epidermis: Adaxial epidermal surface has group of small bulliform cells, not associated with colorless cells in *E. colona* (Fig. 1.64E, arrow), most part made up of large, inflated bulliform cells in *E. crusgalli* and *E. stagnina* (Figs. 1.65D, arrow, 1.66D, arrow); abaxial epidermal cells barrel to rectangular shaped covered with cuticle. Few cells showed papillae which are wide as epidermal cells in *E. crusgalli* and *E. stagnina* (Figs. 1.65D, 1.66C, D, arrow).

Mesophyll: Incompletely radiate chlorenchyma cells continuous between bundles in *E. colona* while irregular, horizontally arranged chlorenchyma cells occupy the major area between the adaxial and abaxial epidermises; chlorenchyma cells continuous between bundles; colorless cells absent.

Vascular bundles: First-order vascular bundles four to five in *E. colona*, five to six in *E. crusgalli* and *E. stagnina* regularly arranged from median to marginal bundles in half lamina.

Five to six third-order and two to three second-order in *E. colona*, seven to eight third-order and three to four second-order in *E. crusgalli*, six to seven third-order and three to four second-order in *E. stagnina* bundles between consecutive first-order bundles present; all the vascular bundles present at same level and placed centrally.

Third-order vascular bundles (Figs. 1.64E, 1.65D, 1.66D): Hexagonal in *E. colona* and *E. crusgalli*, pentagonal in *E. stagnina* with xylem and phloem distinguishable

Second-order vascular bundles: Elliptical

First-order vascular bundles (Figs. 1.64D, 1.65E, 1.66C)

Shape: Circular in all species except in *E. stagnina* elliptical

Relationship of phloem cells and the vascular fibers: Completely surrounded by thick-walled fibers

Nature of lysigenous cavity and protoxylem: Lysigenous cavity present and enlarged protoxylem vessel absent in *E. colona*, present in *E. stagnina* and *E. crusgalli*.

Size of metaxylem vessels in relation to parenchyma sheath cells in cross-section: Wide vessels

Shape of metaxylem vessels: Circular

Midrib vascular bundles

Shape: Circular in all species except elliptical in *E. stagnina*

Relationship of phloem cells and the vascular fibers: Completely surrounded by thick-walled fibers

Nature of lysigenous cavity and protoxylem: Lysigenous cavity present and enlarged protoxylem vessel absent in *E. colona*, present in *E. stagnina* and lysigenous cavity absent and enlarged protoxylem vessel present in *E. crusgalli*.

Size of metaxylem vessels in relation to parenchyma sheath cells in cross-section: Wide vessels

Vascular bundle sheath (Figs. 1.64E, arrowhead, 1.65D, arrowhead, 1.66D, arrowhead)

Third-order vascular bundle's bundle sheath

Shape: Round

Shape of bundle sheath cell: Cells elliptical and elongated

Second-order vascular bundle's bundle sheath

Shape: Circular

Shape of bundle sheath cell: Cells elliptical and elongated

Number of bundle sheath cell: 9–10 in *E. colona* and *E. stagnina*, seven to eight in *E. crusgalli*

Nature of bundle sheath (either complete or incomplete): Complete

Chloroplast position in bundle sheath cell: Chloroplast positioned toward outer tangential wall

Mestome/Inner sheath: Absent

First-order vascular bundle's bundle sheath

Shape: circular in all species except in *E. stagnina* elliptical

Extent of bundle sheath around the vascular bundle: Complete

Extension of bundle sheath: Absent

Number of cells comprising the bundle sheath: 18–19 in *E. colona*, 14–15 in *E. crusgalli*, 15–16 *E. stagnina* parenchyma cells

Bundle sheath cell shape: All cells are elliptical, elongated in shape; chloroplast concentrated near outer tangential wall

Mestome/Inner sheath: Absent

Sclerenchyma

Adaxial sclerenchyma: Absent

Abaxial sclerenchyma: Minute subepidermal strands in *E. colona*, straight horizontal band of strands in *E. stagnina* and absent in *E. crusgalli*

Sclerenchyma between bundles: Absent

Sclerenchyma in leaf margin: Forming very small round fibrous cap; epidermal cells at margin are thin with thick outer tangential wall (Figs. 1.64C, 1.66E)

Leaf-sheath anatomy (Figs. 1.64G–J, 1.65F–H, 1.66F–H)

Outline of lamina: "V" shaped standard angle with straight arms in all species (Figs. 1.64H, 1.65F, 1.66F); ribs and furrows absent

Epidermis

Adaxial epidermal cells: Flat, rectangular
Abaxial epidermal cells: Small, flat, rectangular shaped with thick cuticle
Ground tissue: Parenchymatous with absence of air cavity

Vascular bundle

Arrangement of vascular bundle: Two to three in *E. colona* and *E. crusgalli*, three to four in *E. stagnina* consecutive third-order vascular bundles between first and second-order vascular bundle

Third-order vascular bundle (Figs. 1.64I, 1.65F, 1.66F): Circular in all species except in *E. stagnina* hexagonal with xylem and phloem distinguishable

Second-order vascular bundle: Circular shaped (Fig. 1.64H)

First-order vascular bundle (Figs. 1.64J, 1.65H, 1.66H): Circular with phloem completely surrounded by thick-walled fibers. Lysigenous cavity present and enlarged protoxylem vessel absent in *E. colona*, present in *E. stagnina* and lysigenous cavity absent and enlarged protoxylem vessel present in *E. crusgalli*; circular wide metaxylem vessels element

Vascular bundle sheath (Figs. 1.64J, arrowhead, 1.65H, arrowhead, 1.66H, arrowhead)

Shape: Circular

Extent of bundle sheath around the vascular bundle: Complete

Number of cells comprising the bundle sheath: 18–19 in *E. colona*, 17–18 in *E. crusgalli*, 20–21 in *E. stagnina* parenchyma cells

Bundle sheath cell shape: All cells are elliptical, elongated in shape; cells smaller than the ground tissue

Mestome/Inner sheath: Absent

Sclerenchyma: In form of narrow, straight horizontal abaxial strand

Ligule anatomy (Figs. 1.64K, 1.65I, 1.66I): Ligule absent in *E. colona* and *E. crusgalli*

Shape: Ligule attached to the leaf sheath more than half width and very broad round shaped at the connection point (Fig. 1.66I)

Anatomy: Three layered mesophyll cells between adaxial and abaxial surfaces

DIFFERENTIATING FEATURES

"U"-shaped keel … *E. colona*
"V"-shaped keel
Ligule absent, circular-shaped first-order vascular bundles in leaf lamina …
E. crusgalli
Ligule present, elliptical-shaped first-order vascular bundles in leaf lamina
… *E. stagnina*

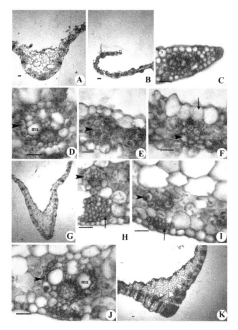

FIGURE 1.64   Leaf anatomy of *Echinichloa colona.*

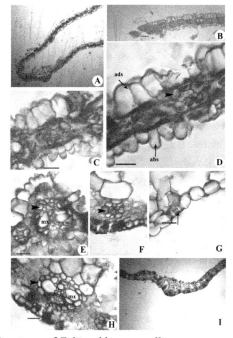

FIGURE 1.65   Leaf anatomy of *Echinochloa crusgalli.*

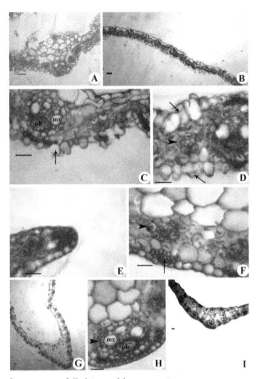

FIGURE 1.66   Leaf anatomy of *Echionochloa stagnina.*

52. *Eremopogon foveolatus*

Leaf lamina (Fig. 1.67A–E)

Outline of the lamina: "V" shaped, wide angle with outwardly curving arms (Fig. 1.67A). Two halves are symmetrical on either side of the median vascular bundle. Ribs and furrows are absent on the leaf surface.

Keel is well developed, V shaped with single median vascular bundle. Small, round parenchyma cells present surrounding the median vascular bundle adaxially. Sclerenchyma present on abaxial surface of keel associated with vascular bundles in form of strands.

Epidermis: Adaxial epidermal surface has most of epidermis made up of large, inflated bulliform cells (Fig. 1.67C, arrow). Abaxial epidermis surface has small, barrel-shaped epidermal cells with thick cuticle. Few epidermal cells have long, balloon-shaped papillae (Fig. 1.67D, arrow).

Mesophyll: Radiate, single layered, isodiametric chlorenchyma cells completely surrounding bundles; compact homogenous chlorenchymatous cells occupy the major area between the adaxial and abaxial epidermises; chlorenchyma cells continuous between bundles; colorless cells absent.

Vascular bundles: Two to three first-order vascular bundles in half lamina regularly arranged from median to marginal bundles; six to seven third-order bundles between consecutive larger bundles and one to two second-order bundles between consecutive first-order bundles present; all the vascular bundles present at same level and placed centrally.

Third-order vascular bundles: Hexagonal with vascular tissue consisting of only few indistinguishable vascular elements (Fig. 1.67B, C)

Second-order vascular bundles: Circular shaped

First-order vascular bundles (Fig. 1.67E)

Shape: Elliptical

Relationship of phloem cells and the vascular fibers: Completely surrounded by thick-walled fibers

Nature of lysigenous cavity and protoxylem: Lysigenous cavity and enlarged protoxylem vessel present

Size of metaxylem vessels in relation to parenchyma sheath cells in cross-section: Wide vessels

Shape of metaxylem vessels: Circular

Median vascular bundles

Shape: Elliptical

Relationship of phloem cells and the vascular fibers: Completely surrounded by thick-walled fibers

Nature of lysigenous cavity and protoxylem: Lysigenous cavity and enlarged protoxylem vessel present

Size of metaxylem vessels in relation to parenchyma sheath cells in cross-section: Wide vessels

Vascular bundle sheath

Third-order vascular bundle's bundle sheath

Shape: Round

Shape of bundle sheath cell: Cells elliptical and elongated

Second-order vascular bundle's bundle sheath

Shape: Round

Shape of bundle sheath cell: Cells elliptical and elongated

Number of bundle sheath cell: 9–10

Nature of bundle sheath (either complete or incomplete): Complete

Chloroplast position in bundle sheath cell: Chloroplast concentrated near outer tangential wall

Mestome/Inner sheath: Absent

First-order vascular bundle's bundle sheath (Fig. 1.67E)

Shape: Round, circular

Extent of bundle sheath around the vascular bundle: Sheath incomplete, abaxially interrupted by sclerenchyma

Extension of bundle sheath: Broad abaxial extension of the bundle sheath

Number of cells comprising the bundle sheath: 13–14 parenchyma cells comprise the sheath

Bundle sheath cell shape: All cells are elliptical, elongated in shape; chloroplast concentrated near outer tangential wall

Mestome/Inner sheath: Absent

Sclerenchyma

Adaxial sclerenchyma: Minute strand consisting of only a few subepidermal fibers

Abaxial sclerenchyma: Well-developed strand, wider than deep and it follows shape of abaxial rib

Sclerenchyma between bundles: Absent

Sclerenchyma in leaf margin: Forming very small pointed fibrous cap; epidermal cells thin with thick outer tangential wall

Leaf-sheath anatomy (Fig. 1.67F–L)

Outline of lamina: V shaped standard angle with inrolled margins (Fig. 1.67F); ribs and furrows absent

Epidermis

Adaxial epidermal cells: Big, rectangular (Fig. 1.67J, arrow)

Abaxial epidermal cells: Small, flat, squarish to round shaped with thick cuticle (Fig. 1.67L, arrow)

Ground tissue: Parenchymatous ground tissue, air cavity absent (Fig. 1.67G)

Vascular bundle

Arrangement of vascular bundle: Four to five consecutive third-order vascular bundles between first and second-order vascular bundle

Third-order vascular bundle: Squarish shaped, vascular tissue consists of only a few distinguishable vascular elements (Fig. 1.67L)

Second-order vascular bundle: Elliptical shaped

First-order vascular bundle: Elliptical shaped, phloem completely surrounded by thick-walled fibers, lysigenous cavity absent and enlarged protoxylem vessel present; circular wide metaxylem vessel elements (Fig. 1.67I)

Vascular bundle sheath

Shape: Round, circular

Extent of bundle sheath around the vascular bundle: Sheath incomplete, interrupted by sclerenchyma cells extending till abaxial epidermis

Number of cells comprising the bundle sheath: 13–14 parenchyma cells comprise the sheath

Bundle sheath cell shape: All cells are elliptical, elongated in shape; cells smaller than the ground tissue

Mestome/Inner sheath: Absent

Sclerenchyma: In form of well-developed broad-anchored girder present on abaxial surface (Fig. 1.67H, I, arrow)

Ligule anatomy (Fig. 1.67M, N)

Shape: Ligule attached up to three-fourth of the lamina and further portion is free and blunt. Ligule broad "V" shaped at the connection (Fig. 1.67M, arrow)

Anatomy: Single layered mesophyll cells between adaxial and abaxial surfaces (Fig. 1.67N)

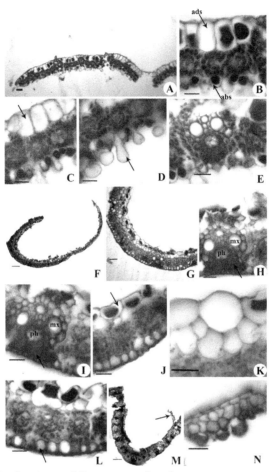

FIGURE 1.67   Leaf anatomy of *Eremopogon foveolatus.*

53. *Eriochloa procera*

Leaf lamina (Fig. 1.68A–G)

Outline of the lamina: Expanded with undulating leaf blade (Fig. 1.68A, B); ribs and furrows present on abaxial surface; shallow and wide furrows present between all vascular bundles; round ribs present over all vascular bundles.

Keel is well-developed, V-shaped keel with single median vascular bundle, three to four third-order vascular bundles; small, round parenchyma cells surrounding the median vascular bundle adaxially; sclerenchyma present on abaxial surface associated with vascular bundles in form of strands.

Epidermis: Adaxial epidermal surface with fan-shaped bulliform cells which occupies one-fourth to half of the leaf thickness (Fig. 1.68E). Abaxial epidermal cells are barrel shaped with thick cuticle.

Mesophyll: Incomplete radiating chlorenchyma cells, continuous between bundles.

Vascular bundles: Three to four first-order vascular bundles in half lamina regularly arranged from median to marginal bundles; seven to eight third-order bundles between consecutive larger bundles and two to three second-order bundles between consecutive first-order bundles present; all the vascular bundles present at same level abaxially.

Third-order vascular bundles: Pentagonal (Fig. 1.68F) or hexagonal or square shaped (Fig. 1.68G) with vascular tissue consisting of only few indistinguishable vascular elements

Second-order vascular bundles: Elliptical

First-order vascular bundles (Fig. 1.68E)

Shape: Elliptical

Relationship of phloem cells and the vascular fibers: Phloem adjoins the inner parenchyma cells

Nature of lysigenous cavity and protoxylem: Lysigenous cavity and enlarged protoxylem vessel present

Size of metaxylem vessels in relation to parenchyma sheath cells in cross-section: narrow vessels

Shape of metaxylem vessels: Circular

Median vascular bundles (Fig. 1.68D)

Shape: Elliptical

Relationship of phloem cells and the vascular fibers: Completely surrounded by thick-walled fibers

Nature of lysigenous cavity and protoxylem: Lysigenous cavity and enlarged protoxylem vessel present

Size of metaxylem vessels in relation to parenchyma sheath cells in cross-section: narrow vessels

Shape of metaxylem vessels: Circular

Vascular bundle sheath

Third-order vascular bundle's bundle sheath (Fig. 1.68F, G, arrow)

Shape: Round

Shape of bundle sheath cell: Radial and inner tangential wall straight while outer tangential wall inflated

Second-order vascular bundle's bundle sheath

Shape: Elliptical

Shape of bundle sheath cell: Radial and inner tangential wall straight while outer tangential wall inflated

Number of bundle sheath cell: 8–9

Nature of bundle sheath (either complete or incomplete): Complete

Chloroplast position in bundle sheath cell: Chloroplast concentrated near outer tangential wall

Mestome/Inner sheath: Absent

First-order vascular bundle's bundle sheath (Fig. 1.68E, D, arrowhead)

Shape: Horseshoe shaped

Extent of bundle sheath around the vascular bundle: Complete

Extension of bundle sheath: Absent

Number of cells comprising the bundle sheath: 9–10 parenchyma cells

Bundle sheath cell shape: Cells varying in size having radial and inner tangential wall straight while outer tangential wall inflated; chloroplast concentrated near outer tangential wall

Mestome/Inner sheath: Complete sheath present with cells smaller than the outer bundle sheath

Sclerenchyma

Adaxial sclerenchyma: Absent

Abaxial sclerenchyma: Minute subepidermal strands present

Sclerenchyma between bundles: Absent

Sclerenchyma in leaf margin: Forming well-developed pointed cap; epidermal cells thin with thick outer tangential wall

Leaf-sheath anatomy (Fig. 1.68H–J)

Outline of lamina: "V" shaped inrolled margins of arms; ribs and furrows present; medium, wide furrows present between all vascular bundles on abaxial surface (Fig. 1.68H)

Epidermis

Adaxial epidermal cells: Flat, rectangular

Abaxial epidermal cells: Flat, rectangular shaped with thick cuticle

Ground tissue: Parenchymatous with absence of air cavity

Vascular bundle

Arrangement of vascular bundle: One to two consecutive third-order vascular bundles between first and second-order vascular bundle

Third-order vascular bundle: Pentagonal with vascular tissue consisting of only few indistinguishable vascular elements (Fig. 1.68I)

Second-order vascular bundle: Elliptical shaped

First-order vascular bundle: Elliptical shaped; phloem completely surrounded by thick-walled fibers; lysigenous cavity absent and enlarged protoxylem vessel present; circular narrow metaxylem vessel elements (Fig. 1.68J)

Vascular bundle sheath (Fig. 1.68I, J, arrowhead)

Shape: Elliptical

Extent of bundle sheath around the vascular bundle: Sheath incomplete, interrupted by sclerenchyma cells abaxially

Number of cells comprising the bundle sheath: 8–9 parenchyma cells

Bundle sheath cell shape: Cells of varying size with radial and inner tangential wall straight while outer tangential wall inflated

Mestome/Inner sheath: Complete sheath present with cells smaller than the outer bundle sheath

Sclerenchyma: In form of well-developed broad, wide, and deep girder on abaxial surface

Ligule anatomy (Fig. 1.68K, L)

Shape: Ligule attached to the midrib and few parts of lamina region because it is hairy ligule (Fig. 1.68K, arrow)

Anatomy: Two layered mesophyll cells between adaxial and abaxial surfaces (Fig. 1.68L)

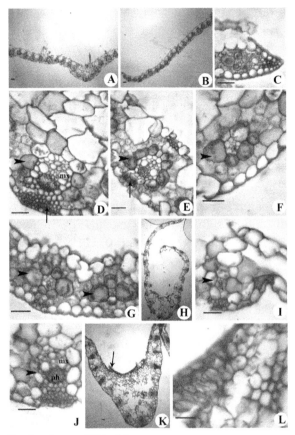

FIGURE 1.68   Leaf anatomy of *Eriochloa procera*.

Oplismenus

54. *Oplismenus burmannii* (Fig. 1.69)

55. *Oplismenus composites* (Fig. 1.70)

Leaf lamina (Figs. 1.69A–H, 1.70A–E)

Outline of the lamina: Expanded with wavy leaf blade (Figs. 1.69A, B, 1.70A, B); two halves are symmetrical on either side of the median vascular bundle; ribs and furrows are present on the leaf surface; shallow, wide furrows present between first and second-order vascular bundles and rounded ribs present on first and second-order vascular bundles.

Keel is well developed, triangular with single median vascular bundle. Small, round parenchyma cells present surrounding the median vascular bundle adaxially. Sclerenchyma present on abaxial surface associated with vascular bundles in form of strands.

Epidermis: Adaxial epidermal surface has two types of bulliform cells (1) fan shaped (Figs. 1.69A, 1.70E) in *O. burmannii*, *O. composites* and (2) restricted group of large, inflated bulliform cells present above the level of epidermis in *O. burmannii* (Fig. 1.69H, arrow); long, slender hairs present with sunken bases (Figs. 1.69F, arrow, 1.70E, arrow); sharp, prickles present (Fig. 1.69D, arrow); abaxial epidermal cells barrel shaped covered with cuticle; short, sharp trichomes (Fig. 1.69G, arrowhead) and prickles (Fig. 1.69G, arrow) present on abaxial surface.

Mesophyll: Irregular, horizontally arranged chlorenchyma cells occupy the major area between the adaxial and abaxial epidermises; chlorenchyma cells continuous between bundles; colorless cells absent.

Vascular bundles: Three to four first-order vascular bundles in half lamina regularly arranged from median to marginal bundles. Four to five third-order bundles between consecutive larger bundles and eight to nine second-order bundles between consecutive first-order bundles present. First and second-order vascular bundles placed abaxially and third-order bundles placed centrally.

Third-order vascular bundles (Fig. 1.69D): Hexagonal with vascular tissue consisting of only few indistinguishable vascular elements

Second-order vascular bundles: Circular shaped

First-order vascular bundles (Figs. 1.69E, 1.70C)

Shape: Circular

Relationship of phloem cells and the vascular fibers: Phloem adjoins the inner or parenchyma sheath

Nature of lysigenous cavity and protoxylem: Lysigenous cavity absent and enlarged protoxylem vessel present

Size of metaxylem vessels in relation to parenchyma sheath cells in cross-section: Narrow vessels

Shape of metaxylem vessels: Circular

Median vascular bundles

Shape: Circular

Relationship of phloem cells and the vascular fibers: Phloem adjoins the inner or parenchyma sheath

Nature of lysigenous cavity and protoxylem: Lysigenous cavity absent and enlarged protoxylem vessel present

Size of metaxylem vessels in relation to parenchyma sheath cells in cross-section: Narrow vessels

Vascular bundle sheath

Third-order vascular bundle's bundle sheath (Fig. 1.69D, arrowhead)

Shape: Round

Shape of bundle sheath cell: Cells elliptical and elongated

Second-order vascular bundle's bundle sheath

Shape: Round

Shape of bundle sheath cell: Cells elliptical and elongated

Number of bundle sheath cell: 8–9

Nature of bundle sheath (either complete or incomplete): Complete

Chloroplast position in bundle sheath cell: Chloroplast small and indistinct

Mestome/Inner sheath: Absent

First-order vascular bundle's bundle sheath (Figs. 1.69E, arrowhead, 1.70C, arrowhead)

Shape: Round, circular

Extent of bundle sheath around the vascular bundle: Complete

Extension of bundle sheath: Absent

Number of cells comprising the bundle sheath: 9–10 parenchyma cells comprise the sheath

Bundle sheath cell shape: All cells are elliptical and elongated in shape; chloroplast small and indistinct

Mestome/Inner sheath: Complete, inner sheath cells smaller than the outer sheath cell

Sclerenchyma

Adaxial sclerenchyma: Minute strand in *O. burmannii* or shallow in *O. composites* consisting of only a few subepidermal fibers

Abaxial sclerenchyma: Well-developed strand following the abaxial rib

Sclerenchyma between bundles: Absent

Sclerenchyma in leaf margin: Forming very small pointed fibrous cap; epidermal cells thin with thick outer tangential wall

Leaf-sheath anatomy (Figs. 1.69I–L, 1.70F–H)

Outline of lamina: Round shaped with inrolled margins (Figs. 1.69I, 1.70F); ribs and furrows present on abaxial surface; shallow, wide furrows present between first and second-order vascular bundles and rounded ribs present on first and second-order vascular bundles.

Epidermis: Adaxial epidermal cells flat and rectangular (Fig. 1.69L). Restricted group of large, inflated bulliform cells present above the level of epidermis (Fig. 1.69K) and long slender hair present (Fig. 1.69K, arrow) in *O. burmannii*

Abaxial epidermal cells: Small, flat, rectangular shaped with thick cuticle

Ground tissue: Parenchymatous with absence of air cavity in *O. burmannii* but oval-shaped air cavities present in *O. composites* (Fig. 1.70H)

Vascular bundle

Arrangement of vascular bundle: One consecutive third-order vascular bundle between first and second-order vascular bundle

Third-order vascular bundle: Hexagonal with vascular tissue consisting of only few indistinguishable vascular elements

Second-order vascular bundle: Circular shaped

First-order vascular bundle (Figs. 1.69J, 1.70G): Circular shaped, phloem adjoins inner or parenchyma sheath, lysigenous cavity and enlarged proto-xylem vessel present; circular wide metaxylem vessel element

Vascular bundle sheath

Shape: Round, circular

Extent of bundle sheath around the vascular bundle: Complete

Number of cells comprising the bundle sheath: 9–10 parenchyma cells

Bundle sheath cell shape: All cells are elliptical, elongated, and smaller than the ground tissue cells

Mestome/Inner sheath: Complete, inner sheath cells smaller than the outer sheath cell

Sclerenchyma: In form of small, four to six fibers strand present on abaxial surface

Ligule anatomy (Figs. 1.69K, L, 1.70I)

Shape: Ligule attached to the leaf sheath more than half sheath region and is round shaped at connection point (Figs. 1.69K, 1.70I)

Anatomy: Single layered mesophyll cells between adaxial and abaxial surfaces

DIFFERENTIATING FEATURES

Air cavities absent in leaf sheath … *O. burmanii*
Oval shaped air cavities in leaf sheath … *O. composites*

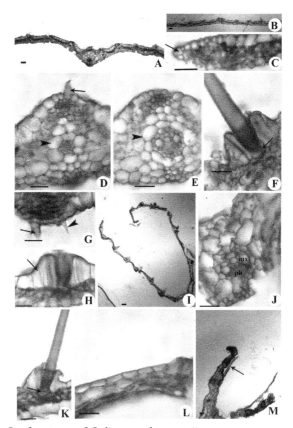

FIGURE 1.69   Leaf anatomy of *Oplismenus burmannii.*

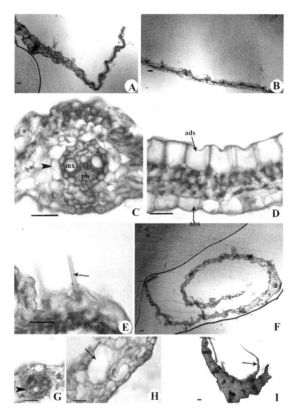

FIGURE 1.70   Leaf anatomy of *Oplismenus composites.*

Panicum

56. *Panicum antidotale* (Fig. 1.71)

57. *Panicum maximum* (Fig. 1.72)

58. *Panicum miliaceum* (Fig. 1.73)

59. *Panicum trypheron* (Fig. 1.74)

Leaf lamina (Figs. 1.71A–J, 1.72A–E, 1.73A–E, 1.74A–H)

Outline of the lamina: Expanded with gently wavy leaf blade (Figs. 1.71A, B, 1.72A, 1.73A) but V shaped, wide angle with convex shape of the arms in *P. trypheron* (Fig. 1.74A, B), two halves are symmetrical on either side of the median vascular bundle. Ribs and furrows are present in all species except absent in *P. trypheron*. Shallow, wide furrows present between all vascular bundles and rounded ribs present on all vascular bundles.

Keel is well developed, semicircular in *P. antidotale*, *P. maximum*, and *P. miliaceum*, "V" shaped in *P. trypheron*, with single median vascular bundle in *P. antidotale*, *P. maximum*, and *P. miliaceum* and four to five third-order vascular bundles present in *P. trypheron*. Small, rounded parenchyma cells are present surrounding the median vascular bundle adaxially. Sclerenchyma is present on abaxial surface associated with vascular bundles in form of well-developed strands.

Epidermis: Adaxial epidermal surface has fan-shaped bulliform cells (Figs. 1.71F, 1.72E, arrow, 1.73F, arrow, 1.74E, arrow) with diamond shaped in *P. antidotale* and *P. maximum*, triangular in *P. miliaceum* and large, round in *P. trypheron*-shaped central cell. Bulliform cells occupy less than half portions in *P. antidotale* and *P. maximum* or one-fourth portions in *P. miliaceum* and *P. trypheron* of leaf thickness; small hook with pointed barb in *P. antidotale* (Fig. 1.71I, arrow) on adaxial surface.

Abaxial epidermis has small round to barrel-shaped epidermal cells with thick cuticle; restricted groups of tall and narrow bulliform cells (Fig. 1.71F) in which long, slender macrohairs (Fig. 1.71G, arrow) with bulbous, sunken base present.

Mesophyll: Radiate, single layered, isodiametric chlorenchyma cells completely surrounding bundles. Compact homogenous chlorenchymatous cells occupy the major area between the adaxial and abaxial epidermises; colorless cells absent in *P. miliaceum* and *P. trypheron* while present in *P. antidotale* and *P. maximum* (Fig. 1.71E, arrow); colorless cells smaller than bulliform cells, forming one extension up to abaxial epidermis in *P. antidotale* while in *P. maximum* partially.

Vascular bundles: Six to seven first-order in *P. antidotale*, eight to nine in *P. maximum* and *P. miliaceum*, three to four in *P. trypheron* regularly arranged from median to marginal bundles in half lamina. Three to four third-order and two to three second-order in *P. miliaceum*, five to six third-order and three to four second-order and four to five third-order and two to three second-order in *P. trypheron* bundles between consecutive first-order bundles present. All the vascular bundles present at same level and placed centrally.

Third-order vascular bundles (Figs. 1.71D, 1.72D, 1.73D, E, 1.74C, D, E, G): Hexagonal in all species except pentagonal (Fig. 1.74C, D, G) and square (Fig. 1.74E) in *P. trypheron*, with vascular tissue consisting of only few indistinguishable vascular elements

Second-order vascular bundles: Circular shaped (Fig. 1.73C)

First-order vascular bundles (Figs. 1.71C, 1.72C, 1.73B, 1.73F)

Shape: Elliptical

Relationship of phloem cells and the vascular fibers: Completely surrounded by thick-walled fibers in all species except in *P. trypheron* where phloem adjoins inner or parenchyma cells

Nature of lysigenous cavity and protoxylem: Enlarged protoxylem vessel present and lysigenous cavity absent in *P. trypheron*

Size of metaxylem vessels in relation to parenchyma sheath cells in cross-section: Narrow vessels

Shape of metaxylem vessels: Circular

Median vascular bundles (Fig. 1.71J)

Shape: Elliptical

Relationship of phloem cells and the vascular fibers: Completely surrounded by thick-walled fibers in all species except in *P. trypheron* where phloem adjoins inner or parenchyma cells

Nature of lysigenous cavity and protoxylem: Enlarged protoxylem vessel present and Lysigenous cavity absent in *P. trypheron*

Size of metaxylem vessels in relation to parenchyma sheath cells in cross-section: Narrow vessels

Vascular bundle sheath

Third-order vascular bundle's bundle sheath (Figs. 1.71D, 1.72D, 1.73D, E, 1.74C, D, G, arrowhead)

Shape: Round

Shape of bundle sheath cell: Fan shaped in all species. *P. trypheron* has cells toward adaxial side larger than the others

Second-order vascular bundle's bundle sheath

Shape: Round

Shape of bundle sheath cell: Fan shaped in all species but *P. trypheron* has cell toward adaxial side is larger than the others

Number of bundle sheath cell: 8–9 in *P. antidotale* and *P. maximum*, six to seven in *P. trypheron* and *P. miliaceum*

Nature of bundle sheath (either complete or incomplete): Complete

Chloroplast position in bundle sheath cell: Chloroplast concentrated near inner tangential wall in *P. antidotale* and *P. miliaceum* and outer tangential wall in *P. trypheron* and *P. maximum*

Mestome/Inner sheath: Absent

First-order vascular bundle's bundle sheath (Figs. 1.71C, 1.72C, 1.73C, 1.74F)

Shape: Elliptical in all species except in *P. antidotale* circular

Extent of bundle sheath around the vascular bundle: Complete

Extension of bundle sheath: Absent

Number of cells comprising the bundle sheath: 10–11 in *P. antidotale*, 13–14 in *P. maximum*, 9–10 in *P. miliaceum*, *P. trypheron* parenchyma cells comprise the sheath

Bundle sheath cell shape: Fan shaped, chloroplast concentrated near inner tangential wall in *P. antidotale* and *P. miliaceum* or near outer tangential wall in *P. trypheron* and *P. maximum*

Mestome/Inner sheath: Complete with cells smaller than the outer sheath cells

Sclerenchyma

Adaxial sclerenchyma: Absent

Abaxial sclerenchyma: Minute strand consisting of only a few subepidermal fibers

Sclerenchyma between bundles: Absent

Sclerenchyma in leaf margin: Forming very small round fibrous cap in *P. antidotale*, *P. miliaceum*, and *P. maximum* while pointed fibrous cap in *P. trypheron*. Epidermal cells thin with thick outer tangential wall

Leaf-sheath anatomy (Figs. 1.71K–P, 1.72G–I, 1.73G–I, 1.74I–K)

Outline of lamina: V shaped with standard angle, inrolled margins in *P. antidotale* and *P. maximum* (Figs. 1.71K, 72G) or round with overlapping inrolled margins in *P. miliaceum* and *P. trypheron* (Figs. 1.73G, 1.74I). Ribs and furrows absent in *P. antidotale*, *P. maximum* while present in *P. miliaceum*, *P. trypheron*. Shallow, wide furrows present between all vascular bundles and rounded ribs present on all vascular bundles.

Epidermis

Adaxial epidermal cells: Flat, rectangular

Abaxial epidermal cells: Big, round cells with thick cuticle; epidermis interrupted by stomata (Fig. 1.71O, arrow)

Ground tissue: Parenchymatous with horizontally elongated oval shaped in *P. antidotale* and *P. maximum* or round shaped in *P. miliaceum* air cavity; air cavity absent in *P. trypheron*

Vascular bundle

Arrangement of vascular bundle: One in *P. antidotale* and *P. maximum*, two in *P. miliaceum* and *P. trypheron* consecutive third-order vascular bundles between first and second-order vascular bundle

Third-order vascular bundle (Figs. 1.71L, 1.72H, 1.74J): Hexagonal in *P. antidotale* and *P. miliaceum*, elliptical in *P. maximum*, and square in *P. trypheron* vascular bundle with vascular tissue consisting of only few indistinguishable vascular elements

Second-order vascular bundle: Circular shaped

First-order vascular bundle (Figs. 1.71N, 1.72G, 1.73I, 1.74K): Circular shaped in all species except in *P. maximum* elliptical with phloem completely surrounded by thick-walled fibers; lysigenous cavity and enlarged protoxylem vessel present in all species except *P. trypheron* with an absence of lysigenous cavity; circular narrow metaxylem vessels elements

Vascular bundle sheath (Fig. 1.74K)

Shape: Round in all species except in *P. maximum* elliptical

Extent of bundle sheath around the vascular bundle: Complete in all species except *P. antidotale* with incomplete vascular bundle, abaxially interrupted by sclerenchyma cells

Number of cells comprising the bundle sheath: 10–11 in *P. antidotale*, 12–13 in *P. maximum*, 7–8 in *P. miliaceum*, and 9–10 in *P. trypheron* parenchyma cells

Bundle sheath cell shape: Fan shaped with cells smaller than the ground tissue

Mestome/Inner sheath: Complete with cells smaller than the outer sheath cell in *P. antidotale* and *P. maximum* or incomplete, interrupted by sclerenchyma cells on abaxial side in *P. miliaceum* and *P. trypheron*

Sclerenchyma: In form of well-developed broad and wide girder in *P. antidotale* and *P. maximum* (Figs. 1.71N, arrow, 1.72I, arrow) while straight horizontal abaxial band in *P. miliaceum* and *P. trypheron* (Fig. 74I, arrow)

Ligule anatomy (Figs. 1.71P, Q, 1.72J, K, 1.73J, K, 1.74L, M)

Shape: Ligule attached to the leaf sheath more than the half width and very broad round connection point (Figs. 1.71Q, 1.72J, 1.73J, 1.74L)

Anatomy: Two layered in all species except in *P. antidotale* single layered, mesophyll cells between adaxial and abaxial surfaces (Figs. 1.71P, 1.72K, 1.73K, 1.74M)

DIFFERENTIATING FEATURES

"V" shaped leaf blade … *P. trypheron*
Expanded leaf blade
Colorless cells absent … *P. miliaceum*
Colorless cells present
Hexagonal-shaped third-order vascular bundles in leaf sheath … *P. antidotalae*
Elliptical-shaped third-order vascular bundles in leaf sheath … *P. maximum*

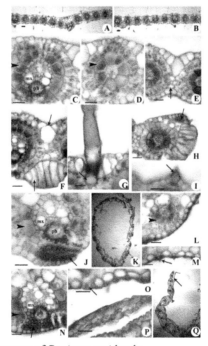

FIGURE 1.71 Leaf anatomy of *Panicum antidotale.*

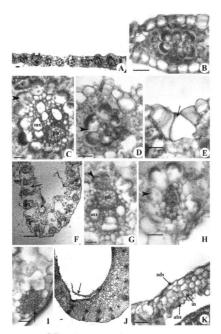

FIGURE 1.72 Leaf anatomy of *Panicum maximum.*

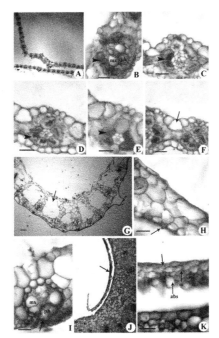

FIGURE 1.73 Leaf anatomy of *Panicum miliaceum.*

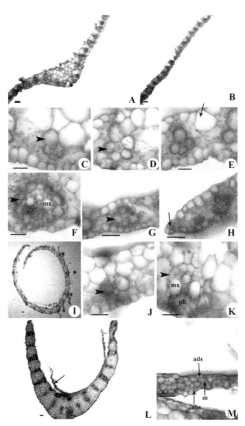

FIGURE 1.74   Leaf anatomy of *Panicum trypheron*.

Paspalidium

60. *Paspalidium flavidum* (Fig. 1.75)

61. *Paspalidium geminatum* (Fig. 1.76)

Leaf lamina (Figs. 1.75A–G, 1.76A–F)

Outline of the lamina: "V" shaped, wide angled in *P. flavidum* or standard angled in *P. geminatum* with outwardly curving, convex arms (Figs. 1.75A, B, 1.76A, B); two halves are symmetrical on either side of the median vascular bundle; ribs and furrows are absent on the leaf surface.

Keel is well developed, V shaped with single median vascular bundle. In *P. flavidum*, small, round parenchyma cells present surrounding the median vascular bundle adaxially. In *P. geminatum*, leaves are terete so parenchyma may be developed but is not specifically in association with the midrib.

Sclerenchyma present on abaxial surface associated with vascular bundles in form of strands.

Epidermis: Adaxial epidermal surface mostly made up of bulliform cells (Fig. 1.75C), epidermal cells shows papillae in which it is wide as epidermal cell (Figs. 1.75D, arrow, 1.76B, arrow); stomata is present (Fig. 1.75E, arrow); abaxial epidermis has big, barrel to square-shaped epidermal cells covered with thick cuticle; long, slender macrohairs present.

Mesophyll: *P. flavidum* has radiate, single layered, isodiametric chlorenchyma cells completely surrounding bundles; compactly arranged homogenous chlorenchyma cells occupy the major area between the adaxial and abaxial epidermises; *P. geminatum* has irregular, chlorenchyma arranged in "U"-shaped manner and cells occupying the major area between the adaxial and abaxial epidermises; chlorenchyma cells continuous between bundles; colorless cells absent

Vascular bundles: Four to five first-order vascular bundle in *P. flavidum* or two to three in *P. geminatum* regularly arranged from median to marginal bundle vascular bundles in half lamina.

Five to seven third-order vascular bundle in *P. flavidum* or four to five in *P. geminatum* arranged between consecutive larger bundles and two to three second-order bundles between consecutive first-order bundles present. All the vascular bundles present at same level and placed centrally in *P. flavidum* while abaxially placed in *P. geminatum*.

Third-order vascular bundles (Figs. 1.75C, D, E, 1.76E): Hexagonal (Fig. 1.75C, D) and elliptical (Fig. 1.76E) in *P. flavidum* or circular in *P. geminatum* (Fig. 1.76E) with vascular tissue consisting of only few indistinguishable vascular elements

Second-order vascular bundles: Elliptical shaped

First-order vascular bundles (Figs. 1.75G, 1.76D)

Shape: Elliptical in *P. flavidum*, circular in *P. geminatum*

Relationship of phloem cells and the vascular fibers: Completely surrounded by thick-walled fibers

Nature of lysigenous cavity and protoxylem: Lysigenous cavity and enlarged protoxylem vessel present

Size of metaxylem vessels in relation to parenchyma sheath cells in cross-section: Wide vessels

Shape of metaxylem vessels: Circular

Median vascular bundles (Fig. 1.75F)

Shape: Elliptical in *P. flavidum*, circular in *P. geminatum*

Relationship of phloem cells and the vascular fibers: Completely surrounded by thick-walled fibers

Nature of lysigenous cavity and protoxylem: Lysigenous cavity and enlarged protoxylem vessel present

Size of metaxylem vessels in relation to parenchyma sheath cells in cross-section: Wide vessels

Vascular bundle sheath

Third-order vascular bundle's bundle sheath (Figs. 1.75C, D, E, 1.76E)

Shape: Round

Shape of bundle sheath cell: Cells elliptical and elongated

Second-order vascular bundle's bundle sheath

Shape: Elliptical

Shape of bundle sheath cell: Cells elliptical and elongated

Number of bundle sheath cell: 9–10

Nature of bundle sheath (either complete or incomplete): Complete

Chloroplast position in bundle sheath cell: Chloroplast positioned toward the outer tangential wall

Mestome/Inner sheath: Absent

First-order vascular bundle's bundle sheath (Figs. 1.75G, 1.76D)

Shape: Elliptical

Extent of bundle sheath around the vascular bundle: Complete

Extension of bundle sheath: Absent

Number of cells comprising the bundle sheath: 15–16 in *P. flavidum*, 20–21 in *P. geminatum* parenchyma cells

Bundle sheath cell shape: All cells are elliptical, elongated in shape; chloroplast positioned toward the outer tangential wall

Mestome/Inner sheath: Absent

Sclerenchyma

Adaxial sclerenchyma: Absent

Abaxial sclerenchyma: Well-developed strand, wider than deep

Sclerenchyma between bundles: Absent

Sclerenchyma in leaf margin: Forming very small with pointed fibrous cap; epidermal cells at margin are thin with thick outer tangential wall

Leaf-sheath anatomy (Figs. 1.75H–L, 1.76G–J)

Outline of lamina: V shaped, narrow angled with inrolled margins in *P. flavidum* (Fig. 1.75H) while round shaped, with inrolled margins in *P. geminatum* (Fig. 1.76G); ribs and furrows absent

Epidermis

Adaxial epidermal cells: Flat, rectangular (Fig. 1.76J)

Abaxial epidermal cells: Small, flat, rectangular shaped with thick cuticle

Ground tissue: Parenchymatous with vertically oval-shaped air cavity in both species (Figs. 1.75I, arrow, 1.76G, arrow)

Vascular bundle

Arrangement of vascular bundle: Three (*P. flavidum*) or three to four (*P. geminatum*) consecutive third-order vascular bundle between first and second-order vascular bundle

Third-order vascular bundle (Fig. 1.75K): Hexagonal in *P. flavidum*, circular in *P. geminatum* with vascular tissue consisting of only a few indistinguishable vascular element.

Second-order vascular bundle: Elliptical shaped

First-order vascular bundle (Figs. 1.75J, 1.76H, I): Obovate in *P. flavidum* or diamond shaped in *P. geminatum* with phloem completely surrounded by thick-walled fibers; lysigenous cavity and enlarged protoxylem vessel present; circular wide metaxylem vessels present

Vascular bundle sheath (Figs. 1.75J, 1.76H)

Shape: Elliptical

Extent of bundle sheath around the vascular bundle: Complete

Number of cells comprising the bundle sheath: 23–24 in *P. flavidum*, 34–35 in *P. geminatum* parenchyma cells

Bundle sheath cell shape: All cells are elliptical and elongated in shape; cells smaller than the ground tissue

Mestome/Inner sheath: Absent

Sclerenchyma: In form of straight horizontal band on abaxial surface

Ligule anatomy (Figs. 1.75M, N, 1.76K, L)

Shape: Ligule attached to the lamina more than half-length of it and blunt narrow "V" shaped at the connection point in *P. flavidum* (Fig. 1.75M) while attached to the midrib and few parts of lamina region because it is hairy ligule in *P. geminatum* (Fig. 1.76K)

Anatomy: Two layered in *P. flavidum* or three layered in *P. geminatum* mesophyll cells between adaxial and abaxial surfaces (Figs. 1.75N, 1.76L)

DIFFERENTIATING FEATURES

"V" shaped in outline of leaf sheath, radiate arrangement of mesophyll cells, two layered mesophyll present in ligule ... *P. flavidum*
Round shaped in outline of leaf sheath, irregularly arranged mesophyll cells, two layered mesophyll present in ligule ... *P. geminatum*

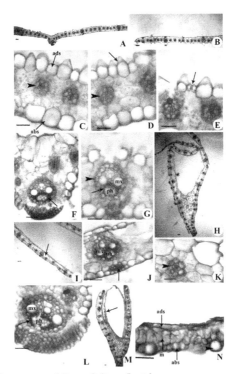

FIGURE 1.75   Leaf anatomy of *Paspalidium flavidum.*

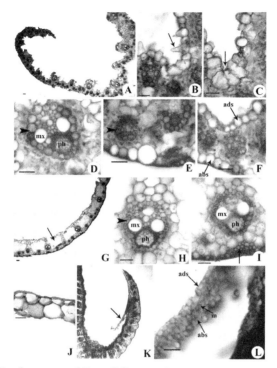

FIGURE 1.76   Leaf anatomy of *Paspalidium geminatum.*

62. *Paspalum scorbiculatum*

Leaf lamina (Fig. 1.77A–G)

Outline of the lamina: Expanded with flat, straight leaf blade (Fig. 1.77A). Two halves are symmetrical on either side of the median vascular bundle. Ribs and furrows present on the leaf surface. Shallow, wide furrows present between all vascular bundles. Keel is inconspicuous.

Epidermis: Adaxial epidermal surface has most part of epidermis made up of bulliform cells (Fig. 1.77C, arrow), epidermal cells shows long, pointed papillae (Fig. 1.77D, arrow); abaxial epidermal cells large, barrel to square shaped covered with thick cuticle.

Mesophyll: Irregular, vertically arranged chlorenchyma cells occupying the major area between the adaxial and abaxial epidermises; chlorenchyma cells continuous between bundles; colorless cells absent.

Vascular bundles: Three to four first-order vascular bundles in half lamina regularly arranged from median to marginal bundles. Four to five third-order bundles between consecutive larger bundles and two to three second-order bundles between consecutive first-order bundles present. All the vascular bundles present at same level and placed centrally

Third-order vascular bundles: Hexagonal (Fig. 1.77E) with vascular tissue consisting of only few indistinguishable vascular elements

Second-order vascular bundles: Elliptical shaped (Fig. 1.77D)

First-order vascular bundles (Fig. 1.77E)

Shape: Elliptical

Relationship of phloem cells and the vascular fibers: Completely surrounded by thick-walled fibers

Nature of lysigenous cavity and protoxylem: Lysigenous cavity and enlarged protoxylem vessel present

Size of metaxylem vessels in relation to parenchyma sheath cells in cross-section: Wide vessels

Shape of metaxylem vessels: Circular

Median vascular bundles

Shape: Elliptical

Relationship of phloem cells and the vascular fibers: Completely surrounded by thick-walled fibers

Nature of lysigenous cavity and protoxylem: Lysigenous cavity and enlarged protoxylem vessel present

Size of metaxylem vessels in relation to parenchyma sheath cells in cross-section: Wide vessels

Vascular bundle sheath

Third-order vascular bundle's bundle sheath (Fig. 1.77E)

Shape: Round

Shape of bundle sheath cell: Cells elliptical and elongated

Second-order vascular bundle's bundle sheath

Shape: Elliptical

Shape of bundle sheath cell: Cells elliptical and elongated

Number of bundle sheath cell: 12–13

Nature of bundle sheath (either complete or incomplete): Complete

Chloroplast position in bundle sheath cell: Chloroplast positioned toward the outer tangential wall

Mestome/Inner sheath: Absent

First-order vascular bundle's bundle sheath (Fig. 1.77E, arrowhead)

Shape: Elliptical

Extent of bundle sheath around the vascular bundle: Complete

Extension of bundle sheath: Absent

Number of cells comprising the bundle sheath: 17–18 parenchyma cells

Bundle sheath cell shape: All cells are elliptical, elongated in shape; chloroplast positioned toward the outer tangential wall

Mestome/Inner sheath: Absent

Sclerenchyma

Adaxial sclerenchyma: Minute subepidermal strands present

Abaxial sclerenchyma: Well-developed strand, wider than deep

Sclerenchyma between bundles: Absent

Sclerenchyma in leaf margin: Forming well-developed pointed fibrous cap (Fig. 1.77B); epidermal cells at margin are thin with thick outer tangential wall

Leaf-sheath anatomy (Fig. 1.77F–G)

Outline of lamina: Round with inrolled margins (Fig. 1.77F); ribs and furrows absent

Epidermis

Adaxial epidermal cells: Flat, rectangular

Abaxial epidermal cells: Small, flat, rectangular shaped with thick cuticle

Ground tissue: Parenchymatous with absence of air cavity

Vascular bundle

Arrangement of vascular bundle: Two consecutive third-order vascular bundles between first and second-order vascular bundle

Third-order vascular bundle: Hexagonal (Fig. 1.77G) with vascular tissue consisting of only few indistinguishable vascular elements

Second-order vascular bundle: Elliptical shaped

First-order vascular bundle: Elliptical shaped with phloem completely surrounded by thick-walled fibers; lysigenous cavity and enlarged proto-xylem vessel present; circular wide metaxylem vessels present

Vascular bundle sheath (Fig. 1.77G, arrowhead)

Shape: Elliptical

Extent of bundle sheath around the vascular bundle: Complete

Number of cells comprising the bundle sheath: 20–21 parenchyma cells

Bundle sheath cell shape: All cells are elliptical and elongated in shape; cells smaller than the ground tissue

Mestome/Inner sheath: Absent

Sclerenchyma: In form of straight horizontal abaxial band

Ligule anatomy (Fig. 1.77H)

Shape: Ligule attached to the half width of the leaf sheath region and it is round shaped at the connection point (Fig. 1.77H, arrow)

Anatomy: Two layered mesophyll cells between adaxial and abaxial surfaces

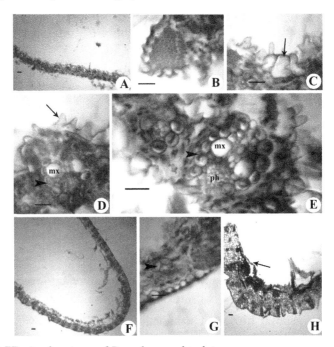

FIGURE 1.77　Leaf anatomy of *Paspalum scorbiculatum.*

63. *Pennisetum setosum*

Leaf lamina (Fig. 1.78A–F)

Outline of the lamina: "U" shaped (Fig. 1.78A); two halves are symmetrical on either side of the median vascular bundle; ribs and furrows are present on abaxial surface of leaf; shallow, wide furrows present between first and second-order vascular bundles and triangular ribs present.

Keel is well developed, U shaped with three median vascular bundles, five to six third-order vascular bundles; parenchyma cells associated with vascular bundles; sclerenchyma abaxially associated with vascular bundles in form of strands.

Epidermis: Adaxial epidermal surface has fan-shaped bulliform cells (Fig. 1.78E, arrow), present at the base of furrows; hook with sharp burb (Fig. 1.78F, arrow); abaxial epidermis has big, barrel to square-shaped epidermal cells covered with thick cuticle.

Mesophyll: Incompletely radiate chlorenchyma present; colorless cells absent

Vascular bundles: Three to four first-order vascular bundles in half lamina regularly arranged from median to marginal bundles; three to four third-order bundles between consecutive larger bundles and one to two second-order bundles between consecutive first-order bundles present; all are placed abaxially

Third-order vascular bundles: Hexagonal shaped (Fig. 1.78D) with xylem and phloem distinguishable

Second-order vascular bundles: Circular shaped

First-order vascular bundles (Fig. 1.78C)

Shape: Obovate

Relationship of phloem cells and the vascular fibers: Completely surrounded by thick-walled fibers

Nature of lysigenous cavity and protoxylem: Lysigenous cavity and enlarged protoxylem vessel present

Size of metaxylem vessels in relation to parenchyma sheath cells in cross-section: Wide vessels

Shape of metaxylem vessels: Circular

Median vascular bundles (Fig. 1.78D)

Shape: Obovate

Relationship of phloem cells and the vascular fibers: Completely surrounded by thick-walled fibers

Nature of lysigenous cavity and protoxylem: Lysigenous cavity and enlarged protoxylem vessel present

Size of metaxylem vessels in relation to parenchyma sheath cells in cross-section: Wide vessels

Vascular bundle sheath

Third-order vascular bundle's bundle sheath (Fig. 1.78D, arrowhead)

Shape: Round

Shape of bundle sheath cell: Cells elliptical and elongated

Second-order vascular bundle's bundle sheath

Shape: Elliptical

Shape of bundle sheath cell: Cells elliptical and elongated

Number of bundle sheath cell: 7–8

Nature of bundle sheath (either complete or incomplete): Complete

Chloroplast position in bundle sheath cell: Chloroplast positioned toward the outer tangential wall

Mestome/Inner sheath: Absent

First-order vascular bundle's bundle sheath (Fig. 1.78C, arrowhead)

Shape: Elliptical

Extent of bundle sheath around the vascular bundle: Complete

Extension of bundle sheath: Absent

Number of cells comprising the bundle sheath: 26–28 parenchyma cells

Bundle sheath cell shape: All cells are elliptical and elongated in shape; chloroplast positioned toward the outer tangential wall

Mestome/Inner sheath: Absent

Sclerenchyma

Adaxial sclerenchyma: Well-developed triangular shaped strand followed ribs

Abaxial sclerenchyma: Well-developed strand, wider than deep

Sclerenchyma between bundles: Absent

Sclerenchyma in leaf margin: Forming pointed very small fibrous cap (Fig. 1.78B); epidermal cells at margin are thin with thick outer tangential wall

Leaf-sheath anatomy (Fig. 1.78G–J)

Outline of lamina: Round shaped with inrolled margins (Fig. 1.78G); ribs and furrows absent

Epidermis

Adaxial epidermal cells: Flat, rectangular (Fig. 1.78J)

Abaxial epidermal cells: Small, flat, rectangular shaped with thick cuticle

Ground tissue: Parenchymatous with vertically oval-shaped air cavity (Fig. 1.78H, J, arrow)

Vascular bundle

Arrangement of vascular bundle: One consecutive third-order vascular bundle between first and second-order vascular bundle

Third-order vascular bundle: Circular shaped, xylem and phloem distinguishable

Second-order vascular bundle: Circular shaped

First-order vascular bundle: Circular shaped with phloem completely surrounded by thick-walled fibers; lysigenous cavity and enlarged proto-xylem vessel present; circular wide metaxylem vessels present (Fig. 1.78I)

Vascular bundle sheath (Fig. 1.78I)

Shape: Elliptical

Extent of bundle sheath around the vascular bundle: Complete

Number of cells comprising the bundle sheath: 20–21 parenchyma cells

Bundle sheath cell shape: All cells are elliptical and elongated in shape; cells smaller than the ground tissue

Mestome/Inner sheath: Absent

Sclerenchyma: In form of straight horizontal band on abaxial surface

Ligule anatomy (Fig. 1.78K)

Shape: Ligule is attached to the midrib and few parts of lamina region because it is hairy ligule (Fig. 1.78K, arrow)

Anatomy: Three layered mesophyll cells between adaxial and abaxial surfaces (Fig. 1.78L)

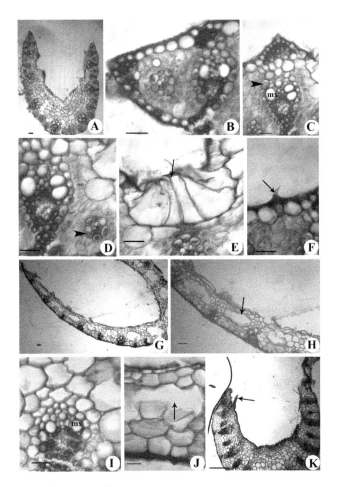

FIGURE 1.78　Leaf anatomy of *Pennisetum setosum*.

Setaria

64. *Setaria glauca* (Fig. 1.79)

65. *Setaria tomentosa* (Fig. 1.80)

66. *Setaria verticillata* (Fig. 1.81)

Leaf lamina (Figs. 1.79A–D, 1.80A–F, 1.81A–H)

Outline of the lamina: Expanded with flat, straight leaf blade in *S. glauca* and *S. tomentosa* (Fig. 1.79A, 1.80A) while distinctly wavy leaf blade in *S. verticillata* (Fig. 1.79A, B); two halves are symmetrical on either side of the median vascular bundle; ribs and furrows are absent on the leaf surface.

Keel is well developed, V shaped with single median vascular bundle in *S. glauca*, single median vascular bundle, five to six third-order vascular bundles in *S. tomentosa* and *S. verticillata*; parenchyma cells associated with vascular bundles; sclerenchyma abaxially associated with vascular bundles in form of strands.

Epidermis: Adaxial epidermal surface has most part of epidermis made up of bulliform cells in *S. glauca* and *S. verticillata* (Figs. 1.79A, 1.80A) while restricted group of large, inflated bulliform cells in *S. tomentosa* (Fig. 1.80B); epidermal cells have long, pointed papillae in *S. glauca* (Fig. 1.79D, arrow); long slender hairs with bulbous base sunken in bulliform cells in *S. tomentosa* (Fig. 1.80C); prickles in *S. verticillata* (Fig. 1.81F, arrow).

Abaxial epidermis has big, barrel-to-square-shaped epidermal cells covered with thick cuticle; restricted group of large, inflated bulliform cells in *S. tomentosa* and *S. verticillata* (Figs. 1.80B, arrow, 1.81G, arrow).

Mesophyll: Irregular, vertically arranged chlorenchyma cells occupying the major area between the adaxial and abaxial epidermises in *S. glauca* and *S. verticillata*, incompletely radiate chlorenchyma in *S. tomentosa*; chlorenchyma cells continuous between bundles; colorless cells absent.

Vascular bundles: Three to four first-order vascular bundles in half lamina regularly arranged from median to marginal bundles. Ten to 11 third-order in *S. glauca*, 12–13 third-order in *S. tomentosa*, 11–12 third-order in *S. verticillata* bundles and three to four second-order bundles between consecutive first-order bundles present. All the vascular bundles present at same level and placed centrally.

Third-order vascular bundles (Figs. 1.79B, 1.80B, 1.81C): Hexagonal with vascular tissue consisting of only few indistinguishable vascular elements

Second-order vascular bundles: Circular in *S. glauca* and *S. tomentosa*, elliptical in *S. verticillata*

First-order vascular bundles (Figs. 1.79C, 1.80E, 1.81D)

Shape: Circular in all species except in *S. verticillata* elliptical

Relationship of phloem cells and the vascular fibers: Phloem adjoins inner or parenchyma sheath cell

Nature of lysigenous cavity and protoxylem: Lysigenous cavity absent and enlarged protoxylem vessel present

Size of metaxylem vessels in relation to parenchyma sheath cells in cross-section: Wide vessels

Shape of metaxylem vessels: Circular

Median vascular bundles (Figs. 1.80D, 1.81E)

Shape: Circular in all species except in *S. verticillata* elliptical

Relationship of phloem cells and the vascular fibers: Phloem adjoins inner or parenchyma sheath cell

Nature of lysigenous cavity and protoxylem: Lysigenous cavity absent and enlarged protoxylem vessel present

Size of metaxylem vessels in relation to parenchyma sheath cells in cross-section: Wide vessels

Vascular bundle sheath

Third-order vascular bundle's bundle sheath (Figs. 1.79B, 1.80F, arrowhead, 1.81C, H, arrowhead)

Shape: Round

Shape of bundle sheath cell: Cells elliptical and elongated

Second-order vascular bundle's bundle sheath

Shape: Elliptical

Shape of bundle sheath cell: Cells elliptical and elongated

Number of bundle sheath cell: 8–9 in *S. glauca* and *S. tomentosa*, 13–14 in *S. verticillata*

Nature of bundle sheath (either complete or incomplete): Complete

Chloroplast position in bundle sheath cell: Chloroplast positioned toward the outer tangential wall

Mestome/Inner sheath: Absent

First-order vascular bundle's bundle sheath (Figs. 1.79C, arrowhead, 1.80E, arrowhead, 1.81K, L, arrowhead)

Shape: Circular in all species except in *S. verticillata* elliptical

Extent of bundle sheath around the vascular bundle: Complete

Extension of bundle sheath: Absent

Number of cells comprising the bundle sheath: 14–15 in *S. glauca*, 19–20 in *S. tomentosa*, *S. verticillata* parenchyma cells

Bundle sheath cell shape: All cells are elliptical and elongated in shape; chloroplast positioned toward the outer tangential wall

Mestome/Inner sheath: Absent

Sclerenchyma

Adaxial sclerenchyma: Absent in *S. glauca* and *S. tomentosa*, minute subepidermal, 4–6 fibers strands present in *S. verticillata*

Abaxial sclerenchyma: Minute subepidermal strands present

Sclerenchyma between bundles: Absent

Sclerenchyma in leaf margin: Forming well developed with pointed fibrous cap in all species (Fig. 187B); epidermal cells at margin are thin with thick outer tangential wall

Leaf-sheath anatomy (Figs. 1.79E–I, 1.80G, K, 1.81I–L)

Outline of lamina: Round-shaped with inrolled margins in *S. glauca* (Fig. 1.79E), V shaped with narrow angle in *S. tomentosa* or standard angle in *S. verticillata* with inrolled margins (Figs. 1.80E, 1.81I); ribs and furrows absent

Epidermis

Adaxial epidermal cells: Flat, rectangular

Abaxial epidermal cells: Small, flat, rectangular shaped with thick cuticle; interrupted by stomata (Figs. 1.79F, arrow, 1.81J, arrow); restricted group of tall, narrow, inflated bulliform cells in *S. tomentosa* (Fig. 1.80I, arrow)

Ground tissue: Parenchymatous with oval-shaped air cavity in *S. glauca* (Fig. 1.79E) or absence of air cavity in *S. tomentosa* and *S. verticillata*

Vascular bundle

Arrangement of vascular bundle: Four to five consecutive third-order vascular bundles between first and second-order vascular bundle

Third-order vascular bundle (Figs. 1.79F, I, 1.80H, 1.81K): Hexagonal in all species except circular in *S. tomentosa* with vascular tissue consisting of only few indistinguishable vascular elements

Second-order vascular bundle: Elliptical shaped (Fig. 1.79H)

First-order vascular bundle (Figs. 1.79G, arrowhead, 1.80J, K, 1.81L): Elliptical with phloem completely surrounded by thick-walled fibers; lysigenous cavity and enlarged protoxylem vessel present; circular wide metaxylem vessels present

Vascular bundle sheath (Figs. 1.79G, arrowhead, 1.80J, K, arrowhead, 1.81L, arrowhead)

Shape: Elliptical

Extent of bundle sheath around the vascular bundle: Incomplete, abaxially interrupted by sclerenchyma cells

Number of cells comprising the bundle sheath: 17–18 in *S. glauca* and *S. verticillata*, 20–22 in *S. tomentosa* parenchyma cells

Bundle sheath cell shape: All cells are elliptical and elongated in shape and cells smaller than the ground tissue

Mestome/Inner sheath: Absent

Sclerenchyma: In form of straight horizontal band girder in *S. glauca* and *S. verticillata* or well developed in *S. tomentosa* on abaxial surface (Figs. 1.79H, arrow, 1.80K)

Ligule anatomy (Figs. 1.79J, K, 1.80L, M, 1.81M, N)

Shape: Ligule attached to the half width of the leaf sheath region and it is round shaped at the connection point in all species (Figs. 1.79J, 1.80L, 1.81M)

Anatomy: Two layered in *S. tomentosa* and *S. verticillata*, three layered in *S. glauca* mesophyll cells between adaxial and abaxial surfaces (Figs. 1.79K, 1.80M, 1.81N); few cells of abaxial surface showed sclerenchyma cells in *S. glauca* and *S. tomentosa* (Figs. 1.79K, arrowhead, 1.80M)

DIFFERENTIATING FEATURES

Oval shaped air cavities in leaf sheath ... *S. glauca*
Air cavities absent
Elliptical-shaped first-order vascular bundles ... *S. verticillata*
Circular-shaped first-order vascular bundles ... *S. tomentosa*

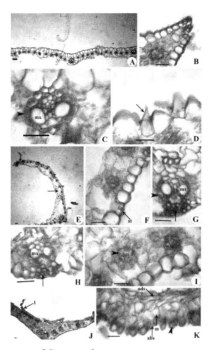

FIGURE 1.79   Leaf anatomy of *Setaria glauca*.

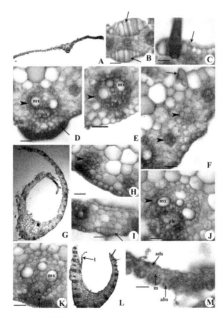

FIGURE 1.80   Leaf anatomy of *Setaria tomentosa*.

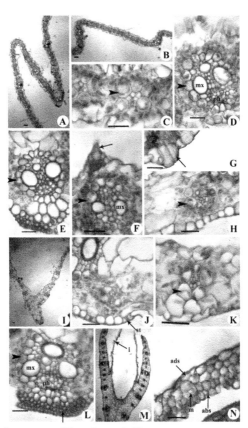

FIGURE 1.81   Leaf anatomy of *Setaria verticillata.*

67. *Isachne globosa*

Leaf lamina (Fig. 1.82A–D)

Outline of the lamina: Expanded with distinctly wavy leaf blade (Fig. 1.82A); two halves symmetrical on either side of the median vascular bundle; ribs and furrows are present on the leaf surface; deep, wide furrows present between all vascular bundles and triangular ribs with pointed apex present over all vascular bundles; keel is inconspicuous.

Epidermis: Adaxial epidermal surface has no bulliform cells (Fig. 1.82A), epidermal cells are small, barrel shaped covered with cuticle. Abaxial epidermis has small, barrel to square-shaped epidermal cells covered with thick cuticle.

Mesophyll: Incompletely radiating chlorenchyma cells present; colorless cells absent.

Vascular bundles: Two to three first-order vascular bundles regularly arranged from median to marginal bundle in half lamina; four to five third-order bundles between consecutive larger bundles and one to two second-order bundles between consecutive first-order bundles present; all the vascular bundles present at same level and placed centrally.

Third-order vascular bundles: Circular shaped (Fig. 1.82B) with vascular tissue consisting of only few indistinguishable vascular elements

Second-order vascular bundles: Elliptical shaped (Fig. 1.82C)

First-order vascular bundles (Fig. 1.82D)

Shape: Elliptical

Relationship of phloem cells and the vascular fibers: Phloem adjoins inner or parenchyma sheath cell

Nature of lysigenous cavity and protoxylem: Lysigenous cavity and enlarged protoxylem vessel present

Size of metaxylem vessels in relation to parenchyma sheath cells in cross-section: Wide vessels

Shape of metaxylem vessels: Circular

Vascular bundle sheath

Third-order vascular bundle's bundle sheath (Fig. 1.82B, arrrowhead)

Shape: Round

Shape of bundle sheath cell: Cells are large with outer tangential wall thick

Second-order vascular bundle's bundle sheath (Fig. 1.82C, arrow)

Shape: Elliptical

Shape of bundle sheath cell: Cells are large with outer tangential wall thick

Number of bundle sheath cell: 8–9

Nature of bundle sheath (either complete or incomplete): Complete

Chloroplast position in bundle sheath cell: Translucent

Mestome/Inner sheath: Absent

First-order vascular bundle's bundle sheath (Fig. 1.82D)

Shape: Elliptical

Extent of bundle sheath around the vascular bundle: Complete

Extension of bundle sheath: Absent

Number of cells comprising the bundle sheath: 13–14 parenchyma cells

Bundle sheath cell shape: Cells are large with outer tangential wall thick; chloroplast translucent

Mestome/Inner sheath: Absent

Sclerenchyma

Adaxial sclerenchyma: Minute subepidermal, four to six fibers strands present

Abaxial sclerenchyma: Minute subepidermal strands present

Sclerenchyma between bundles: Absent

Sclerenchyma in leaf margin: Forming well developed with very pointed fibrous cap. Epidermal cells at margin are thin with thick outer tangential wall

Leaf-sheath anatomy (Fig. 1.82E, F)

Outline of lamina: V shaped, standard angle with inrolled margins (Fig. 1.82E); ribs and furrows present; shallow, wide furrows present between all vascular bundles

Epidermis

Adaxial epidermal cells: Flat, rectangular

Abaxial epidermal cells: Small, flat, rectangular shaped with thick cuticle

Ground tissue: Parenchymatous with oval-shaped air cavity (Fig. 1.82E, arrow)

Vascular bundle

Arrangement of vascular bundle: One consecutive third-order vascular bundle between first and second-order vascular bundle

Third-order vascular bundle: Circular shaped with vascular tissue consisting of only few indistinguishable vascular elements

Second-order vascular bundle: Elliptical shaped

First-order vascular bundle: Elliptical shaped with phloem completely surrounded by thick-walled fibers; lysigenous cavity and enlarged proto-xylem vessel present; circular wide metaxylem vessels present (Fig. 1.82F)

Vascular bundle sheath (Fig. 1.82F, arrow)

Shape: Elliptical

Extent of bundle sheath around the vascular bundle: Incomplete, abaxially interrupted by sclerenchyma cells

Number of cells comprising the bundle sheath: 15–16 parenchyma cells

Bundle sheath cell shape: Cells are large with outer tangential wall thick; cells smaller than the ground tissue

Mestome/Inner sheath: Absent

Sclerenchyma: In form of straight horizontal abaxial band

Ligule anatomy (Fig. 1.82G, H)

Shape: Ligule attached to the half width of the leaf sheath region and it is round shaped at the connection point (Fig. 1.82G, arrow)

Anatomy: Three layered mesophyll cells between adaxial and abaxial surfaces (Fig. 1.82H)

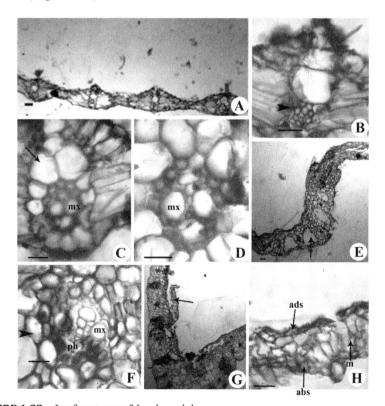

FIGURE 1.82   Leaf anatomy of *Isachne globosa*.

Aristida

68. *Aristida adscensionis* (Fig. 1.83)

69. *Aristida funiculata* (Fig. 1.84)

Leaf lamina (Figs. 1.83A–D, 1.84A–D)

Outline of the lamina: "V" shaped, standard angle with concave arms forming heart shape in both species (Figs. 1.83A, 1.84A). Two halves are symmetrical on either side of the median vascular bundle.

Ribs and furrows are present on surface of leaf; shallow, wide furrows with rounded ribs present over all vascular bundles.

Keel is well developed, V shaped with single median vascular bundle. Small, parenchyma associated with median vasucalr bundle. Sclerenchyma present on abaxial surface associated with vascular bundles in form of strands.

Epidermis: Adaxial epidermal surface has no bulliform cells in *A. adscensionis* or fan-shaped bulliform cells in *A. funiculata* (Fig. 1.84C, arrow). Small, barrel-shaped epidermal cells covered with thick cuticle. Abaxial epidermal cells are barrel to square shaped covered with thick cuticle.

Mesophyll: Irregular, horizontally arranged chlorenchyma cells occupying the major area between the adaxial and abaxial epidermises; chlorenchyma cells continuous between bundles; colorless cells absent.

Vascular bundles: One to two first-order vascular bundles in half lamina regularly arranged from median to marginal bundles. Two to three third-order bundles between consecutive larger bundles and one to two second-order bundles between consecutive first-order bundles present. Vascular bundles arranged at different levels. First and second-order vascular bundles placed centrally while third-order vascular bundles placed abaxially.

Third-order vascular bundles (Fig. 1.83D, 1.84C): Circular with xylem and phloem distinguishable

Second-order vascular bundles: Circular shaped

First-order vascular bundles (Figs. 1.83B, 1.84D)

Shape: Circular

Relationship of phloem cells and the vascular fibers: Phloem adjoins inner or parenchyma sheath

Nature of lysigenous cavity and protoxylem: Lysigenous cavity and enlarged protoxylem vessel present

Size of metaxylem vessels in relation to parenchyma sheath cells in cross-section: Wide vessels

Shape of metaxylem vessels: Circular

Median vascular bundles (Fig. 1.83C)

Shape: Circular

Relationship of phloem cells and the vascular fibers: Phloem adjoins inner or parenchyma sheath

Nature of lysigenous cavity and protoxylem: Lysigenous cavity and enlarged protoxylem vessel present

Size of metaxylem vessels in relation to parenchyma sheath cells in cross-section: Wide vessels

Vascular bundle sheath

Third-order vascular bundle's bundle sheath (Figs. 1.83D, arrowhead, 1.84C, arrowhead)

Shape: Round

Shape of bundle sheath cell: Cells elliptical and elongated

Second-order vascular bundle's bundle sheath

Shape: Elliptical

Shape of bundle sheath cell: Cells elliptical and elongated

Number of bundle sheath cell: 11–12 in *A. adscensionis*, 7–8 in *A. funiculata*

Nature of bundle sheath (either complete or incomplete): Complete

Chloroplast position in bundle sheath cell: Translucent

Mestome/Inner sheath: Absent

First-order vascular bundle's bundle sheath (Fig. 1.83B)

Shape: Circular

Extent of bundle sheath around the vascular bundle: Incomplete, abaxially interrupted by sclerenchyma cells

Extension of bundle sheath: Broad abaxial extension present

Number of cells comprising the bundle sheath: 11–12 in *A. adscensionis*, 9–10 in *A. funiculata* parenchyma cells

Bundle sheath cell shape: All cells are elliptical and elongated in shape; chloroplast translucent

Mestome/Inner sheath: Complete and cells smaller than outer sheath cell

Sclerenchyma

Adaxial sclerenchyma: Absent

Abaxial sclerenchyma: Well-developed strand (Figs. 1.83C, D, arrow, 1.84D, arrow)

Sclerenchyma between bundles: Absent

Sclerenchyma in leaf margin: Forming very small with very pointed fibrous cap; epidermal cells at margin are with thick outer tangential wall

Leaf-sheath anatomy (Figs. 1.83E–G, 1.84E–G)

Outline of lamina: Round shaped with inrolled margins (Figs. 1.83E, 1.84E); ribs and furrows absent in *A. adscensionis* while present in *A. funiculata* in which shallow, wide furrows present over first and second-order vascular bundles.

Epidermis

Adaxial epidermal cells: Flat, rectangular

Abaxial epidermal cells: Small, flat, rectangular shaped with thick cuticle

Ground tissue: Parenchymatous with round-shaped air cavity present in *A. adscensionis* (Fig. 1.83F, arrow) while absence of air cavity in *A. funiculata*

Vascular bundle

Arrangement of vascular bundle: Three to four consecutive third-order vascular bundles between first and second-order vascular bundle

Third-order vascular bundle: Circular with xylem and phloem distinguishable

Second-order vascular bundle: Elliptical shaped

First-order vascular bundle (Figs. 1.83G, 1.84H): Circular with phloem completely surrounded by thick-walled fibers; lysigenous cavity and enlarged protoxylem vessel present; circular wide metaxylem vessels present

Vascular bundle sheath (Figs. 1.83G, 1.84H)

Shape: Elliptical

Extent of bundle sheath around the vascular bundle: Incomplete, abaxially interrupted by sclerenchyma cells

Number of cells comprising the bundle sheath: 14–15 in *A. adscensionis*, 10–11 in *A. funiculata* parenchyma cells

Bundle sheath cell shape: All cells are elliptical and elongated in shape and cells smaller than the ground tissue

Mestome/Inner sheath: Complete and cells smaller than outer sheath cell

Sclerenchyma: In form of straight horizontal girder on abaxial surface (Fig. 1.83G, arrow)

Ligule anatomy (Figs. 1.83H, I, 1.84H, I)

Shape: Ligule is attached to the midrib and few parts of lamina region because it is hairy ligule in both species (Figs. 1.83H, 1.84H)

Anatomy: Single layered mesophyll cells between adaxial and abaxial surfaces in both species (Figs. 1.83I, 1.84I)

DIFFERENTIATING FEATURES

Round-shaped air cavities present in leaf sheath, ribs and furrows absent in leaf sheath … *A. adscensionis*
Air cavities absent, shallow furrows present in leaf sheath … *A. funiculata*

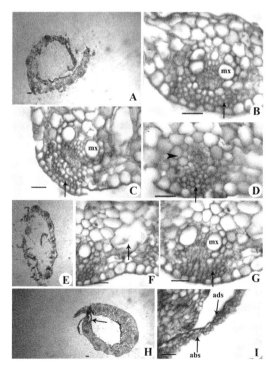

FIGURE 1.83   Leaf anatomy of *Aristida adscensionis*.

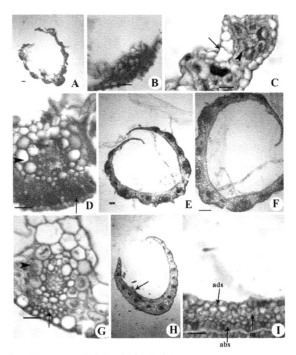

FIGURE 1.84   Leaf anatomy of *Aristida funiculata.*

70. *Perotis indica*

Leaf lamina (Fig. 1.85A–E)

Outline of the lamina: Expanded with flat leaf blade (Fig. 1.85A). Two halves are symmetrical on either side of the median vascular bundle. Ribs and furrows are present on the leaf surface. Very shallow furrows present on abaxial surface. Keel is inconspicuous.

Epidermis: Adaxial epidermal surface has fan-shaped bulliform cells (Fig. 1.85E, arrow), epidermal cells are small, barrel shaped covered with cuticle. Abaxial epidermal cells are small, barrel to square shaped covered with thick cuticle.

Mesophyll: Incompletely radiate chlorenchyma cells present; colorless cells absent.

Vascular bundles: Four to five first-order vascular bundles regularly arranged from median to marginal bundle in half lamina. Six to seven third-order bundles between consecutive larger bundles and one to two second-order bundles between consecutive first-order bundles present. All the vascular bundles present at same level and placed centrally.

Third-order vascular bundles: Pentagonal (Fig. 1.85C) with vascular tissue consisting of only few indistinguishable vascular elements

Second-order vascular bundles: Elliptical shaped

First-order vascular bundles (Fig. 1.85D)

Shape: Elliptical

Relationship of phloem cells and the vascular fibers: Phloem adjoins inner or parenchyma sheath cell

Nature of lysigenous cavity and protoxylem: Lysigenous cavity and enlarged protoxylem vessel present

Size of metaxylem vessels in relation to parenchyma sheath cells in cross-section: Narrow vessels

Shape of metaxylem vessels: Angular

Vascular bundle sheath

Third-order vascular bundle's bundle sheath (Fig. 1.85D, arrowhead)

Shape: Horseshoe shaped

Shape of bundle sheath cell: Fan shaped

Second-order vascular bundle's bundle sheath

Shape: Horseshoe shaped

Shape of bundle sheath cell: Fan shaped

Number of bundle sheath cell: 6–7

Nature of bundle sheath (either complete or incomplete): Incomplete, abaxially interrupted by sclerenchyma cells

Chloroplast position in bundle sheath cell: Chloroplast fills entire lumen of cell

Mestome/Inner sheath: Absent

First-order vascular bundle's bundle sheath

Shape: Horseshoe shaped

Extent of bundle sheath around the vascular bundle: Incomplete, abaxially interrupted by sclerenchyma cells

Extension of bundle sheath: Absent

Number of cells comprising the bundle sheath: 6–7 parenchyma cells

Bundle sheath cell shape: Fan shaped

Mestome/Inner sheath: Complete and cells are smaller than the outer sheath cell

Sclerenchyma

Adaxial sclerenchyma: Absent

Abaxial sclerenchyma: Well-developed wide girder present (Fig. 1.85C, D, B)

Sclerenchyma between bundles: Absent

Sclerenchyma in leaf margin: Forming very pointed cap; epidermal cells at margin are thin with thick outer tangential wall

Leaf-sheath anatomy (Fig. 1.85F–H)

Outline of lamina: Round shaped with inrolled margins (Fig. 1.85F); ribs and furrows absent

Epidermis

Adaxial epidermal cells: Flat, rectangular

Abaxial epidermal cells: Small, flat, rectangular shaped with thick cuticle

Ground tissue: Parenchymatous with round-shaped air cavity (Fig. 1.85F, arrow)

Vascular bundle

Arrangement of vascular bundle: Two consecutive third-order vascular bundles between first and second-order vascular bundle

Third-order vascular bundle: Pentagonal and vascular tissue consisting of only a few distinguishable vascular elements (Fig. 1.85H, arrowhead)

Second-order vascular bundle: Elliptical shaped

First-order vascular bundle: Elliptical shaped with phloem completely surrounded by thick-walled fibers; lysigenous cavity and enlarged proto-xylem vessel present; circular narrow metaxylem vessels present (Fig. 1.85G)

Vascular bundle sheath (Fig. 1.85G)

Shape: Horseshoe shaped

Extent of bundle sheath around the vascular bundle: Incomplete, abaxi-ally interrupted by sclerenchyma cells

Number of cells comprising the bundle sheath: 7–8 parenchyma cells

Bundle sheath cell shape: Fan-shaped cells smaller than the ground tissue

Mestome/Inner sheath: Complete and cells are smaller than the outer sheath cell

Sclerenchyma: In form of straight horizontal band on abaxial surface (Fig. 1.85G, arrow)

Ligule anatomy (Fig. 1.85H, I)

Shape: Ligule attached to the lamina up to one-third length of it and free, blunt, narrow, "V" shaped at the point of connection (Fig. 1.85H, arrow)

Anatomy: Single layered mesophyll cells between adaxial and abaxial surfaces (Fig. 1.85I)

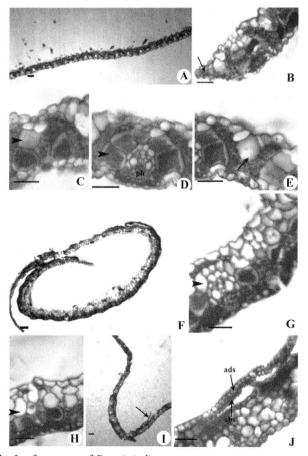

FIGURE 1.85 Leaf anatomy of *Perotis indica.*

Chloris

71. *Chloris barbata* (Fig. 1.86)

72. *Chloris montana* (Fig. 1.87)

73. *Chloris virgata* (Fig. 1.88)

Leaf lamina (Figs. 1.86A–H, 1.87A–E, 1.88A–F)

Outline of the lamina: "V" shaped in all species (Figs. 1.86A, 1.87A, 1.88A), broad angled with concave arms in *C. barbata*, convex arms in *C. montana* or standard angled with straight arms in *C. virgata*.

Two halves are symmetrical on either side of the median vascular bundle; ribs and furrows absent in *C. montana*, *C. virgata* while present in *C. barbata*; shallow, wide furrows with rounded ribs present over all vascular bundles.

Keel is well developed, V shaped with single median vascular bundle, five to six third-order vascular bundles. Small, parenchyma associated with median vascular bundle. Sclerenchyma present on abaxial surface associated with vascular bundles in form of strands.

Epidermis: Adaxial epidermal surface has fan-shaped bulliform cells in all species (Figs. 1.86F, arrow, 1.87C, arrow, 1.88B, arrow) in which central cell is diamond shape in *C. barbata* (Fig. 1.86G, arrow), inflated round in *C. montana* or shield shape in *C. virgata*.

Abaxial epidermal cells are barrel to square shaped covered with thick cuticle. Epidermal cells have papillae in all species (Figs. 1.86D, arrow, 1.87D, arrow, 1.88E, arrow).

Mesophyll: Incompletely radiate chlorenchyma cells present in *C. barbata* or radiate, single layered, isodiametric cells completely surrounding bundles in *C. montana* and *C. virgata*; compactly arranged homogenous chlorenchymatous cells occupying the major area between the adaxial and abaxial epidermises in *C. montana* and *C. virgata*; chlorenchyma cells continuous between bundles; colorless cells absent.

Vascular bundles: Three to four first-order vascular bundles in *C. barbata* and *C. virgata* or two to three in *C. montana*, regularly arranged from median to marginal bundle in half lamina. Four to five third-order bundles between consecutive larger bundles and two to three second-order bundles between consecutive first-order bundles present. Vascular bundles placed at different levels. First and second-order vascular bundles placed centrally while third-order vascular bundles placed abaxially.

Third-order vascular bundles (Figs. 1.86E, 1.87B, C, 1.88C): Hexagonal in *C. barbata* and *C. virgata* while pentagonal in *C. montana* with xylem and phloem distinguishable

Second-order vascular bundles: Circular shaped

First-order vascular bundles (Figs. 1.86B, 1.88D)

Shape: Elliptical

Relationship of phloem cells and the vascular fibers: Completely surrounded by sclerenchyma fibers

Nature of lysigenous cavity and protoxylem: Enlarged protoxylem vessel present in all species and lysigenous cavity present only in *C. virgata*

Size of metaxylem vessels in relation to parenchyma sheath cells in cross-section: Narrow vessels

Shape of metaxylem vessels: Circular

Median vascular bundles (Fig. 1.87D)

Shape: Circular in all species except in *C. barbata* elliptical

Relationship of phloem cells and the vascular fibers: Completely surrounded by sclerenchyma fibers

Nature of lysigenous cavity and protoxylem: Lysigenous cavity absent and enlarged protoxylem vessel present

Size of metaxylem vessels in relation to parenchyma sheath cells in cross-section: Narrow vessels

Vascular bundle sheath

Third-order vascular bundle's bundle sheath (Figs. 1.86E, arrowhead, 1.87B, arrowhead, 1.88C, E, arrowhead)

Shape: Round

Shape of bundle sheath cell: Outer tangential wall inflated; radial and inner tangential wall straight

Second-order vascular bundle's bundle sheath

Shape: Circular in all species except in *C. barbata* elliptical

Shape of bundle sheath cell: Outer tangential wall inflated; radial and inner tangential wall straight

Number of bundle sheath cell: 7–8 in *C. barbata* and *C. montana*, 8–9 in *C. virgata*

Nature of bundle sheath (either complete or incomplete): Complete

Chloroplast position in bundle sheath cell: Chloroplast fills lumen of cell

Mestome/Inner sheath: Absent

First-order vascular bundle's bundle sheath (Figs. 1.86B, 1.88D, arrowhead)

Shape: Circular in all species except in *C. barbata* elliptical

Extent of bundle sheath around the vascular bundle: Incomplete, abaxially interrupted by sclerenchyma cells in *C. barbata* and *C. montana* or complete in *C. virgata*

Extension of bundle sheath: Broad abaxial extension present in *C. barbata* and *C. montana* or absent in *C. virgata*

Number of cells comprising the bundle sheath: 9–10 in *C. barbata* and *C. montana*, 15–16 in *C. virgata* parenchyma cells

Bundle sheath cell shape: Outer tangential wall inflated; radial and inner tangential wall straight; chloroplast fills lumen of cell

Mestome/Inner sheath: Complete and cells smaller than outer sheath cell; gradation in cell size toward abaxial side seen only in *C. virgata*

Sclerenchyma

Adaxial sclerenchyma: Absent

Abaxial sclerenchyma: Well-developed strand in *C. barbata*, minute subepidermal strands in *C. montana* and *C. virgata*

Sclerenchyma between bundles: Absent

Sclerenchyma in leaf margin: Forming very small with very pointed fibrous cap in *C. barbata* (Fig. 1.86C, arrow), only pointed cap in *C. montana*, *C. virgata* (Figs. 1.87D, 1.88F); epidermal cells at margin are thin with thick outer tangential wall

Leaf-sheath anatomy (Figs. 1.86I–K, 1.87F–H, 1.88G–K)

Outline of lamina: Round shaped with inrolled margins in *C. barbata* (Fig. 1.86I), "V" shaped with inrolled margins in *C. montana* and *C. virgata* (Figs. 1.87F, 1.88G); ribs and furrows absent

Epidermis

Adaxial epidermal cells: Flat, rectangular

Abaxial epidermal cells: Small, flat, rectangular shaped with thick cuticle

Ground tissue: Parenchymatous with absence of air cavity

Vascular bundle

Arrangement of vascular bundle: Two to three consecutive third-order vascular bundles between first and second-order vascular bundle

Third-order vascular bundle (Figs. 1.86K, 1.87G, 1.88I): Hexagonal with xylem and phloem distinguishable

Second-order vascular bundle: Elliptical in all species except in *C. virgata* circular

First-order vascular bundle (Figs. 1.86J, 1.87H, 1.88H, K): Elliptical in all species except in *C. virgata* circular with phloem completely surrounded by thick-walled fibers; lysigenous cavity and enlarged protoxylem vessel present; circular wide metaxylem vessels present

Vascular bundle sheath (Figs. 1.86J, K, arrowhead, 1.87H, G, arrowhead, 1.88H, K, arrowhead)

Shape: Circular in all species except in *C. barbata* elliptical

Extent of bundle sheath around the vascular bundle: Incomplete, abaxially interrupted by sclerenchyma cells

Number of cells comprising the bundle sheath: 10–11 in *C. montana* and *C. barbata*, 14–15 in *C. virgata* parenchyma cells

Bundle sheath cell shape: Outer tangential wall inflated; radial and inner tangential wall straight and cells smaller than the ground tissue

Mestome/Inner sheath: Complete and cells smaller than outer sheath cell

Sclerenchyma: In form of straight horizontal girder in *C. montana* and *C. barbata* (Fig. 1.86K, arrow) or well-developed girder in *C. virgata* (Fig. 1.88M) on abaxial surface

Ligule anatomy (Figs. 1.86L, M, 1.87I, J, 1.88L, M)

Shape: Ligule is attached to the half width of the leaf sheath region and it is round shaped at the connection point in *C. barbata* (Fig. 1.86H) or attached to the lamina up to the half-length of it and is free and "V" shaped at the connection point in *C. montana* and *C. virgata* (Figs. 1.87I, 1.88L)

Anatomy: Two layered in *C. barbata* and *C. montana* or three layered in *C. virgata* mesophyll cells between adaxial and abaxial surfaces (Figs. 1.86M, 1.87J, 1.88M)

DIFFERENTIATING FEATURES

Round-shaped ligule … *C. barbata*
"V"-shaped ligule

Central cell of bulliform cell is inflated round … *C. montana*
Central cell of bulliform cell is shield shape … *C. virgata*

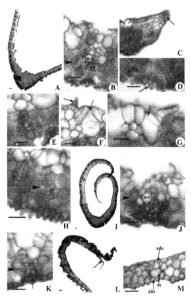

FIGURE 1.86   Leaf anatomy of *Chloris barbata*.

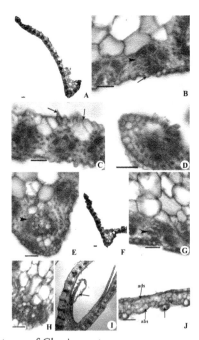

FIGURE 1.87   Leaf anatomy of *Choris montana*.

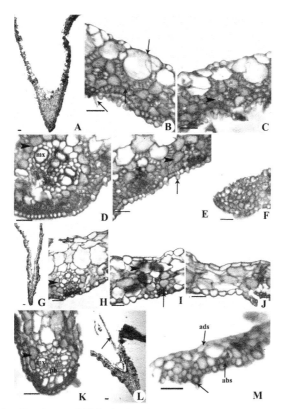

FIGURE 1.88   Leaf anatomy of *Chloris virgata.*

74. *Cynodon dactylon*

Leaf lamina (Fig. 1.89A–E)

Outline of the lamina: Expanded with undulating leaf blade (Fig. 1.89A). Two halves are symmetrical on either side of the median vascular bundle; ribs and furrows are present on the leaf surface; shallow and narrow furrows between all vascular bundles and flat ribs present over all vascular bundles. Keel is inconspicuous.

Epidermis: Adaxial epidermal surface has fan-shaped bulliform cells (Fig. 1.89B, arrow) in which central cell is elongated with parallel sides; epidermal cells showed pointed prickles (Fig. 1.89E, arrow); abaxial epidermal cells are small, barrel to square shaped covered with thick cuticle.

Mesophyll: Incompletely radiate chlorenchyma cells present; colorless cells present (Fig. 1.89E) which is smaller than bulliform cells and with one extension.

Vascular bundles: One to two first-order vascular bundles in half lamina regularly arranged from median to marginal bundle. Three to four third-order bundles between consecutive larger bundles and one to two second-order bundles between consecutive first-order bundles present. All the vascular bundles present at same level and placed centrally.

Third-order vascular bundles: Triangular (Fig. 1.89B) with vascular tissue consisting of only few indistinguishable vascular elements

Second-order vascular bundles: Circular shaped (Fig. 1.89D)

First-order vascular bundles (Fig. 1.89C)

Shape: Elliptical

Relationship of phloem cells and the vascular fibers: Completely covered by sclerenchyma cells

Nature of lysigenous cavity and protoxylem: Lysigenous cavity absent and enlarged protoxylem vessel present

Size of metaxylem vessels in relation to parenchyma sheath cells in cross-section: Narrow vessels

Shape of metaxylem vessels: Angular

Vascular bundle sheath

Third-order vascular bundle's bundle sheath (Fig. 1.89B, arrrowhead)

Shape: Triangular

Shape of bundle sheath cell: Fan shaped

Second-order vascular bundle's bundle sheath (Fig. 1.89D, arrowhead)

Shape: Triangular

Shape of bundle sheath cell: Fan shaped

Number of bundle sheath cell: 8–9

Nature of bundle sheath (either complete or incomplete): Complete

Chloroplast position in bundle sheath cell: Translucent

Mestome/Inner sheath: Complete and cells are smaller than outer sheath cells

First-order vascular bundle's bundle sheath (Fig. 1.89C, arrowhead)

Shape: Triangular

Extent of bundle sheath around the vascular bundle: Complete

Extension of bundle sheath: Absent

Number of cells comprising the bundle sheath: 13–14 parenchyma cells

Bundle sheath cell shape: Fan shaped

Mestome/Inner sheath: Complete and cells are smaller than the outer sheath cell

Sclerenchyma

Adaxial sclerenchyma: Minute subepidermal strands present

Abaxial sclerenchyma: subepidermal strands present

Sclerenchyma between bundles: Absent

Sclerenchyma in leaf margin: Forming very pointed cap; epidermal cells at margin are thin with thick outer tangential wall

Leaf-sheath anatomy (Fig. 1.89F–I)

Outline of lamina: Round with inrolled margins (Fig. 1.89F); ribs and furrows absent

Epidermis

Adaxial epidermal cells: Flat, rectangular
Abaxial epidermal cells: Small, flat, rectangular shaped with thick cuticle
Ground tissue: Parenchymatous ground tissue, parenchyma cells contain starch grains (Fig. 1.89I, arrow), air cavity absent

Vascular bundle

Arrangement of vascular bundle: One consecutive third-order vascular bundle between first and second-order vascular bundle

Third-order vascular bundle: Pentagonal with vascular tissue consisting of only few indistinguishable vascular elements

Second-order vascular bundle: Elliptical shaped

First-order vascular bundle: Elliptical with phloem completely surrounded by thick-walled fibers; lysigenous cavity and enlarged protoxylem vessel present; circular narrow metaxylem vessels present (Fig. 1.89H)

Vascular bundle sheath (Fig. 1.89G, H, arrowhead)

Shape: Elliptical shaped

Extent of bundle sheath around the vascular bundle: Incomplete, abaxially interrupted by sclerenchyma cells

Number of cells comprising the bundle sheath: 8–9 parenchyma cells

Bundle sheath cell shape: Fan shaped and cells smaller than the ground tissue

Mestome/Inner sheath: Complete and cells are smaller than the outer sheath cell

Sclerenchyma: In form of straight horizontal abaxial band

Ligule anatomy (Fig. 1.89J, K)

Shape: Ligule attached to the lamina up to one-third length of it and then free, blunt, narrow, "V" shaped at the point of connection (Fig. 1.89J, arrow)

Anatomy: Three layered mesophyll cells between adaxial and abaxial surfaces (Fig. 1.89K)

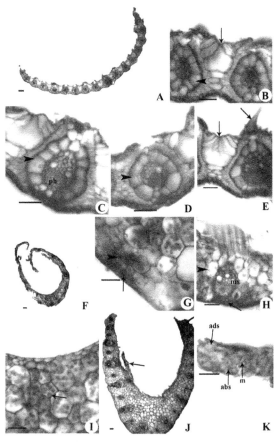

FIGURE 1.89   Leaf anatomy of *Cynodon dactylon*.

75. *Melanocenchris jaequemontii*

Leaf lamina (Fig. 1.90A–F)

Outline of the lamina: Expanded with undulating leaf blade (Fig. 1.90A). Two halves are symmetrical on either side of the median vascular bundle. Ribs and furrows are present on the leaf surface. Medium, narrow furrows between all vascular bundles and rounded ribs present over all vascular bundles. Keel is inconspicuous.

Epidermis: Adaxial epidermal surface has fan-shaped bulliform, a cell (Fig. 1.90B, arrow) in which central cell is diamond shaped. Epidermal cells showed pointed, curved prickles (Fig. 1.90D). Abaxial epidermal cells are small, barrel to square shaped covered with thick cuticle.

Mesophyll: Incompletely radiate chlorenchyma cells present; colorless cells present (Fig. 1.90E) which is smaller than bulliform cells and with one extension.

Vascular bundles: One to two first-order vascular bundles in half lamina regularly arranged from median to marginal bundle. Three to four third-order bundles between consecutive larger bundles and one to two second-order bundles between consecutive first-order bundles present. All the vascular bundles present at same level and placed centrally.

Third-order vascular bundles: Hexagonal (Fig. 1.90E) with vascular tissue consisting of only few indistinguishable vascular elements

Second-order vascular bundles: Circular shaped

First-order vascular bundles (Fig. 1.90C)

Shape: Round

Relationship of phloem cells and the vascular fibers: Completely covered by sclerenchyma

Nature of lysigenous cavity and protoxylem: Lysigenous cavity absent and enlarged protoxylem vessel present

Size of metaxylem vessels in relation to parenchyma sheath cells in cross-section: Narrow vessels

Shape of metaxylem vessels: Angular

Vascular bundle sheath

Third-order vascular bundle's bundle sheath

Shape: Round

Shape of bundle sheath cell: Fan shaped

Second-order vascular bundle's bundle sheath

Shape: Round

Shape of bundle sheath cell: Fan shaped

Number of bundle sheath cell: 8–9

Nature of bundle sheath (either complete or incomplete): Complete

Chloroplast position in bundle sheath cell: Translucent

Mestome/Inner sheath: Complete and cells are smaller than outer sheath cells

First-order vascular bundle's bundle sheath (Fig. 1.90C, arrowhead)

Shape: Triangular

Extent of bundle sheath around the vascular bundle: Complete

Extension of bundle sheath: Absent

Number of cells comprising the bundle sheath: 8–9 parenchyma cells

Bundle sheath cell shape: Fan shaped

Mestome/Inner sheath: Complete and cells are smaller than the outer sheath cell

Sclerenchyma

Adaxial sclerenchyma: Minute subepidermal strands present

Abaxial sclerenchyma: Subepidermal strands present

Sclerenchyma between bundles: Absent

Sclerenchyma in leaf margin: Forming very pointed cap; epidermal cells at margin are thin with thick outer tangential wall (Fig. 1.90B)

Leaf-sheath anatomy (Fig. 1.90G, H)

Outline of lamina: Round shaped with inrolled margins (Fig. 1.90G); ribs and furrows absent

Epidermis

Adaxial epidermal cells: Flat, rectangular

Abaxial epidermal cells: Small, flat, rectangular shaped with thick cuticle

Ground tissue: Parenchymatous with absence of air cavity

Vascular bundle

Arrangement of vascular bundle: One consecutive third-order vascular bundle between first and second-order vascular bundle

Third-order vascular bundle: Hexagonal with vascular tissue consisting of only few indistinguishable vascular elements

Second-order vascular bundle: Circular shaped

First-order vascular bundle: Circular shaped with phloem completely surrounded by thick-walled fibers; lysigenous cavity and enlarged proto-xylem vessel present; circular narrow metaxylem vessels present

Vascular bundle sheath

Shape: Circular shaped

Extent of bundle sheath around the vascular bundle: Complete

Number of cells comprising the bundle sheath: 8–9 parenchyma cells

Bundle sheath cell shape: Fan shaped and cells smaller than the ground tissue

Mestome/Inner sheath: Complete and cells are smaller than the outer sheath cell

Sclerenchyma: In form of straight horizontal band on abaxial surface

Ligule anatomy (Fig. 1.90I)

Shape: Ligule attached to the midrib and few parts of lamina region because it is hairy ligule

Anatomy: Two layered mesophyll cells between adaxial and abaxial surfaces (Fig. 1.90I)

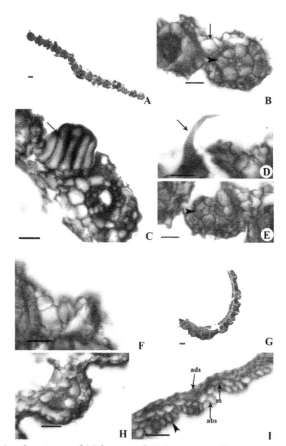

FIGURE 1.90   Leaf anatomy of *Melanocenchris jaequemontii.*

76. *Oropetium villosulum*

Leaf lamina (Fig. 1.91A–F)

Outline of the lamina: Expanded with undulating leaf blade (Fig. 1.91A). Two halves are symmetrical on either side of the median vascular bundle. Ribs and furrows are present on the leaf surface. Medium, narrow furrows between all vascular bundles and rounded ribs present over all vascular bundles. Keel is inconspicuous.

Epidermis: Adaxial epidermal surface has fan-shaped bulliform cell (Fig. 1.91E, arrow) in which central cell is diamond shaped; epidermal cells have pointed, curved prickles (Fig. 1.91F); few epidermal cells have papillae, one papillae per cell (Fig. 1.91E, arrow); abaxial epidermal cells are small, barrel to square shaped covered with thick cuticle.

Mesophyll: Incompletely radiate chlorenchyma cells present; colorless cells present (Fig. 1.91E) which is smaller than bulliform cells and with one extension.

Vascular bundles: One to two first-order vascular bundles in half lamina regularly arranged from median to marginal bundle. Two to three third-order bundles between consecutive larger bundles and one to two second-order bundles between consecutive first-order bundles present. All the vascular bundles present at same level and placed centrally.

Third-order vascular bundles: Hexagonal (Fig. 1.91C) with vascular tissue consisting of only few indistinguishable vascular elements

Second-order vascular bundles: Circular shaped

First-order vascular bundles (Fig. 1.91D)

Shape: Elliptical

Relationship of phloem cells and the vascular fibers: Completely covered by sclerenchyma cells

Nature of lysigenous cavity and protoxylem: Lysigenous cavity absent and enlarged protoxylem vessel present

Size of metaxylem vessels in relation to parenchyma sheath cells in cross-section: Narrow vessels

Shape of metaxylem vessels: Circular

Vascular bundle sheath

Third-order vascular bundle's bundle sheath (Fig. 1.91C, F, arrowhead)

Shape: Round

Shape of bundle sheath cell: Fan shaped

Second-order vascular bundle's bundle sheath

Shape: Round

Shape of bundle sheath cell: Fan shaped

Number of bundle sheath cell: 7–8

Nature of bundle sheath (either complete or incomplete): Complete

Chloroplast position in bundle sheath cell: Translucent

Mestome/Inner sheath: Absent

First-order vascular bundle's bundle sheath (Fig. 1.91D, arrowhead)

Shape: Circular

Extent of bundle sheath around the vascular bundle: Complete

Extension of bundle sheath: Absent

Number of cells comprising the bundle sheath: 8–9 parenchyma cells

Bundle sheath cell shape: Fan shaped

Mestome/Inner sheath: Complete and cells are smaller than the outer sheath cell

Sclerenchyma

Adaxial sclerenchyma: Minute subepidermal strands present

Abaxial sclerenchyma: subepidermal strands present

Sclerenchyma between bundles: Absent

Sclerenchyma in leaf margin: Forming very pointed cap. Epidermal cells at margin are thin with thick outer tangential wall (Fig. 1.91B)

Leaf-sheath anatomy (Fig. 1.91G–I)

Outline of lamina: V shaped, narrow angle with inrolled margins (Fig. 1.91G); ribs and furrows absent

Epidermis

Adaxial epidermal cells: Flat, rectangular

Abaxial epidermal cells: Small, flat, rectangular shaped with thick cuticle

Ground tissue: Parenchymatous with absence of air cavity

Vascular bundle (Fig. 1.91H, I, arrowhead)

Arrangement of vascular bundle: One consecutive third-order vascular bundle between first and second-order vascular bundle

Third-order vascular bundle: Hexagonal with vascular tissue consisting of only few indistinguishable vascular elements (Fig. 1.91H)

Second-order vascular bundle: Circular shaped

First-order vascular bundle: Elliptical with phloem completely surrounded by thick-walled fibers; lysigenous cavity and enlarged protoxylem vessel present; circular narrow metaxylem vessels present (Fig. 1.91I)

Vascular bundle sheath

Shape: Circular shaped

Extent of bundle sheath around the vascular bundle: Complete

Number of cells comprising the bundle sheath: 10–11 parenchyma cells

Bundle sheath cell shape: Fan shaped and cells smaller than the ground tissue

Mestome/Inner sheath: Complete and cells are smaller than the outer sheath cell

Sclerenchyma: In form of straight horizontal abaxial band

Ligule anatomy (Fig. 1.91J, K)

Shape: Ligule attached to the midrib and few parts of lamina region because it is hairy ligule (Fig. 1.91J, arrow)

Anatomy: Two layered mesophyll cells between adaxial and abaxial surfaces (Fig. 1.91K)

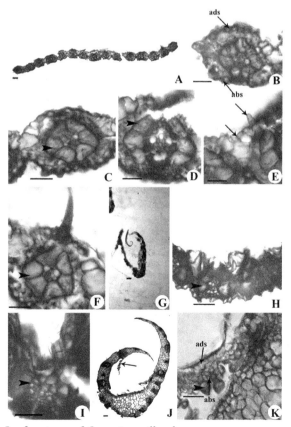

FIGURE 1.91   Leaf anatomy of *Oropetium villosulum.*

77. *Schoenefeldia gracilis*

Leaf lamina (Fig. 1.92A–E)

Outline of the lamina: V shaped, wide angle with straight arms (Fig. 1.92A). Two halves are symmetrical on either side of the median vascular bundle. Ribs and furrows are present on the leaf surface. Medium, narrow furrows between all vascular bundles and rounded ribs present over all vascular bundles.

Keel is well-developed, V-shaped keel with single median vascular bundle, three to four third-order vascular bundles. Small, round parenchyma cells are present surrounding the median vascular bundle adaxially. Sclerenchyma on abaxial surface associated with vascular bundles in form of girders.

Epidermis: Adaxial epidermal surface has fan-shaped bulliform cells (Fig. 1.92E, arrow) in which central cell is shield shaped; epidermal cells showed pointed, curved prickles (Fig. 1.92D); abaxial epidermal cells are small, barrel to square shaped covered with thick cuticle.

Mesophyll: Radiate, single layered, isodiametric chlorenchyma cells completely surrounding bundles. Compact homogenous chlorenchymatous cells occupy the major area between the adaxial and abaxial epidermises; chlorenchyma cells continuous between bundles; colorless cells present (Fig. 1.92E) which are smaller than bulliform cells and with one extension.

Vascular bundles: One to two first-order vascular bundles in half lamina regularly arranged from median to marginal bundle; four to five third-order bundles between consecutive larger bundles and two to three second-order bundles between consecutive first-order bundles present; all the vascular bundles present at same level and placed abaxially.

Third-order vascular bundles: Hexagonal (Fig. 1.92D) with vascular tissue consisting of only few indistinguishable vascular elements

Second-order vascular bundles: Circular shaped

First-order vascular bundles (Fig. 1.92C)

Shape: Elliptical

Relationship of phloem cells and the vascular fibers: Completely covered by sclerenchymatous fibers

Nature of lysigenous cavity and protoxylem: Lysigenous cavity and enlarged protoxylem vessel present

Size of metaxylem vessels in relation to parenchyma sheath cells in cross-section: Wide vessels

Shape of metaxylem vessels: Circular

Median vascular bundle

Shape: Elliptical

Relationship of phloem cells and the vascular fibers: Completely covered by sclerenchymatous fibers

Nature of lysigenous cavity and protoxylem: Lysigenous cavity and enlarged protoxylem vessel present

Size of metaxylem vessels in relation to parenchyma sheath cells in cross-section: Wide vessels

Vascular bundle sheath

Third-order vascular bundle's bundle sheath (Fig. 1.92D, arrowhead)

Shape: Round

Shape of bundle sheath cell: Fan shaped

Second-order vascular bundle's bundle sheath

Shape: Round

Shape of bundle sheath cell: Fan shaped

Number of bundle sheath cell: 6–7

Nature of bundle sheath (either complete or incomplete): Complete

Chloroplast position in bundle sheath cell: Chloroplast positioned toward inner tangential wall

Mestome/Inner sheath: Absent

First-order vascular bundle's bundle sheath (Fig. 1.92C, arrowhead)

Shape: Circular

Extent of bundle sheath around the vascular bundle: Incomplete, adaxially interrupted by parenchyma cell and abaxially interrupted by sclerenchyma cells

Extension of bundle sheath: Broad well-developed girder present toward abaxial side

Number of cells comprising the bundle sheath: 8–9 parenchyma cells

Bundle sheath cell shape: Fan shaped

Mestome/Inner sheath: Complete and cells are smaller than the outer sheath cell

Sclerenchyma

Adaxial sclerenchyma: Minute subepidermal strands present

Abaxial sclerenchyma: Well-developed girder, wider than deep

Sclerenchyma between bundles: Absent

Sclerenchyma in leaf margin: Forming very small rounded cap; epidermal cells at margin are thin with thick outer tangential wall (Fig. 1.92B)

Leaf-sheath anatomy (Fig. 1.92F–I)

Outline of lamina: U shaped with inrolled margins (Fig. 1.92F); ribs and furrows absent

Epidermis

Adaxial epidermal cells: Flat, rectangular

Abaxial epidermal cells: Small, flat, rectangular shaped with thick cuticle

Ground tissue: Parenchymatous with absence of air cavity

Vascular bundle

Arrangement of vascular bundle: Two to three consecutive third-order vascular bundles between first and second-order vascular bundle

Third-order vascular bundle: Hexagonal with vascular tissue consisting of only few indistinguishable vascular elements

Second-order vascular bundle: Elliptical shaped (Fig. 1.92I)

First-order vascular bundle: Elliptical with phloem completely surrounded by thick-walled fibers; lysigenous cavity and enlarged protoxylem vessel present; circular narrow metaxylem vessels present (Fig. 1.92G)

Vascular bundle sheath (Fig. 1.92G, H, I, arrowhead)

Shape: Horseshoe shaped

Extent of bundle sheath around the vascular bundle: Incomplete, abaxially interrupted by sclerenchyma cells

Number of cells comprising the bundle sheath: 8–9 parenchyma cells

Bundle sheath cell shape: Fan shaped and cells smaller than the ground tissue

Mestome/Inner sheath: Complete and cells are smaller than the outer sheath cell

Sclerenchyma: In form of well-developed girder on abaxial surface (Fig. 1.92H, arrow)

Ligule anatomy (Fig. 1.92J, K)

Shape: Ligule attached to the midrib and few parts of lamina region because it is hairy ligule (Fig. 1.92J, arrow)

Anatomy: Two layered mesophyll cells between adaxial and abaxial surfaces, few cells have presence of sclerenchyma (Fig. 1.92K, arrowhead)

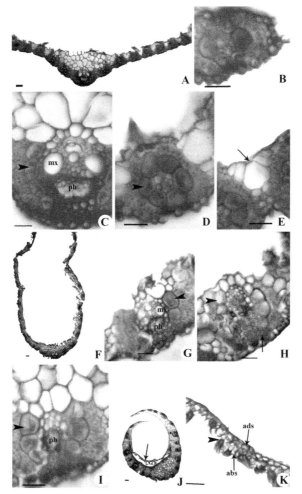

FIGURE 1.92   Leaf anatomy of *Schoenefeldia gracilis*.

Tetrapogon

78. *Tetrapogon tenellus* (Fig. 1.93)

79. *Tetrapogon villosus* (Fig. 1.94)

Leaf lamina (Figs. 1.93A–G, 1.94A–G)

Outline of the lamina: Expanded in both species with straight arms in *T. tenellus* or wavy arms in *T. villosus* (Figs. 1.93A, B, 1.94A). Two halves are symmetrical on either side of the median vascular bundle. Ribs and furrows are present on the leaf surface. Medium, narrow furrows between all vascular bundles and rounded ribs present over all vascular bundles.

Keel is well developed, V shaped with single median vascular bundle. Small sclerenchyma cells are present surrounding adaxial to the median vascular bundle; sclerenchyma on abaxial surface associated with vascular bundles in form of girders.

Epidermis: Adaxial epidermal surface has fan-shaped bulliform cells in both species in which central cell is inflated (Figs. 1.93F, arrow, 1.94D, arrow). Abaxial epidermal cells are small, barrel to square shaped covered with thick cuticle. Epidermal cells have papillae in both species (Figs. 1.93G, arrow, 1.94E, arrow).

Mesophyll: Radiate, single layered, isodiametric chlorenchyma cells surrounding bundles; compactly arranged homogenous chlorenchyma cells occupying the major area between the adaxial and abaxial epidermises; chlorenchyma cells continuous between bundles; colorless cells present (Fig. 1.93E) which are smaller than bulliform cells and present near the bulliform cells.

Vascular bundles: Two to three first-order vascular bundles in half lamina regularly arranged from median to marginal bundle. Four to five third-order bundles in *T. tenellus* or six to seven third-order bundles in *T. villosus* between consecutive larger bundles and two to three second-order bundles between consecutive first-order bundles present. All the vascular bundles present at same level and placed abaxially.

Third-order vascular bundles (Figs. 1.93E, 1.94C, E): Pentagonal with vascular tissue consisting of only few indistinguishable vascular elements

Second-order vascular bundles: Elliptical shaped

First-order vascular bundles (Figs. 1.93D, 1.94B)

Shape: Elliptical

Relationship of phloem cells and the vascular fibers: Completely covered by sclerenchymatous fibers

Nature of lysigenous cavity and protoxylem: Lysigenous cavity absent and enlarged protoxylem vessel present

Size of metaxylem vessels in relation to parenchyma sheath cells in cross-section: Narrow vessels

Shape of metaxylem vessels: Circular

Median vascular bundle (Fig. 1.93C)

Shape: Elliptical

Relationship of phloem cells and the vascular fibers: Completely covered by sclerenchymatous fibers

Nature of lysigenous cavity and protoxylem: Lysigenous cavity absent and enlarged protoxylem vessel present

Size of metaxylem vessels in relation to parenchyma sheath cells in cross-section: Narrow vessels

Vascular bundle sheath

Third-order vascular bundle's bundle sheath (Figs. 1.93E, arrowhead, 1.94C, arrowhead)

Shape: Triangular

Shape of bundle sheath cell: Fan shaped

Second-order vascular bundle's bundle sheath

Shape: Horseshoe shaped

Shape of bundle sheath cell: Fan shaped

Number of bundle sheath cell: 6–7

Nature of bundle sheath (either complete or incomplete): Incomplete, abaxially interrupted by sclerenchyma cells

Chloroplast position in bundle sheath cell: Chloroplast fills entire cell lumen in *T. tenellus* or positioned toward inner tangential wall in *T. villosus.*

Mestome/Inner sheath: Absent

First-order vascular bundle's bundle sheath (Figs. 1.93D, arrowhead, 1.94B, arrowhead)

Shape: Circular

Extent of bundle sheath around the vascular bundle: Incomplete, adaxially interrupted by parenchyma cells and abaxially interrupted by sclerenchyma cells

Extension of bundle sheath: Broad well-developed girder present toward abaxial side

Number of cells comprising the bundle sheath: 7–8 parenchyma cells

Bundle sheath cell shape: Fan shaped and chloroplast fill entire cell lumen in *T. tenellus* or positioned toward inner tangential wall in *T. villosus*

Mestome/Inner sheath: Complete and cells are smaller than the outer sheath cell

Sclerenchyma

Adaxial sclerenchyma: Minute subepidermal strands present

Abaxial sclerenchyma: Well-developed girder, wider than deep

Sclerenchyma between bundles: Absent

Sclerenchyma in leaf margin: Forming very small rounded cap; epidermal cells at margin are thin with thick outer tangential wall

Leaf-sheath anatomy (Figs. 1.93H–L, 1.94F–I)

Outline of lamina: "V" shaped with inrolled margins in both species (Figs. 1.93H, 1.94F); ribs and furrows absent

Epidermis

Adaxial epidermal cells: Flat, rectangular (Fig. 1.93L)

Abaxial epidermal cells: Small, flat, rectangular shaped with thick cuticle

Ground tissue: Parenchymatous with absence of air cavity

Vascular bundle

Arrangement of vascular bundle: Two to three in *T. tenellus*, one to two in *T. villosus* consecutive third-order vascular bundle between first and second-order vascular bundle

Third-order vascular bundle (Figs. 1.93J, 1.94I): Hexagonal in *T. tenellus* and pentagonal in *T. villosus* with vascular tissue consisting of only few indistinguishable vascular elements

Second-order vascular bundle: Elliptical shaped

First-order vascular bundle (Figs. 1.93I, 1.94G): Elliptical with phloem completely surrounded by thick-walled fibers; enlarged protoxylem vessel present in both and lysigenous cavity present only in *T. tenellus*; circular narrow metaxylem vessels present

Vascular bundle sheath (Fig. 1.93I, J, K, arrowhead)

Shape: Horseshoe shaped

Extent of bundle sheath around the vascular bundle: Incomplete, abaxially interrupted by sclerenchyma cells

Number of cells comprising the bundle sheath: 8–9 in *T. tenellus*, 7–8 in *T. villosus* parenchyma cells

Bundle sheath cell shape: Fan shaped and cells smaller than the ground tissue

Mestome/Inner sheath: Complete and cells are smaller than the outer sheath cell

Sclerenchyma: In form of straight horizontal abaxial strands

Ligule anatomy (Figs. 1.93M, N, 1.94J, K)

Shape: Ligule attached to the midrib and few parts of lamina region because it is hairy ligule in both species (Figs. 1.93M, 1.94J)

Anatomy: Single in *T. tenellus* or two in *T. villosus* layered mesophyll cells between adaxial and abaxial surfaces (Figs. 1.93N, 1.94K)

DIFFERENTIATING FEATURES

Chloroplast fill entire full lumen, straight arms of leaf blade, single layered mesophyll in ligule … *T. tenellus*
Chloroplast concentrated near inner tangential wall, wavy arms of leaf blade, two layered mesophyll in ligule … *T. villosus*

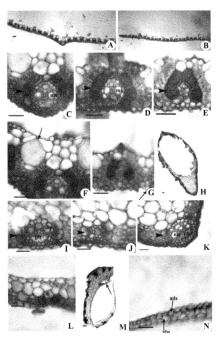

FIGURE 1.93   Leaf anatomy of *Tetrapogon tenellus*.

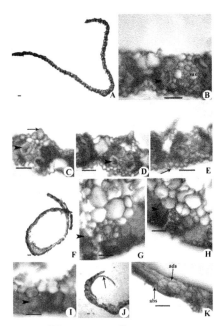

FIGURE 1.94   Leaf anatomy of *Tetrapogon villosus*.

80. *Acrachne racemosa*

Leaf lamina (Fig. 1.95A–G)

Outline of the lamina: "V" shaped, broad angle with straight arms (Fig. 1.95A); two halves are symmetrical on either side of the median vascular bundle; ribs and furrows are absent on the leaf surface.

Keel is well-developed, V-shaped keel with single median vascular bundle, three to four third-order vascular bundles. Small, sclerenchyma cells are present surrounding adaxial to the median vascular bundle. Sclerenchyma on abaxial surface associated with vascular bundles in form of girders.

Epidermis: Adaxial epidermal surface has fan-shaped bulliform cells in which central cell is shield shaped (Fig. 1.95E, arrow). Abaxial epidermal cells are small, barrel to square shaped covered with thick cuticle. Epidermal cells have papillae (Fig. 1.95C, arrow) and papillae are long and as wide as epidermal cell width.

Mesophyll: Radiate, single layered, isodiametric chlorenchyma cells surrounding bundles; compact homogenous chlorenchymatous cells occupying the major area between the adaxial and abaxial epidermises; chlorenchyma cells continuous between bundles; colorless cells present which are smaller than bulliform cells and present near the bulliform cells.

Vascular bundles: Three to four first-order vascular bundles in half lamina regularly arranged from median to marginal bundle; six to seven third-order bundles between consecutive larger bundles and two to three second-order bundles between consecutive first-order bundles present; all the vascular bundles present at same level and placed centrally.

Third-order vascular bundles: Hexagonal (Fig. 1.95D, E) vascular tissue consisting of only few indistinguishable vascular elements

Second-order vascular bundles: Elliptical shaped

First-order vascular bundles (Fig. 1.95C)

Shape: Elliptical

Relationship of phloem cells and the vascular fibers: Completely covered by sclerenchymatous fibers

Nature of lysigenous cavity and protoxylem: Lysigenous cavity absent and enlarged protoxylem vessel present

Size of metaxylem vessels in relation to parenchyma sheath cells in cross-section: Narrow vessels

Shape of metaxylem vessels: Circular

Median vascular bundle

Shape: Elliptical

Relationship of phloem cells and the vascular fibers: Completely covered by sclerenchymatous fibers

Nature of lysigenous cavity and protoxylem: Lysigenous cavity absent and enlarged protoxylem vessel present

Size of metaxylem vessels in relation to parenchyma sheath cells in cross-section: Narrow vessels

Vascular bundle sheath

Third-order vascular bundle's bundle sheath (Fig. 1.95D, E, arrowhead)

Shape: Circular

Shape of bundle sheath cell: Cells with tangential wall inflated, two cells large which are placed in sides of vascular bundle

Second-order vascular bundle's bundle sheath

Shape: Circular

Shape of bundle sheath cell: Cells with tangential wall inflated, two cells large which are placed in sides of vascular bundle

Number of bundle sheath cell: 8–9

Nature of bundle sheath (either complete or incomplete): Complete

Chloroplast position in bundle sheath cell: Chloroplast positioned toward outer tangential wall

Mestome/Inner sheath: Absent

First-order vascular bundle's bundle sheath (Fig. 1.95C, arrowhead)

Shape: Elliptical

Extent of bundle sheath around the vascular bundle: Complete

Extension of bundle sheath: Absent

Number of cells comprising the bundle sheath: 12–13 parenchyma cells

Bundle sheath cell shape: Cells with tangential wall inflated, two large cells which are placed in sides of vascular bundle; chloroplast positioned toward outer tangential wall

Mestome/Inner sheath: Complete and cells are smaller than the outer sheath cell

Sclerenchyma

Adaxial sclerenchyma: Minute subepidermal strands present (Fig. 1.95C, arrow)

Abaxial sclerenchyma: Well-developed girder, wider than deep

Sclerenchyma between bundles: Absent

Sclerenchyma in leaf margin: Forming very small rounded cap; epidermal cells at margin are thin with thick outer tangential wall

Leaf-sheath anatomy (Fig. 1.95F–I)

Outline of lamina: "V" shaped with inrolled margins (Fig. 1.95F); ribs and furrows absent

Epidermis

Adaxial epidermal cells: Flat, rectangular

Abaxial epidermal cells: Small, flat, rectangular shaped with thick cuticle

Ground tissue: Parenchymatous with absence of air cavity

Vascular bundle

Arrangement of vascular bundle: One to two consecutive third-order vascular bundles between first and second-order vascular bundle

Third-order vascular bundle: Hexagonal with vascular tissue consisting of only few indistinguishable vascular elements (Fig. 1.95G)

Second-order vascular bundle: Elliptical shaped

First-order vascular bundle: Elliptical with phloem completely surrounded by thick-walled fibers; lysigenous cavity absent and enlarged protoxylem vessel present; circular narrow metaxylem vessels present (Fig. 1.95I)

Vascular bundle sheath (Fig. 1.95G, I, arrowhead)

Shape: Circular

Extent of bundle sheath around the vascular bundle: Incomplete, abaxially interrupted by sclerenchyma cells

Number of cells comprising the bundle sheath: 7–8 parenchyma cells

Bundle sheath cell shape: Cells with tangential wall inflated, two large cells which are placed in sides of vascular bundle and cells smaller than the ground tissue

Mestome/Inner sheath: Complete and cells are smaller than the outer sheath cell

Sclerenchyma: In form of straight horizontal abaxial band

Ligule anatomy (Fig. 1.95J, K)

Shape: Ligule attached to the midrib and broad round shaped at connection point (Fig. 1.95J, arrow)

Anatomy: Two layered mesophyll cells between adaxial and abaxial surfaces (Fig. 1.95K)

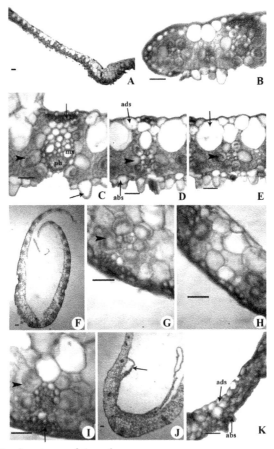

FIGURE 1.95   Leaf anatomy of *Acrachne racemosa.*

Dactyloctenium

81. *Dactyloctenium aegyptium* (Fig. 1.96)

82. *Dactyloctenium scindicus* (Fig. 1.97)

Leaf lamina (Figs. 1.96A–E, 1.97A–G)

Outline of the lamina: Expanded with straight arms in both species (Figs. 1.96A, 1.97A); two halves are symmetrical on either side of the median vascular bundle; ribs and furrows are present on the leaf surface; medium, narrow furrows between all vascular bundles and round ribs present over all vascular bundles.

Keel is well developed, V shaped with single median vascular bundle. Small, sclerenchyma cells are present surrounding adaxial to the median vascular bundle. Sclerenchyma on abaxial surface associated with vascular bundles in form of girders.

Epidermis: Adaxial epidermal surface has fan-shaped bulliform cells in both species, inflated central cell is in *D. aegyptium* and shield-shaped central cell in *D. scindicus* (Figs. 1.96D, arrow, 1.97, arrow).

Abaxial epidermal cells are small, barrel to square shaped covered with thick cuticle. Epidermal cells have large, balloon-shaped papillae (Figs. 1.96E, arrow, 1.97E). Abaxial epidermis has restricted group of long, inflated bulliform cells present in *D. scindicus* (Fig. 1.97C).

Mesophyll: Radiate, single layered, isodiametric chlorenchyma cells completely surrounding bundles; compactly arranged homogenous chlorenchyma cells occupying the major area between the adaxial and abaxial epidermises; chlorenchyma cells continuous between bundles; colorless cells absent.

Vascular bundles: Three to four first-order vascular bundles in half lamina regularly arranged from median to marginal bundle; six to seven third-order bundles between consecutive larger bundles and two to three second-order bundles between consecutive first-order bundles present; all the vascular bundles present at same level and placed abaxially

Third-order vascular bundles (Figs. 1.96E, 1.97G): Hexagonal with vascular tissue consisting of only few indistinguishable vascular elements

Second-order vascular bundles: Elliptical shaped (Fig. 1.97F)

First-order vascular bundles (Figs. 1.96D, 1.97E)

Shape: Elliptical

Relationship of phloem cells and the vascular fibers: Completely covered by sclerenchymatous fibers

Nature of lysigenous cavity and protoxylem: Lysigenous cavity absent and enlarged protoxylem vessel present

Size of metaxylem vessels in relation to parenchyma sheath cells in cross-section: Narrow vessels

Shape of metaxylem vessels: Circular

Median vascular bundle (Figs. 1.96C, 1.97D)

Shape: Elliptical

Relationship of phloem cells and the vascular fibers: Completely covered by sclerenchymatous fibers

Nature of lysigenous cavity and protoxylem: Lysigenous cavity absent and enlarged protoxylem vessel present

Size of metaxylem vessels in relation to parenchyma sheath cells in cross-section: Narrow vessels

Vascular bundle sheath

Third-order vascular bundle's bundle sheath (Figs. 1.96E, arrowhead, 1.97G, arrowhead)

Shape: Circular

Shape of bundle sheath cell: Outer tangential wall inflated; radial walls straight

Second-order vascular bundle's bundle sheath (Fig. 1.97F)

Shape: Circular

Shape of bundle sheath cell: Outer tangential wall inflated; radial walls straight

Number of bundle sheath cell: 8–9 in *D. aegyptium*, 6–7 in *D. scindicus*

Nature of bundle sheath (either complete or incomplete): Incomplete, abaxially interrupted by sclerenchyma cells

Chloroplast position in bundle sheath cell: Chloroplast centrally placed

Mestome/Inner sheath: Absent

First-order vascular bundle's bundle sheath (Figs. 1.96D, arrowhead, 1.97E, arrowhead)

Shape: Circular

Extent of bundle sheath around the vascular bundle: Incomplete, adaxially interrupted by parenchyma cells and abaxially interrupted by sclerenchyma cells

Extension of bundle sheath: Broad well-developed girder present toward abaxial side

Number of cells comprising the bundle sheath: 7–8 in *D. aegyptium*, 9–10 in *D. scindicus* parenchyma cells

Bundle sheath cell shape: Outer tangential wall inflated; radial walls straight; chloroplast centrally placed

Mestome/Inner sheath: Complete and cells are smaller than the outer sheath cell

Sclerenchyma

Adaxial sclerenchyma: Minute subepidermal strands present

Abaxial sclerenchyma: Well-developed girder, wider than deep

Sclerenchyma between bundles: Absent

Sclerenchyma in leaf margin: Forming crescent shaped in *D. aegyptium* or pointed shaped cap in *D. scindicus*; epidermal cells at margin are thin with thick outer tangential wall and showed papillae

Leaf-sheath anatomy (Figs. 1.96F–I, 1.97H–J)

Outline of lamina: Round shaped with inrolled margins in both species (Figs. 1.96F, 1.97H); ribs and furrows absent

Epidermis

Adaxial epidermal cells: Flat, rectangular

Abaxial epidermal cells: Small, flat, rectangular shaped with thick cuticle

Ground tissue: Parenchymatous with absence of air cavity

Vascular bundle

Arrangement of vascular bundle: One to two consecutive third-order vascular bundles between first and second-order vascular bundle

Third-order vascular bundle (Figs. 1.96I, 1.97I): Hexagonal with vascular tissue consisting of only few indistinguishable vascular elements

Second-order vascular bundle: Elliptical shaped

First-order vascular bundle (Figs. 1.96H, 1.97J): Elliptical with phloem completely surrounded by thick-walled fibers; lysigenous cavity absent and enlarged protoxylem vessel present; circular narrow metaxylem vessels present

Vascular bundle sheath (Figs. 1.96H, arrowhead, 1.97J, arrowhead)

Shape: Horseshoe shaped

Extent of bundle sheath around the vascular bundle: Incomplete, abaxially interrupted by sclerenchyma cells

Number of cells comprising the bundle sheath: 9–10 parenchyma cells

Bundle sheath cell shape: Fan shaped and cells smaller than the ground tissue

Mestome/Inner sheath: Complete and cells are smaller than the outer sheath cell

Sclerenchyma: In form of straight horizontal abaxial strand

Ligule anatomy (Figs. 1.96J, K, 1.97K, L)

Shape: Ligule attached to the midrib and parts of lamina region and forms round shaped in *D. aegyptium* or "V" shaped in *D. scindicus* at connection of point (Figs. 1.96J, 1.97J).

Anatomy: Single in *D. aegyptium* or two in *D. scindicus*-layered mesophyll cells between adaxial and abaxial surfaces (Figs. 1.96K, 1.97L)

DIFFERENTIATING FEATURES

Crescent shaped leaf margin, round-shaped ligule, single layered mesophyll in ligule ... *D. aegyptium*
Pointed leaf margin, "V"-shaped ligule, two layered mesophyll in ligule ... *D. sindicus*

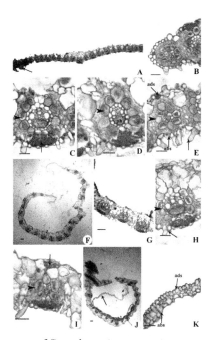

FIGURE 1.96   Leaf anatomy of *Dactyloctenium aegyptium*.

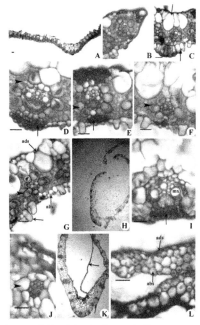

FIGURE 1.97   Leaf anatomy of *Dactyloctenium scindicus*.

83. *Desmostachya bipinnata*

Leaf lamina (Fig. 1.98A–G)

Outline of the lamina: Expanded with straight arms (Fig. 1.98A); two halves are symmetrical on either side of the median vascular bundle; ribs and furrows are present on the leaf surface; shallow, narrow furrows between all vascular bundles and round ribs present over all vascular bundles.

Keel is well-developed, V-shaped keel with single median vascular bundle, three to four third-order vascular bundles; small, sclerenchyma cells are present surrounding adaxial to the median vascular bundle; sclerenchyma on abaxial surface associated with vascular bundles in form of girders.

Epidermis: Adaxial epidermal surface has fan-shaped bulliform cells in which central cell is inflated, large shaped (Fig. 1.98E, arrow). Abaxial epidermal cells are small, barrel to square shaped covered with thick cuticle.

Mesophyll: Radiate, single layered, isodiametric chlorenchyma cells completely surrounding bundles; compact homogenous chlorenchymatous cells occupying the major area between the adaxial and abaxial epidermises; chlorenchyma cells discontinuous by colorless cells between bundles; colorless cells present (Fig. 1.98E) and smaller than bulliform cells with one extension.

Vascular bundles: Two to three first-order vascular bundles in half lamina regularly arranged from median to marginal bundle; five to six third-order bundles between consecutive larger bundles and two to three second-order bundles between consecutive first-order bundles present; all the vascular bundles present at same level and placed centrally.

Third-order vascular bundles: Triangular (Fig. 1.98D) with vascular tissue consisting of only few indistinguishable vascular elements (Fig. 1.98D, arrow)

Second-order vascular bundles: Elliptical shaped (Fig. 1.98C)

First-order vascular bundles (Fig. 1.98B)

Shape: Elliptical

Relationship of phloem cells and the vascular fibers: Phloem divided by intrusion of small fibers resulting in sclerosed phloem (Fig. 1.98B, arrow)

Nature of lysigenous cavity and protoxylem: Lysigenous cavity absent and enlarged protoxylem vessel present

Size of metaxylem vessels in relation to parenchyma sheath cells in cross-section: Narrow vessels

Shape of metaxylem vessels: Circular

Median vascular bundle (Fig. 1.98D)

Shape: Elliptical

Relationship of phloem cells and the vascular fibers: Completely covered by sclerenchymatous fibers

Nature of lysigenous cavity and protoxylem: Lysigenous cavity absent and enlarged protoxylem vessel present

Size of metaxylem vessels in relation to parenchyma sheath cells in cross-section: Narrow vessels

Vascular bundle sheath

Third-order vascular bundle's bundle sheath (Fig. 1.98D)

Shape: Circular

Shape of bundle sheath cell: Outer tangential wall inflated; radial walls straight

Second-order vascular bundle's bundle sheath (Fig. 1.98C)

Shape: Circular

Shape of bundle sheath cell: Fan shaped

Number of bundle sheath cell: 11–12

Nature of bundle sheath (either complete or incomplete): Complete

Chloroplast position in bundle sheath cell: Chloroplast positioned toward inner tangential wall

Mestome/Inner sheath: Complete and cells are smaller than the outer sheath cell

First-order vascular bundle's bundle sheath (Fig. 1.98B, arrowhead)

Shape: Circular

Extent of bundle sheath around the vascular bundle: Complete

Extension of bundle sheath: Absent

Number of cells comprising the bundle sheath: 12–14 parenchyma cells

Bundle sheath cell shape: Fans shaped; chloroplast positioned toward inner tangential wall

Mestome/Inner sheath: Complete and cells are smaller than the outer sheath cell

Sclerenchyma

Adaxial sclerenchyma: Minute subepidermal strands present

Abaxial sclerenchyma: Well-developed girder, wider than deep

Sclerenchyma between bundles: Absent

Sclerenchyma in leaf margin: Forming very small pointed cap; epidermal cells at margin are thin with thick outer tangential wall and showed papillae

Leaf-sheath anatomy (Fig. 1.98F–H)

Outline of lamina: Round shaped with inrolled margins; ribs and furrows absent

Epidermis

Adaxial epidermal cells: Flat, rectangular

Abaxial epidermal cells: Small, flat, rectangular shaped with thick cuticle

Ground tissue: Parenchymatous with absence of air cavity

Vascular bundle

Arrangement of vascular bundle: One to two consecutive third-order vascular bundles between first and second-order vascular bundle

Third-order vascular bundle: Hexagonal shaped with vascular tissue consists of only a few indistinguishable vascular elements (Fig. 1.98G)

Second-order vascular bundle: Circular shaped

First-order vascular bundle: Circular with phloem completely surrounded by thick-walled fibers; lysigenous cavity absent and enlarged protoxylem vessel present; circular narrow metaxylem vessels present (Fig. 1.98H)

Vascular bundle sheath (Fig. 1.98G, H, arrowhead)

Shape: Circular shaped

Extent of bundle sheath around the vascular bundle: Incomplete, abaxially interrupted by sclerenchyma cells

Number of cells comprising the bundle sheath: 9–10 parenchyma cells

Bundle sheath cell shape: Fan shaped and cells smaller than the ground tissue

Mestome/Inner sheath: Complete and cells are smaller than the outer sheath cell

Sclerenchyma: In form of straight horizontal abaxial strand (Fig. 1.98H, arrow)

Ligule anatomy (Fig. 1.98I, J)

Shape: Ligule attached to the midrib and parts of lamina region and broad "V" shaped at connection of point (Fig. 1.98I, arrow)

Anatomy: Single layered mesophyll cells between adaxial and abaxial surfaces (Fig. 1.98J)

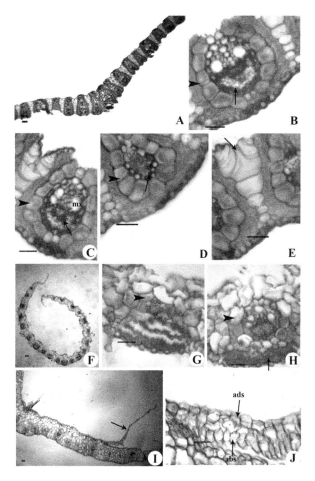

FIGURE 1.98   Leaf anatomy of *Desmostachya bipinnata.*

84. *Dinebra retroflexa*

Leaf lamina (Fig. 1.99A–F)

Outline of the lamina: Expanded with straight arms (Fig. 1.99A). Two halves are symmetrical on either side of the median vascular bundle. Ribs and furrows are present on the leaf surface. Medium, narrow furrows between all vascular bundles and rounded ribs present over all vascular bundles. Keel is inconspicuous.

Epidermis: Adaxial epidermal surface has fan-shaped bulliform cells in which central cell is inflated, large shaped (Fig. 1.99F, arrow); prickles present (Fig. 1.99C, arrow); abaxial epidermal cells are small, barrel to square shaped covered with thick cuticle.

Mesophyll: Irregularly, vertically arranged chlorenchyma cells continuous between vascular bundles; colorless cells absent.

Vascular bundles: Four to five first-order vascular bundles in half lamina regularly arranged from median to marginal bundle; five to six third-order bundles between consecutive larger bundles and two to three second-order bundles between consecutive first-order bundles present; all the vascular bundles present at same level and placed centrally.

Third-order vascular bundles: Hexagonal (Fig. 1.99F) with vascular tissue consisting of only few indistinguishable vascular elements

Second-order vascular bundles: Hexagonal shaped (Fig. 1.99E)

First-order vascular bundles (Fig. 1.99D)

Shape: Circular

Relationship of phloem cells and the vascular fibers: Completely covered by sclerenchymatous fibers

Nature of lysigenous cavity and protoxylem: Lysigenous cavity and enlarged protoxylem vessel present

Size of metaxylem vessels in relation to parenchyma sheath cells in cross-section: Narrow vessels

Shape of metaxylem vessels: Circular

Vascular bundle sheath

Third-order vascular bundle's bundle sheath (Fig. 1.99F)

Shape: Circular

Shape of bundle sheath cell: Elliptical and elongated

Second-order vascular bundle's bundle sheath (Fig. 1.99E, arrowhead)

Shape: Circular

Shape of bundle sheath cell: Elliptical and elongated

Number of bundle sheath cell: 6–7

Nature of bundle sheath (either complete or incomplete): Complete

Chloroplast position in bundle sheath cell: Chloroplast positioned toward outer tangential wall

Mestome/Inner sheath: Absent

First-order vascular bundle's bundle sheath (Fig. 1.99D, arrowhead)

Shape: Circular

Extent of bundle sheath around the vascular bundle: Incomplete, adaxially interrupted by sclerenchyma cells

Extension of bundle sheath: Absent

Number of cells comprising the bundle sheath: 10–11

Bundle sheath cell shape: Elliptical shaped; chloroplast positioned toward inner tangential wall

Mestome/Inner sheath: Complete and cells are smaller than the outer sheath cell

Sclerenchyma

Adaxial sclerenchyma: Minute subepidermal strands present (Fig. 1.99D, arrow)

Abaxial sclerenchyma: Well-developed girder, wider than deep

Sclerenchyma between bundles: Absent

Sclerenchyma in leaf margin: Forming very small pointed cap; epidermal cells at margin are thin with thick outer tangential wall and showed papillae

Leaf-sheath anatomy (Fig. 1.99G–J)

Outline of lamina: Round shaped with inrolled margins (Fig. 1.99G); ribs and furrows absent

Epidermis

Adaxial epidermal cells: Flat, rectangular

Abaxial epidermal cells: Small, flat, rectangular shaped with thick cuticle

Ground tissue: Parenchymatous with absence of air cavity

Vascular bundle

Arrangement of vascular bundle: One consecutive third-order vascular bundle between first and second-order vascular bundle

Third-order vascular bundle: Hexagonal with vascular tissue consisting of only few indistinguishable vascular elements (Fig. 1.99I)

Second-order vascular bundle: Circular shaped

First-order vascular bundle: Circular with phloem completely surrounded by thick-walled fibers; lysigenous cavity and enlarged protoxylem vessel present; circular narrow metaxylem vessels present (Fig. 1.99H)

Vascular bundle sheath (Fig. 1.99H, arrowhead)

Shape: Circular shaped

Extent of bundle sheath around the vascular bundle: Incomplete, abaxially interrupted by sclerenchyma cells

Number of cells comprising the bundle sheath: 12–13 parenchyma cells

Bundle sheath cell shape: Fan shaped; cells smaller than the ground tissue

Mestome/Inner sheath: Complete and cells are smaller than the outer sheath cell

Sclerenchyma: In form of straight horizontal band girder on abaxial surface

Ligule anatomy (Fig. 1.99K, L)

Shape: Ligule attached to the midrib and parts of lamina region and broad round shaped at connection of point (Fig. 1.99K, arrow)

Anatomy: Two layered mesophyll cells between adaxial and abaxial surfaces (Fig. 1.99L)

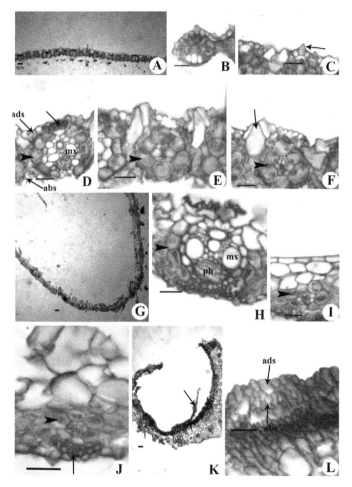

FIGURE 1.99   Leaf anatomy of *Dinebra retroflexa.*

85. *Eleusine indica*

Leaf lamina (Fig. 1.100A–E)

Outline of the lamina: V shaped, broad angle with straight arms (Fig. 1.100A); two halves are symmetrical on either side of the median vascular bundle; ribs and furrows are absent on the leaf surface.

Keel is well developed, V shaped with single median vascular bundle; small, sclerenchyma cells are present surrounding adaxial to the median vascular bundle; sclerenchyma on abaxial surface associated with vascular bundles in form of girders.

Epidermis: Adaxial epidermal surface has fan-shaped bulliform cells (Fig. 1.100C, arrow); abaxial epidermal cells are small, barrel to square shaped covered with thick cuticle; epidermal cells have papillae (Fig. 1.100D).

Mesophyll: Radiate, single layered, isodiametric chlorenchyma cells completely surrounding bundles; compact homogenous chlorenchymatous cells occupying the major area between the adaxial and abaxial epidermises; chlorenchyma cells continuous between bundles; colorless cells present which are smaller than bulliform cells and present near the bulliform cells.

Vascular bundles: Three to four first-order vascular bundles in half lamina regularly arranged from median to marginal bundle; five to six third-order bundles between consecutive larger bundles and two to three second-order bundles between consecutive first-order bundles present; all the vascular bundles present at same level and placed abaxially.

Third-order vascular bundles: Hexagonal (Fig. 1.100E) with vascular tissue consisting of only few indistinguishable vascular elements

Second-order vascular bundles: Elliptical shaped

First-order vascular bundles (Fig. 1.100D)

Shape: Circular

Relationship of phloem cells and the vascular fibers: Completely covered by sclerenchymatous fibers

Nature of lysigenous cavity and protoxylem: Lysigenous cavity absent and enlarged protoxylem vessel present

Size of metaxylem vessels in relation to parenchyma sheath cells in cross-section: Narrow vessels

Shape of metaxylem vessels: Circular

Median vascular bundle (Fig. 1.100C)

Shape: Circular

Relationship of phloem cells and the vascular fibers: Completely covered by sclerenchymatous fibers

Nature of lysigenous cavity and protoxylem: Lysigenous cavity absent and enlarged protoxylem vessel present

Size of metaxylem vessels in relation to parenchyma sheath cells in cross-section: Narrow vessels

Vascular bundle sheath

Third-order vascular bundle's bundle sheath (Fig. 1.100E, arrowhead)

Shape: Circular

Shape of bundle sheath cell: Fan shaped

Second-order vascular bundle's bundle sheath

Shape: Circular

Shape of bundle sheath cell: Fan shaped

Number of bundle sheath cell: 8–9

Nature of bundle sheath (either complete or incomplete): Complete

Chloroplast position in bundle sheath cell: Chloroplast positioned toward inner tangential wall

Mestome/Inner sheath: Absent

First-order vascular bundle's bundle sheath (Fig. 1.100D, arrowhead)

Shape: Elliptical

Extent of bundle sheath around the vascular bundle: Complete

Extension of bundle sheath: Absent

Number of cells comprising the bundle sheath: 12–13 parenchyma cells

Bundle sheath cell shape: Fan shaped; chloroplast positioned toward inner tangential wall

Mestome/Inner sheath: Complete and cells are smaller than the outer sheath cell

Midrib vascular bundle (Fig. 1.100C)

Shape: Elliptical

Extent of bundle sheath around the vascular bundle: Complete

Extension of bundle sheath: Absent

Number of cells comprising the bundle sheath: 12–13 parenchyma cells

Bundle sheath cell shape: Fan shaped; chloroplast positioned toward inner tangential wall

Mestome/Inner sheath: Complete and cells are smaller than the outer sheath cell

Sclerenchyma

Adaxial sclerenchyma: Minute subepidermal strands present

Abaxial sclerenchyma: Well-developed girder, wider than deep

Sclerenchyma between bundles: Absent

Sclerenchyma in leaf margin: Forming very small rounded cap; epidermal cells at margin are thin with thick outer tangential wall

Leaf-sheath anatomy (Fig. 1.100F–I)

Outline of lamina: "U" shaped with inrolled margins (Fig. 1.100F); ribs and furrows absent

Epidermis

Adaxial epidermal cells: Flat, rectangular

Abaxial epidermal cells: Small, flat, rectangular shaped with thick cuticle; stomata present (Fig. 1.100I, arrow)

Ground tissue: Parenchymatous with absence of air cavity

Vascular bundle

Arrangement of vascular bundle: One to two consecutive third-order vascular bundles between first and second-order vascular bundle

Third-order vascular bundle: Pentagonal with vascular tissue consisting of only few indistinguishable vascular elements (Fig. 1.100H)

Second-order vascular bundle: Elliptical shaped

First-order vascular bundle: Elliptical with phloem completely surrounded by thick-walled fibers; lysigenous cavity absent and enlarged protoxylem vessel present; circular narrow metaxylem vessels present (Fig. 1.100G)

Vascular bundle sheath (Fig. 1.100G, arrowhead)

Shape: Elliptical

Extent of bundle sheath around the vascular bundle: Incomplete, abaxially interrupted by sclerenchyma cells

Number of cells comprising the bundle sheath: 11–12 parenchyma cells

Bundle sheath cell shape: Fan shaped and cells smaller than the ground tissue

Mestome/Inner sheath: Complete and cells are smaller than the outer sheath cell

Sclerenchyma: In form of straight horizontal band of girder on abaxial surface (Fig. 1.100G, H, arrow)

Ligule anatomy (Fig. 1.100J, K)

Shape: Ligule attached to the midrib and broad round shaped at connection point (Fig. 1.100J, arrow)

Anatomy: Two layered mesophyll cells between adaxial and abaxial surfaces (Fig. 1.100K)

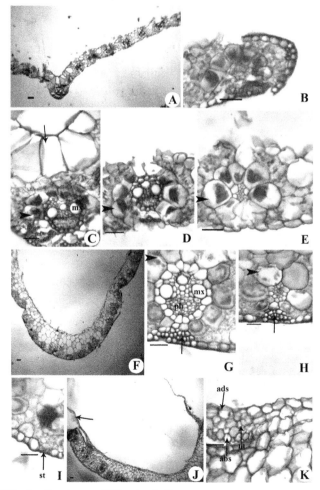

FIGURE 1.100   Leaf anatomy of *Eleusine indica.*

Eragrostiella

86. *Eragrostiella bachyphylla* (Fig. 1.101)

87. *Eragrostiella bifaria* (Fig. 1.102)

Leaf lamina (Figs. 1.101A–E, 1.102A–F)

Outline of the lamina: "V" shaped, wide angled with straight arms (Figs. 1.101A and 1.102A); two halves are symmetrical on either side of the median vascular bundle; ribs and furrows are present on the leaf surface; shallow, narrow furrows present between all vascular bundles with flat ribs present over all vascular bundles; keel is inconspicuous.

Epidermis: Adaxial epidermal surface has fan-shaped bulliform cells, elongated with parallel side in both species (Figs. 1.101C, arrow, 1.102C, arrow). Adaxial surface has small, barrel-shaped cells with thick cuticle. Sharp macrohairs present in base of bulliform cells in *E. bachyphylla* (Fig. 1.102D, arrow). Abaxial epidermal cells are round to barrel shaped covered with thick cuticle.

Mesophyll: Incompletely radiate chlorenchyma cells present between abaxial and adaxial epidermises; colorless cells present, associated with bulliform cells and smaller than bulliform cells with single extension, extended till opposite epidermis in both species (Figs. 1.101C, arrow, 1.102C, arrow).

Vascular bundles: One to two first-order in *E. bachyphylla* or three to four first-order in *E. bifaria* vascular bundles regularly arranged from median to marginal bundle in half lamina. One to two third-order in *E. bachyphylla* or two to three third-order in *E. bifaria* bundles between consecutive larger bundles and one to two second-order bundles between consecutive first-order bundles present. All the vascular bundles present at same level and placed centrally.

Third-order vascular bundles (Figs. 1.101B, 1.102D): Circular in *E. bachyphylla* or elliptical in *E. bifaria* with vascular tissue consisting of only few indistinguishable vascular elements

Second-order vascular bundles (Fig. 1.101E): Circular in *E. bachyphylla* and elliptical in *E. bifaria*

First-order vascular bundles (Figs. 1.101F, 1.102F)

Shape: Circular in *E. bachyphylla* or elliptical in *E. bifaria*

Relationship of phloem cells and the vascular fibers: Completely surrounded by thick-walled fibers

Nature of lysigenous cavity and protoxylem: Lysigenous cavity and enlarged protoxylem vessel present

Size of metaxylem vessels in relation to parenchyma sheath cells in cross-section: Narrow vessels

Shape of metaxylem vessels: Circular

Vascular bundle sheath

Third-order vascular bundle's bundle sheath (Figs. 1.101B, 1.102E)

Shape: Circular in *E. bachyphylla* and elliptical in *E. bifaria*

Shape of bundle sheath cell: Fan shaped

Second-order vascular bundle's bundle sheath (Fig. 1.102E)

Shape: Circular in *E. bachyphylla* and elliptical *E. bifaria*

Shape of bundle sheath cell: Fan shaped

Number of bundle sheath cell: 7–8

Nature of bundle sheath (either complete or incomplete): Incomplete, adaxially and abaxially interrupted by sclerenchyma cells in *E. bachyphylla*, only on abaxially interrupted by sclerenchyma cells in *E. bifaria*

Chloroplast position in bundle sheath cell: Chloroplast positioned toward inner tangential wall

Mestome/Inner sheath: Complete and inner sheath cells are smaller than the outer bundle sheath cell

First-order vascular bundle's bundle sheath (Figs. 1.101D, 1.102F)

Shape: Circular in *E. bachyphylla* and elliptical in *E. bifaria*

Extent of bundle sheath around the vascular bundle: Incomplete, adaxially and abaxially interrupted by sclerenchyma cells

Extension of bundle sheath: Anchored shaped well-developed girder present presented toward adaxially and abaxially in *E. bachyphylla*, only on abaxially interrupted by sclerenchyma cells in *E. bifaria*

Number of cells comprising the bundle sheath: 8–9 in *E. bachyphylla* and 9–10 *E. bifaria*

Bundle sheath cell shape: Fan shaped; chloroplast positioned toward inner tangential wall

Mestome/Inner sheath: Complete in *E. bifaria* and incomplete, adaxially and abaxially interrupted by sclerenchyma cells in *E. bachyphylla*; inner sheath cells are smaller than the outer bundle sheath cell

Sclerenchyma

Adaxial sclerenchyma: Well-developed arched shaped girder present over vascular bundles in *E. bachyphylla* while equidimensional girder present over vascular bundles in *E. bifaria*

Abaxial sclerenchyma: Well-developed arched shaped girder present below vascular bundle

Sclerenchyma between bundles: Absent

Sclerenchyma in leaf margin: Forming well-developed pointed fibrous cap curved toward adaxial surface (*E. bachyphylla*) and not curved in *E. bifaria*; epidermal cells are small, thin walled (Fig. 1.102B)

Leaf-sheath anatomy (Figs. 1.101F–I, 1.102G–J)

Outline of lamina: "V" shaped, wide angled in *E. bachyphylla* and round shaped in *E. bifaria* with inrolled margins (Figs. 1.101F, 1.102G); ribs and furrows absent on leaf sheath surface

Epidermis

Adaxial epidermal cells: Flat, rectangular

Abaxial epidermal cells: Small, barrel shaped with thick cuticle

Ground tissue: Parenchymatous with absence of air cavity

Vascular bundle

Arrangement of vascular bundle: One to two in *E. bachyphylla* or two to three in *E. bifaria* consecutive third-order vascular bundle between first and second-order vascular bundle

Third-order vascular bundle (Figs. 1.101H, 1.102I): Circular with vascular tissue consisting of only few indistinguishable vascular elements

Second-order vascular bundle: Circular shaped

First-order vascular bundle (Figs. 1.101G, 1.102H)

Circular with phloem completely surrounded by thick-walled fibers; lysigenous cavity and enlarged protoxylem vessel present; circular narrow metaxylem vessel elements

Vascular bundle sheath (Figs. 1.101G, arrowhead, 1.102H, arrowhead)

Shape: Circular

Extent of bundle sheath around the vascular bundle: Sheath incomplete, abaxially interrupted by sclerenchyma cells

Number of cells comprising the bundle sheath: 8–9 parenchyma cells

Bundle sheath cell shape: Fan shaped

Mestome/Inner sheath: Complete made up of by sclerenchyma cells, inner bundle sheath smaller than outer bundle sheath

Sclerenchyma: In form of well-developed anchor-shaped girder on abaxial surface in both species (Figs. 1.101I, arrow, 1.102J, arrow)

Ligule anatomy (Figs. 1.101J, K, 1.102K, L)

Shape: Ligule attached to the half width of the leaf sheath region and it is round shaped at the connection point in both species (Figs. 1.101J, 102K)

Anatomy: Single in *E. bachyphylla* or two in *E. bifaria*-layered mesophyll cells between adaxial and abaxial surfaces (Figs. 1.101K, 1.102L)

DIFFERENTIATING FEATURES

Circular shaped vascular bundles, "V"-shaped leaf sheath … *E. bachyphylla*
Elliptical shaped vascular bundles, round-shaped leaf sheath … *E. bifaria*

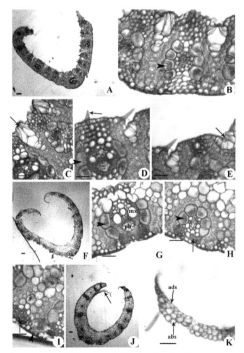

FIGURE 1.101   Leaf anatomy of *Eragrostiella bachyphylla.*

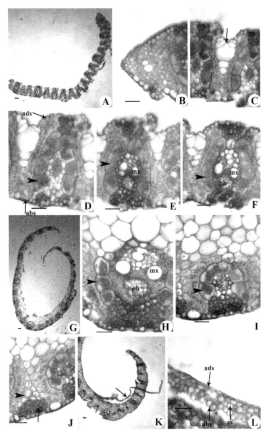

FIGURE 1.102 Leaf anatomy of *Eragrostiella bifaria.*

Eragrostis

88. *Eragrostis cilianensis* (Fig. 1.103)

89. *Eragrostis ciliaris* (Fig. 1.104)

90. *Eragrostis japonica* (Fig. 1.105)

91. *Eragrostis nutans* (Fig. 1.106)

92. *Eragrostis pilosa* (Fig. 1.107)

93. *Eragrostis tenella* (Fig. 1.108)

94. *Eragrostis tremula* (Fig. 1.109)

95. *Eragrostis unioloides* (Fig. 1.110)

96. *Eragrostis viscosa* (Fig. 1.111)

Leaf lamina (Figs. 1.103A–F, 1.104A–D, 1.105A–D, 1.106A–D, 1.107A–F, 1.108A–F, 1.109A–F, 1.110A–G, 1.111A–E)

Outline of the lamina: Expanded with gently straight arms in *E. cilianensis* and *E. ciliaris* (Figs. 1.103A, 1.104A), "V" shaped wide angled with straight arms in *E. japonica, E. nutans, E. pilosa, E. tenella,* and *E. unioloides* (1.105A, 1.106A, 1.107A, 1.108A, 1.109A) while inrolled in *E. tremula* and *E. viscosa.*

Two halves are asymmetrical on either side of the median vascular bundle; ribs and furrows are present on the leaf surface; shallow, wide furrows in *E. cilianensis, E. ciliaris, E. japonica,* and *E. nutans*; narrow, medium furrows in *E. pilosa, E. tenella, E. tremula, E. unioloides,* and *E. viscosa* present between all vascular bundles. Flat ribs present in *E. pilosa* and *E. tenella* while other species have round ribs present over all vascular bundles.

Keel is inconspicuous in *E. tremula* while other species showed well developed, V shaped with single median vascular bundle in *E. cilianensis, E. ciliaris, E. japonica, E. nutans, E. pilosa, E. tenella,* and *E. viscosa* while in *E. unioloides* round shaped. Small, round parenchyma cells are present surround the median vascular bundle adaxially. Sclerenchyma is present on abaxial surface associated with vascular bundles in form of narrow girders in all species except in *E. unioloides, E. viscosa.*

Epidermis: Adaxial epidermal surface has fan-shaped bulliform cells in all species (Figs. 1.103C, arrow, 1.104B, 1.105C, 1.106C, arrow, 1.107E, 1.108F, 1.107F). Abaxial epidermis has round to barrel shaped covered with thick cuticle. Short straight hook is present in *E. tenella* (Fig. 1.108E); short, slender hair present in *E. tremula* (Fig. 1.109E); long, slender macrohairs present in *E. unioloides*; short macrohairs present in *E. viscosa* (Fig. 1.111B, arrow); prickles present in *E. japonica* (Fig. 1.105D, arrow).

Abaxial epidermal cells have barrel shaped with thick cuticle. In *E. unioloides,* cuticle is equal to or greater than the depth of the average epidermal cells.

Mesophyll: Radiate, single layered, tubular shaped chlorenchyma cells present between abaxial and adaxial epidermis in *E. cilianensis, E. tremula, E. unioloides,* and *E. viscosa* or indistinctly or incompletely radiate chlorenchyma continuous between bundles and in form of horizontally elongated strap in *E. nutans, E. pilosa,* and *E. tenella* while irregular, vertically arranged chlorenchyma in *E. ciliaris* and *E. japonica.*

Colorless cells absent in *E. ciliaris, E. japonica, E. nutans, E. tremula,* and *E. viscosa* while it was present in *E. cilianensis, E. unioloides,* and *E.*

tenella. They are associated with bulliform cells and smaller than bulliform cells with single extension.

Vascular bundles: Three to four in *E. cilianensis*, *E. ciliaris*, and *E. tenella*, two to three in *E. pilosa*, *E. japonica*, and *E. nutans*, three in *E. tremula* and *E. viscosa*, four in *E. unioloides* first-order vascular bundles regularly arranged from median to marginal bundle in half lamina.

Two to three in *E. tremula* and *E. viscosa*, three to four in *E. japonica* and *E. nutans*, three in *E. unioloides*, four to five in *E. ciliaris* and *E. tenella*, five to six in *E. pilosa*, seven to eight in *E. cilianensis* third-order bundles between consecutive larger bundles.

Four to five in *E. cilianensis*, two to three in *E. pilosa*, *E. viscosa*, *E. ciliaris*, *E. japonica*, and *E. nutans*, three to four in *E. tenella*, one to two in *E. tremula*, three in *E. unioloides* second-order bundles between consecutive first-order bundles. All the vascular bundles present abaxially at same level.

Third-order vascular bundles (Figs. 1.103E, 1.105C, 1.106B, C, 1.107D, 1.108F, 1.109C, 1.111D): Elliptical in *E. cilianensis* and *E. unioloides*, pentagonal in *E. viscosa*, hexagonal in *E. nutans* and *E. pilosa*, circular in *E. ciliaris*, *E. japonica*, and *E. tenella*, triangular in *E. tremula* with vascular tissue consisting of only few indistinguishable vascular elements

Second-order vascular bundles (Figs. 1.106E, 1.107F): Circular in *E. cilianensis*, *E. ciliaris*, *E. japonica*, *E. nutans*, *E. pilosa*, and *E. tenella*, triangular in *E. tremula*, elliptical in *E. unioloides* and *E. viscosa*

First-order vascular bundles (Figs. 1.103D, 1.104B, 1.105B, 1.106F, 1.107C, 1.108E, 1.109D, 1.111E)

Shape: Elliptical

Relationship of phloem cells and the vascular fibers: Completely surrounded by thick-walled fibers

Nature of lysigenous cavity and protoxylem: Lysigenous cavity and enlarged protoxylem vessel present

Size of metaxylem vessels in relation to parenchyma sheath cells in cross-section: Wide vessels in all species except in *E. pilosa*, *E. viscosa* showed narrow vessels

Shape of metaxylem vessels: Circular

Median vascular bundle (Figs. 1.103C, 1.104C, 1.106D, 1.107C)

Shape: Elliptical in all species except round in *E. unioloides*

Relationship of phloem cells and the vascular fibers: Completely surrounded by thick-walled fibers

Nature of lysigenous cavity and protoxylem: Lysigenous cavity and enlarged protoxylem vessel present

Size of metaxylem vessels in relation to parenchyma sheath cells in cross-section: Wide vessels in all species except in *E. pilosa*, *E. viscosa* showed narrow vessels

Vascular bundle sheath

Third-order vascular bundle's bundle sheath (Figs. 1.103E, 1.105C, 1.106B, C, 1.107D, 1.108F, 1.109C, 1.111D)

Shape: Horseshoe shaped in *E. cilianensis* and *E. pilosa*, round in *E. ciliaris*, *E. japonica*, and *E. nutans*, triangular in *E. tenella*, *E. tremula*, and *E. viscosa*, elongated in *E. unioloides*

Shape of bundle sheath cell: Fan shaped in *E. cilianensis*, elliptical in *E. ciliaris*, *E. japonica*, *E. nutans*, *E. tremula*, and *E. unioloides*, radial walls straight, tangential walls inflated in *E. pilosa* and *E. tenella*, elliptical and elongated cells with two basal cells toward abaxial surface are large in *E. viscosa*.

Second-order vascular bundle's bundle sheath

Shape: Horseshoe shaped in *E. cilianensis*, *E. nutans*, *E. pilosa*, and *E. tenella*, triangular in *E. ciliaris*, *E. tremula* and *E. viscosa*, round in *E. japonica*, *E. unioloides*

Shape of bundle sheath cell: Fan shaped in *E. cilianensis*, elliptical and elongated in *E. ciliaris*, *E. japonica*, *E. unioloides*, and *E. tremula*, elliptical and elongated cells with two basal cells toward abaxial surface are large in *E. viscosa*, radial walls straight, tangential walls inflated in *E. tenella*, *E. pilosa*, and *E. nutans*.

Number of bundle sheath cell: 7–8 in *E. pilosa*, *E. japonica*, and *E. nutans*, 9–10 in *E. tenella* and *E. cilianensis*, 8–9 in *E. ciliaris*, *E. tremula*, and *E. viscosa*, 16 in *E. unioloides*, 11–12 in *E. japonica*

Nature of bundle sheath (either complete or incomplete): Complete in *E. cilianensis E. ciliaris*, *E. japonica*, *E. viscosa*, and *E. unioloides* while incomplete, abaxially interrupted by sclerenchyma in *E. pilosa*, *E. tenella*, *E. tremula*, and *E. nutans*

Chloroplast position in bundle sheath cell: Chloroplast positioned toward inner tangential wall in *E. cilianensis*, translucent chloroplast in *E. ciliaris*, chloroplast fills entire lumen in *E. unioloides*, chloroplast placed centrally in *E. japonica*, *E. nutans*, *E. pilosa*, *E. tenella*, *E. tremula*, and *E. viscosa.*

Mestome/Inner sheath: Absent in *E. cilianensis*, *E. ciliaris*, *E. japonica*, *E. tenella*, and *E. viscosa* while present in other species; complete, inner bundle sheath smaller than outer bundle sheath in *E. pilosa*, *E. nutans*, and *E. unioloides*

First-order vascular bundle's bundle sheath (Figs. 1.103D, 1.104B, 1.105B, 1.106F, 1.107C, 1.108E, 1.109D, 1.111E)

Shape: Horseshoe shaped in *E. cilianensis*, *E. pilosa*, and *E. tenella*, round in *E. ciliaris* and *E. japonica*, elliptical in *E. nutans*, *E. tremula*, *E. unioloides*, *E. cilianensis*, and *E. ciliaris*, triangular in *E. viscosa*

Extent of bundle sheath around the vascular bundle: Complete in *E. cilianensis*, *E. japonica*, and *E. viscosa* while incomplete, abaxially interrupted by sclerenchyma cells in *E. ciliaris*, *E. nutans*, *E. pilosa*, *E. tenella*, and *E. tremula* or abaxially interrupted by sclerenchyma, adaxially interrupted by parenchyma cells in *E. unioloides.*

Extension of bundle sheath: Absent in *E. cilianensis*, *E. japonica*, and *E. viscosa.* Narrow, small girder present on abaxial side in *E. ciliaris*, *E. nutans*, *E. pilosa*, and *E. tenella* while broad abaxial extension of the bundle sheath in *E. tremula* in *E. unioloides.*

Number of cells comprising the bundle sheath: 11–12 in *E. cilianensis* and *E. ciliaris*, 10–11 in *E. pilosa* and *E. tenella*, 13–14 in *E. tremula* and *E. unioloides*, 12–13 in *E. viscosa* and *E. japonica*, and 16–17 *in E. nutans.*

Bundle sheath cell shape: Fan shaped in *E. cilianensis*, elliptical and elongated cells in *E. japonica*, *E. nutans*, *E. tremula*, and *E. unioloides.* Elliptical cells and cells toward adaxial side are inflated and bigger in size than the others in *E. ciliaris*; cells are with straight radial walls, inflated tangential walls in *E. pilosa* and *E. tenella*, cells elliptical and elongated with two basal cells toward abaxial surface are large in *E. viscosa.*

Chloroplast positioned toward inner tangential wall in *E. cilianensis*, chloroplast translucent in *E. ciliaris*, chloroplast placed centrally in *E. japonica*, *E. nutans*, *E. pilosa*, *E. tenella*, *E. tremula*, and *E. viscosa*, chloroplast fills entire lumen of cell in *E. unioloides.*

Mestome/Inner sheath: Present in all species except in *E. tremula* it is absent. Complete, inner bundle sheath smaller than outer bundle sheath in

E. cilianensis, *E. nutans*, *E. pilosa*, *E. tenella*, *E. unioloides*, and *E. viscosa* while incomplete, abaxially interrupted by sclerenchyma and inner bundle sheath smaller than outer bundle sheath in *E. ciliaris* and *E. japonica*.

Sclerenchyma

Adaxial sclerenchyma: Absent in *E. japonica*. Narrow strands present over vascular bundles in *E. cilianensis*, subepidermal strands present in *E. ciliaris*, minute strand consisting of only a few subepidermal fibers in *E. nutans*, *E. pilosa*, *E. tenella*, *E. tremula*, and *E. viscosa* and well-developed girder, wider than deep and it follows shape of abaxial rib in *E. unioloides*

Abaxial sclerenchyma: Well-developed straight horizontal strands present in *E. cilianensis*, *E. pilosa*, and *E. tremula*, well-developed girder, wider than deep girder present below first-order vascular bundle in *E. ciliaris*, *E. japonica*, *E. nutans*, *E. pilosa*, *E. tenella*, and *E. unioloides*, as wide as tall in *E. tremula* and well-developed narrow strand in *E. viscosa*

Sclerenchyma between bundles: Absent in all species

Sclerenchyma in leaf margin: Forming very small pointed fibrous cap in all species except round shaped in *E. tremula*, *E. unioloides*; epidermal cells are thin walled, elongated, arranged side by side

Leaf-sheath anatomy (Figs. 1.103H–J, 1.104E–H, 1.105G–I, 1.106G–I, 1.107G–J, 1.108G–J, 1.109G–J, 1.110H–K, 1.111F–G)

Outline of lamina: V shaped, standard angle with overlapping inrolled margins in *E. cilianensis*, *E. nutans*, *E. pilosa*, *E. tenella*, and *E. unioloides* (Figs. 1.103H, 1.106G, 1.107G, 1.108G, 1.110H) or round shaped with inrolled margin in *E. ciliaris*, *E. japonica*, *E. tremula*, and *E. viscosa* (Figs. 1.104E, 1.105E, 1.109G, 1.111F).

Ribs and furrows present in all species except in *E. unioloides* and *E. viscosa* showed it is absent. Shallow, wide furrows present with rounded ribs present over all vascular bundles except medium, wide furrows present between all vascular bundles in *E. japonica*.

Epidermis

Adaxial epidermal cells: Flat, rectangular

Abaxial epidermal cells: Small, barrel shaped with thick cuticle

Ground tissue: Parenchymatous with absence of air cavity in *E. cilianensis*, *E. ciliaris*, *E. tenella*, and *E. tremula*; round-shaped air cavities in

E. unioloides and *E. viscosa* or oval-shaped air cavities in *E. pilosa*, *E. japonica*, and *E. nutans*

Vascular bundle

Arrangement of vascular bundle: One to two consecutive third-order vascular bundles between first and second-order vascular bundle

Third-order vascular bundle: Elliptical in *E. cilianensis* and *E. unioloides*, round in *E. ciliaris*, *E. japonica*, *E. nutans*, *E. pilosa*, and *E. viscosa*, hexagonal in *E. tenella*, pentagonal in *E. tremula* with vascular tissue consisting of only few indistinguishable vascular elements

Second-order vascular bundle: Circular shaped in *E. cilianensis* and *E. ciliaris*, elliptical in *E. japonica*, *E. nutans*, *E. pilosa*, *E. tenella*, *E. tremula*, *E. unioloides*, and *E. viscosa*

First-order vascular bundle: Circular in all species except in *E. pilosa* and *E. viscosa* it is elliptical; phloem completely surrounded by thick-walled fibers; lysigenous cavity and enlarged protoxylem vessel present; circular narrow metaxylem vessel elements in *E. cilianensis*, *E. pilosa*, *E. tenella*, and *E. unioloides* while other species showed wide metaxylem vessel elements

Vascular bundle sheath

Shape: Circular in all species except elliptical in *E. unioloides* and horseshoe shaped in *E. viscosa*

Extent of bundle sheath around the vascular bundle: Sheath incomplete, abaxially interrupted by sclerenchyma cells in all species except in *E. unioloides* it is complete

Number of cells comprising the bundle sheath: 10–11 in *E. cilianensis* and *E. ciliaris*, 8–9 in *E. pilosa*, 17–18 in *E. tremula*, 9–10 in *E. tenella*, 11–13 in *E. japonica*, *E. nutans*, and *E. viscosa*, 22–23 in *E. unioloides* parenchyma cells

Bundle sheath cell shape: Fan shaped in *E. cilianensis*; elliptical cells in *E. ciliaris*, *E. japonica*, *E. nutans*, *E. tremula*, *E. unioloides*, and *E. viscosa*; cells with straight radial walls; inflated tangential walls in *E. pilosa* and *E. tenella*

Mestome/Inner sheath: Present in all species except in *E. tremula* it is absent; complete, inner bundle sheath smaller than outer bundle sheath in *E. cilianensis*, *E. nutans*, *E. pilosa*, *E. tenella*, and *E. viscosa* while incomplete, abaxially interrupted by sclerenchyma and inner bundle sheath smaller than outer bundle sheath in *E. ciliaris*, *E. japonica*, and *E. unioloides*.

Sclerenchyma: In form of straight, horizontal band in *E. cilianensis*, *E. pilosa*, well developed, wider than deep girder in *E. ciliaris*, *E. japonica*, and *E. nutans*, straight horizontal well-developed girder in *E. tenella*, triangular shaped girder in *E. tremula*, anchored shaped girder present below vasucalr bundle present in *E. unioloides*, well-developed narrow girder in *E. viscosa*

Ligule anatomy (Figs. 1.103K, L, 1.104I, J, 1.105J, K, 1.106K, L, 1.107K, L, 1.108K, L, 1.109L, M, 1.110L, M, 1.111H, I)

Shape: Ligule attached to the lamina more than half-length of it and blunt narrow "V" shaped at the connection point in *E. cilinensis* and *E. japonica*, ligule attached to the lamina till the three-fourth than is free and blunt, broad "V" shaped at the connection in *E. ciliaris* (Fig. 1.104I) while ligule attached to the lamina only one-fourth portions and at connection point it was very almost like straight "V" shaped with small sharp angled in *E. nutans* (Fig. 1.106K).

Ligule attached to the midrib region and partially with lamina region, and the connection, it is broad round shaped in *E. pilosa*, *E. tenella*, and *E. unioloides*, ligule attached only to the midrib region and broad round shaped at connections point in *E. tremula* (Fig. 1.109J) or ligule attached to the half width of the leaf sheath region and it is round shaped at the connection point in *E. viscosa* (Fig. 1.111H).

Anatomy: Two in *E. cilinensis*, three in *E. nutans*, single in *E. pilosa* and *E. tenella* or more than three in *E. ciliaris*, *E. japonica*, *E. tremula*, *E. unioloides*, and *E. viscosa*-layered mesophyll cells between adaxial and abaxial surfaces. Sclerenchyma cells present toward abaxial surface in *E. viscosa* (Fig. 1.111I, arrowhead)

DIFFERENTIATING FEATURES

Inrolled leaf blade
Colorless cells present, centrally placed chloroplast in mesophyll cells … *E. tremula*
Colorless cells absent, chloroplast concentrated toward outer tangential wall of mesophyll cells … *E. viscosa*
Expanded leaf blade
Round-shaped outline of leaf sheath … *E. ciliaris*

"V"-shaped outline of leaf sheath

"V"-shaped ligule…….…………….…………………….…… *E. cilianensis*

Round-shaped ligule

Air cavities absent in leaf sheath………......…..……………..... *E. tenella*

Round-shaped air cavities present in leaf sheath……..………… *E. unioloides*

"V" shaped leaf blade

Vascular bundles are arranged toward abaxial side of leaf…......…..*E. pilosa*

Vascular bundles are placed center of leaf

Lysigenous cavity absent, chloroplast placed centrally of mesophyll cells….

……………………………………………………………….…… *E. japonica*

Lysigenous cavity present, chloroplast concentrated near inner wall of tangential wall of mesophyll cells…………………………………… *E. nutans*

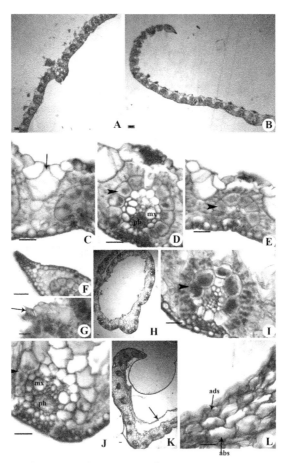

FIGURE 1.103   Leaf anatomy of *Eragrostis cilianensis.*

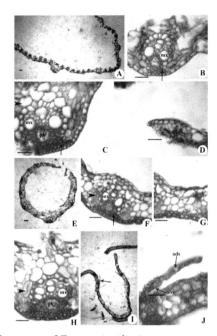

FIGURE 1.104   Leaf anatomy of *Eragrostis ciliaris.*

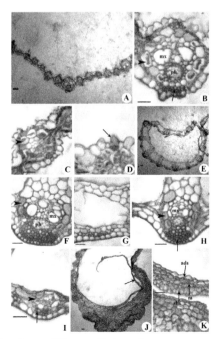

FIGURE 1.105   Leaf anatomy of *Eragrostis japonica.*

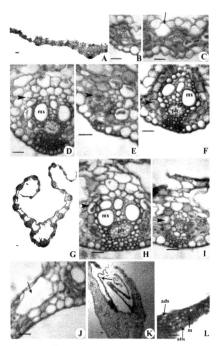

FIGURE 1.106   Leaf anatomy of *Eragrostis nutans.*

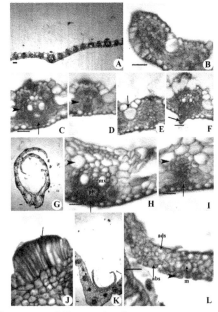

FIGURE 1.107   Leaf anatomy of *Eragrostis pilosa.*

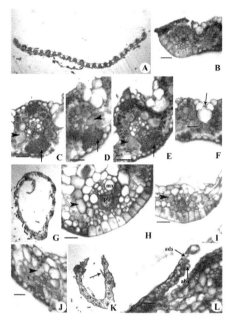

FIGURE 1.108   Leaf anatomy of *Eragrostis tenella*.

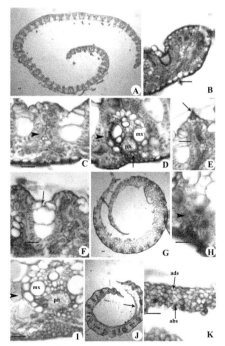

FIGURE 1.109   Leaf anatomy of *Eragrostis tremula*.

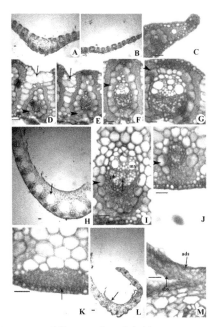

FIGURE 1.110   Leaf anatomy of *Eragrostis unioloides.*

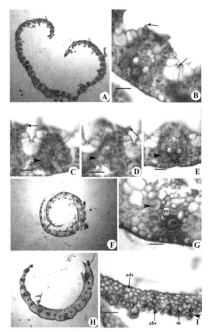

FIGURE 1.111   Leaf anatomy of *Eragrostis viscosa.*

Sporobolus

97. *Sporobolus coromandelianus* (Fig. 1.112)

98. *Sporobolus diander* (Fig. 1.113)

99. *Sporobolus indicus* (Fig. 1.114)

Leaf lamina (Figs. 1.112A–E, 1.113A–E, 1.114A–F)

Outline of the lamina: Expanded with flat, straight leaf blade in *S. coro-mandelianus* (Fig. 1.112A). "V" shaped with wide angle and outwardly curving, convex-shaped arms in *S. diander* (Fig. 1.113A) or "V" shaped with broad angle and distinctly undulating with rounded wave in *S. indicus* (Fig. 1.114A).

Two halves are symmetrical on either side of the median vascular bundle. Ribs and furrows are present on the leaf surface. Narrow, wide furrows present between all vascular bundles and ribs are flat in *S. coromandelianus*, on adaxial surface medium, deep furrow present between all vascular bundle in *S. diander*. Round ribs with obtuse apex situated over the vascular bundles placed adaxially in *S. indicus*.

Keel is inconspicuous in *S. coromandelianus* and *S. diander* but well developed in *S. indicus*, semicircular with single median vascular bundle.

Small, round parenchyma cells surround the median vascular bundle abaxially. Sclerenchyma in form of narrow adaxial strands and on abaxial surface associated with vascular bundles in form of subepidermal strands.

Epidermis: Adaxial epidermis has bulliform cells which resembles fan shaped due to central large cell in all species (Figs. 1.112C, 1.113F, 1.114C, arrow). Epidermal cells are small, round, and highly cuticularized. Adaxial epidermal cells in *S. diander* frequently papillate with length of papillae much lesser than half the width of the cell and one papillae per cell (Fig. 1.113E, arrow). Bulliform cells of *S. indicus* papillate with swollen tip (Fig. 1.114C, arrow). Abaxial epidermal cells small, barrel shaped in *S. coromandelianus*, round to square in *S. diander* or round in *S. indicus* covered with cuticle.

Mesophyll: Indistinctly or incompletely radiate chlorenchyma cells forming horizontal strap continuous between bundles in *S. coromandelianus* and *S. diander* or radiate, single layered, isodiametric chlorenchyma cells completely surrounding bundles in *S. indicus*. Compactly arranged homogenous chlorenchyma cells occupy the major area between the adaxial and abaxial epidermises. Well-defined group of colorless cells in *S. indicus* and

S. coromandelianus (Figs. 1.112B, arrow, 1.114D, arrow) which are closely associated with bulliform cells and girders extend to the opposite epidermis in *S. indicus*, smaller than bulliform cells with one extension in *S. coromandelianus*; colorless cells absent in *S. diander.*

Vascular bundles: Four first-order in *S. coromandelianus*, three first-order in *S. diander* or two first-order in *S. indicus* vascular bundles regularly arranged from median to marginal bundle in half lamina.

Three to four third and two to three second-order in *S. coromandelianus*, two third and two second-order in *S. diander*, two to three third and second-order in *S. indicus* bundles between consecutive larger bundles between consecutive first-order bundles present. All the vascular bundles present at same level and placed centrally.

Third-order vascular bundles (Figs. 1.112D, 1.113C, 1.114F): Circular in all species except in *S. indicus* hexagonal with vascular tissue consisting of only few indistinguishable vascular elements

Second-order vascular bundles: Circular shaped

First-order vascular bundles (Figs. 1.112E, 1.113B, 1.114E)

Shape: Elliptical

Relationship of phloem cells and the vascular fibers: Completely surrounded by thick-walled fibers

Nature of lysigenous cavity and protoxylem: Enlarged protoxylem vessel present in all species and lysigenous cavity present only in *S. indicus*

Size of metaxylem vessels in relation to parenchyma sheath cells in cross-section: Narrow vessels

Shape of metaxylem vessels: Circular

Vascular bundle sheath

Third-order vascular bundle's bundle sheath (Figs. 1.112D, 1.113C, arrowhead, 1.114E)

Shape: Round

Shape of bundle sheath cell: All cells are similar in size with straight radial walls and inflated tangential walls in *S. coromandelianus* and *S. diander* or fan shaped in *S. indicus*.

Second-order vascular bundle's bundle sheath

Shape: Round

Shape of bundle sheath cell: Radial walls straight, tangential walls inflated in *S. coromandelianus* and *S. diander* or fan shaped in *S. indicus*

Number of bundle sheath cell: 8–9

Nature of bundle sheath (either complete or incomplete): Complete

Chloroplast position in bundle sheath cell: Chloroplast fills entire lumen of cell in *S. coromandelianus* and *S. diander* or positioned toward outer tangential wall in *S. indicus*

Mestome/Inner sheath: Absent

First-order vascular bundle's bundle sheath (Figs. 1.112E, 1.113B, arrowhead, 1.114E, arrowhead)

Shape: Elliptical

Extent of bundle sheath around the vascular bundle: Complete

Extension of bundle sheath: Narrow abaxial extension of the bundle sheath in *S. coromandelianus* and *S. diander* or absent in *S. indicus*

Number of cells comprising the bundle sheath: 10 in *S. coromandelianus*, 15–16 in *S. diander*, 12–13 in *S. indicus* parenchyma cells

Bundle sheath cell shape: All cells are similar in size with straight radial walls and inflated tangential walls and chloroplast fill entire lumen of cell in *S. coromandelianus* and *S. diander* while fan shaped with chloroplast positioned toward outer tangential wall in *S. indicus*

Mestome/Inner sheath: Absent

Sclerenchyma

Adaxial sclerenchyma: Minute strand consisting of only a few subepidermal fibers in *S. coromandelianus* and *S. indicus*, well-developed girder, wider than deeper in *S. diander*

Abaxial sclerenchyma: Well-developed girder, wider than deep in *S. coromandelianus*, minute small strand in *S. diander* or minute strand consisting of only a few subepidermal fibers in *S. indicus*

Sclerenchyma between bundles: Absent

Sclerenchyma in leaf margin: Forming well-developed pointed fibrous cap in *S. coromandelianus* and *S. diander*, only pointed cap in *S. indicus*. Epidermal cells at margin are with thick outer tangential wall (Fig. 1.112B)

Leaf-sheath anatomy (Figs. 1.112F–H, 1.113H–J, 1.114G–K)

Outline of lamina: Round with inrolled margins in *S. coromandelianus* and *S. diander* or "V" shaped broad angle with inrolled margins in *S. indicus* (Figs. 1.112F, 1.113H, 1.114G); ribs and furrows absent

Epidermis

Adaxial epidermal cells: Flat, rectangular

Abaxial epidermal cells: Small, flat, rectangular in *S. coromandelianus* and *S. indicus* or round shaped in *S. diander* with thick cuticle

Ground tissue: Parenchymatous with absence of air cavity

Vascular bundle

Arrangement of vascular bundle: Two third-order vascular bundles in *S. coromandelianus* and *S. indicus* and one third-order in *S. diander* between first and second-order vascular bundle

Third-order vascular bundle (Figs. 1.112G, 1.113J, 1.114H, I): Elliptical in *S. coromandelianus*, circular in *S. diander*, and hexagonal in *S. indicus* with vascular tissue consisting of only few indistinguishable vascular elements in *S. coromandelianus* and *S. indicus* while xylem and phloem groups distinguishable in *S. diander*

Second-order vascular bundle: Elliptical shaped

First-order vascular bundle (Figs. 1.112H, 1.113I, 1.114J): Elliptical in *S. coromandelianus* and *S. diander* and circular in *S. indicus* with phloem completely surrounded by thick-walled fibers; lysigenous cavity and enlarged protoxylem vessel present; circular narrow metaxylem vessel in *S. coromandelianus* and *S. diander*; wide metaxylem vessel in *S. indicus*

Vascular bundle sheath (Figs. 1.112H, 1.114K)

Shape: Elliptical in *S. coromandelianus* and *S. diander* and circular in *S. indicus*

Extent of bundle sheath around the vascular bundle: Sheath incomplete and abaxially interrupted by sclerenchyma cells in *S. coromandelianus* and *S. indicus* while complete in *S. diander*

Number of cells comprising the bundle sheath: 8 parenchymatous cells in *S. coromandelianus*, 14–15 celled in *S. diander* and *S. indicus*

Bundle sheath cell shape: All cells similar in size and shape with straight radial walls, inflated tangential walls in *S. coromandelianus*; cells toward abaxial surface are smaller than other cells in *S. diander* (Fig. 1.113I, arrowhead); fan shaped and adaxially interrupted by parenchyma cells and abaxially interrupted by sclerenchyma cells in *S. indicus*

Mestome/Inner sheath: Absent in *S. coromandelianus* and *S. indicus* while in *S. diander* it forms incomplete sheath with cells abaxially sclerenchymatous (Fig. 1.113I)

Sclerenchyma: In form of straight horizontal well developed in *S. coromandelianus* and *S. diander* or small in *S. indicus* abaxial girder

Ligule anatomy (Figs. 1.112I, J, 1.113K, L, 1.114L, M)

Shape: Ligule attached to the midrib region and partially with lamina with the connection being round and broad in *S. coromandelianus* and *S. diander* (Figs. 1.112I, 1.113K) or attached to half the width of the leaf sheath region with a round connection point in *S. indicus* (Fig. 1.114L)

Anatomy: Single in *S. coromandelianus*, two in *S. diander*, three in *S. indicus* layered mesophyll cells between adaxial and abaxial surfaces (Figs. 1.112J, 1.113L, 1.114M)

DIFFERENTIATING FEATURES

Expanded shaped leaf lamina … *S. coromardelianus*
"V" shaped leaf lamina
Chloroplast fill entire full lumen, bundle sheath cells with inflated tangential walls … *S. diander*
Chloroplast concentrated near outer tangential wall, fan shaped vascular bundle sheath cells … *S. indicus*

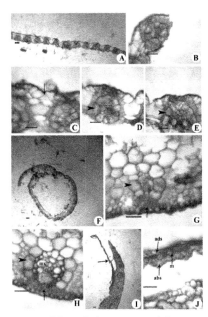

FIGURE 1.112   Leaf anatomy of *Sporobolus coromandelianus*.

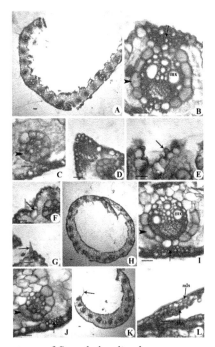

FIGURE 1.113   Leaf anatomy of *Sporobolus diander*.

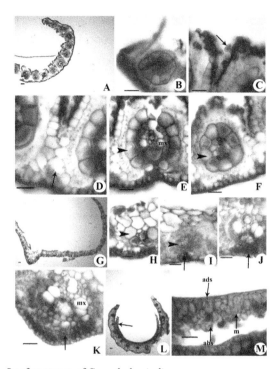

FIGURE 1.114   Leaf anatomy of *Sporobolus indicus.*

100. *Tragrus biflorus*

Leaf lamina (Fig. 1.115A–D)

Outline of the lamina: Expanded with flat, straight leaf blades (Fig. 1.115A, B); ribs and furrows are absent; midrib bundle not distinguishable; keel is inconspicuous.

Epidermis: Adaxial epidermal surface with narrow group of bulliform cells occupying less than half of the leaf thickness (Fig. 1.115C, arrow); bulliform cells with small papillae (Fig. 1.115D, arrow); abaxial epidermal cells round which have thick cuticle.

Mesophyll: Indistinctly or incompletely radiate chlorenchyma continuous between bundles; colorless cells absent.

Vascular bundles: Five first-order vascular bundles in half lamina regularly arranged from median to marginal bundles; two to three third-order bundles between consecutive larger bundles and two to three second-order bundles between consecutive first-order bundles present; all the vascular bundles centrally placed and at same level.

Third-order vascular bundles: Hexagonal and vascular tissue consisting of only few indistinguishable vascular elements (Fig. 1.115C)

Second-order vascular bundles: Circular shaped

First-order vascular bundles (Fig. 1.115D)

Shape: Elliptical

Relationship of phloem cells and the vascular fibers: Phloem adjoins the inner or parenchyma sheath

Nature of lysigenous cavity and protoxylem: Lysigenous cavity present protoxylem vessel absent

Size of metaxylem vessels in relation to parenchyma sheath cells in cross-section: Narrow vessels

Shape of metaxylem vessels: Circular

Vascular bundle sheath

Third-order vascular bundle's bundle sheath (Fig. 1.115C, arrowhead)

Shape: Round

Shape of bundle sheath cell: Radial and inner tangential walls straight; outer tangential wall inflated; fan-shaped, all cells similar in shape and size and chloroplasts concentrated near the inner tangential wall

Second-order vascular bundle's bundle sheath

Shape: Round

Shape of bundle sheath cell: Radial and inner tangential walls straight; outer tangential wall inflated; fan-shaped, all cells similar in shape and size and chloroplasts positioned toward the inner tangential wall

Number of bundle sheath cell: 9–10

Nature of bundle sheath (either complete or incomplete): Incomplete, abaxially interrupted by sclerenchyma

Chloroplast position in bundle sheath cell: Chloroplasts concentrated near the inner tangential wall

Mestome/Inner sheath: Absent

First-order vascular bundle's bundle sheath (Fig. 1.115D, arrowhead)

Shape: Round, circular

Extent of bundle sheath around the vascular bundle: Sheath incomplete, abaxially partially interrupted caused by a narrow girder of one to three fibers wide

Extension of bundle sheath: Absent

Number of cells comprising the bundle sheath: 10–11 parenchyma cells

Bundle sheath cell shape: Radial and inner tangential walls straight; outer tangential wall inflated; fan-shaped, all cells similar in shape and size and chloroplasts positioned toward the inner tangential wall

Mestome/Inner sheath: Sheath complete; completely surrounding the xylem and phloem and adaxially situated cells larger than lateral cells of the sheath

Sclerenchyma

Adaxial sclerenchyma: Minute strand consisting of only a few subepidermal fibers

Abaxial sclerenchyma: Minute strand consisting of only a few subepidermal fibers

Sclerenchyma between bundles: Absent

Sclerenchyma in leaf margin: Forming very small fibrous pointed cap; epidermal cells at margin with outer thick tangential wall

Leaf-sheath anatomy (Fig. 1.115E–G)

Outline of lamina: Round with inrolled margins (Fig. 1.115E); ribs and furrows absent

Epidermis

Adaxial epidermal cells: Flat, rectangular

Abaxial epidermal cells: Small, barrel shaped with thick cuticle

Ground tissue: Parenchymatous with absence of air cavity

Vascular bundle

Arrangement of vascular bundle: One consecutive third-order vascular bundle between first and second-order vascular bundle

Third-order vascular bundle: Hexagonal with vascular tissue consisting of only few indistinguishable vascular elements (Fig. 1.115F)

Second-order vascular bundle: Oval shaped

First-order vascular bundle: Elliptical with phloem adjoins the inner or parenchyma sheath; lysigenous cavity present and protoxylem vessel absent; circular narrow metaxylem vessel elements

Vascular bundle sheath (Fig. 1.115F, G, arrowhead)

Shape: Round, circular

Extent of bundle sheath around the vascular bundle: Sheath incomplete, partially interrupted by a narrow girder of one to three fibers wide

Number of cells comprising the bundle sheath: 10–11 parenchyma cells

Bundle sheath cell shape: Radial and inner tangential walls straight; outer tangential wall inflated; fan-shaped, all cells similar in shape and size

Mestome/Inner sheath: Sheath complete; completely surrounding the xylem and phloem

Sclerenchyma: In form of minute strand consisting of only a few subepidermal fibers present on abaxial surface

Ligule anatomy (Fig. 1.115H, I)

Shape: Ligule attached only to the midrib region and broad round shaped at connections point (Fig. 1.115H, arrow).

Anatomy: Single layered mesophyll cells between adaxial and abaxial surfaces (Fig. 1.115I)

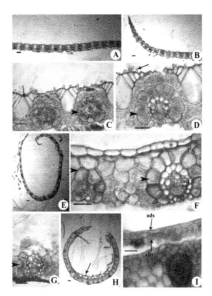

FIGURE 1.115   Leaf anatomy of *Tragrus biflorus*.

1.3 DISCUSSION AND IDENTIFICATION KEY

Initially, grasses were classified by their morphological characteristic features of spike and inflorescence. Prat (1932, 1936) used the microscopic characters of leaf epidermal cells for identification of grasses. Later on, different authors studied aspects like chromosomes, root hairs, stem apices, first seedling leaf, embryo structure, physiology, nucleoli, geographical distribution leaf epidermis, leaf anatomy, and characterized the taxa within a particular family (Avdulov, 1931; Yakovlev, 1950; Stebbins, 1956; Brown and Emery, 1957; Brown et al., 1957; Reeder 1957; Row and Reeder, 1957).

Grass leaf anatomical study has been emphasized as a very fundamental character. Transverse sections of grass leaves are also helpful in the identification and taxonomic delimitation of grasses (Kumar and Nautiyal, 2017). Leaf anatomy of different grass species has been examined by various authors (Renvoize, 2002; Keshavarzi et al., 2007; Abbasi et al., 2010; Abbasi and Asadi, 2014). Duval-Jouve (1875), used character like presence or absence of bulliform cells in both abaxial and adaxial epidermises and position of bulliform cells in respect to vascular bundle. He concludes that if bulliform cells are present over tertiary nerves this was character observed in both Paniceae and Andropogoneae. In Paniceae also, it may be present in both epidermises. This feature was not found in Andropogoneae and hence a specific character of Paniceae. Avdulov (1931) also described two basic types of leaf anatomy: Type I and Type II. The species with characters like thick-walled mestome sheath, connected by sclerenchyma to both epidermises, a poorly developed parenchyma sheath and irregular arrangement of chlorenchyma, were of Type I and the species with characters like the large size of the parenchyma sheath cells which separate the xylem from the sclerenchyma next to upper epidermis, the radial arrangement of chlorenchyma cells belongs to the Type II. Prat (1932) designated Type I as Festucoid type and Type II as Panicoid type of leaf anatomy. Tribes Paniceae, Maydeae, Andropogoneae, Chlorideae, and Zoysieae in the present study belong to the Type II leaf anatomy.

In the present study, the characteristic features used for description are adapted from Ellis (1976) which includes a very detailed description of the different tissues of the leaf, its sheath, and ligule. Characteristic features of the cells which form each tissue (epidermal, mesophyll and vascular) have been critically examined and presented.

Brown (1958) studied leaf blades of 101 species in 72 genera and categorized them into 6 categories based on their plastid structure, presence and absence of mestome, arrangement of mesophyll chlorenchyma cells, and character of outer bundle sheath.

Festucoid type: It consists of well-developed, thick-walled endodermis surrounded by a very indistinct parenchyma sheath, the cells of which are small and very thin walled and contain chloroplasts similar to those of loose, irregularly arranged, spongy mesophyll cells.

Bambusoide type: Characterized by an endodermis, some specialization of the parenchyma sheath cells and modified chlorenchyma tissue.

Arundinoid type: Characterized by the complete lack of chloroplasts in the enlarged parenchyma sheath cells. The chlorenchyma cells are densely packed and the endodermis is poorly developed.

Panicoid type: Characterized by radial arrangement of the adjacent chlorenchyma cells, some species showed absence of mestome or endodermis while some showed presence of it.

Aristicoid type: It is unique type with no mestome sheath but having a double parenchyma sheath, the cells of both containing specialized plastids.

Chloridoid type: Characterized by an endodermis around the large bundles by a single parenchyma sheath, chlorenchyma of long, narrow, radially arranged cells forming one layer and containing few chloroplasts.

Renvoize studied leaf anatomy of different tribes like Andropogoneae (1982c), Arundinelleae (1982b), Garnotideae (1982a), Eragrostideae (1983), Arundinoideae (1986a), Centothecoideae (1986b), Bambuseae (1987b), Paniceae (1987a), and only described the leaf micromorphological character and leaf anatomical characters which includes long cells, short cells, stomata, outline of lamina in transverse section, vascular bundles, epidermis, and mesophyll cells arrangement for the different tribes. Calderón and Soderstrom (1973) studied morphological and anatomical features of *Maclurolyra* belonging to Bambusoideae.

Metcalfe (1960) in his book studied on anatomy of the monocotyledons and included around 315 species of grasses. The 315 grass species studied have been treated in general. He studied and described the micromorphology and anatomy of leaf blade. Some of the observations varied from the present study. Individual species have been characterized anatomically but the different features have not been critically analyzed as done in the present study. *Perotis indica* has anchor-shaped girder which is made up of by sclerenchyma cells, whereas in present study, it showed minute subepidermal strands on abaxial surface. In present study also, species like *Apluda mutica, Chloris virgata, Coix, Dactyloctenium aegyptium, Desmostachya bipinnata, Eleusine indica, Panicum maximum, Panicum miliaceum, Rottbelia exaltata*, *Vetiveria zinzanoides* showed large sclerenchyma bundle on abaxial side. Moreover, only selective diagrammatic representation and no microphotographs have been given. In the present study, each species has been characterized individually and represented with microphotographs.

Patel (1976) examined morphotaxonomic characters in 51 grass species from Ahmedabad districts. He also examined the leaf anatomy of 24 species. However, only 17 species from the present study have been included in his work. Moreover, these species have not been described anatomically in detail and are supplemented with diagrams unlike the present study which describes leaf, leaf sheath, and ligule anatomy of 100 grass species in detail as per Ellis (1976). Martínez-Sagarra et al. (2017) studied leaf anatomy in the Iberian species of *Festuca*. According to them, leaf anatomy has certain limitations for the delimitation of species, although it has taxonomic value for the separation of some groups. Few characters were found to be over-lapping to a great extent, especially number of outer/inner bundle sheath cells, number of bulliform cells, thickness of sclerenchyma, abundance, and length of trichomes. The reason proposed for existence of these characters may be due to their responses to environmental conditions. The anatomical differences between the species of the two major clades are evident and there are many features that distinguish them. Kok and Van der Schiff (1973) proposed a key for the identification of 60 grass species on the basis of leaf anatomical features.

Aliscioni et al. (2016) studied morphology and leaf anatomy of *Setaria* in which characters like well-developed chlorenchyma with irregular or incompletely radiate arrangement, presence of colorless cells, presence of sclerenchyma in form of girders either on adaxial or abaxial surface, presence of features like papillae and mestome has been considered. In the present study, *Setaria* contradicted by showing absence of mestome and papillae, well-developed sclerenchyma in form of girder, irregularly arranged chlorenchyma cells and well-developed, semicircular-shaped keel. *Setaria glauca* and *Setaria tomentosa* had straight arms of leaf blade while *Setaria verticiallata* had gently wavy arms of leaf blade.

Ghahraman et al. (2006) studied anatomy of 25 Iranin species of *Bromus*. They considered 81 total characters were selected for variable and constant species characters, among them 28 characters are variable, whereas 40 characters constant in all species. Variable anatomical quantitative and qualitative characters like shape of lamina in transverse section, width of one-half lamina, thickness, size of adaxial rib, number of adaxial ribs and furrows, depth of adaxial furrow, thickness of keel, length and width of midrib, distance of midrib from abaxial and adaxial surface, number of vascular bundles, arrangement of different orders of vascular bundles, shape of midrib, number of protoxylem, number of lysigenous cavity, size of metaxylem, number of outer sheath cells, radial wall of outer sheath cells, macrohairs density, shape and distribution, and number of bulliform cells. When they evaluate the anatomical characters in the *Bromus* apart from separating and categorizing the perennial and the annuls, no other grouping is shown and the remaining anatomic attributes have only taxonomic value among the species. Kumar and Nautiyal (2017) studied leaf anatomy of two genera, *Eragrostis* and *Eleusine* belonging to tribe Eragrostideae from Uttarakhand. They said that different species exhibit variations in different anatomical characters, which are valuable in their identification and differentiation, some characters, and similar in all species of the tribe. *Eleusine indica* having V-shaped keel, radially arranged chlorenchyma, fan-shaped bulliform cells, or irregular groups deeply penetrating into the mesophyll.

Olsen et al. (2013) studied the environmental and inherited variation in leaf anatomy of *Andropogon gerardii*. They concluded that leaf anatomy was influenced by that environment and genetic source population. Leaves from plants grown in mesic sites had thicker midribs, larger veins, fewer trichomes, and a larger proportion of bulliform cells compared to plants grown in drier situations. Bor (1960) noticed that the blades of grass leaves may be penned up, convoluted, or pleated in the bud. Thus, the arrangements of leaves in a bud may simply be related to the major types of involution exhibited in mature leaves. In the present study, all the members of Maydeae showed expanded leaf lamina. In Andropogoneae, members showed either expanded or "V" shaped leaf lamina except for *Triplopogon ramosissimus* which showed an inrolled leaf lamina. Tribe Paniceae members also generally showed either expanded or "V"-shaped leaf lamina except *Pennnisetum setosum* showed "U"-shaped leaf lamina. Among the tribe Chlorideae, only *Chloris* showed "V" shaped.

Longitudinal ribs and furrows were present either on the adaxial or abaxial surface of a leaf. Depth, shape, spacing, and location of the ribs and furrows varied in the different species. The ribs were usually developed in association with the large vascular bundles and were more developed on the adaxial than the abaxial surface (Metcalfe, 1960). Most of the studied species showed presence of ribs and furrows except *Chionachnae, Coix, Andropogon, Bothriochloa, Chrysopogon, Cymbopogon, Ophiuros, Rottboellia, Saccharum, Sorghum, Thelepogon, Vetiveria, Digitaria, Eremopogon, Paspalidium, Setaria, Acrachne, Eleusine,* and *Tragus. Heteropogon triticeus*, *Triplopogon ramosissimus*, *Alloteropsis cimicina*, *Cynodon*, *Chloris*, *Melanocenchris*, and so on showed smooth ribs on the adaxial surface of the leaf. The ribs were present on first-order vascular bundle and furrows were present below the other vascular bundles. *Cenchrus biflorus* showed prominent smooth ribs abaxially. *Panicum miliaceum*, *Eragrostis* species, and *Sporobolus* species also have smooth ribs on both the surfaces of leaf lamina. *Capillipedium hugelii*, *Sehima ischaemoides*, *Heteropogon contortus* var. *geninus* subvar. *typicus*, *Themeda cymbaria*, *Oplismenus*, and so on possess triangular shaped, sharp ribs present on abaxial surface of leaf lamina. *Iseilema laxum* and *Isachne globosa* have triangular-shaped ribs on the adaxial surface of leaf lamina.

Keel is an important diagnostic feature to recognize and easily distinguish the species. Presence and absence of keel, its size and shape, number of vascular bundles within the keel are some of the important features which help in distinguishing the species. In the studied species, variation was observed in the number of vascular bundles in the keel. The variation was observed in the different genus of the same tribe as well as in different tribes. *Chionachnae koenigii* and *Coix lachryma-jobi* of tribe Maydeae showed many vascular bundles within keel. Species like *Bothriochloa pertusa*, *Chrysopogon fulvus*, *Ischaemum*, *Saccharum spontanum*, *Thelopogon elegans*, *Themeda*, and so on, of tribe Andropogoneae also had many vascular bundles in the keel. But other species from Andropogoneae like *Apluda mutica*, *Capillipedium huegelli*, *Heteropogon*, *Sehima*, *Sorghum*, and so on had a single vascular bundle in keel. *Chloris virgata*, *Chloris montana*, *Paspalidium flavidum*, *Digitaria longiflora*, *Cenchrus ciliaris*, *Alloteropsis cimiciana*, and so on also had single vascular bundle in keel. Species like *Oplismenus*, *Panicum trypheron*, *Setaria tomentosa*, *Setaria verticillata*, *Chloris*, *Eleusine*, and so on had many vascular bundles in keel region. The keel was protruded abaxially either because of sclerenchyma or because of the vascular bundle which was located closer to abaxial surface. Shape of the keel also varied, it was either "V" shaped or "U"/semicircular shaped.

Generally, epidermal cells in the leaves of all the grass species were barrel, square, rectangular, or round shaped which was covered with either thin or thick cuticle. Jauhar and Joshi (1967) studied the epidermal features of four species of *Panicum*. He pointed out a wide range of variation existed in respect to the length of the long cells and the extent of rippling and cuticularization of their anticlinal walls, shape of short cells, cuticular appendages, frequencies, and size of stomata in the different species. In the present study, the characteristic features of cuticularization have been considered in detail.

In some species, few epidermal cells showed papilla which were either sharp, blunt, or balloon shaped and broad. Width of papillae was equal to the width of epidermal cell in some species like *Bothriochloa pertusa*, *Heteropogon* sp., *Sehima* sp., *Digitaria ciliaris*, *Echinochloa crusgalli*, *Chloris montana*, *Chloris virgata*, and *Acrachne ramosa* showed blunt end/round papillae which were as wide as the epidermal cells. *Capillepedium hugleii*, *Ischaemum molle*, *Ischaemum rugosum*, *Themeda cymbaria*, *Themeda triandra* showed sharp papillated epidermal cells. Very few species showed *Eremopogon faveolatus*, *Dactyloctenium aegyptium*, and *Dactyloctenium scindicus* balloon-shaped long papillae.

Presence of bulliform cells in most of the grass species is a well-known feature. Features like the shape/form of the bulliform cells, its arrangement, presence on both epidermises or on a single epidermis, number of bulliform cells served as a characteristic diagnostic feature for identification of the species. Their morphology combined with enlarged mesophyll, colorless cells have been used as taxonomic characteristics (Metcalfe, 1960). Function of bulliform cells is different according to different authors (Haberlandt, 1914; Mauseth, 1988; Moulia, 2000). It is mainly considered for water storage (Eleftheriou and Noitsakis, 1978; Vecchia et al., 1998), or it can participate in the young leaf expansion that is rolled in the apex, due to water stress (Shields, 1951; Jane and Chiang, 1991). Esau (1965) said that during excessive water loss, the bulliform cells become flaccid and enabled the leaf either to fold or to roll, in conjugation with or without colorless cells. Alvarez et al. (2008) studied structure of bulliform cells in relation to its function in *Tristachya leiostachya* and *Loudetiopsis chrysothrix*. According to their study, bulliform cells appeared to play an active role during the leaf movement. Thus, involving the bulliform cells in the involution mechanism of young and mature leaves. Clayton and Renvoize (1986) suggested that bulliform cells favored the light enhance in the hygroscopic leaf movement because they accumulated large amount of silicon and their outermost walls might thicken and cutinize becoming stiff (Ellis, 1976). Longhi-Wagner (2001) studied two native grasses *Loudetiopsis chrysothrix* and *Tristachya leiostachya* and showed that leaf rolling occurred under water stress and features like epicuticular waxes, silica bodies in the coastal zone and stomata in furrows can be related to water saving (Alvarez et al., 2005). Alvarez et al. (2008) studied the anatomical and ultrastructural characterization of the bulliform cells in these plants and described the main characteristic features as the periclinal walls being thinner than the adjacent epidermal wall, abundance of pectic substance in cuticular layer, sinuous anticlinal walls with ramified plasmodesmata, vacuum formed by a developed vacuole or in numerous small vacuoles, abundance of phenolic substances and oil drops, all of these characteristics suggesting its involvement in the mechanism of foliar involution. Considering the taxonomic and likely ecophysiological role of bulliform cells, their structural details have been studied in the 100 grass species selected in the present study. Fisher (1939) found the keel-size, amount of parenchymatous and sclerenchymatous tissue and bulliform cells to be a valuable diagnostic character in separating four species of *Chloris*. In the present study, *Chloris* species showed fan-shaped bulliform cells which were similar to observation made by Fisher (1939) but shape of central cell differed. In *Chloris barbata*, it was diamond shaped, *Chloris virgata* shield shaped and *Chloris montana* inflated round shaped.

Breakwell (1914) on the basis of the leaf anatomy divided six grass species into three groups, based on their vascular bundle character, sclerenchyma mass developed under different bundles, epidermal cells, and bulliform cells; group I had three species—*A. interiuedius, A. affinis, A. cericeus*; group II had *A. ischaemum* and group III had *A. refractus* and *A. intermedius*. In the present study, *Andropogon plumis* also shared few similar characters like fan-shaped bulliform cells, a well-developed sclerenchymatous girder on abaxial surface and thick cuticle. Most of the studied species showed fan-shaped bulliform cells, for example, *Eragrostis, Sporobolus diander, Andropogon, Imperata, Iseilema, Sehima, Oplismenus, Panicum, Perotis, Tragus, Cynodon, Melanocenchris, Eragrostis, Eragrostiella, Dinebra, Tetrapogon, Desmostachya* and *Eleusine. Soporoblus coromandelians, Ischemum pilosum, Iseilema laxum*, and *Eragroestiella* showed bulliform cells in furrows. Arrangement of the bulliform cells was found to be varying in different species. In some species like *Coix, Bothriochloa, Capillipedium, Chrysopogon, Heteropogon, Ischaemum, Ophiuros, Rottboellia, Sorghum, Thelepogon, Triplopogon, Cenchrus, Digitaria, Eremopogon, Paspalidium, Paspalum*, and *Setaria* showed large, inflated bulliform cells while species like *Chionachane, Apluda, Dicanthium, Saccharum, Sehima*, and *Oplismenus* have restricted group of bulliform cells. Few species showed very small or narrow group of bulliform cells, for example, *Vetiveria, Alloteropsis, Echinochloa.*

Mesophyll cells and its characteristic features were also found to be a diagnostic identifying feature. The mesophyll can be subdivided into the assimilatory chlorenchyma and the colorless parenchyma which consists of translucent cells. In some species, the adaxial chlorenchyma cells are more regular and vertically arranged than the residual. Metcalfe (1960) correctly stresses that this distinction is rare and at best unclear. According to him *Apluda mutica, Cenchrus biflorus, Cenchrus ciliaris, Chloris virgata, Cymbopogon martini, Cynodon dactylon, Echinochloa colona, Echinochloa crusgalli, Eleusine indica, Heteropogon contortus, Imperata cylindrica, Panicum miliaceum, Perotis indica, Setaria glauca, Sorghum halepense,* and *Themeda triandra* had radiate arrangement of chlorenchyma cells, *Chloris barbata, Dacyloctenium aegyptium, Desmostachya bipinnata, Eragrostis unioloides, Panicum maximum, Paspalidium geminatum, Rottboellia exaltata, Sporobolus diander* had incomplete radiate arrangement of chlorenchyma cells while *Coix* had irregular arrangement of chlorenchyma cells. But in present study, *Cenchrus biflorus, Cenchrus ciliaris, Echinochloa colona, Echinochloa crusgalli, Setaria glauca,* and *Paspalidium geminatum* showed irregular arrangement of chlorenchyma cells, *Cynodon dactylon, Coix, Chloris barbata, Sporobolus diander,* and *Perotis indica* showed incomplete radiate arrangement of chlorenchyma cells while *Rottboellia exaltata, Panicum maximum, Eragrostis unioloides, Dacyloctenium aegyptium, Desmostachya bipinnata, Apluda mutica, Chloris virgata, Cymbopogon martini, Eleusine indica, Heteropogon contortus, Panicum miliaceum,* and *Sorghum halepense* showed radiate arrangement of chlorenchyma cells.

    Chonan (1978) studied the shape of mesophyll cell and divided it into and three categories: (1) tubular palisade, (2) longitudinally elongated arm palisade cell, (3) transversely elongated arm palisade cell. Garnier and Laurent (1994) proposed that leaf thickness was similar in annual and perennial grass species but density was different in all and it was higher in perennials. The area occupied by mesophyll in leaf was higher in annuals, at the expense of the three other tissue, that is, epidermis, sclerenchyma, vascular tissue. He also on the basis of the anatomy of mesophyll in leaves of Gramineae classified the grasses into four subfamily groups: Festucoideae, Bambusoideae, Panicoideae, and Eragrostideae based on structure of mesophyll. According to Ellis (1976), the species which showed radiate arrangement of chlorenchyma cells belonged to Panicoid group while those who showed irregular arrangement belonged to festucoid group. *Ophiuros exaltatus, Oplismenus, Vetiveria zinzanoides, Andropogon plumis, Themeda triandra, Bothriochloa pertusa, Arthraxon lanceolatus, Dicanthium annulatum, Dicanthium caricosum, Cymbopogon martini, Ischaemum,* and *Rottboellia exaltata* showed radiate arrangement of chlorenchyma, that is, they are belong to panicoid group while *Setaria glauca, Setaria verticillata, Setaria tomentosa, Eragrostis japonica, Eragrostis nutans, Digitaria ciliaris, Digitaria longiflora, Tragus biflorus, Sporobolus,* and *Aristida* species had irregular chlorenchyma arrangement and belonged to festucoid group. Few species showed incomplete radial arrangement of chlorenchyma cells, for example, *Coix*, *Chionachnae*, *Echinochloa colona*, *Echinochloa crusgalli*, *Echinochloa stagnina*, *Dactyloctenium aegyptium*, *Dactyloctenium scindicus*, and *Acrachne racemosa*.

There was evidence documented for the taxonomic distribution of C_3 and C_4 photosynthetic pathways in the subfamilies of the Poaceae (Kanai and Kashiwagi, 1975; Dengler et al., 1994; Sage and Sage, 2009; Lundgren et al., 2014). In tribe Bambusoideae and group Pooideae with tribe Arundinoidae (from Panicoideae group), being C_3 only few genera have C_4 pathway. Panicoideae and Chlorideae had the C_4 pathway with only few Panicoid genera recorded to being C_3. Makino and Ueno (2018) studied structural and physiological responses of the C_4 pathway in *Sorghum bicolor* to nitrogen limitation. Nitrogen is one of the major nutrients influencing photosynthesis and productivity of C_4 plants as well as C_3 plants. This study revealed the structural response of *Sorghum* leaves to N limitation in relation to the physiological responses of photosynthesis. C_4 photosynthesis is a highly regulated biochemical mechanism achieved by close coordination between mesophyll cells and bundle sheath cells. They proved that structural traits like leaf thickness, stomatal density, chloroplast number and arrangement, cell wall thickness, density of plsmodesmata, vascular bundle sheath are altered along with physiological traits of C_4 photosynthesis. Most changes would be associated with cellular allocation of N, light use, CO_2 diffusion and leakiness, and metabolite transport under N limitation. Hattersley (1984) studied on the characterization of C_4 leaf type anatomy in grasses and described the leaf anatomy with relation to physiology. He observed that evolution of C_4 suitable anatomy might require relatively few changes in plant lineges with anatomical traits across C_3 and C_4 taxa has important implications for the functional diversity of C_4 lineges and for the approaches used to identify genetic determinants of C_4 anatomy. In studied species, most of the genera showed C_4 leaf anatomy except the species *Oplismenus* and *Chionachnae*.

In grasses vascular bundles are arranged centrally within the vertical plane of the blade. In leaf blades of *Ophiuros*, *Sorghum*, *Apluda*, *Haceklochloa*, *Paspalidium flavidum*, *Eragrostis*, *Brachiaria*, *Panicum*, *Iseilema*, *Setaria*, *Themeda*, *Eremopogon*, *Capillipedium huegelii*, *Digitaria*, *Echinochloa*, *Tragus*, and so on had a position of vascular bundles were centrally positioned. But in some grass species, vascular bundles were observed to be located closer to the abaxial or to the adaxial surface. Species like *Chionachnae*, *Coix*, *Heteropogon triticeus*, *Chloris montana*, *Schoenefeldia gracilis*, *Tetrapogon*, *Dactyloctenium*, *Eragrostis*, and so on had vascular bundles arranged closer to abaxial surface. Irregular arrangements of vascular bundles of different orders situated at different levels within the mesophyll were also observed. This positioning of the vascular bundles in the blade appears to be a useful diagnostic character above the genus level that has been largely overlooked in the past. In the present study, few grass species showed irregular arrangement of vascular bundles, in which third-order vascular bundles placed closer to abaxial surface and first-order vascular bundles placed centrally of leaf blade. Species like *Vetiveria zinzanoides*, *Themeda cymbaria*, *Cenchrus biflorus*, *Paspalidium geminatum*, *Chloris barbata*, and so on showed irregular arrangement of vascular bundles.

It was observed that wider leaves do not necessarily have more vascular bundles than narrower leaves. Studied grass species showed large variation in arrangement and alternation of vascular bundles of different orders. For example, *Coix* had 2 second and 2 third-order vascular bundles between 2 consecutive first-order vascular bundles, *Andropogon plumis* had 4 second and 2 third-order vascular bundles between 2 consecutive first-order vascular bundles, *Bothriochloa pertusa* had 2 second and 4 third-order vascular bundles between 2 consecutive first-order vascular bundles, *Cymbopogon martini* had 2 second and 1 third-order vascular bundles between 2 consecutive first-order vascular bundles, *Dicanthium caricosum*, *Heteropogon contortus*, *Heteropogon ritchei* had 3 second and 3 third-order vascular bundles between 2 consecutive first-order vascular bundles, *Heteropogon triticeus* had 1 second and 6 third-order vascular bundles between 2 consecutive first-order vascular bundles, *Ischaemum* showed had 4 second and three third-order vascular bundles between 2 consecutive first-order vascular bundles, *Vetivaria* had 1 second and 3 third-order vascular bundles between 2 consecutive first-order vascular bundles, *Alloteropsis* had 3 second and 10 third-order vascular bundles between two consecutive first-order vascular bundles, *Capillipedium* had 10 vascular bundles between two consecutive first-order vascular bundles, *Ophiuros* had 9 third-order vascular bundles between 2 consecutive first-order vascular bundles but second-order vascular bundles absent, *Apluda mutica*, *Chloris* showed alternate arrangement of second and third-order vascular bundles between 2 consecutive first-order vascular bundles, *Eragrostis* had 1–2 second and 3–4 third-order vascular bundles between 2 consecutive first-order vascular bundles, *Echinochloa colona* had 2 second and 3 third-order vascular bundles between 2 consecutive first-order vascular bundles, *Echinochloa crusgalli* had 4 second and 7 third-order vascular bundles between 2 consecutive first-order vascular bundles. This relative proportion and the alternation of the various sizes of vascular bundle along leaves of any one species of grass are remarkably constant. The distribution patterns of bundles of different orders and their variation are important diagnostic characters.

Generally, three different orders of vascular bundles are present in grasses, which differ in their sizes and presence of vascular tissue type. The first-order bundles according to Metcalfe (1960) are characterized by having a metaxylem vessel on either side of the protoxylem, and as the protoxylem was nonfunctional it was replaced by a lysigenous cavity. A few well-developed protoxylem vessels may be present in addition to the lysigenous cavity, or both cavity and vessels may be absent. In second-order bundles, the xylem and phloem are easily distinguishable, but the large metaxylem vessels are lacking and xylem and phloem is reduced to a few elements or even may be indistinguishable. Vickery (1935) and Goossens (1938) have pointed out that second-order bundles may show a tendency to develop into first-order bundles. In the third-order vascular bundles, metaxylem vessel elements are always lacking and in the smallest bundles of this order, the xylem and phloem elements may be indistinguishable. Third-order vascular bundles were angular, circular, or elliptical in outline. This feature was taxonomically very important. Angular bundles were a characteristic feature of the panicoids and nonangular bundles were the characteristic feature of festucoid grasses. Amongst the studied species most of Panicoideae members had angular or elliptical-shaped third-order vascular bundles and most of Pooideae members had circular-shaped third-order vascular bundles. *Chionachnae, Coix, Andropogon, Chrysopogon, Cymbopogon, Capillepedium, Heteropogon*, and so on of Panicoideae had angular-shaped third-order vascular bundles and species like *Cynodon dactylon, Eragrostis tremula* of Panicoideae had a triangular-shaped third-order vascular bundle, whereas *Brachiaria, Eragrostis, Sporobolus, Dactyloctenium*, and so on of festucoid showed circular-shaped third-order vascular bundle. *Desmostachya, Eragrostiella*, and so on of festucoid showed an elliptical shape of third-order vascular bundles. Second-order vascular bundles also showed shapes like third-order vascular bundle, but most of the species had circular-shaped second-order vascular bundle, that is, the same as festucoid.

Vascular bundles were surrounded by bundle sheath either completely or partially. Sometimes two layers of bundle sheaths were present, with outer one known as outer bundle sheath and inner one as inner bundle sheath or mestome. Bundle sheath distribution and type of cells were very important character for the taxonomic study. Ellis (1974) reported anomalous bundle sheath in *Alloteropsis semialata*. It showed double-bundle sheath with the inner bundle sheath composed of large cells with specialized chloroplast and outer bundle sheath of smaller cells with or without chloroplast. Schwendener (1890) studied the distribution of sclerenchyma between the vascular bundle and epidermises to be of systematic importance and he classified grasses into two groups, based on the presence and absence of mestome sheath. Members of Maydeae, Andropogoneae, and few members of Paniceae showed absence of mestome sheath while members of Pooideae and few members of Paniceae showed presence of mestome. The species with single bundle sheath was characteristic feature of panicoid, whereas the species with double sheaths was the characteristic of festucoid members. Most of the Panicoideae members showed absence of mestome, that is, single bundle sheath except *Brachiaria*, *Eriochloa*, and *Panicum*. From the tribe Pooideae, *Aristida*, *Perotis*, *Chloris*, *Tetrapogon*, *Dactyloctenium*, *Elesusine*, *Eragrostis*, *Ergrostiella*, *Sporobolus*, and *Tragus* showed presence of mestome, that is, double-bundle sheath within the first-order vascular bundle. Brown (1958) differentiates the genus *Aristida* from all other grasses in having each vascular bundle surrounded by two parenchymatous sheaths and with an absence of sclerenchymatous, thick-walled, or mestome sheath being absent. But Metcalfe (1960) is of the opinion that the bundle may be surrounded by three sheaths the innermost being mestome sheath.

Shape of bundle sheath varied with the change in the shape of vascular bundles. The third-order vascular bundles are generally angular, circular, or elliptical shape of bundle sheath was also either circular or elliptical. *Chionachnae*, *Coix*, *Apluda*, *Andropogon*, *Chrysopogon*, *Cymbopogon*, *Dicanthium*, *Hackelochloa*, *Ischaemum*, *Sehima*, *Sorghum*, *Thelepogon*, *Brachiaria*, *Cenchrus*, *Digitaria*, *Eremopogon*, *Eriochloa*, *Oplismenus*, *Setaria*, *Chloris*, *Eragrostis* from group Panicoideae or Pooideae and so on showed round/circular-shaped bundle sheath, *Bothriochloa pertusa*, *Imperata*, *Ophiuros*, *Sachharum*, *Vetiveria*, *Pennisetum*, and *Eragrostis tenella* showed elliptical-shaped bundle sheath while *Perotis*, *Tetrapogon*, *Tragus* showed horseshoe-shaped bundle sheath. Bundle sheath is further also designated as complete or incomplete and if incomplete then it is because of interruption by sclerenchyma or parenchyma, either from abaxial or adaxial or both sides. Species like *Chionachnae*, *Coix*, *Apluda*, *Cymbopogon*, *Hackelochloa*, *Thelepogon*, *Cenchrus*, *Panicum*, *Setaria*, and so on showed complete bundle sheath while species like *Andropogon*, *Bothriochloa*, *Chrysopogon*, *Dicanthium*, *Iseilema*, *Echinochloa*, *Eremopogon*, *Eriochloa*, *Tragus*, and so on showed incomplete bundle sheath because sclerenchyma interrupted from abaxial side. But *Ophiuros*, *Eragrostiella*, and *Vetiveria* showed incomplete bundle sheath from the abaxial and adaxial side. *Eragrostiella* in fact showed abaxial interruption because of sclerenchyma and adaxially because of parenchyma while other two species showed sclerenchyma on both sides.

In the grass leaf blade, the sclerenchymatous tissue includes all fibers as well as other thick-walled cells in certain instances. The sclerenchyma is commonly found in association with the vascular bundles, the midrib, or keel and in the margin. Species like *Hackelochloa granularis, Panicum antidotale, Brachiaria ramosa, Chlrois montana, Capillipedium hugelii, Sporobolus coromandenlianus*, and so on showed absence of sclerenchyma in the margin. *Apluda, Paspalidium flavidum, Eragrostis nutans, Pennisetum setosum, Iseilema laxum, Oplismenus compositus, Panicum maximum, Vetiveria, Themeda triandra, Eremopogon foveolatus, Desmostachya, Tragus*, and so on had a small cap of sclerenchyma a margin while *Chionachnae koenigii, Coix, Ophiuros, Sorghum halepense, Brachiaria eruciformis, Bothriochloa pertusa, Eragrostiella bifaria*, and so on had a well-developed sclerenchyma in the margin. The sclerenchyma associated with the vascular bundles is in the form of subepidermal longitudinal bands following the course of either first-order vascular bundle or each vascular bundle. They were in the form of either strands or girders; the former is usually associated with the third-order vascular bundles and the latter with larger bundles. The distribution and arrangement of the sclerenchyma associated with the bundles were found to be characters of taxonomic interest.

Changes in leaf biochemistry, ultrastructure and plant biomass allocation, and leaf morphological and anatomical modifications are frequent components of plant adjustment to contrasting habitat light conditions (Mendes et al., 2001; Oguchi et al., 2003; Lambers et al., 2004; Mojzes and Kalapos, 2005). The species, which were grown under different habitats, showed variations in their anatomical structure. Species like *Melanocenchris, Eragrostiella bachyphylla*, and *Eragrostiella bifaria* grown in stony or rocky area showed unique characters like incomplete outer bundle sheath which was interrupted from both the sides by sclerenchyma cells, presence of colorless cells which follows bulliform cells and ends till opposite end of epidermis, mestome was either parenchymatous or sclerenchymatous which was either complete or incomplete. But the species like *Coix, Brachiaria* grown in moist area showed characters like presence of colorless cells which generally occupied either half or one-fourth or less than that portion of the leaf thickness, mestome was always made up of parenchyma cells, outer bundle sheath was complete or incomplete, and if incomplete, it was due to interruption of parenchyma or sclerenchyma cells from abaxial surface.

Leaf-sheath anatomy was more or less same as leaf lamina. Most of the characters like adaxial and abaxial epidermis, ground tissue, distribution of vascular bundles, vascular bundle arrangement, vascular bundle structure, vascular bundle sheath structure, and outline of leaf sheath are similar to the structure of leaf lamina. Leaf sheath showed variation in outline, that is, "V"/"U" or round shaped. Among studied species, *Chionachnae*, *Hetero-pogon triticeus*, *Eragrostis pilosa*, *Paspalidium flavidum*, *Eragrostis nutans*, *Pennisetum setosum*, *Brachiaria eruciformis*, *Panicum antidotale*, *Vetiveria zinzanoides*, *Chloris montana*, *Digitaria ciliaris*, *Sachoenefeldia gracilis*, *Bothriochloa pertusa*, and so on showed "V"-shaped outline of leaf sheath while species like *Coix*, *Hackelochloa*, *Eleusine indica*, *Tragus*, *Panicum miliaceum*, *Paspalum scorbiculatum*, *Aristida funiculata*, and so on showed "U"-shaped outline of leaf sheath. Species like *Ophiuros*, *Sorghum*, *Apluda*, *Iseilema*, *Setaria glauca*, *Eragrostis japonica*, *Eragrostis tremula*, *Eragrostis viscosa*, *Eragrostis unioloides*, *Cymbopogon*, *Dicanthium*, and so on showed round-shaped outline of leaf sheath. Unlike leaf lamina, in the ground tissue of leaf sheath, air cavities were present which could be categorized into four different shapes. Some of the species showed an absence of air cavities. *Chionachnae*, *Coix*, *Ophiuros*, *Sorghum*, *Panicum antidotale*, *Vetivaria*, and so on showed vertically oval-shaped air cavities, whereas *Hackelochloa*, *Hetropogon triticeus*, *Paspalidium flavidum*, *Penni-setum setosum*, *Brachiaria eruciformis*, *Setaria glauca*, *Eragrostis japonica*, *Triplopogon*, *Brachiaria distachya*, *Brachiaria ramosa*, and so on showed horizontally oval-shaped air cavities. Species like *Eragrostis unioloides*, *Panicum miliaceum*, *Arthraxon lanceolatus*, *Ischanae globosa*, and so on showed round-shaped air cavities, whereas *Chloris montana*, *Andropogon*, *Themeda triandra*, *Capillipedium*, *Digitaria ciliaris*, *Echinochloa colona*, *Bothriochloa*, *Schoenefeldia*, *Tetrapogon*, *Dactyloctenium*, *Eragrostiella*, *Eragrostis ciliaris*, *Acrachanae ramosa*, *Paspalum*, *Eragrostis tremula*, *Eragrostis viscosa*, and so on showed absence of air cavities.

Grasses with ligules have higher fitness (Moreno et al., 1997; Korzun et al., 1997). Ligule may be membranous or hairy. It may be mainly fully or partially attached to leaf lamina or leaf sheath. The main role of the ligule is to protect the inner side of the leaf sheath from water, dust, and spore but in a few species, it also had a function of secretion (Chaffey, 2000). Chaffey (1984) studied epidermal structure in the ligules of four species of *Poa.* He described long cells, short cells, and unicellular prickle hairs over the surface, with unicellular hairs and papillate cells on the free edges of the ligules. Ligule anatomy is important for identifying individuals living in nonoptional environments (Neumann, 1938). Chaffey (1994) demonstrated the functional anatomy of ligules in 49 grass species from 10 tribes. As chloroplast was found to be present in the ligules studied he has deduced it as a photosynthetic leaf organ. Structurally ligule anatomy is simpler than the leaf lamina. Sometimes, prickle hairs are observed and are present on abaxial surface. Ligules consist of three layers: the mesophyll and two epidermal cell layers with absence of intercellular spaces. Mesophyll is generally present throughout the ligule except at the edges which consisted of epidermal cells only (Neumann, 1938). Neumann also suggested that ligule character is also vital for identifying individual and species living in nonoptimal environments (Pathak et al., 2013). Ligule cells are generally much smaller than other grass plant parts. Generally, the abaxial epidermis appears to consist of parenchyma like cells. Szabo et al. (2006) studied ligule anatomy and morphology of five species of *Poa*, both by light and scanning electron microscopy. They have described the ligule in detail emphasizing interspecific differences and habitat-dependent variation of shape, length, and prickle hair density. They observed that ligules of *Poa anustifloia* are decurrent on the leaf sheath margin and the connection of the ligule blade and the adaxial epidermis of the leaf sheath is a straight line forming a right angle to the longitudinal axis of the leaf sheath while the ligule of *P. pratensis* describes a curve in its connection to the adaxial epidermis of the leaf sheath. This provides a good discriminating character between the two closely related taxa, irrespective of modifications caused by habitat conditions. Among the studied species, mesophyll layers are either one, two, three, or more. Sometimes prickles and sclerenchyma cells were present on abaxial surface. Adaxial cells of the ligule were usually flat, square to rectangular shaped and abaxial cells were square shaped with outer tangential wall round. Species like *Andropogon, Brachiaria reptans, Digitaria granularis, Digitaria longiflora, Eremopogon, Eriochloa, Panicum antidotale, Perotis, Eleusine indica, Eragrostis ciliaris*, and so on showed single layered mesophyll, species like *Setaria, Chloris virgata, Imperata, Ischaemum molle, Iesilema, Ophiuros, Rottboellia, Themeda laxa, Triplopogon, Vetiveria*, and so on showed two layered mesophyll and species like *Apluda, Chrysopogon, Cymbopogon, Sorghum, Alloteropsis, Cenchrus, Dinebra*, and so on showed three or more layered of mesophyll. Few studied species showed presence of sclerenchyma on abaxial surface, that is, *Apluda, Arthraxon, Capillipedium, Dicanthium caricosum, Dicanthium annulatum, Hackelochloa granularis, Ischaemum molle, Ischaemum rugosum, Saccharum, Sorghum halepense, Paspalidium geminatum, Setaria glauca, Setaria verticillata*, and so on. Species like *Heteropogon contortus, Cenchrus setigerus* show presence of prickles on abaxial surface.

Based on the characteristic features of a dichotomous key for the studied species has been prepared.

1. Outline of leaf lamina is "U" shaped ....***Pennisetum setosum***

1. Outline of leaf lamina is "V"/expanded/inrolled shaped **2**

2. Outline of lamina inrolled...**3**

2. Outline of leaf lamina is "V"/expanded shaped.....................................**5**

3. Vascular bundle positioned centrally within
   the leaf blade ....***Triplopogon ramosissimus***

3. Vascular bundle located close to abaxial surface of leaf blade................**4**

4. Vascular bundle sheath cells with chloroplast
   in the center..***Eragrostis tremula***

4. Vascular bundle sheath cell with chloroplast
   adpressed near the outer tangential wall ....***Eragrostis viscosa***

5. Outline of lamina "V" shaped..**6**

5. Outline of lamina expanded...**56**

6. Ribs and furrows present **7**

6. Ribs and furrows absent...**37**

7. Ligule absent..**8**

7. Ligule present...**9**

8. "U"-shaped keel... ***Echinochloa colona***

8. "V"-shaped keel..***Echinochloa crusgalli***

9. "V"-shaped ligule...**10**

9. Round-shaped ligule **18**

10. Well-developed sclerenchyma present at the margin of leaf lamina....**11**

10. Only 4–6 sclerenchyma cells present at the margin of leaf lamina **13**

11. Vertically oval-shaped air cavities present
    in leaf sheath ....***Alloteropsis cimicina***

11. Air cavities absent in leaf sheath..**12**

12. Phloem adjoins with the inner or parenchyma sheath
    cells in first-order vascular bundle............................***Cenchrus biflorus***

12. Phloem completely surrounded by
    thick-walled fibers .... ***Cenchrus ciliaris***

13. Mestome present in first-order vascular bundle.........***Eragrostis nutans***

53. Xylem and phloem distinguishable elements in
    third-order vascular bundles .... *Vetivaria zinzanoides*

53. Vascular tissue consisting of a few indistinguishable
    vascular elements in third-order vascular bundles..............................**54**

54. Square-shaped third-order vascular bundle of
    leaf sheath .... *Eremopogon foveolatus*

54. Pentagonal-shaped third-order vascular bundle of leaf sheath **55**

55. Circular-shaped first-order vascular bundle.......*Themeda qurqdrivalvis*

55. Elliptical-shaped first-order vascular bundle.............*Themeda triandra*

56. Mestome present in first-order vascular bundle.................................**57**

56. Mestome absent in first-order vascular bundle.................................**75**

57. "V"-shaped outline of leaf sheath..**58**

57. Round-shaped outline of leaf sheath...**64**

58. Lysigenous cavity absent in first-order vascular bundle.....................**59**

58. Lysigenous cavity present in first-order vascular bundle **62**

59. Pointed leaf lamina margin **60**

59. Round leaf lamina margin..**61**

60. Radiate arrangement of mesophyll cells............*Brachiaria eruciformis*

60. Irregularly arrangement of mesophyll cells ....*Oropetium villosulum*

61. Chloroplast fill entire full lumen, straight arms of
    leaf blade, single layered mesophyll in ligule.........*Tetrapogon tenellus*

61. Chloroplast concentrated near inner tangential wall, wavy arms
    of leaf blade, two layered mesophyll in ligule........ *Tetrapogon villosus*

62. "V"-shaped ligule... *Echinochloa stagnina*

62. Round-shaped ligule **63**

63. Air cavities absent in leaf sheath.............................. *Eragrostis tenella*

63. Round-shaped air cavities present in leaf sheath . *Eragrostis unioloides*

64. Colorless cells absent **65**

64. Colorless cells present..**73**

65. Ribs and furrows absent...*Tragrus biflorus*

65. Ribs and furrows present **66**

66. Irregular arrangement of mesophyll cells **67**

66. Radiate/incomplete radiate arrangement of mesophyll cells **68**

93. Colorless cells present............................ *Sporobolus coromandelianus*

93. Colorless cells absent...**94**

94. Irregularly arranged mesophyll cells **95**

94. Radiate/incomplete radiate arranged mesophyll cells.........................**98**

95. Inconspicuous keel...*Paspalum scrobiculatum*

95. Well-developed keel...**96**

96. Wide vessels present ....*Setaria glauca*

96. Narrow vessels present **97**

97. Air cavities absent in leaf sheath........................*Oplismenus burmannii*

97. Oval shaped air cavities absent in leaf sheath... *Oplismenus composites*

98. Incomplete radiate arrangement of
    mesophyll cells.. *Arthraxon lanceolatus*

98. Radiate arrangement of mesophyll cells...**99**

99. Chloroplast placed toward outer tangential
    wall of vascular bundle sheath................................*Panicum trypheron*

99. Chloroplast placed at center of vascular
    bundle sheath .... *Iseilema laxum*

1.4 CLUSTER ANALYSIS

The software used displays a single tree among possible ones (Fig. 1.116). In the dendrogram based on the anatomical features of leaf lamina, leaf sheath, and ligule:

I.   Outline of lamina (four criteria)
      (1) Expanded, (2) V shaped, (3) U shaped, and (4) inrolled
II.   Presence and absence of ribs and furrows (two criteria)
      (1) Presence and (2) absence
III. Position of vascular bundles within the leaf blade (five criteria)
      (1) Center of blade, (2) closer to abaxial surface, (3) closer to adaxial surface, (4) first-order bundles central and third-order bundles abaxial, and (5) third-order bundles central and first-order bundles displaced adaxially in ribs
IV.   Bulliform cells (nine criteria)
      (1) Absent, (2) most part of epidermis made up of bulliform cells, (3) groups of bulliform cells present in epidermis, (4) *Zea* type, (5) restricted groups of tall and narrow bulliform cells, (6) fan-shaped bulliform cells, (7) *Sporobolus* type, (8) Arundo type, and (9)

bulliform cells and closely associated colorless cells forming an extensive column or girder extending from the base of an adaxial furrows deep into leaf

V.   Arrangement of chlorenchyma cells (three criteria)
(1) Radiate, (2) indistinctly or incompletely radiating chlorenchyma, and (3) irregular chlorenchyma

VI.   Colorless cells (two criteria)
(1) Present and (2) absent

VII.   Relationship of phloem cells and vascular fibers (three criteria)
(1) Phloem adjoins the inner or parenchyma sheath, (2) phloem completely surrounded by thick-walled fibers, and (3) phloem divided by intrusion of small fibers resulting in sclerosed phloem

VIII.   Nature of lysigenous cavity and protoxylem (four criteria)
(1) Lysigenous cavity and enlarged protoxylem vessel present, (2) lysigenous cavity but no protoxylem vessel present, (3) enlarged protoxylem vessel present but no lysigenous cavity, and (4) no lysigenous cavity or protoxylem vessel present

IX.   Size of metaxylem vessel (three criteria)
(1) Narrow vessel, (2) wide vessel, and (3) very wide vessel

X.   Chloroplast structure of the bundle sheath cells (six criteria)
(1) Translucent, (2) small chloroplast, not distinct, (3) chloroplast fill entire cell lumen, (4) chloroplast concentrated near the outer tangential wall, (5) chloroplast concentrated near the inner tangential wall, and (6) chloroplasts centrally situated

XI.   Mestome (two criteria)
(1) Present and (2) absent

XII.   Sclerenchyma in the margin (three criteria)
(1) Absent, (2) small cap, and (3) well developed

XIII.   Shape of sclerenchymatous cap (six criteria)
(1) Rounded, (2) pointed, (3) narrow, very long pointed, (4) crescent shaped, (5) curved—extending along adaxial side, and (6) curved—extending along abaxial side

XIV.   Outline of leaf sheath (three criteria)
(1) V shaped, (2) U shaped, and (3) round shaped

XV.   Shape of air cavities present in leaf sheath (four criteria)
(1) Vertically oval, (2) horizontally oval, (3) oval shape, and (4) round shape

XVI.   Outline of ligule anatomy (two criteria)
(1) V shaped and (2) round shaped

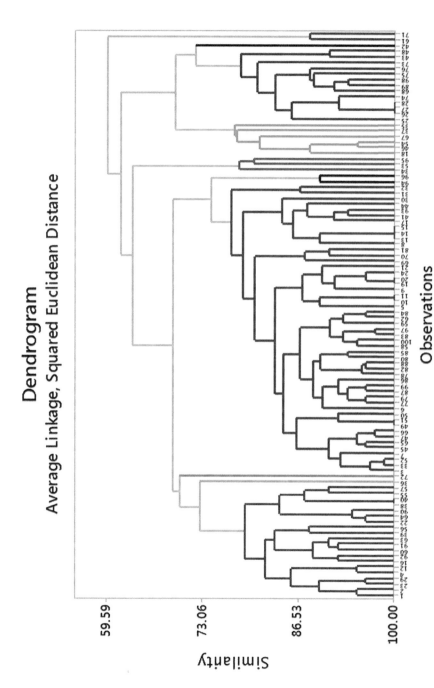

FIGURE 1.116 A dendrogram showing clustering of grass species on the basis of leaf anatomy.

One hundred studied species showed 59.59% similarity with each other. Prepared dendrogram showed in Fig. 1.116 and different values from centroid were shown in Table 1.1. Two major clusters were formed in dendrogram. One major cluster consisted single cluster 8, which contained only two species, that is, *Chloris barbata, Paspalidium geminatum*. Cluster 8 showed 88.16% similarity with other clusters and 1.500 distance from centroid. Other major clusters contain cluster 1, 2, 3, 4, 5, 6, 7, 9, and 10. Among them cluster 6, 7, and 9 were *simplicifolius* clusters, that is, single species in cluster. Cluster 6 had species *Vetiveria zinzanoides*, which showed 73.70% similarity in characters with other species and had 0.00 distances from centroid. Cluster 7 had species *Cenchrus biflorus*, which showed 72.07% similarity with other clusters. Cluster 9 had species *Chloris montana*, which showed 69.86% similarity with other species.

Cluster 10 contained two species, like *Eragrostis tremula* and *Eragrostis viscosa*. Both the species showed 89.72% similarity in characters and had 1.414 maximum value from centroid. Cluster 5 had three species, *Themeda quadrivalvis, Eriochloa procera*, and *Eragrostis unioloides*. They showed 78.39% similarity with other clusters and had 2.380 maximum value form centroid.

TABLE 1.1 Distribution of Leaf into Clusters on the Basis of Their Qualitative Features.

Cluster number	Number of observation (grass species) per cluster	Maximum value from centroid	Average distance from centroid
1	20	2.867	2.527
2	51	3.637	2.503
3	6	2.853	2.432
4	13	2.949	2.377
5	3	2.380	2.308
6	1	0.00	0.00
7	1	0.00	0.00
8	2	1.500	1.500
9	1	0.00	0.00
10	2	1.414	1.414

Cluster 3 contained six species like *Triplopogon ramosissimus, Ischaemum indicum, Digitaria granularis, Oplismenus burmanii, Isachne globosa*, and *Alloteropsis cimicina*. They all showed 77.96% similarity with other species and had 2.853 maximum values from centroid. Species like

Saccharum spontaneum, Sehima ischaemoides, Sehima nervosum, Sehima sulcatum, Cynodon dactylon, Aristida, Eragrostis ciliaris, Sporobolus diander, Melanocenchris, and so on forming cluster 4. This cluster had total 13 species and showed 78.41% similarity with others. Cluster 4 was closer to cluster 7, that is, more affinity toward each other.

Cluster 1 contained 20 species, such as *Chionachnae koengii, Coix, Ophiorus exaltatus, Sorghum halepense, Apluda mutica, Hackelochloa granularis, Heteropoogn triticeous, Eragrostis pilosa, Panicum antidotale, Iseilema laxum, Brachiaria ramose, Brachiaria distachya, Panicum maximum, Oplismenus comositus*, and so on. This cluster showed 79.05% similarity and 2.826 maximum value from centroid.

Cluster 2 was large cluster among all. It contained 51 species and showed similarity 77.02% with the 3.367 maximum value form centroid, for example, species like *Andropogon plumis, Themeda triandra, Digitaria ciliais, Capillepedium hugelii, Setaria tomentosa, Setaria verticillata, Echinochloa colonum, Sachoenefeldia gracilis*, and so on.

KEYWORDS

- **leaf lamina**
- **leaf sheath**
- **ligule**
- **anatomy**
- **panicoids**
- **festucoids**

REFERENCES

Abbasi, M.; Assadi, M. Evaluation of Generic Concept in *Colpodium* Trin. and *Catabrosella* Tzvel. (Tzvel.): A New Subspecies in Poaceae Family in Iran. *Int. J. Biosci.* **2014,** *5,* 50–57.

Abbasi, M.; Assadi, M.; Attar, F.; Nejadsattari, T. Micromorphological and Anatomical Features of *Puccinellia dolicholepis* (Poaceae), a New Record for the Flora of Iran. *Iran. J. Bot.* **2010,** *16,* 267–272.

Aliscioni, S. S.; Ospina, J. C.; Gomiz, N. E. Morphology and Leaf Anatomy of *Setaria* (Poaceae: Panicoideae: Paniceae) and Its Taxonomic Significance. *Plant Syst. Evol.* **2016,** *302* (2), 173–185.

Alvarez, J. M.; Rocha, J. F.; Machado, S. R. Estrutura Foliar de *Loudetiopsis chrysothrix* (Nees) Conert and *Tristachya leiostachya* Nees (Poaceae). *Rev. Bras. Bot.* **2005,** *28,* 23–37.

Alvarez, J. M.; Rocha, J. F.; Machado, S. R. Bulliform Cells in *Loudetiopsis chrysothrix* (Nees) Conert and *Tristachya leiostachya* Nees (Poaceae): Structure in Relation to Function. *Braz. Arch. Biol. Technol.* **2008**, *51* (1), 113–119.

Arber, A. The Phyllode Theory of the Monocotyledonous Leaf, with Special Reference to Anatomical Evidence. *Ann. Bot.* **1918**, *32* (128), 465–501.

Arber, A. Leaves of the Gramineae. *Bot. Gaz.* **1923**, *76*, 374–388.

Avdulov, N. P. Karyo-systematische Untersuchungen der Familie Gramineen (in Russian; German Summary). *Bull. Appl. Bot. Suppl.* **1931**, *44*, 428.

Blackman, E. Observations on the Development of the Silica Cells of the Leaf Sheath of Wheat (*Triticum aestivum*). *Can. J. Bot.* **1969**, *47* (6), 827–838.

Bor, N. L. *The Grasses of Burma, Ceylon, India and Pakistan*; Pergamon Press: London, 1960; pp 767.

Breakwell, E. A Study of the Leaf Anatomy of Some Native Species of the Genus *Andropogon*, N.O. Gramineae. *Proc. Linn. Soc. N.S.W.* **1914**, *39*, 385–394.

Brown, W. V. Leaf Anatomy in Grass Systematics. *Bot. Gaz.* **1958**, *119* (3), 170–178.

Brown, W. V. A Cytological Difference between the Eupanicoideae and the Chloridoideae (Gramineae). *Southwest. Nat.* **1960**, *5*, 7–11.

Brown, W. V. Another Cytological Difference among the Kranz Subfamilies of the Gramineae. *Bull. Torrey Bot. Club* **1974a**, *101* (3), 120–123.

Brown, W. V. Morphological Variation and Chromosome Numbers of North American Populations of *Koeleria cristata*. *Bull. Torrey Bot. Club* **1974b**, *101* (3), 124–128.

Brown, W. V. Variations in Anatomy, Associations, and Origins of Kranz Tissue. *Am. J. Bot.* **1975**, *62*, 395–402.

Brown, W. V. The Kranz Syndrome and Its Subtypes in Grass Systematics. *Mem. Torrey Bot. Club* **1977**, *23* (3), 1–97.

Brown, W. V.; Emery, W. H. Persistent Nucleoli and Grass Systematics. *Am. J. Bot.* **1957**, *44*, 585–590.

Brown, W. V.; Heimsch, C.; Emery, W. H. P. The Organization of the Grass Shoot Apex and Systematics. *Am. J. Bot.* **1957**, *44*, 590–595.

Bugnon, P. *La feuille chez les Graminees*. Theses, University of Paris: Paris, 1921.

Calderón, C. E.; Soderstrom, T. R. Morphological and Anatomical Considerations of the Grass Subfamily Bambusoideae Based on the Genus *Maclurolyra*. *Smith Contr. Bot.* **1973**, *11*, 1–54.

Cerros-Tlatilpa, R. *Estudio sistemático del género Chloris Sw. en México. Tesis de Maestría en Ciencias*. Facultad de Ciencias, Universidad Nacional Autónoma de México: México, DF, 1999; 165 pp.

Chaffey, N. J. Epidermal Structure in the Ligules of Four Species of the Genus *Poa* L. (Poaceae). *Bot. J. Linn. Soc.* **1984**, *89*, 341–354.

Chaffey, N. J. Structure and Function of the Membranous Grass Ligule—A Comparative Study. *Bot. J. Linn. Soc.* **1994**, *116*, 53–69.

Chaffey, N. J. Research Review: Physiological Anatomy and Function of the Membranous Grass Ligule. *New Phytol.* **2000**, *146*, 5–21.

Chakravarty, A. K.; Verma, C. M. Anatomical Study in Arid Zone Grasses of Western Rajasthan. *J. Ind. Bot. Soc.* **1966**, *44*, 506–511.

Chonan, N. Comparative Anatomy of Mesophyll among the Leaves of Gramineous Crops. *Jpn. Agric. Res. Q.* **1978**, *12*, 128–131.

Clayton, W. D.; Renvoize, S. A. *Genera Graminum. Kew Bulletin Adicional Series XIII*, London. 1986; pp 389.

Columbus, J. T. *Lemma Micromorphology, Leaf Anatomy and Phylogenetics of Bouteloua, Hilaria and Relatives (Gramineae: Chlorideae: Boutelouinae).* Ph.D. Dissertation. Univeristy of California, Berkeley, CA, 1996; pp 258.

Crookston, R. K.; Moss, D. N. A Variation of C4 Leaf Anatomy in *Arundinella hirta* (Gramineae). *Plant Physiol.* **1973,** *52* (5), 397–402.

De Wet, J. M. J. Leaf Anatomy in Six South African Grass Genera. *Bothalia* **1960,** *7,* 299–301.

Dengler, N. G.; Dengler, R. E.; Donnelly, P. M.; Hattersley, P. W. Quantitative Leaf Anatomy of C3 and C4 Grasses (Poaceae): Bundle Sheath and Mesophyll Surface Area Relationships. *Ann. Bot.* **1994,** *73* (3), 241–255.

Deshpande, B. D.; Sarkar, S. Nodal Anatomy and the Vascular System of Some Members of the Gramineae. *Proc. Nat. Inst. Sci. India Bull.* **1962,** *28,* 1–12.

Duval-Jouve, J. Histotaxie des Feuilles de Graminees. *Ann. Sci. Nat. Bot. Ser.* **1875,** *6* (1), 294–371.

Eleftheriou, E. P.; Noitsakis, B. A Comparative Study of the Leaf Anatomy of the Grasses *Andropogon ischaemum* and *Chrysopogon gryllus. Phyton* **1978,** *19,* 27–36.

Ellis, R. P. Anomalous Vascular Bundle Sheath Structure in *Alloteropsis semialata* Leaf Blades. *Bothalia* **1974,** *11* (3), 273–275.

Ellis, R. P. A Procedure for Standardizing Comparative Leaf Anatomy in the Poaceae. I. The Leaf-Blade as Viewed in Transverse Section. *Bothalia* **1976,** *12* (1), 65–109.

Esau, K. *Plant Anatomy,* 2nd ed.; John Wiley and Sons: New York, 1965; pp 767.

Fisher, B. S. A Contribution to the Leaf Anatomy of Natal Grasses, Series I: *Chloris* Sw.;. and *Eustachys* Desv. *Ann. Natal Mus.* **1939,** *9,* 245–267.

Fuente, V.; Ortunez, E. Festuca sect. Eskia (Poaceae) in the Iberian Peninsula. *Folia Geobot.* **2001,** *36,* 385–421.

Garnier, E.; Laurent, G. Leaf Anatomy, Specific Mass and Water Content in Congeneric Annual and Perennial Grass Species. *New Phytol.* **1994,** *128* (4), 725–736.

Ghahraman, A.; Alemi, M.; Atar, F.; Hamzeh, B.; Columbus, T. Anatomical Studies in Some Species of *Bromus* L. (Poaceae) in Iran. *Iran. J. Bot.* **2006,** *12* (1), 1–14.

Giussani, L. M.; Cota-Sanchez, J. H.; Zuloaga, F. O.; Kellogg, E. A. A Molecular Phylogeny of the Grass Subfamily Panicoideae (Poaceae) Shows Multiple Origins of C4 Photosynthesis. *Am. J. Bot.* **2001,** *88,* 1993–2012.

Goossens, A. P. A Study of the South African Species of *Sporobolus* with Special Reference to Leaf Anatomy. *Trans. R. Soc. S. Afr.* **1938,** *26,* 173–223.

Haberlandt, G. *Physiological Plant Anatomy*; Macmillan Co.: London, 1914.

Hattersley, P. W. Characterization of C4 Type Leaf Anatomy in Grasses (Poaceae). Mesophyll: Bundle Sheath Area Ratios. *Ann. Bot.* **1984,** *53* (2), 163–180.

Hitch, P. A.; Sharman, B. C. Initiation of Procambial Strands in Leaf Primordia of *Dactylis glomerata* L. as an Example of a Temperate Herbage Grass. *Ann. Bot.* **1968,** *32,* 153–164.

Hitch, P. A.; Sharman, B. C. The Vascular Pattern of Festucoid Grass Axes, with Particular Reference to Nodal Plexi. *Bot. Gaz.* **1971,** *132* (1), 38–56.

Hitchcock, A. S. *A Textbook of Grasses*; Macmillan Co.: New York, 1922.

Hsu, C. C. The Classification of Panicum (Gramineae) and Its Allies, with Special Reference to the Characters of the Lodicule, Style-Base and Lemma. *J. Fac. Sci., Univ. Tokyo, Sect. 3* **1965,** *9* (3), 43–150.

Hubbard, C. E. Gramineae. J. Hutchinson. *Fam. Flower Plants* **1934,** *2,* 199–299.

Jane, W. N.; Chiang, S. H. T. Morphology and Development of Bulliform Cells in *Arundo formosana* Hack. *Taiwania* **1991,** *36,* 85–97.

Jauhar, P. P.; Joshi, A. B. Cytotaxonomic Investigations in the *Panicum maximum* Jacq. Complex II. Studies on Epidermal Pattern. *Nelumbo* **1967**, *9* (1–4), 59–62.

Judziewicz, E. J.; Clark, L. G. The South-American Species of *Arthrostylidium* (Poaceae, Bambusoideae, Bambuseae). *Syst. Bot.* **1993**, *15*, 80–99.

Kanai, R.; Kashiwagi, M. *Panicum milioides*, a Gramineae Plant Having Kranz Leaf Anatomy without C4-Photosynthesis. *Plant Cell Physiol.* **1975**, *16* (4), 669–679.

Kaufman, B. P.; McDonald, M. R.; Bernstein, M. H. Cytological Studies of Changes Induced in Cellular Materials by Ionizing Radiations. *Ann. N. Y. Acad. Sci.* **1956**, *59*, 1–53.

Kellogg, E. A. The Grasses: A Case Study in Macroevolution. *Annu. Rev. Ecol. Syst.* **2000**, *31*, 217–238.

Keshavarzi, M.; Rahiminejad, M. R.; Kheradmandnia, M. Anatomical and Morphological Variation of *Aegilops triuncialis* L. of Iran. *Pagohesh Sazandegi* **2002**, *5*, 14–20.

Keshavarzi, M.; Khaksar, M.; Seifali, M. Systematic Study of Annual Weed *Phalaris minor* Retz (Poaceae) in Iran. *Pak. J. Biol. Sci.* **2007**, *10* (8), 1336–1342.

Kok, P. D. F.; Van Der Schijff, H. P. A Key Based on Epidermal Characteristics for the Identification of Certain Highveld Grasses. *Koedoe* **1973**, *16* (1), 27–43.

Korzun, V.; Malyshev, S.; Voylokov, A.; Borner, A. RFLP-Based Mapping of Three Mutant Loci in Rye (*Secale cereale*) and Their Relation to Homoeologous Loci within the Gramineae. *Theor. Appl. Genet.* **1997**, *95*, 468–473.

Kumar, N.; Nautiyal, S. Leaf Anatomy of Two Genera of Tribe Eragrostideae (Poaceae) from Mandal Forest of Kedarnath Wildlife Sanctuary, Uttarakhand, India. *Int. J. Bot. Stud.* **2017**, *2* (5), 50–55.

Kuo, J.; O'Brien, T. P.; Canny, M. J. Pit-field Distribution, Plasmodesmatal Frequency, Assimilate Flux in the Mestome Sheath Cells of Wheat Leaves. *Planta* **1974**, *121*, 97–118.

Lambers, J. H. R.; Harpole, W. S.; Tilman, D.; Knops, J.; Reich, P. B. Mechanisms Responsible for the Positive Diversity–Productivity Relationship in Minnesota Grasslands. *Ecol. Lett.* **2004**, *7* (8), 661–668.

Longhi-Wagner, H. M. Tribo Arundinelleae. In *Flora fanerogâmica do Estado de São Paulo, Poaceae*; Wanderley, M. G. L., Shepherd, G. J., Giulietti, A. M.; Fapesp-Hucitec: São Paulo, 2001; Vol 1, pp 119–123.

Lundgren, M. R.; Osborne, C. P.; Christin, P. A. Deconstructing Kranz Anatomy to Understand C4 Evolution. *J. Exp. Bot.* **2014**, *65* (13), 3357–3369.

Makino, Y.; Ueno, O. Structural and Physiological Responses of the C_4 Grass Sorghum Bicolor to Nitrogen Limitation. *Plant Prod. Sci.* **2018**, *21* (1), 39–50.

Martínez-Sagarra, G.; Abad, P.; Devesa, J. A. Study of the Leaf Anatomy in Cross-Section in the Iberian Species of *Festuca* L. (Poaceae) and Its Systematic Significance. *PhytoKeys* **2017**, *83*, 43–74.

Mauseth, J. D. *Plant Anatomy*; Benjamin Cummings: California, 1988.

Mendes, M. M.; Gazarini, L. C.; Rodrigues, M. L. Acclimation of *Myrtus communis* to Contrasting Mediterranean Light Environments Effects on Structure and Chemical Composition of Foliage and Plant Water Relations. *Environ. Exp. Bot.* **2001**, *45*, 165–178.

Metcalfe, C. R. *Anatomy of the Monocotyledons. 1. Gramineae*; Clarendon Press: Oxford, 1960; pp 731.

Mojzes, A.; Kalapos, T. Leaf Anatomical Plasticity of *Brachypodium pinnatum* (L.) Beauv. Growing in Contrasting Microenvironments in a Semiarid Loess Forest-Steppe Vegetation Mosaic. *Comm. Ecol.* **2005**, *6* (1), 49–56.

Moreno, M. A.; Harper, L. C.; Krueger, R. W.; Dellaporta, S. L.; Freeling, M. Liguleless 1 Encodes a Nuclear-Localized Protein Required for Induction of Ligules and Auricles During Maize Leaf Organogenesis. *Genes Dev.* **1997**, *11*, 616–628.

Moulia, B. Leaves as Shell Structures: Double Curvature, Auto-stresses, and Minimal Mechanical Energy Constraints on Leaf Rolling in Grasses. *J. Pl. Growth Reg.* **2000**, *19*, 19–30.

Neumann, H. Zur Kenntnis der Anatomie und ersten Anlage der Graminenligula. *Beitr. Biol. Pflanzen* **1938**, *25*, 1–22.

Oguchi, R.; Hikosaka, K.; Hirose, T. Does the Photosynthetic Light-Acclimation Need Change in Leaf Anatomy? *Plant, Cell Environ.* **2003**, *26* (4), 505–512.

Ogundipe, O. T.; Olatunji, O. A. Systematic Anatomy of *Brachiaria* (Trin.) Griseb. (Poaceae). *Feddes Repert.* **1992**, *103* (1–2), 19–30.

Olsen, J. T.; Caudle, K. L.; Johnson, L. C.; Baer, S. G.; Maricle, B. R. Environmental and Genetic Variation in Leaf Anatomy among Populations of *Andropogon gerardii* (Poaceae) along a Precipitation Gradient. *Am. J. Bot.* **2013**, *100* (10), 1957–1968.

Paliwal, G. S. Stomatal Ontogeny and Phylogeny. I. Monocotyledons. *Acta Bot Neerland.* **1969**, *18* (5), 654–668.

Patel, S. R. *Morphotaxonomic Studies in Some Gramineae*. Thesis. Sardar Patel University: Anand, 1976.

Pathak, S.; Kar, S.; Singh, P. Ligules as Aid to Identification of Grasses. *Pleione* **2013**, *7* (1), 241–246.

Picket-Heaps, J. D.; Northcote, D. H. Cell Division in the Formation of the Stomatal Complex of the Young Leaves of Wheat. *J. Cell. Sci.* **1966**, *1*, 121–128.

Prat, H. L'epiderme des Graminées. Ètude Anatomique et Systematique. *Ann. Sci. Nat. Bot., Ser.* **1932**, *10* (14), 117–324.

Prat, H. La Sistématique des Graminées. *Ann. Sci. Nat. Bot., Ser.* **1936**, *10* (18), 165–257.

Reeder, J. R. The Embryo in Grass Systematics. *Am. J. Bot.* **1957**, *44* (9), 756–768.

Renvoize, S. A. A Survey of Leaf-Blade Anatomy in Grasses. III. Garnotideae. *Kew Bull.* **1982a**, *37* (3), 497–500.

Renvoize, S. A. A Survey of Leaf-Blade Anatomy in Grasses. II. Arundinelleae. *Kew Bull.* **1982b**, *37* (3), 489–495.

Renvoize, S. A. A Survey of Leaf-Blade Anatomy in Grasses. I. Andropogoneae. *Kew Bull.* **1982c**, *37* (2), 315–321.

Renvoize, S. A. A Survey of Leaf-Blade Anatomy in Grasses. IV. Eragrostideae. *Kew Bull.* **1983**, *38* (3), 469–478.

Renvoize, S. A. A Survey of Leaf-Blade Anatomy in Grasses. VIII. Arundinoideae. *Kew Bull.* **1986a**, *41* (2), 323–338.

Renvoize, S. A. A Survey of Leaf-Blade Anatomy in Grasses. IX. Centothecoideae. *Kew Bull.* **1986b**, *41* (2), 339–342.

Renvoize, S. A. A Survey of Leaf-Blade Anatomy in Grasses. XI. Paniceae. *Kew Bull.* **1987a**, *42*, 739–768.

Renvoize, S. A. A Survey of Leaf-Blade Anatomy in Grasses. X. Bambuseae. *Kew Bull.* **1987b**, *42* (1), 201–207.

Row, H. C.; Reeder, J. R. Root-Hair Development as Evidence of Relationships among Genera of Gramineae. *Am. J. Bot.* **1957**, *44*, 596–601.

Sage, T. L.; Sage, R. F. The Functional Anatomy of Rice Leaves: Implications for Refixation of Photorespiratory CO_2 and Efforts to Engineer C4 Photosynthesis into Rice. *Plant Cell Physiol.* **2009**, *50* (4), 756–772.

Sangster, A. G.; Parry, D. W. Some Factors in Relation to Bulliform Cell Silicification in the Grass Leaf. *Ann. Bot.* **1969**, *33* (2), 315–323.

Schwendener, S. Die Mestomscheiden der Gramineenblatter. *S. B. Akad. Wiss. Berl.* **1890**, *22*, 405–426.

Shields, L. M. The Involution Mechanism in Leaves of Certain Grasses. *Phytomorphology* **1951**, *1*, 225–251.

Siqueiros Delgado, M. E.; Herrera Arrieta, Y. Taxonomic Value of Culm Anatomical Characters in the Species of *Bouteloua lagasca* (Poaceae: Eragrostideae). *Phytologia* **1996**, *81* (2), 124–141.

Stebbins, G. L. Taxonomy and the Evolution of Genera, with Special Reference to the Family Gramineae. *Evolution* **1956**, *10* (3), 235–245.

Stebbins, G. L.; Jain, S. K. A Study of Stomatal Development in *Allium*, *Rhoeo* and *Commelina*. *Dev. Biol.* **1960**, *1*, 409–426.

Stebbins, G. L.; Khush, G. S. Variations in the Organization of the Stomatal Complex in the Leaf Epidermis of Monocotyledons and Its Bearing on Their Phylogeny. *Am. J. Bot.* **1961**, *48*, 51–59.

Stebbins, G. L.; Shah, S. S. Developmental Studies of Cell Ditferentiation in the Epidermis of Monocotyledons. II. Cytological Figures of Stomatal Development in the Gramineae. *Dev. Biol.* **1960**, *2*, 477–500.

Szabo, Z. K.; Papp, M.; Daroczi, L. Ligule Anatomy and Morphology of Five *Poa* Species. *Acta Biol. Cracovien. Ser. Bot.* **2006**, *48* (2), 83–88.

Tateoka, T. The Place of the Genus *Phyllorachis* in the System of Gramineae. *Bot. Mag.* **1956**, *69*, 83–86.

Tateoka, T. Miscellaneous Papers on the Phylogeny of Poaceae, Proposition of a New Phylogenetic System of Poaceae. *J. Jap. Bot.* **1957**, *32*, 275–287.

Vecchia, F. D.; Asmar, T. E.; Calamassi, R.; Rascio, N.; Vazzana, C. Morphological and Ultrastructural Aspects of Dehydration and Rehydration in Leaves of *Sporobolus stapfianus*. *J. Pl. Growth Reg.* **1998**, *24*, 219–228.

Vickery, J. W. Leaf Anatomy and Vegetative Characters of the Indigenous Grasses of New South Wales. *Proc. Linn. Soc. N.S.W.* **1935**, *60*, 340–373.

Yakovlev, M. S. Structure of Endosperm and Embryo in Cereals as a Systematic Feature. *Inst. Akad. Sci. S.S.S.R.* **1950**, *1*, 121–218.

Zuloaga, F.; Morrone, O.; Dubcovsky, J. Exomorphological, Anatomical, and Cytological Studies in *Panicum validum* (Poaceae: Panicoideae: Paniceae): Its Systematic Position within the Genus. *Syst. Bot.* **1989**, *14*, 220–230.

CHAPTER 2

Culm Anatomy

ABSTRACT

Anatomical features of culm are also a very important feature for taxonomic identification. Internally the internodes in grasses may be solid throughout development, or may become hollow. Characteristic features of identification are represented with photographs. Both qualitative and quantitative features have been taken into consideration. Qualitative features include shape in cross section, epidermis, hypodermis, ground tissue, type of sclerenchyma, I and II IVBs, III PVB, kranz anatomy, shape of kranz arc, shape of kranz cell, chloroplast type, mesotme while quantitative features include features like cross sectional area, number of vascular bundles, size of vascular bundles, size of metaxylem, number of kranz cells were considered in study. Also based on the presence or absence of kranz anatomy, the studied species have been divided into two categories.

Five different major types of epidermis were identified among the studied grass species on the basis of their shape and cuticularization: barrel shaped, rectangular shaped, square shaped, elongated shaped, and round shaped.

Vascular bundles are found irregularly scattered in the ground tissue. Toward the periphery, the bundles are smaller in size while toward the center, they are larger in size. The smaller bundles are younger, while the larger ones are older. Each vascular bundle has a covering called bundle sheath formed by a single layer of sclerenchyma cells. The cambium is absent. Hence the vascular bundles are described as conjoint, collateral, and closed.

Three types of vascular bundles are present: (1) Internal vascular bundle (IVB)—I and II vascular bundles are present toward the center so they are known as IVB. I and II IVB are organized in one to three concentric circles. (2) Peripheral vascular bundles (PVB)—vascular bundles are present toward the periphery. Third-order peripheral vascular bundles (III PVB)—inconspicuous proto and metaxylem cells. Nine types of I IVB are present, seven types of II IVB are present, one type of III PVB is present. The culms of a few grass species show the presence of chlorenchyma.

Chlorenchyma surrounds the III PVB. It is in the form of arc, so it is known as Kranz arc. Kranz anatomy is observed only in species like *Eragrostis, Dinebra, Cenchrus setigerus, Dactyloctenium sps., Tragus biflors, Perotis indica, Tetrapogon sp.,* etc. This chlorenchyma is radial chlorenchyma. The shape of Kranz arc was: straight, half circle, horseshoe shaped, and circular shaped.

Kranz arc, radial cholrenchyma, culm outline and chloroplast shape and position, are some of the culm anatomical features suggested to be useful for inferring phylogenies. Based on all above characters a dichotomous key has been prepared and statically analyzed using cluster analysis.

2.1 INTRODUCTION

Generally, grasses have been stated to have a hollow internode (Hitchcock, 1914; Bews, 1929; Arber, 1934; Armstrong, 1937). However, the members of Andropogoneae and Paniceae internodes are or may be solid (Brown et al., 1959). Canfield (1934) reported, 74% of the grasses examined from the Jornada experimental Range station, New Mexico, to have solid internodes relating the solid stem structure to the environment and concluded that hollow stemmed grasses are not well adapted to arid regions. He studied nearly 133 species in 80 genera. Their study revealed that a large number of Panicoideae grasses have thick walls and small hollows where the festucoid grasses have thin walls and large hollows. He arranged tribes with respect to % of species with solid internodes. For example, Maydeae have 100% solid culm while Paniceae have 49% solid culm, Andropogoneae and Chlorideae have 78% solid stem.

For the culm anatomy, features like incomplete bundle sheath which is forming are external to the vascular bundles, Kranz arc, radial chlorenchyma, culm outline, and chloroplast shape and position are suggested to be useful for inferring phylogenies (Siqueiros and Herrera, 1996).

Except in bamboos, culm anatomy in grasses (Grosor and Liese, 1971; Liese, 1980, 1998; Sekar and Balsubramanian, 1994; Agrasar and Rodriguez, 2002; Londoño et al., 2002; Yao et al., 2002) has been little explored, as it is very useful in phylogenetics (Cenci et al., 1984; Gasser et al., 1994; Siqueiros and Herrera, 1996; Ramos et al., 2002).

Siqueiros and Herrea (1996) pointed out that culm anatomy at the epinodal culm region in *Bouteloua* has restricted phylogenetic value. Ramos et al. (2002) found most important variations in the culm components in *Bromus aleuticus* are cortical and medular parenchyma, the development and position of the vascular bundles and the development of the scleren-chyma ring associated with these bundles.

Delgado (2007) studied culm anatomy of 55 taxa of *Boutelouinae* (Chlorideae) and conclude that culm anatomy does not display the same degree of variation as leaf anatomy does. There are several culm antatomy characters that can be useful for inferring relationships at higher levels. And the most important phylogenetic culm anatomy features are Kranz structures, number and position of the vascular bundles, and sclerenchyma girders.

2.2 ANATOMICAL FEATURES

Grasses are monocotyledonous flowering plants having hollow stem known as the culm which are divided into sections by solid nodes which bear the plant leaves. Variation in the internal anatomy of the internode may be taxonomically diagnostic and is most useful at the level of genus and species. Described below are the internal features of grass culm. Characteristic diagnostic anatomical features of internodal region observed in the studied grasses have been described and quatitative data represented in Table 2.1.

General grass culm anatomy

Internally the internodes may be solid throughout development, or may become hollow.

Epidermis

Epidermis is the outermost covering of the stem represented by a single layer of compactly arranged, barrel-shaped parenchyma cells. Intercellular spaces are absent. A cuticle is present. The epidermis contains stomata and silica.

Hypodermis

Hypodermis is a region that lies immediately below the epidermis. It is represented by a few layers of compactly arranged sclerenchyma or parenchyma cells.

Ground tissue

Ground tissue is a component of the stem. It is undifferentiated. The ground tissue is represented by several layers of loosely arranged parenchyma cells enclosing prominent intercellular spaces. The ground tissue is meant for storage of food.

Vascular bundles

Vascular bundles are found irregularly scattered in the ground tissue. Toward the periphery, the bundles are smaller in size while toward the center, they are larger in size. The smaller bundles are younger, while the larger ones are older. Each vascular bundle has a covering called bundle sheath formed by a single layer of sclerenchyma cells. The vascular bundle encloses both xylem and phloem. Xylem is found toward the inner surface and phloem toward the outer surface. Cambium is absent. Hence, the vascular bundles are described as conjoint, collateral, and closed. There are two metaxylem and protoxylem vessels arranged in the shape of "Y." The lower protoxylem vessel is nonfunctional and remains as a water-filled cavity called lysigenous cavity or protoxylem cavity. In the phloem, only sieve tubes, companion cells and phloem fibers are present. Culm anatomy has been described by a few authors. Grosser and Liese (1971) histologically compared 52 bamboo species. Metcalfe (1960) studied 206 genera and 413 species examined and described culm anatomy of them. Description of the internal features of the studied culm internode has been done with help of characteristic features described by Grosser and Liese (1971) and Mecalfe (1960).

A general detailed description of the different characteristic features and their types noted in the 100 different species studied has been described in the first part followed by which the type of characteristic feature observed in each grass species has been discussed separately.

The anatomical structure of the culms is characterized exclusively by the collateral vascular bundles embedded in parenchymatous ground tissue which exhibits four basic types (Diagram 1):

Type I-consisting of one part (central vascular strand, i.e., CVB); supporting tissue only as sclerenchyma sheaths; intercellular space with tyloses.

Type II - consisting of one part (CVB); supporting tissue only as sclerenchyma sheaths; sheath at the intercellular space (protoxylem) strikingly larger than the other ones; intercellular space without tyloses.

Type III - consisting of two parts (CVB and one FS, i.e., fiber strand); fiber strand inside the central strand; sheath at the intercellular space (protoxylem) generally smaller than the other ones.

Type IV - consisting of three parts (CVB and two FS); fiber strands outside and inside the central strand.

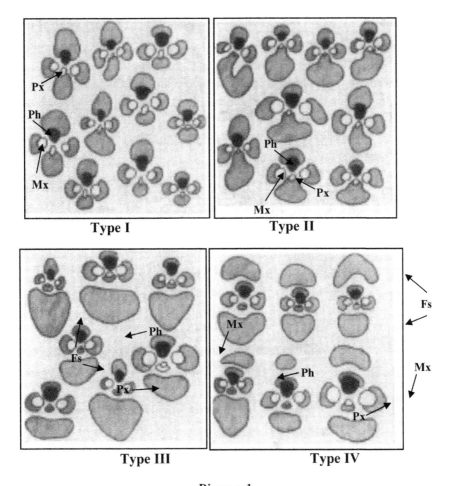

Diagram 1
(Adapted from Grosser and Liese, 1971)
(Fs= fiber strand, Mx= metaxylem, Px= protoxylem, Ph= phloem)

The culms of grasses are typically hollow cylinders interrupted at intervals by transverse partitions. The hollow units represent the internodes and transverse partition is at nodes. The diameter of the hollow center may be large or small in relation to the total diameter of the culm. Sometimes few grass species have solid culm, that is, internodal region is occupied by ground tissues and culm is not hollow. Few grasses show an intermediate state between the hollow and solid culm, in which tissue becomes loose and spongy without actually breaking down sufficiently to become hollow (Metcalfe, 1960).

Epidermis

It is the outermost layer and is covered by cuticle. Epidermal cells are round to square shaped. Sometimes stomata and silica bodies are also present. Five different major types of epidermises were identified among the studied grass species on the basis of their shape and cuticularization which has been described with the help of diagrams below. Microphotographs depicting these characteristic features are represented in Figure 2.1.

1. Barrel shaped (BS)
    I. Big cells, tangential wall covered with evenly thickened outer and inner tangential wall, and only one of the radial walls is thickened (Fig. 2.1A).
    II. Flattened cells with outer tangential wall coveredwith even thick cuticle which penetrate between radially two epidermal cells. Inner tangential wall is thin (Fig. 2.1B).
    III. Outer tangential wall of epidermal cells covered with layer of thin cuticle (Fig. 2.1C).
    IV. Small-sized cells with outer tangential wall covered with layer of cuticle (Fig. 2.1D).
    V. Tangential wall covered with very thick cuticle (Fig. 2.1E).
2. Rectangular shaped (RS)
    I. Tangential wall covered with very thin cuticle (Fig. 2.1F).
    II. Flat cells, tangential wall with slightly thick cuticle (Fig. 2.1G).
    III. Tangential wall of cells covered with cuticle which penetrate 1/3rd portion between cells (Fig. 2.1H).
3. Square shaped (SS)
    I. Small cells, outer tangential wall and radial walls covered with thick cuticle (Fig. 2.1I).
    II. Big cells, tangential wall covered with very thin cuticle (Fig. 2.1J).

III. Small cells, covered with thick cuticle and it penetrates between epidermal cells; thickness of cuticle is larger than the cell wall thickness (Fig. 2.1K).

4. Elongated shaped (ES)

I. Big cells, tangential wall covered with thick cuticle, inner tangential wall arched or round (Fig. 2.1L).

II. Small cells, tangential wall covered with thin cuticle, inner tangential wall angular (Fig. 2.1M).

5. Round shaped epidermal cells, outer tangential wall covered with thick cuticle, cuticle penetrate between cell and form triangular-shaped thickening at the junction of two adjacent cells (Fig. 2.1N).

Ground tissue (Metcalfe, 1960)

Transverse sections of culm showed that the parenchymatous tissue is the main component tissue which forms ground tissue. Sometimes few cells contain chloroplasts and serve for photosynthesis; few cells become elongated and thickened walls and provide mechanical support. Within the ground tissue, vascular bundles and other cells are arranged in different patterns.

Mechanical tissues are restricted toward the peripheral part of the culm where they provide effective support in resisting strains and stresses. They form a subepidermal ring in the form of a cylinder in which peripheral vascular bundles are present. Sclerenchymatous cylinder may be single termed **"unicylindrical"** which is very commonly observed in grasses or it may be in two concentric rings separated by ground tissue and termed as **"bicylindrical."** Sometimes sclerenchyma forms **girder**. Other than this, the individual vascular bundles even in the inner part of the culm and are accompanied by strands of thick walled fibers.

The distribution of sclerenchyma could be categorized as:

1. **Type A**: Where fibers are of medium thickness (Fig. 2.1O, *arrow*).

2. **Type B**: Fibers are very thick walled, lumen almost completely occluded (Fig. 2.1P, *arrow*)

3. **Type C:** Fiber walls thick, walled interrupted with pits (Fig. 2.1Q, *arrow*).

4. **Type D**: Not markedly thickened, penta to hexagonal-shaped cells (Fig. 2.1R).

5. **Type E**: Thick cell wall, with big lumen (Fig. 2.1S).

Intercellular cavities or canal were observed in the grasses which are aquatic or which are growing in moist area. They develop intercellular air cavities or canals which vary in size, shape, and distribution in different species.

Mestome is formed by small, thick-walled cells, similar to the sclerenchyma, surrounding the vascular bundles and occurring usually in a single layer. Mestome cells constitute the internal bundle sheath, and their size and shape are variable.

Vascular tissue:

Vascular bundles like any other monocot stem were found to be scattered and widely distributed irregularly throughout the cross sectional area of the culm. In few grass species, the vascular bundles are in a single circle or a few concentric circles. The number of circles that are present is largely bound up with the diameter of the culm, and the proportion of its cross-sectional area that is hollow. The vascular bundles nearest to the centers of the culms are much larger than those present next to the epidermis. The individual vascular bundles are usually circular or oblong to elliptical. Shapes vary considerably in grasses. Vascular bundles are collateral (phloem is present at the pole of the bundle toward the periphery of the culm), conjoint, endarch, and closed.

Three types of vascular bundles are present:

(1) **Internal vascular bundle (IVB):** I and II vascular bundles are present toward the center so they are known as IVB. I and II IVB are organized in one to three concentric circles.

    (a) **First order internal vascular bundles (I IVB):** Proto and metaxylem cells are well developed. I IVB are smaller in size and metaxylem vessels are not conspicuously larger in diameter than the neighboring cells.

(b) Second order internal vascular bundles (II IVB): Only metaxylem cells are developed.

(2) Peripheral vascular bundles (PVB): Vascular bundles are present toward the periphery. Third order peripheral vascular bundles (III PVB); inconspicuous proto and metaxylem cells.

Vascular bundles could be differentiated on the basis of shapes in all three first and second order internal vascular bundles and the peripheral bundles. The shapes could be categorized as given below. Diagrammatic representations of the shapes are also given along with the description and figures 59A–X depict the microphotographs of the same. Pitted portion in the diagram represents sclerenchymatous cells.

1. **Types of I IVB:**

Type A: Diamond-shaped vascular bundle with round shaped metaxylem, sclerenchymatous sheath present toward phloem and small patch of sclerenchyma present toward protoxylem (Fig. 2.2 A).

Type B: Diamond-shaped vascular bundle with elliptical shaped metaxylem, sclerenchymatous sheath present toward phloem (Fig. 2.2B).

Type C: Triangular diamond shaped (Fig. 2.2C).

Type D: Obovate-shaped vascular bundle with a broad top and narrow base (Fig. 2.2D).

Type E: Top shaped/inverted triangular shaped vascular bundle (Fig. 2.2E).

Type F: Vertically elliptical-shaped vascular bundle with sclerenchymatous patch present at both the polar ends of the bundle (Fig. 2.2F).

Type G: Inverted cork or bulb-shaped vascular bundle (Fig. 2.2G).

Type H: Round-shaped vascular bundle (Fig. 2.2H).

2. Types of II IVB:

Type I: Oval-shaped vascular bundle with broad base and narrow top (Fig. 2.2I).

Type J: Oblong-shaped vascular bundle (Fig. 2.2J).

Type K: Oval-shaped vascular bundle with very broad base and abrupt narrow tip (Fig. 2.2K).

Type L: Triangular shaped vascular bundle with sclerenchymatous sheath around xylem (Fig. 2.2L).

Type M: Round-shaped vascular bundle (Fig. 2.2M).

Type N: Inverted triangular shaped vascular bundle with sclerenchyma cap ensheathing the entire vascular bundle except at the region of lysigenous cavity (*arrow*) (Fig. 2.2N).

Type O: Round-shaped, 2–3 layers of sclerenchymatous sheath encircling the entire vascular bundle (Fig. 2.2O).

Type P: Diamond-shaped vascular bundle.

3. Type of III PVB:

Type Q: Round-shaped vascular bundle with single-layered sclerenchymatous sheath (Fig. 2.2P).

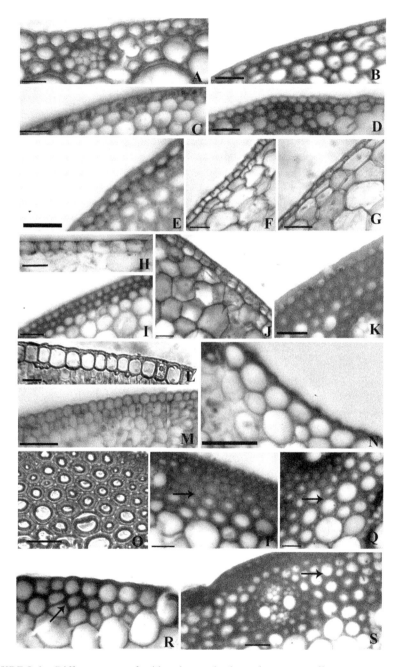

FIGURE 2.1   Different types of epidermises and sclerenchymatous cells.

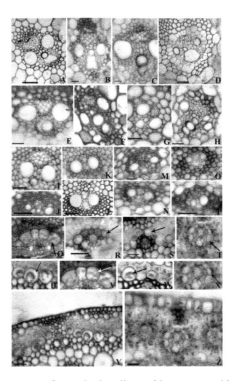

FIGURE 2.2   Different types of vascular bundles and kranz arc and kranz cell.

Chlorenchyma tissue

The culms of few grass species show presence of chlorenchyma. Where the peripheral tissues are largely fibrous it is usual to find that the chlorenchyma is in the form of longitudinal columns bounded externally by the epidermis and otherwise surrounded by or embedded in the peripheral sclerenchyma. Chlorenchyma cells are irregular in shape and arrangement. In some grasses, chlorenchyma may form a continuous cylinder subjacent to the epidermis or the cylinder may be traversed at intervals by girders of sclerenchyma, one of which often lies opposite to each of the outermost vascular bundles and extends from these to the epidermis. In certain plants, a layer of radically oriented mesophyll cells which surrounds the vascular bundles and it called this layer "Kranz" (wreath) and the type of anatomy Kranz type. Kranz type of anatomy includes both the mesophyll and bundle sheath and mostly found in C_4 plants. In culm anatomy, chlorenchyma present surrounds the third order peripheral vascular bundle (III PVB), but it is not present in the form of a ring. It is in the form of an arc so it is known as Kranz arc. This chlorenchyma is radial chlorenchyma.

The shape of Kranz arc was:

(i)   Straight (Fig. 2.2Q, *arrow*)
(ii) Half circle (Fig. 2.2R, *arrow*)
(iii) Horseshoe (Fig. 2.2S, *arrow*)
(iv) Circular shaped (Fig. 2.2T, *arrow*)

Shapes of the Kranz cells are:

(i) Round to oval (Fig. 2.2U, *arrow*)
(ii) Square or trapezoid (Fig. 2.2V, *arrow*)

Chloroplasts in the Kranz cells were arranged in two ways:

(i) Centripetal chloroplasts (chlorophyll present toward inner side of cell) (Fig. 2.2W, *arrow*).
(ii) Centrifugal chloroplasts (chlorophyll present toward the outer side/periphery of cell) (Fig. 2.2X, *arrow*).

Morphometric/Histoarchitechtural Analysis Of Culm Anatomy

1. *Chionachnae koenigii* (Fig. 2.3(1) A–E)

Round in transaction with slightly undulating surface (Fig. A)

Size: 295 × 275 ± 24.87 µm, hairs/prickles absent

Solid internode with parechymatous pith

Epidermis: Square Shaped III (Fig. C)

Hypodermis: 3–4 layered sclerenchymatous

With III PVB embedded

Type-B sclerenchyma present

Ground tissue: Round to oval-shaped parenchymatous (Fig. E)

Vascular bundles arranged in ring form

Vascular system: All present

I IVB: Type I, sclerenchymatous cap present (Fig. D)

Size: 14.68 × 12.5 ±3.87 µm

II IVB: Type P (Fig. C)

Size: 7 × 9.67 ± 1.73 µm

III PVB: Type Q (Fig. B)

Size: 4.5 × 3 ± 0.84 µm

Bar size: A = 50 µm, B–E = 5 µm

2. *Coix lachryma-jobi* (Fig. 2.3(2) A–E)

Oval in transaction with smooth surface (Fig. A)

Size: 923.07 × 646.15 ± 31.49 μm, hairs/prickles absent

Solid internode with parechymatous pith

Epidermis: Rectangular shaped II (Fig. B)

Hypodermis: 5–6 layered parenchymatous

With III PVB embedded, below it 2–3 layered

Sclerenchyma present, Type-D sclerenchyma present

Ground tissue: Irregular-shaped parenchymatous (Fig. C)

Vascular bundles arranged in ring form

Vascular system: All present

I IVB: Type D, sclerenchymatous cap present (Fig. E)

Size: 17.78 × 15.56 ± 2.32 μm

II IVB: Type K (Fig. D)

Size: 11.07 × 10.36 ± 3.72 μm

III PVB: Type Q

Size: 7.34 × 6.27 ± 1.47 μm

Bar size: A = 50 μm, B–E = 5 μm

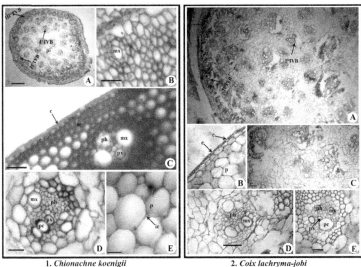

1. *Chionachne koenigii*   2. *Coix lachryma-jobi*

3. *Andropogon pumilus* (Fig. 2.4(3) A–D)

Oval in transaction with slightly undulating surface (Fig. A)

Size: 161.17 × 91.67 ± 18.37 µm, hairs/prickles absent

Hollow internode, that is, hollow pith

Epidermis: Barrel Shaped II (Fig. B)

Hypodermis: 2–3 layered sclerenchymatous

With III PVB embedded

Type-E sclerenchyma present (Fig. C)

Ground tissue: Oval-shaped parenchymatous

Vascular bundles arranged in ring form

Vascular system: All present

I IVB: Type I (Fig. D)

Lysigenous cavity oval shaped

Size: 13.22 × 16.37 ± 3.28 µm

II IVB: Type K

Size: 12.34 × 15.34 ± 2.98 µm

III PVB: Type Q

Size: 3.62 × 4.82 ± .73 µm

Bar size: A = 50 µm, B–D = 5 µm

4. *Apluda mutica* (Fig. 2.4(4) A–G)

Round to oval in transaction with smooth surface (Fig. A)

Size: 358.25 × 433.38 ± 28.73 µm, hairs/prickles present

Solid internode with parechymatous pith

Epidermis: Elongated Shaped II (Fig. C)

Hypodermis: Angular parenchymatous zone

Interrupted by sclerenchymatous girder

Type-D sclerenchyma present

Ground tissue: Round to oval-shaped parenchymatous (Fig. B, D)

Vascular bundles arranged in ring form

Vascular system: All present

I IVB: Type C, scatterly arranged (Fig. F)

Size: 35.71 × 31.43 ± 6.42 µm

II IVB: Type K (Fig. E)

Size: 15 × 18.21 ± 3.72 µm

III PVB: Type Q (Fig. C)

Size: 7.08 × 14.17 ± 1.38 µm

Bar size: A & B= 50 µm, C–G = 5 µm

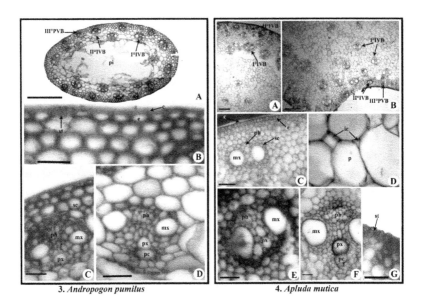

3. *Andropogon pumilus*   4. *Apluda mutica*

5. *Arthraxon lanceolatus* (Fig. 2.5(5) A–F)

Round in transaction with concave distal end (Fig. A)

Size: 116.67 × 138.09 ± 17.82 µm, hairs/prickles absent

Solid internode with parechymatous pith

Epidermis: Barrel shaped III (Fig. C)

Silica bodies present

Hypodermis: 3–4 layered parenchymatous

Interrupted by sclerenchymaotus cells, Type-C sclerenchyma present

With III PVB embedded

Ground tissue: Round to oval-shaped parenchymatous (Fig. B)

Vascular bundles arranged in ring form

Vascular system: All present

I IVB: Type G (Fig. F)

Lysigenous cavity bigger than the protoxylem

Size: $14 \times 11.34 \pm 4.83$ µm

II IVB: Type M (Fig. E)

Size: $11.07 \times 12.86 \pm 2.63$ µm

III PVB: Type Q (Fig. D)

Size: $6.67 \times 7.08 \pm 1.72$ µm

Bar size: A = 50 µm, B–F = 5 µm

6. *Bothriochloa pertusa* (Fig. 2.5(5) A–F)

Round in transaction with smooth surface (Fig. A)

Size: $155.6 \times 138.9 \pm 3.78$µm, hairs/prickles absent

Solid internode with parechymatous pith

Epidermis: Barrel Shaped V (Fig. D)

Hypodermis: 3–4 layered sclerenchymatous

With III PVB-embedded protoxylem

Ground tissue: Round to oval-shaped parenchymatous (Fig. B, C)

Vascular bundles arranged in ring form

Vascular system: All present

I IVB: Type I (Fig. F)

Lysigenous cavity equal to Type-B and E sclerenchyma present

Size: $11 \times 12.67 \pm 3.29$ µm

II IVB: Type M (Fig. E)

Size: $7.5 \times 9.32 \pm 2.85$ µm

III PVB: Size:-

Bar size: A = 50 µm, B–F = 5 µm

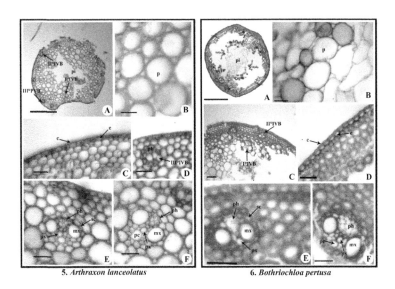

5. *Arthraxon lanceolatus*   6. *Bothriochloa pertusa*

7. *Capillipedium huegelii* (Fig. 2.6(7) A–F)

Round in transaction with concave distal end (Fig. A)

Size: 152.94 × 138.24 ± 18.63 μm, hairs/prickles absent

Solid internode with parechymatous pith

Epidermis: Barrel Shaped IV (Fig. B)

Hypodermis: 3–4 layered parenchymatous

Interrupted by sclerenchyma With III PVB embedded

Type-C sclerenchyma present

Ground tissue: Round to oval-shaped parenchymatous (Fig. C)

Vascular bundles arranged in ring form

Vascular system: All present

I IVB: Type I, scatterly arranged (Fig. F)

Size: 16 × 17 ± 6.37 μm

II IVB: Type P (Fig. E)

Size: 11.36 × 13.18 ± 2.84 μm

III PVB: Type Q (Fig. D)

Size: 5.67 × 7.34 ± 1.35 μm

Bar size: A = 50 μm, B–F = 5 μm

8. *Chrysopogon fulvus* (Fig. 2.6(8) A–F)

Squarish round in transaction (Fig. A)

Size: 277.78 × 283.34 ± 35.22 μm, hairs/prickles absent

Solid internode with parechymatous pith

Epidermis: Square Shaped III (Fig. C)

Hypodermis: 8–10 layered sclerenchymatous

With III PVB embedded

Type-A sclerenchyma present (Fig. D)

Ground tissue: Round to oval-shaped parenchymatous (Fig. B)

Vascular bundles arranged in ring form

Vascular system: All present

I IVB: Type C, partially covered by sclerenchymatous cells (Fig. F)

Size: 28.89 × 21.67 ± 5.18 μm

II IVB: Type K (Fig. E)

Size: 15.47 × 20.16 ± 6.18 μm

III PVB: Type Q

Size: 6.30 × 10 ± 2.73 μm

Bar size: A = 50 μm, B–F = 5 μm

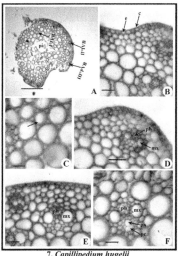

7. *Capillipedium hugelii*

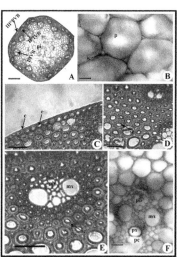

8. *Chrysopogon fulvus*

9. *Cymbopogon martini* (Fig. 2.7(9) A–F)

Round to oval in transaction with slightly undulating surface (Fig. A)

Size: 759 × 900 ± 48.22 µm, hairs/prickles absent

Solid internode with parechymatous pith

Epidermis: Rectangular shaped II

Hypodermis: 10–12 layered sclerenchymatous

With III PVB embedded

Type-E, C sclerenchyma present (Fig. C)

Ground tissue: Round to oval-shaped parenchymatous (Fig. B)

Vascular bundles arranged in ring form

Vascular system: All present

I IVB: Type B scatterly arranged (Fig. F)

Size: 24.5 × 27.5 ± 3.73 µm

II IVB: Type P (Fig. D, E)

Size: 14.09 x 14.09 ± 4.62 µm

III PVB: Type Q   Size: 3.86 × 7.95 ± 1.37 µm

Bar size: A = 50 µm, B–F = 5 µm

Dichanthium

10. *Dichanthium annulatum.* 11. *Dichanthium caricosum*

10. *Dichanthium annulatum* (Fig. 2.7(10) A–G)

Round in transaction (Fig. A)

Size: 253.72 × 241.67 ± 27.56 µm, hairs prickles absent

Solid internode with parechymatous pith

Epidermis: Rectangular shaped III (Fig. C)

Hypodermis: 5–6 layered sclerenchymatous

With III PVB embedded

Type-E sclerenchyma present (Fig. B)

Ground tissue: Round to oval-shaped parenchymatous,

Vascular bundles arranged in ring form

Vascular system: All present

I IVB: Type B (Fig. G)

Lysigenous cavity is irregularly oblong shaped

Size: 21 × 24.5 ± 7.29 μm

II IVB: Type P (Fig. F)

Size: 12 × 16.5 ± 3.81 μm

III PVB: ×shaped (Fig. E)   Size: 7.5 × 11 ± 1.38 μm

Bar size: A = 50 μm, B–G = 5 μm

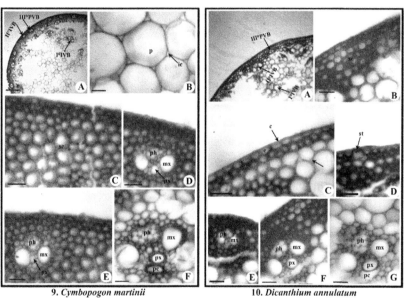

9. *Cymbopogon martinii*   10. *Dicanthium annulatum*

11. *Dichanthium caricosum* (Fig. 2.8(11) A–F)

Round to oval in transaction with smooth surface (Fig. A)

Size: 527.2 × 613.23 ± 61.78 μm, hairs/prickles absent

Solid internode with parechymatous pith

Epidermis: Rectangular shaped III (Fig. C)

Hypodermis: 3–4 layered sclerenchymatous

With III PVB embedded

Type-E sclerenchyma present (Fig. D)

Ground tissue: Round to oval-shaped parenchymatous

Vascular bundles arranged in ring form

Vascular system: All present

I IVB: Type B, sclerenchymatous cap present (Fig. F)

Scatterly arranged

Size: $22 \times 23.5 \pm 4.83$ μm

II IVB: Type M (Fig. E)

Size: $13.18 \times 11.82 \pm 1.92$ μm

III PVB: Type Q (Fig. B)   Size: $7.1 \times 9.13 \pm 2.73$ μm

Bar size: A = 50 μm, B = 10 μm, C–F = 5 μm

DIFFERENTIATING FEATURES:

Diamond shaped II IVB … *D. annulatum*

Round shaped II IVB … *D. caricosum*

12. *Hackelochloa granularis* (Fig. 2.8(12) A–F)

Oval in transaction with smooth surface (Fig. A)

Size: $218.18 \times 254.54 \pm 31.37$ μm, hairs/prickles absent.

Solid internode with parechymatous pith.

Epidermis: Barrel shaped I (Fig. D)

Hypodermis: Single layered sclerenchymatous

Type-B sclerenchyma present

Ground tissue: Angular-shaped parenchymatous (Fig. B, C)

Vascular bundles arranged in ring form

Vascular system: III PVB absent

I IVB: Type F, sclerenchymatous cap present (Fig. F)

Size: $16 \times 23 \pm 4.73$ μm

II IVB: Type N (Fig. E)

Size: $8.08 \times 5.85 \pm 1.36$ μm

III PVB: -Size:-

Bar size: A = 50 μm, B = 10μm, C-F = 5 μm

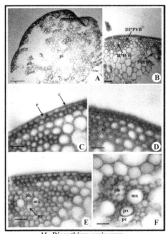 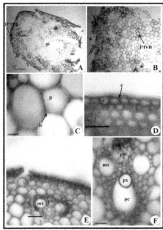

11. *Dicanthium caricosum*   12. *Hackelochloa granularis*

Heteropogon

13. *Heteropogon contortus var. genuinus sub var. typicus*

14. *Heteropogon contortus var. genuinus sub var. hispidissimus*

15. *Heteropogon ritcheii*

16. *Heteropogon triticeus*

13. *Heteropogon contortus var. genuinus sub var. typicus* (Fig. 2.9(13) A–F)

Oblong in transaction with smooth surface (Fig. A)

Size: 241.67 × 316.67 ± 28.82 µm, hairs/prickles absent.

Solid internode with parechymatous pith.

Epidermis: Barrel shaped V (Fig. D)

Hypodermis: 5-6 layered sclerenchymatous

With III PVB embedded

Type-B, E sclerenchyma present

Ground tissue: Round to oval-shaped parenchymatous (Fig. B, C)

Vascular bundles arranged in ring form

Vascular system: All present

I IVB: Type I, sclerenchymatous cap present (Fig. F)

Size: 16.82 × 17.73 ± 4.39 µm

II IVB: Type M (Fig. E)

Size: 12.92 x 16.25 ± 3.82 µm

III PVB: Type Q (Fig. D)   Size: 9.17 × 9.58 ± 3.78 µm

Bar size: A = 50 µm, B = 10 µm, C-F = 5 µm

14. *Heteropogon contortus var. genuinus sub var. hispidissimus* (Fig. 2.9(14) A–E)

Oblong in transaction with smooth surface (Fig. A)

Size: 256.25 × 384.37 ± 28.84 µm, hairs/prickles absent

Solid internode with parechymatous pith

Epidermis: Square shaped I (Fig. B)

Hypodermis: Single-layered sclerenchymatous

With III PVB embedded

Type-B sclerenchyma present

Ground tissue: Round to oval-shaped parenchymatous (Fig. C)

Vascular bundles arranged in ring form

Vascular system: All present

I IVB: Type E (Fig. E)

Size: 15 × 18.75 ± 4.25 µm

II IVB: Type N (Fig. D)

Size: 10.36 × 14.93 ± 4.72 µm

III PVB: Type Q Size: 7.5 × 9.87 ± 4.83 µm

Bar size: A = 50 µm, B–E = 5 µm

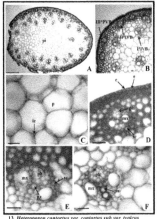
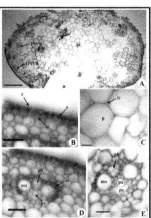

13. *Heteropogon contortus var. contortus sub var. typicus*   14. *Heteropogon contortus var. contortus sub var. genuinus*

15. *Heteropogon ritchiei* (Fig. 2.10(15) A–F)

Oblong in transaction with slightly undulating surface (Fig. A)

Size: 260 × 335 ± 35.59 μm, hairs/prickles absent.

Solid internode with parechymatous pith

Epidermis: Rectangular Shaped III (Fig. D)

Hypodermis: 5–6 layered sclerenchymatous (Fig. B)

With III PVB embedded

Type-B, E sclerenchyma present

Bicylindrical pattern

Ground tissue: Round to oval-shaped Parenchymatous (Fig. C)

Vascular bundles arranged in ring form

Vascular system: All present

I IVB: Type I (Fig. F)

Size: 19.61 × 18.07 ± 3.28 μm

II IVB: Type M (Fig. E)

Size: 10.84 × 13.34 ± 4.18 μm

III PVB: Type Q

Size: 9.23 × 10 ± 3.75 μm

Bar Size: A = 50 μm, B = 10μm, C–F = 5 μm

16. *Heteropogon triticeus* (Fig. 2.10(16) A–F)

Oval to round in transaction with smooth surface (Fig. A)

Size: 1155.56 × 1444.45 ± 47.28 μm, hairs/prickles absent

Solid internode with parechymatous pith

Epidermis: Barrel Shaped IV (Fig. C)

Hypodermis: 3–4 layered sclerenchymatous

With III PVB embedded

Type-B sclerenchyma present

Ground tissue: Round to oval-shaped parenchymatous (Fig. B)

Vascular bundles arranged in ring form

Vascular system: All present

I IVB: Type B, sclerenchymatous cap present (Fig. F)

Size: 25 × 27.27±5.18 μm

II IVB: Type N (Fig. E)

Size: 16.67 × 16.78 ± 5.38 μm

III PVB: Type Q (Fig. D)

 Size: 10.5 × 14±3.28 μm

Bar size: A = 50 μm, B–F = 5 μm

DIFFERENTIATING FEATURES:

Sclerenchyma in form of bicylinder … *H. ritchiei*
Sclerenchyma in form of unicylinder
Square-shaped epidermal cell … *H. contortus var. genuinus sub var. hispidissimus*
Barrel-shaped epidermal cell
Sclerenchyma type E present … *H. contortus* var. *genuinus* sub var. *typicus*
Sclerenchyma type B present … *H. triticeous*

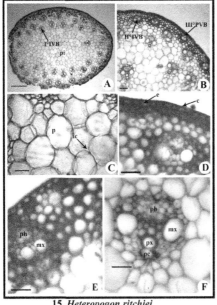

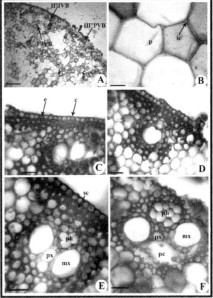

15. *Heteropogon ritchiei*   16. *Heteropogon triticeus*

17. *Imperata cylindrica* **(Fig. 2.11 (17) A–F)**

Oblong in transaction with slightly undulating surface (Fig. A)

Size: 183 × 118.27 ± 34.25 µm, hairs/prickles absent

Solid internode with parechymatous pith

Epidermis: Barrel Shaped V (Fig. D)

Hypodermis: 5–6 layered sclerenchymatous (Fig. B)

With III PVB embedded

Type-B, E sclerenchyma present (Fig. E)

Ground tissue: Round to oval-shaped parenchymatous (Fig. C)

Vascular bundles arranged in ring form

Vascular system: All present

I IVB: Type H, sclerenchymatous cap present (Fig. F)

Size: 22.5 × 20 ± 4.27 µm

II IVB: Type O (Fig. E)

Size: 11.12 × 13.89 ± 2.78 µm

III PVB: Type Q

Size: 6 × 5.71 ± 1.39 µm

Bar size: A = 50 µm, B = 10 µm, C–F = 5 µm

Ischaemum

18. *Ischaemum indicus*

19. *Ischaemum molle*

20. *Ischaemum pilosum*

21. *Ischaemum rugosum*

18. *Ischaemum indicus* **(Fig. 2.11 (18) A–F)**

Elliptical in transaction with smooth surface (Fig. A)

Size: 280 × 420 ± 21.38 µm, hairs/prickles absent

 Solid internode with parechymatous pith

Epidermis: Barrel Shaped IV (Fig. C)

Hypodermis: 3–4 layered sclerenchymatous (Fig. B)

With III PVB embedded

Type-E sclerenchyma present

Ground tissue: Round to oval-shaped parenchymatous (Fig. D),

Vascular bundles arranged in ring form

Vascular system: All present

I IVB: Type D (Fig. F)

Size: 27.5 × 23 ± 6.28 μm

II IVB: Type P (Fig. E)

Size: 9.6 × 11.6 ± 2.74 μm

III PVB: Type Q

Size: 6.26 × 7.93 ± 1.83 μm

Bar size: A = 50 μm, B = 10 μm, C–F = 5 μm

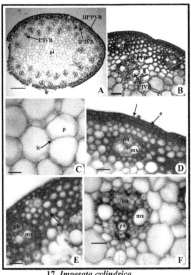

17. *Imperata cylindrica*

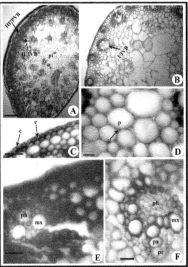

18. *Ischaemum indicum*

19. *Ischaemum molle* (Fig. 2.12 (19) A–F)

Heart in transaction with smooth surface (Fig. A)

Size: 216 × 224 ± 34.87 μm, hairs/prickles absent

Solid internode with parechymatous pith

Epidermis: Barrel Shaped IV (Fig. B, D)

Hypodermis: 2–3 layered sclerenchymatous

With III PVB embedded

Type-D sclerenchyma present (Fig. B)

Ground tissue: Round to oval-shaped parenchymatous (Fig. C)

Vascular bundles arranged in ring form

Vascular system: All present

I IVB: Type B, sclerenchymatous cells present (Fig. F)

Size: $15.5 \times 16 \pm 3.8$ μm

II IVB: Type K (Fig. E)

Size: $8.6 \times 12.4 \pm 3.19$ μm

III PVB: Type Q

Size: $4.38 \times 6.15 \pm 1.83$ μm

Bar size: A = 50 μm, B–F = 5 μm

20. *Ischaemum pilosum* (Fig. 2.12 (20) A–F)

Oblong in transaction with smooth surface (Fig. A)

Size: $250 \pm 331.82 \pm 41.73$ μm, hairs/prickles absent

Solid internode with parechymatous pith. Kranz anatomy present

Epidermis: Elongated Shaped II (Fig. C)

Hypodermis: 3–4 layered chlorenchymatous

Interrupted by sclerenchymatous cells

Type-B sclerenchyma present

Ground tissue: Round to oval-shaped parenchymatous (Fig. B)

Vascular bundles arranged in ring form

Vascular system: All present

I IVB: Type I (Fig. F, D)

Size: $250 \pm 331.82 \pm 41.73$ μm

II IVB: Type M

Size: $12.73 \times 13.82 \pm 3.28$ μm

III PVB: Type Q (Fig. C)

Size: $7.27 \times 6.36 \pm 2.73$ μm

Kranz arc: Half circle (Fig. E)

Chloroplast type: Centrifugal

Shape of Kranz cell: Round to oval (Fig. E)

Mestome: Single layered

Bar size: A = 50 μm, B–F = 5 μm

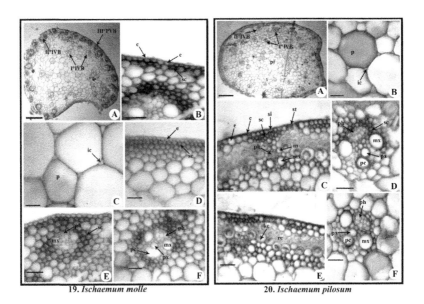

19. *Ischaemum molle*   20. *Ischaemum pilosum*

21. *Ischaemum rugosum* (Fig. 2.13 (21) A–F)

Round in transaction with smooth surface (Fig. A)

Size: 330.43 × 321.74 ± 15.12 μm, hairs/prickles absent

Solid internode with parechymatous pith

Epidermis: Barrel shaped V (Fig. C)

Hypodermis: Single-layered parenchymatous

Than 4–5 layered sclerenchymatous (Fig. B),

With III PVB embedded

Type-B sclerenchyma present (Fig. D)

Ground tissue: Round to oval-shaped parenchymatous

Vascular bundles arranged in ring form

Vascular system: All present

I IVB: Type B, sclerenchymatous cap present, scatterly arranged

Size: $24.63 \times 22.37 \pm 4.62$ μm

II IVB: Type P (Fig. E)

Size: $9.55 \times 8.64 \pm 2.16$ μm

III PVB: Type Q (Fig. F)

Size: $8.89 \times 10 \pm 1.63$ μm

Bar size: A = 50 μm, B = 10 μm, C–F = 5 μm

DIFFERENTIATING FEATURES:

Presence of Kranz arc … *I. pilosum*
Absence of Kranz arc
Sclerenchyma type B present … *I. rugosum*
Sclerenchyma type D present
Elliptical shaped in transaction … *I. indicus*
Heart shaped in transaction … *I. molle*

22. *Iseilema laxum* (Fig. 2.13 (22) A–G)

Oval in transaction with distal concave end, smooth surface (Fig. A)

Size: $333.34 \times 416.67 \pm 24.38$ μm, hairs/prickles absent

Solid internode with parechymatous pith

Epidermis: Square shaped I (Fig. D)

Hypodermis: 2-layered sclerenchymatous (Fig. B)

With III PVB embedded

Type-D sclerenchyma present

Ground tissue: Round to oval-shaped parenchymatous (Fig. C)

Vascular bundles arranged in ring form

Vascular system: All present

I IVB: Type C, sclerenchymatous cap present (Fig. F, G)

Size: $35.45 \times 24.54 \pm 7.19$ μm

II IVB: Type M (Fig. E)

Size: $16.25 \times 18.18 \pm 4.17$ μm

III PVB: Type Q

Size: $10.91 \times 13.18 \pm 2.74$ μm

Bar size: A = 50 μm, B = 10 μm, C–G = 5 μm

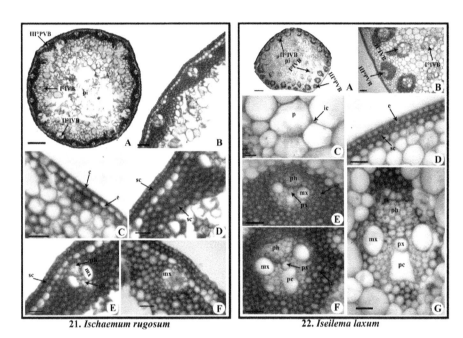

21. *Ischaemum rugosum*   22. *Iseilema laxum*

23. *Ophiuros exaltatu* (Fig. 2.14 (23) A–H)

Oval in transaction with slightly undulating surface (Fig. A)

Size: $480 \times 545 \pm 38.17$ μm, hairs/prickles absent

Solid internode with parechymatous pith. Kranz anatomy is present.

Epidermis: Elongated shaped I (Fig. E)

Hypodermis: 5–6 layered chlorenchymatous (Fig. B)

Interrupted by sclerenchymatous girder

With III PVB embedded

Type-E sclerenchyma present

Ground tissue: Round to oval-shaped parenchymatous (Fig. C)

Vascular bundles arranged in ring form

Vascular system: All present

I IVB: Type B (Fig. H, G), scatterly arranged

Size: 22.78 × 18.34 ± 5.28 μm

II IVB: Type P

Size: 17.5 × 17 ± 3.97 μm

III PVB: Type Q (Fig. E)

Size: 3.82 × 4.12 ± 1.82 μm

Kranz arc: Circle (Fig. E)

Chloroplast type: Centrifugal

Shape of Kranz cell: Round to oval (Fig. D)

Mestome: Single layered

Bar size: A = 50 μm, B = 10 μm, C–H = 5 μm

24. *Rottboellia exaltata* (Fig. 2.14 (24) A–I)

Oval in transaction (Fig. A)

Size: 367.28 × 227.18 ± 24.37 μm, hairs/prickles absent

Solid internode with parechymatous pith

Epidermis: Barrel shaped III (Fig. F)

Hypodermis: 2–3 layered sclerenchymatous

With III PVB embedded

Type-D sclerenchyma present (Fig. C)

Ground tissue: Angular-shaped parenchymatous (Fig. B, I)

Vascular bundles arranged in ring form

Vascular system: All present

I IVB: Type G (Fig. H), scatterly arranged

Size: 24.37 × 21.25 ± 3.17 μm

II IVB: Type M (Fig. G)

Size: 8.27 × 10.17 ± 3.19 μm

III PVB: Type Q (Fig. D)

Size: 7.83 × 8.69 ± 3.71 μm

Bar size: A–C = 50 μm, D–I = 5 μm

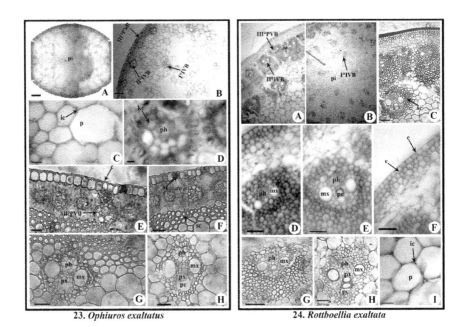

23. *Ophiuros exaltatus*   24. *Rottboellia exaltata*

25. *Saccharum spontanum* (Fig. 2.15 (25) A–F)

Round in transaction with slightly undulating surface (Fig. A)

Size: 305 × 355 ± 21.79 µm, hairs/prickles absent

Solid internode with parechymatous pith

Epidermis: Rectangular shaped III (Fig. E)

Hypodermis: 1–2 layered sclerenchymatous

And parenchyma cells with III PVB embedded

Type-D sclerenchyma present

Ground tissue: Round to oval-shaped parenchymatous (Fig. B, C)

Vascular bundles arranged in ring form

Vascular system: All present

I IVB: Type C sclerenchymatous cap present (Fig. F)

Size: 17.5 × 25 ± 5.72 µm

II IVB: Type M Size: 15.03 × 15.04 ± 3.16 µm

III PVB: Type Q (Fig. D)

Size: 7.69 × 4.62 ± 1.92 μm

Bar size: A = 50 μm, B = 10 μm, C–F = 5 μm

Sehima

26. *Sehima ischaemoides*

27. *Sehima nervosum*

28. *Sehima sulcatum*

26. *Sehima ischaemoides* (Fig. 2.15 (26) A–F)

Round to oval in transaction with slightly undulating surface (Fig. A)

Size: 203.57 × 214.28 ± 18.36 μm, hairs/prickles absent

Solid internode with parechymatous pith

Epidermis: Barrel shaped V (Fig. C)

Hypodermis: 5–6 layered sclerenchymatous

With III PVB embedded

Type-B sclerenchyma present (Fig. D)

Ground tissue: Round to oval-shaped parenchymatous

Vascular bundles arranged in ring form

Vascular system: All present

I IVB: Type E (Fig. E, F)

Size: 17.39 × 25.22 ± 4.28 μm

II IVB: Type P (Fig. B)

Size: 8.26 × 7.93 ± 2.71 μm

III PVB: Type Q

Size: 5.19 × 7.41 ± 2.71 μm

Bar size: A = 50 μm, B–F = 5 μm

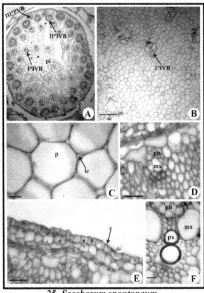

25. *Saccharum spontaneum*

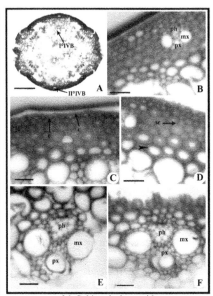

26. *Sehima ischaemoides*

27. *Sehima nervosum* (Fig. 2.16 (27) A–G)

Round in transaction with smooth surface (Fig. A)

Size: 257.41 × 259.26 ± 24.63 μm, hairs/prickles absent

Solid internode with parechymatous pith

Epidermis: Barrel shaped V (Fig. B)

Hypodermis: 5–6 layered sclerenchymatous

With III PVB embedded

Type-B sclerenchyma present (Fig. C)

Ground tissue: Round to oval-shaped parenchymatous (Fig. D)

Vascular bundles arranged in ring form

Vascular system: All present

I IVB: Type C, sclerenchymatous cap present (Fig. G)

Size: 21.74 × 20.22 ± 6.26 μm

II IVB: Type P (Fig. F)

Size: 17.08 × 21.25 ± 3.71 μm

III PVB: Type Q (Fig. E)

Size: $3.94 \times 5.76 \pm 1.52$ µm

Bar size: A = 50 µm, B–G = 5 µm

28. *Sehima sulcatum* (Fig. 2.16 (28) A–F)

Round in transaction with slightly undulating surface (Fig. A)

Size: $127.27 \times 102.27 \pm 11.72$ µm, hairs/prickles absent

Solid internode with parechymatous pith

Epidermis: Barrel shaped III

Hypodermis: 2–3 layered sclerenchymatous in form of girder (Fig. B)

Type-D sclerenchyma present

Ground tissue: Round to oval-shaped parenchymatous (Fig. F)

Vascular bundles arranged in ring form

Vascular system: All present

I IVB: Type C, sclerenchymatous cells present (Fig. D, E)

Size: $11.67 \times 13.94 \pm 2.78$ µm

II IVB: Type P (Fig. C)

Size: $6.34 \times 8.34 \pm 2.19$ µm

III PVB: Type Q

Size: $4.27 \times 4.07 \pm 1.18$ µm

Bar size: A = 50 µm, B = 10 µm, C–F = 5 µm

DIFFERENTIATING FEATURES:

Sclerenchyma in form of girder … *S. sulcatum*
Sclerenchyma in form of unicylinder
Inverted triangular shaped I IVB … *S. ischaemoides*
Triangular diamond shaped I IVB … *S. nervosum*

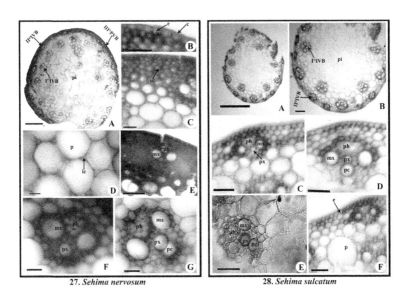

27. *Sehima nervosum*   28. *Sehima sulcatum*

29. *Sorghum halepense* (Fig. 2.17 (29) A–F)

Round in transaction with slightly undulating surface (Fig. A)

Size: 1765 × 1877.5 ± 32.78 μm, hairs/prickles absent

Solid internode with parechymatous pith

Epidermis: Rectangular shaped I (Fig. B)

Hypodermis: 1–2 layered sclerenchymatous in form of girder

Type-D sclerenchyma present

Ground tissue: Round to oval-shaped parenchymatous (Fig. C)

Vascular bundles arranged in ring form

Vascular system: All present

I IVB: Type I, sclerenchymatous cells present (Fig. E, F)

Size: 19.13 × 25.22 ± 5.27 μm

II IVB: Type K (Fig. D)

Size: 17.27 × 21.82 ± 4.87 μm

III PVB: Type Q

Size: 4.55 × 4.31 ± 1.64 μm

Bar size: A = 50 μm, B–F = 5 μm

30. *Thelepogn elegans* (Fig. 2.17 (30) A–F)

Round in transaction with slightly undulating surface (Fig. A)

Size: 500 × 550 ± 16.62 μm, hairs/prickles absent

Solid internode with parechymatous pith

Epidermis: Elongated shaped I(Fig. C)

Hypodermis: 2–3 layered sclerenchymatous interrupted by parenchyma cells

Type-D sclerenchyma present

Ground tissue: Angular parenchymatous (Fig. B)

Vascular bundles arranged in ring form

Vascular system: All present

I IVB: Type B, sclerenchymatous cells present (Fig. F)

Size: 16.67 × 21.43 ± 3.17 μm

II IVB: Type P (Fig. E)

Size: 7.67 × 10 ± 2.73 μm

III PVB: Type Q

Size: 6.82 × 5.93 ± 2.23 μm

Bar size: A = 50 μm, B–F = 5 μm

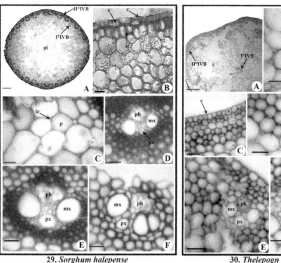

29. *Sorghum halepense*   30. *Thelepogn elegans*

Themeda

31. *Themeda cymbaria*

32. *Themeda laxa*

33. *Themeda triandra*

34. *Themeda quadrivalvis*

31. *Themeda cymbaria* **(Fig. 2.18 (31) A–F)**

Round in transaction with slightly undulating surface (Fig. A)

Size: 269.23 × 369.23 ± 31.73 μm, hairs/prickles absent

Solid internode with parechymatous pith

Epidermis: Square shaped I (Fig. D)

Hypodermis: 6–7 layered sclerenchymatous

With III PVB embedded

Type-E sclerenchyma present (Fig. C)

Ground tissue: Round to oval-shaped parenchymatous (Fig. B)

Vascular bundles arranged in ring form

Vascular system: All present

I IVB: Type B, sclerenchymatous cap present (Fig. F)

Size: 20 × 19.58 ± 2.73 μm

II IVB: Type K (Fig. E)

Size: 8.28 × 5.84 ± 2.74 μm

III PVB: Type Q

Size: 6.86 × 10.29 ± 1.37 μm

Bar size: A = 50 μm, B–F = 5 μm

32. *Themeda laxa* **(Fig. 2.18 (32) A–E)**

Round in transaction with slightly undulating surface (Fig. A)

Size: 275 × 318.75 ± 25.38 μm, hairs/prickles absent

Solid internode with parechymatous pith

Epidermis: Square shaped II (Fig. B)

Hypodermis: 1–2 layered sclerenchymatous

Than angular parenchyma present

Type-D sclerenchyma present

Ground tissue: Angular parenchymatous (Fig. C)

Vascular bundles arranged in ring form

Vascular system: III PVB absent

I IVB: Type I, sclerenchymatous cap present (Fig. E)

Size: $25 \times 24.17 \pm 8.78$ μm

II IVB: Type P (Fig. D)

Size: $9.63 \times 11.36 \pm 3.18$ μm

III PVB: Size

Bar size: A = 50 μm, B–E = 5 μm

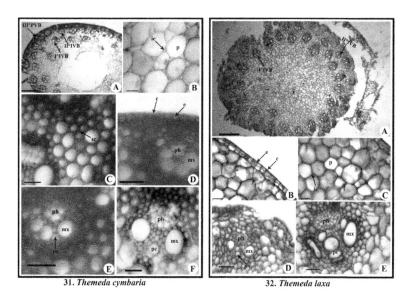

31. *Themeda cymbaria*										32. *Themeda laxa*

33. *Themeda triandra* (Fig. 2.19 (33) A–F)

Round in transaction with smooth surface (Fig. A)

Size: $305 \times 362 \pm 11.28$ μm, hairs/prickles absent

Solid internode with parechymatous pith

Epidermis: Rectangular shaped III (Fig. C)

Hypodermis: 7–8 layered sclerenchymatous

Type-B, E sclerenchyma present (Fig. D)

With III PVB embedded

Ground tissue: Angular parenchymatous (Fig. B)

Vascular bundles arranged in ring form

Vascular system: all present

I IVB: Type E, sclerenchymatous cells present (Fig. F)

Size: 19.77 × 27.5 ± 7.21 μm

II IVB: Type P (Fig. E)

Size: 13.46 × 16.15 ± 4.26 μm

III PVB: Type Q

Size: 8.27 × 8.92 ± 2.81 μm

Bar size: A = 50 μm, B–F = 5 μm

34. *Themeda quadrivalvis* (Fig. 2.19 (34) A–G)

Round in transaction with slightly undulating surface (Fig. A)

Size: 390.91 × 363.64 ± 25.81 μm, hairs/prickles absent

Hollow internode

Epidermis: Elongated shaped II (Fig. D)

Hypodermis: 3–4 layered parenchymatous than 7–8 layered sclerenchyma

With III PVB embedded

Type-B sclerenchyma present (Fig. C)

Ground tissue: Round to oval-shaped parenchymatous (Fig. B)

Vascular bundles arranged in ring form

Vascular system: All present

I IVB: Type A, sclerenchymatous cap present surrounded by sclerenchyma (Fig. G)

Size: 24.71 × 24.71 ± 6.17 μm

II IVB: Type P surrounded by sclerenchyma (Fig. F)

Size: 12.95 × 16.36 ± 3.19 μm

III PVB: Type Q (Fig. E)

Size: 9.5 × 10.5 ± 3.47 μm

Bar size: A = 50 μm, B–G = 5 μm

DIFFERENTIATING FEATURES:

Sclerenchyma type D present
Rectangular-shaped epidermal cell … *T. laxa*
Elongated-shaped epidermal cell … *T. qurqdrivalvis*
Sclerenchyma type E present
Square-shaped epidermal cell … *T. cymbaria*
Rectangular-shaped epidermal cell … *T. triandra*

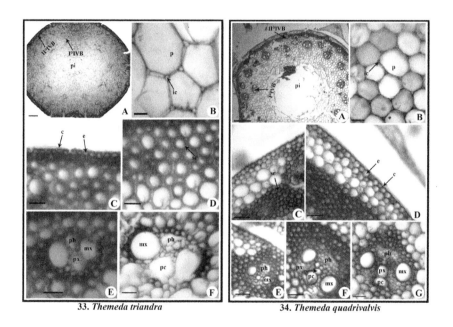

33. *Themeda triandra*   34. *Themeda quadrivalvis*

35. *Triplopogon ramosissimus* (Fig. 2.20 (35) A–D)

Squarish round in transaction with smooth surface (Fig. A)

Size: 1050.14 × 1108.16 ± 5.28 µm, hairs/prickles absent

Solid internode with parechymatous pith

Epidermis: Square shaped III (Fig. B)

Hypodermis: Parenchymatous interrupted by sclerenchyma cells (Fig. A)

Type-E sclerenchyma present

Ground tissue: Round to oval-shaped parenchymatous (Fig. C)

Vascular bundles arranged in ring form

Vascular system: All present

I IVB: Type C (Fig. D)

Size: 58.34 × 91.08 ± 2.73 µm

II IVB: Type L (Fig. B)

Size: 50.34 × 51.63 ± 0.83 µm

III PVB: Type Q (Fig. A)

Size: 33.45 × 29.35 ± 0.21 µm

Bar size: A = 50 µm, B–D = 5 µm

36. *Vetivaria zinzanoides* (Fig. 2.20 (36) A–F)

Oval in transaction with smooth surface (Fig. A)

Size: 1133.34 × 1483.34 ± 34.16 µm, hairs/prickles absent

Solid internode with parechymatous pith

Epidermis: Square shaped III (Fig. D)

Hypodermis: 2–3 layered sclerenchymatous

Than parenchyma cells present

Air spaces present (Fig. C)

Type-D sclerenchyma present

Ground tissue: Round to oval-shaped parenchymatous (Fig. F)

Vascular bundles arranged in ring form

Vascular system: All present

I IVB: Type B, sclerenchymatous cap present (Fig. F)

Size: 36.12 × 27.78 ± 8.15 µm

II IVB: Type L, surrounded by sclerenchyma (Fig. E)

Size: 13.89 × 16.67 ± 4.18 µm

III PVB: Type Q (Fig. A)

Size: 8.34 × 8.34 ± 4.28 µm

Bar size: A = 50 µm, B, D–F= 5 µm, C = 10 µm

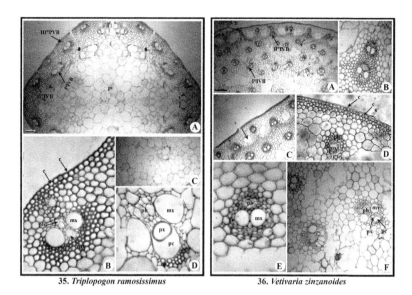

35. *Triplopogon ramosissimus*　　　36. *Vetivaria zinzanoides*

37. *Alloteropsis cimicina* (Fig. 2.21 (37) A–E)

Oval in transaction with undulating surface (Fig. A)

Size: 159.13 × 153.13 ± 16.28 µm, hairs/prickles absent

Hollow internode

Epidermis: Rectangular shaped III (Fig. B)

Hypodermis: 2–3 layered parenchyma cells

3–4 layered sclerenchymatous (Fig. B)

With III PVB embedded

Type-D sclerenchyma present

Ground tissue: Round to oval-shaped parenchymatous (Fig. C)

Vascular bundles arranged in ring form

Vascular system: All present

I IVB: Type B, sclerenchymatous cells present (Fig. E)

Size: 10.56 × 15 ± 2.91 µm

II IVB: Type P (Fig. D)

Size: 13.34 × 16.67 ± 3.18µm

III PVB: Type Q (Fig. C)

Size: 4.5 × 6.5 ± 1.24µm

Bar size: A = 50 µm, B, D–E = 5 µm, C = 10 µm

Brachiaria

38. *Brachiaria distachya*

39. *Brachiaria eruciformis*

40. *Brachiaria ramose*

41. *Brachiaria reptans*

38. *Brachiaria distachya* (Fig. 2.21 (38) A–F)

Oval in transaction with undulating surface (Fig. A)

Size: 141.17 × 170.59 ± 32.34 µm, hairs/prickles present (Fig. E, *arrow*)

Hollow internode

Epidermis: Rectangular shaped III (Fig. B)

Bulliform cells present (Fig. C)

Hypodermis: 2–3 layered parenchyma cells (Fig. D)

3–4 layered sclerenchymatous

Type-E sclerenchyma present (Fig. D)

With III PVB embedded

Ground tissue: Round to oval-shaped parenchymatous (Fig. D)

Vascular bundles arranged in ring form

Vascular system: All present

I IVB: Type C (Fig. F)

Size: 15.45 × 21.36 ± 2.15 µm

II IVB: Type P (Fig. A)

Size: 8.57 × 13.39 ± 3.71 µm

III PVB: Type Q (Fig. A)

Size: 5.88 × 7.35 ± 1.62 µm

Bar size: A = 50 µm, B–F = 5 µm

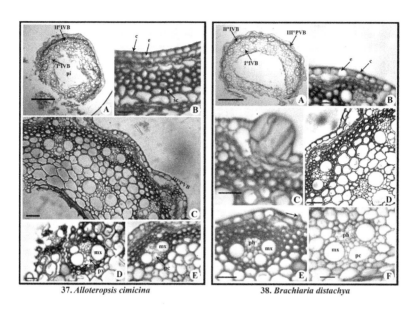

37. *Alloteropsis cimicina*   38. *Brachiaria distachya*

39. *Brachiaria eruciformis* (Fig. 2.22 (39) A–F)

Square to oval in transaction with slightly undulating surface (Fig. A)

Size: 112.5 × 137.5 ± 22.08 μm, hairs/prickles absent

Hollow internode

Epidermis: Barrel shaped III (Fig. F)

Hypodermis: 2–3 layered parenchyma cells (Fig. E)

3–4 layered sclerenchymatous

With III PVB embedded

Type-E sclerenchyma present

Ground tissue: Round to oval-shaped parenchymatous (Fig. B)

Vascular bundles arranged in ring form

Vascular system: All present

I IVB: Type A, sclerenchymatous cells present (Fig. D)

Size: 14.09 × 17.27 ± 2.17 μm

II IVB: Type L (Fig. C)

Size: 8 × 10 ± 1.27 μm

III PVB: Type Q

Size: 3.34 × 4.17 ± 1.34 μm

Bar size: A = 50 μm, B = 10 μm, C–F = 5 μm

40. *Brachiaria ramosa* (Fig. 2.22 (40) A–G)

Oblong in transaction with smooth surface (Fig. A)

Size: 340.91 × 572.73 ± 32.72 µm, hairs/prickles absent

Solid internode with parechymatous pith

Epidermis: Rectangular shaped I (Fig. B)

Hypodermis: 3–4 layered parenchyma cells

5–6 layered sclerenchymatous (Fig. D)

With III PVB embedded

Type-E sclerenchyma present

Ground tissue: Round to oval-shaped parenchymatous

Vascular bundles arranged in ring form

Vascular system: All present

I IVB: Type B, sclerenchymatous cells covered VB (Fig. E)

Size: 26.67 × 22.78 ± 5.16 µm

II IVB: Type P (Fig. G)

Size: 11.25 × 11.56 ± 2.18 µm

III PVB: Type Q (Fig. F)

Size: 7.69 × 11.54 ± 3.18 µm

Bar size: A = 50 µm, B, C, E–G = 5 µm, D = 10 µm

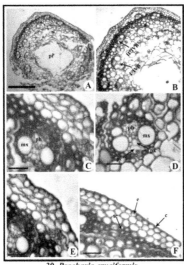

39. *Bracharia eruciformis*

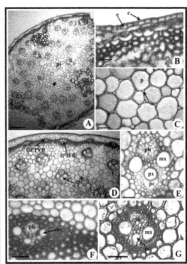

40. *Bracharia ramosa*

41. *Brachiaria reptans* (Fig. 2.23 (41) A–F)

Horseshoe in transaction with slightly undulating surface (Fig. A)

Size: 167.86 × 164.28 ± 21.91 μm, hairs/prickles absent

Hollow internode. Kranz anatomy present (Fig. D)

Epidermis: Barrel shaped III (Fig. F)

Hypodermis: 3–4 layered chlorenchymatous, 2–3 layered Type-D sclerenchymatous cells (Fig. E)

With III PVB embedded

Ground tissue: Round to oval-shaped parenchymatous, vascular bundles arranged in ring form

Vascular system: All present

I IVB: Type H (Fig. B)

Size: 15.24 × 17.14 ± 3.16 μm

II IVB: Type M, Sclerenchyma surrounded (Fig. C)

Size: 10.68 × 9.88 ± 3.16 μm

III PVB: Type Q (Fig. D)

Size: 7.69 × 5 ± 1.26 μm

Kranz arc: Half circle (Fig. F)

Chloroplast type: Centripetal

Shape of Kranz cell: Round to oval (Fig. F)

Mestome: absent

Bar size: A = 50 μm, B–F = 5 μm

DIFFERENTIATING FEATURES:

Presence of Kranz arc ... *B. reptans*

Absence of Kranz arc

Barrel-shaped epidermal cell ... *B. eruciformis*

Rectangular-shaped epidermal cell

Triangular diamond shaped I IVB, oval shaped in transaction......*B. distachya*
Diamond shaped I IVB, oblong shaped in transaction.....…...… *B. ramosa*

Cenchrus

42. *Cenchrus biflorus*

43. *Cenchrus ciliaris*

44. *Cenchrus setigerus*

42. *Cenchrus biflorus* (Fig. 2.23 (42) A–F)

Oblong in transaction with smooth surface (Fig. A)

Size: 165 × 480 ± 14.86 µm, hairs/prickles absent

Solid internode with parechymatous pith

Epidermis: Barrel shaped IV (Fig. C)

Hypodermis: 3–4 layered parenchymatous cell, than 2–3 layered Type-B sclerenchymatous cells

With III PVB embedded

Ground tissue: Round to oval-shaped parenchyma (Fig. D), vascular bundles arranged in ring form

Vascular system: All present

I IVB: Type I, phloem surrounded by sclerenchymatous cells (Fig. F)

Size: 23.18 × 18.18 ± 7.16 µm

II IVB: Type P (Fig. B)

Size: 11.5 × 14.5 ± 2.18µm

III PVB: Type Q (Fig. E)

 Size: 5.56 × 7.78 ± 1.72 µm

Bar size: A = 50 µm, B–F = 5 µm

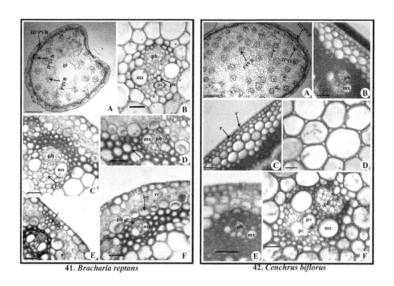

41. *Bracharia reptans*   42. *Cenchrus biflorus*

43. *Cenchrus ciliaris* (Fig. 2.24 (43) A–E)

Round in transaction with smooth surface (Fig. A)

Size: 600 × 591.67 ± 47.61 µm, hairs/prickles absent

Solid internode with parechymatous pith

Epidermis: Rectangular shaped I (Fig. B)

Hypodermis: 3–4 layered parenchymatous (Fig. A)

With III PVB embedded

Type-D sclerenchyma present in form of girder

Ground tissue: Round to oval-shaped parenchymatous (Fig. A)

Vascular bundles arranged in ring form

Vascular system: All present

I IVB: Type B, sclerenchymatous cap absent (Fig. E)

Size: 36.67 × 28 ± 6.17 µm

II IVB: Type M (Fig. D)

Size: 18.63 × 15.91 ± 4.76 µm

III PVB: Type Q (Fig. C)

Size: 15.2 × 13.6 ± 5.15 µm

Bar size: A = 50 µm, B–E = 5 µm

44. *Cenchrus setigerus* (Fig. 2.24 (44) A–G)

Horseshoe in transaction with slightly undulating surface (Fig. A)

Size: 333.34 × 311.12 ± 16.24 µm, hairs/prickles absent

Hollow internode. Kranz anatomy present (Fig. D)

Epidermis: Round shaped (Fig. B)

Hypodermis: 6–7 layered chlorenchymatous interrupted by sclerenchymatous girder

After ring of Type-B sclerenchyma present (Fig. D)

With III PVB embedded

Ground tissue: Round to oval-shaped parenchymatous, starch grains present (Fig. E)

Vascular bundles arranged in ring form

Vascular system: All present

I IVB: Type H (Fig. F)

Size: 28.18 × 32.73 ± 4.16 µm

II IVB: Type P (Fig. G)

Size: 10.88 × 13.82 ± 3.28 µm

III PVB: Type Q

Size: 4.68 × 4.37 ± 1.62 µm

Kranz arc: Half circle (Fig. D)

Chloroplast type: Centrifugal

Shape of Kranz cell: Round to oval (Fig. D)

Mestome: Single layered

Bar size: A = 50 µm, B–G = 5 µm

DIFFERENTIATING FEATURES:

Sclerenchyma in form of girder ... *C. ciliaris*
Sclerenchyma in form of unicylinder
Sclerenchyma type B present ... *C. biflorus*
Sclerenchyma type E present ... *C. setigerus*

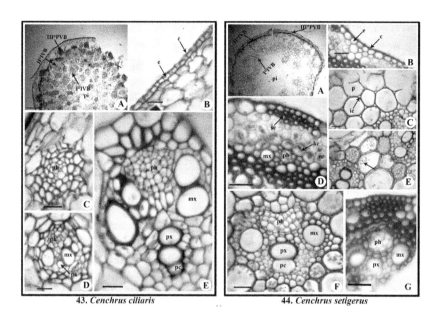

43. *Cenchrus ciliaris*   44. *Cenchrus setigerus*

Digitaria

45. *Digitaria ciliaris*

46. *Digitaria granularis*

47. *Digitaria longiflora*

48. *Digitaria stircta*

45. *Digitaria ciliaris* (Fig. 2.25 (45) A–F)

Round in transaction with smooth surface (Fig. A)

Size: 320 × 305 ± 21.63 µm, hairs/prickles absent

Hollow internode

Epidermis: Barrel shaped III (Fig. B)

Hypodermis: 5–6 layered sclerenchymatous

With III PVB embedded

Type-D sclerenchyma present (Fig. C)

Ground tissue: Round to oval-shaped parenchymatous (Fig. D)

Vascular bundles arranged in ring form

Vascular system: All present

I IVB: Type B, sclerenchymatous cap absent (Fig. F)

Size: 21.54 × 23.85 ± 6.18 μm

II IVB: Type L (Fig. E)

Size: 9.47 × 11.58 ± 3.71 μm

III PVB: Type Q

Size: 6.15 × 5.72 ± 2.16 μm

Bar size: A = 50 μm, B–F = 5 μm

46. *Digitaria granularis* (Fig. 2.25 (46) A–F)

Round to oval in transaction with smooth surface (Fig. A)

Size: 258.34 × 294.26 ± 13.24 μm, hairs/prickles absent

Hollow internode

Epidermis: Barrel shaped II (Fig. B, E)

Hypodermis: 3–4 layered parenchymatous interrupted by sclerenchymatous girder

With III PVB embedded (Fig. C)

Type-D sclerenchyma present

Ground tissue: Round to oval-shaped parenchymatous (Fig. C)

Vascular bundles arranged in ring form

Vascular system: III PVB absent

I IVB: Type D, sclerenchymatous cap present (Fig. F), big square lysigenous cavity

Size: 43.75 × 40 ± 6.16 μm

II IVB: Type O (Fig. D, E)

Size: 16.67 × 25.34 ± 4.14μm

III PVB: Size:

Bar size: A = 50 μm, C & D = 10 μm, B, E–F = 5 μm

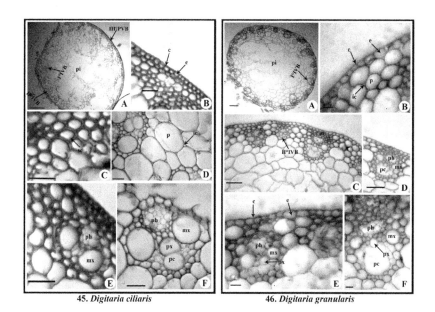

45. *Digitaria ciliaris*   **46. *Digitaria granularis***

47. *Digitaria longiflora* (Fig. 2.26 (47) A–E)

Round to triangular with blunt angle in transaction (Fig. A)

Size: 409.09 × 463.64 ± 72.22 µm, hairs/prickles absent

Hollow internode

Epidermis: Barrel-shaped I (Fig. D)

Hypodermis: 3–4 layered sclerenchymatous

With III PVB embedded

Type-D sclerenchyma present

Ground tissue: Round to oval-shaped parenchymatous (Fig. B)

Vascular bundles arranged in ring form

Vascular system: All present

I IVB: Type F, sclerenchymatous cells surrounded VB (Fig. E)

Size: 36 × 30 ± 9.26 µm

II IVB: Type J (Fig. C)

Size: 16.24 × 21.29 ± 3.28 µm

III PVB: Type Q (Fig. C, D)

Size: 3 × 3.67 ± 1.62 µm

Bar size: A = 50 μm, C = 10 μm, B, D–E = 5 μm

48. *Digitaria stricta* (Fig. 2.26 (48) A–E)

Oval to oblong in transaction with undulating surface (Fig. A)

Size: 250 × 308.34 ± 24.25 μm, hairs/prickles absent

Hollow internode

Epidermis: Barrel shaped II (Fig. E)

Hypodermis: 1–2 layered parenchyma

Than 3–4 layered sclerenchymatous

With III PVB embedded

Type-E sclerenchyma present (Fig. D)

Ground tissue: Irregular-shaped parenchymatous

Vascular bundles arranged in ring form

Vascular system: All present

I IVB: Type B, sclerenchymatous cap present (Fig. C)

Size: 9.56 × 11.36 ± 3.16 μm

II IVB: Type P (Fig. E)

Size: 5 × 5.67 ± 1.25 μm

III PVB: Type Q (Fig. B)

Size: 2 × 4 ± 0.34 μm

Bar size: A = 50 μm, C = 10 μm, B, D–E = 5 μm

DIFFERENTIATING FEATURES:

Sclerenchyma type E present … *D. stricta*
Sclerenchyma type D present
Sclerenchyma in form of unicylinder … *D. ciliaris*
Sclerenchyma in form of girder
III IVB absent, obovate shaped I IVB … *D. granularis*
III IVB present, vertically elliptical shaped I IVB … *D. longiflora*

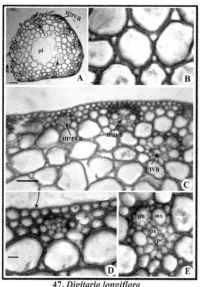

47. *Digitaria longiflora*

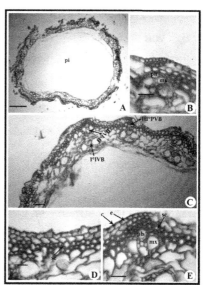

48. *Digitaria stricta*

Echinichloa

49. *Echinichloa colona*

50. *Echinochloa crusgalli*

51. *Echionochloa stagnina*

49. *Echinochloa colona* (Fig. 2.27 (49) A–G)

Round in transaction with smooth surface (Fig. A)

Size: 1123.08 × 1184.62 ± 82.63 µm, hairs/prickles absent

Solid internode with parechymatous pith

Epidermis: Barrel shaped III (Fig. C)

Hypodermis: 3–4 layered sclerenchymatous (Fig. D)

With III PVB embedded

Type-D sclerenchyma present

Ground tissue: Round to oval-shaped parenchymatous (Fig. B)

Vascular bundles arranged in ring form

Vascular system: All present

I IVB: Type I and E, sclerenchymatous cap present (Fig. F, G)

Size: 24.17 × 31.67 ± 7.25 μm

II IVB: Type K

Size: 20.56 × 25.56 ± 6.24 μm

III PVB: Type Q (Fig. E)

Size: 14 × 11.5 ± 4.26 μm

Bar size: A = 50 μm, B–C, E–G = 5 μm, D = 10 μm

50. *Echinochloa crusgalli* (**Fig. 2.27 (50) A–E**)

Round in transaction with slightly undulating surface (Fig. A)

Size: 1584.61 × 1638.46 ± 69.18 μm, hairs/prickles absent

Solid internode with parechymatous pith

Epidermis: Rectangular shaped III (Fig. B)

Hypodermis: 7–8 layered parenchyma

Than 6–7 layered sclerenchymatous

With III PVB embedded

Type-D sclerenchyma present (Fig. C)

Ground tissue: Round to oval-shaped parenchymatous

Vascular bundles arranged in ring form

Vascular system: All present

I IVB: Type I, sclerenchymatous cap present (Fig. E)

Size: 14.64 × 14.29 ± 3.12 μm

II IVB: Type M

Size: 0.09 × 9.55 ± 3.17 μm

III PVB: Type Q (Fig. D)

Size: 4.09 × 3.64 ± 2.13 μm

Bar size: A = 50 μm, B–E = 5 μm

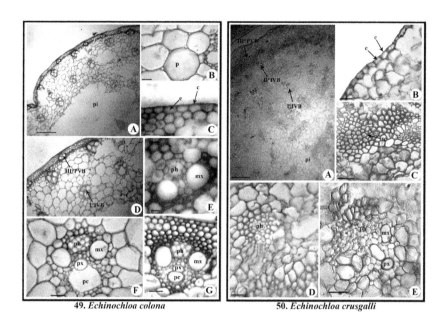

49. *Echinochloa colona*   **50. *Echinochloa crusgalli***

51. *Echinochloa stagnina* (Fig. 2.28 (51) A–E)

Round in transaction with slightly undulating surface (Fig. A)

Size: 1050 × 1163 ± 62.58 μm, hairs/prickles absent

Solid internode with parechymatous pith

Epidermis: Rectangular shaped II (Fig. B)

Hypodermis: 1–2 layered Type-B sclerenchymatous

Than thickend Type E sclerenchyma 4–5 layered present (Fig. C)

With III PVB embedded

Ground tissue: Angular-shaped parenchymatous (Fig. A)

Vascular bundles arranged in ring form

Vascular system: All present

I IVB: Type B, sclerenchymatous cell present surrounded VB (Fig. E)

Size: 29.17 × 25 ± 7.14 μm

II IVB: Type O (Fig. A)

Size: 23.34 × 25 ± 5.12 μm

III PVB: Type Q (Fig. D)

Size: 8.34 × 12.5 ± 3.62 μm

Bar size: A = 50 μm, B–E = 5 μm

DIFFERENTIATING FEATURES:

Barrel-shaped epidermal cell … *E. colona*
Rectangular-shaped epidermal cell
Sclerenchyma in form of unicylinder … *E. crusgalli*
Sclerenchyma in form of bicylinder … *E. stagnina*

52. *Eremopogon foveolatus* (Fig. 2.28 (52) A–F)

Round to oval in transaction with smooth surface (Fig. A)

Size: 1210 × 1560 ± 102.13 μm, hairs/prickles absent

Solid internode with parechymatous pith

Epidermis: Barrel shaped V (Fig. B)

Hypodermis: 5–6 layered sclerenchymatous

With III PVB embedded

Type-E sclerenchyma present

Ground tissue: Round to oval-shaped parenchymatous (Fig. C)

Vascular bundles arranged in ring form

Vascular system: All present

I IVB: Type F, sclerenchymatous cell present around phloem (Fig. F)

Size: 26.12 × 16.24 ± 6.15 μm

II IVB: Type O (Fig. E)

Size: 21.92 × 16.15 ± 7.36 μm

III PVB: Type Q (Fig. D)

Size: 7 × 14 ± 3.16 μm

Bar size: A = 50 μm, B–F = 5 μm

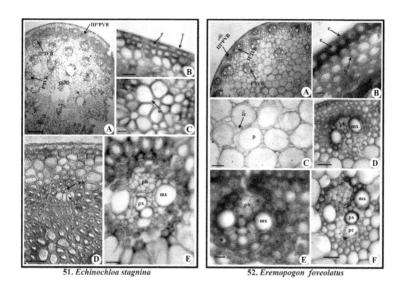

51. *Echinochloa stagnina*   52. *Eremopogon foveolatus*

53. *Eriochloa procera* (Fig. 2.29 (53) A–G)

Round in transaction with smooth surface (Fig. A)

Size: 261.90 × 304.76 ± 36.41 μm, hairs/prickles absent

Solid internode with parechymatous pith. Kranz anatomy present

Epidermis: Round shaped (Fig. D)

Hypodermis: 4–5 layered chloenchymatous (Fig. B)

Than 1–2 layered sclerenchyma with III PVB embedded

Type-B sclerenchyma present

Ground tissue: Round to oval-shaped parenchymatous (Fig. C)

Vascular bundles arranged in ring form

Vascular system: All present

I IVB: Type B, sclerenchymatous cap absent (Fig. G)

Size: 12.72 × 16.82 ± 4.52 μm

II IVB: Type K (Fig. F)

Size: 7.73 × 11.36 ± 3.17 μm

III PVB: Type Q (Fig. E)

Size: 6.07 × 8.03 ± 2.76 μm

Kranz arc: Half circle (Fig. E)

Chloroplast type: Centrifugal

Shape of Kranz cell: Round to oval

Mestome: Single layered

Bar size: A = 50 µm, B = 10 µm, C–G = 5 µm

Oplismenus

54. *Oplismenus burmannii*

55. *Oplismenus composites*

54. *Oplismenus burmannii* (Fig. 2.29 (54) A–F)

Irregular round in transaction with undulating surface (Fig. A)

Size: 173.34 × 193.34 ± 36.29 µm, hairs/prickles absent

Solid internode with parechymatous pith

Epidermis: Barrel shaped IV (Fig. C)

Hypodermis: 3–4 layered parenchymatous (Fig. C)

Interrupted by sclerenchymatous girder

With III PVB embedded

Type-E sclerenchyma present

Ground tissue: Round to oval-shaped parenchymatous (Fig. D)

Vascular bundles arranged in ring form

Vascular system: All present

I IVB: Type E (Fig. F)

Size: 14.34 × 13.34 ± 5.26 µm

II IVB: Type M (Fig. E)

Size: 12.34 × 11.34 ± 4.26 µm

III PVB: Type Q (Fig. A)

Size: 4 × 5 ± 2.16 µm

Bar size: A = 50 µm, B = 10 µm, C–F = 5 µm

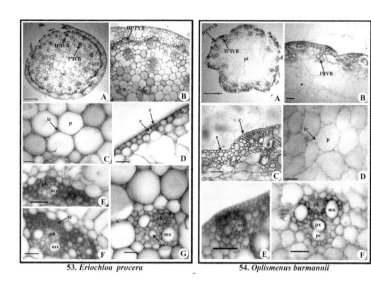

53. *Eriochloa procera*   54. *Oplismenus burmannii*

55. *Oplismenus composites* (Fig. 2.30 (55) A–F)

Oval in transaction with undulating surface (Fig. A)

Size: 214.28 × 257.14 ± 37.34 μm, hairs/prickles absent

Solid internode with parechymatous pith

Epidermis: Barrel-shaped V (Fig. B)

Bulliform cell present (Fig. D)

Hypodermis: 4–5 layered sclerenchymatous (Fig. C)

With III PVB embedded

Type-D sclerenchyma present

Ground tissue: Round to oval-shaped parenchymatous (Fig. C)

Vascular bundles arranged in ring form

Vascular system: III PVB absent

I IVB: Type D, sclerenchymatous cap absent (Fig. F)

Size: 12.5 × 14.75 ± 3.71 μm

II IVB: Type P (Fig. E)

Size: 5.22 × 6.52 ± 2.53 μm

III PVB:

Bar size: A = 50 μm, B–F = 5 μm

DIFFERENTIATING FEATURES:

Sclerenchyma type E present ... *O. burmanii*
Sclerenchyma type D present ... *O. composites*

Panicum

56. *Panicum antidotale*

57. *Panicum maximum*

58. *Panicum miliaceum*

59. *Panicum trypheron*

56. *Panicum antidotale* (Fig. 2.30 (56) A–E)

Oval in transaction with slightly undulating surface (Fig. A)

Size: 480 × 1146.67 ± 62.25 µm, hairs/prickles absent

Hollow internode

Epidermis: Elongated shaped-I (Fig. C)

Hypodermis: 3–4 layered parenchymaotus , than 1–2 layered Type-D sclerenchymatous cells

With III PVB embedded

Ground tissue: Round to oval-shaped parenchymatous (Fig. B)

Vascular bundles arranged in ring form

Vascular system: All present

I IVB: Type B, sclerenchymatous cap absent (Fig. E)

Size: 19.55 × 20.91 ± 7.21 µm

II IVB: Type P (Fig. D)

Size: 10 × 12.61 ± 4.14 µm

III PVB: Type Q

Size: 6.67 × 10 ± 3.61 µm

Bar size: A = 50 µm, B–E = 5 µm

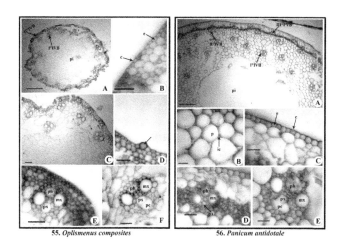

55. *Oplismenus composites*   56. *Panicum antidotale*

57. *Panicum maximum* (Fig. 2.31 (57) A–H)

Round in transaction with smooth surface (Fig. A)

Size: 533.34 × 438.09 ± 36.26 μm, hairs/prickles absent

Hollow internode

Epidermis: Rectangular shaped III (Fig. D)

Hypodermis: 6–7 layered parenchymatous

Than 3–4 layered sclerenchymatous cells

With III PVB embedded

Type-D sclerenchyma present

Ground tissue: Round to oval-shaped parenchymatous (Fig. C, D)

Vascular bundles arranged in ring form

Vascular system: All present

I IVB: Type C (Fig. H, G)

Size: 22.5 × 23.5 ± 7.25 μm

II IVB: Type N (Fig. F)

Size: 8.69 × 16.95 ± 2.47 μm

III PVB: Type Q (Fig. E)

Size: 5.16 × 4.17 ± 2.13 μm

Bar size: A = 50 μm, B = 10 μm, C–H = 5 μm

58. *Panicum miliaceum* (Fig. 2.31(58) A–F)

Horseshoe in transaction with smooth surface (Fig. A)

Size: 315.38 × 323.07 ± 25.16 µm, hairs/prickles absent

Hollow internode

Epidermis: Round shaped (Fig. B)

Hypodermis: 2–3 layered parenchymatous cells interrupted by sclerenchymatous cells (Fig. B).

With III PVB embedded

Type-B sclerenchyma present

Ground tissue: Round to oval-shaped parenchymatous (Fig. C)

Vascular bundles arranged in ring form

Vascular system: All present

I IVB: Type E (Fig. F)

Size: 19.68 × 23.22 ± 5.27 µm

II IVB: Type M (Fig. E)

Size: 17.23 × 19.45 ± 6.16 µm

III PVB: Type Q (Fig. D)

Size: 6.07 × 9.28 ± 2.72 µm

Bar size: A = 50 µm, B–F = 5 µm

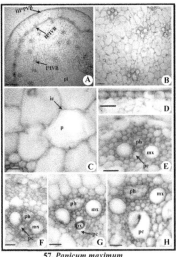

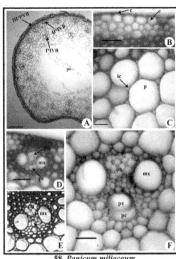

57. *Panicum maximum*   58. *Panicum miliaceum*

59. *Panicum trypheron* (Fig. 2.32 (59) A–F)

Horseshoe in transaction with smooth surface (Fig. A). Hairs/prickles absent

Size: $316.67 \times 308.34 \pm 51.56$ µm, hollow internode. Kranz anatomy present (Fig. C)

Epidermis: Elongated shaped II (Fig. B)

Hypodermis: 3–4 layered chlorenchymatous cells

Interrupted by Type-E sclerenchymatous (Fig. C) cells with III PVB embedded

Ground tissue: Round to oval-shaped parenchyma (Fig. E), vascular bundles arranged in ring form

Vascular system: all present

I IVB: Type E (Fig. F)

Size: $15.94 \times 19.69 \pm 6.17$ µm

II IVB: Type P (Fig. D)

Size: $8.34 \times 12.5 \pm 3.17$ µm

III PVB: Type Q (Fig. C)

Size: $6.39 \times 10 \pm 2.15$ µm

Kranz arc: Half circle (Fig. C)

Chloroplast type: Centrifugal

Shape of Kranz cell: Round to oval

Mestome: Single layered

Bar size: A = 50 µm, B–F = 5 µm

DIFFERENTIATING FEATURES:

Presence of Kranz arc … *P. trypheron*
Absence of Kranz arc
Sclerenchyma type B present … *P. miliaceum*
Sclerenchyma type D present
Elongated-shaped epidermal cell … *P. antidotalae*
Rectangular-shaped epidermal cell … *P. maximum*

Paspalidium

60. *Paspalidium flavidum* 61. *Paspalidium geminatum*

60. *Paspalidium flavidum* (Fig. 2.32 (60) A–G)

Oval with distal end flat in transaction with slightly undulating surface (Fig. A), hairs/prickles absent

Size: 361.90 × 280.95 ± 47.25 µm, solid internode with parenchyma pith. Kranz anatomy (Fig. E)

Epidermis: Rectangular shaped III (Fig. C)

Hypodermis: 3–4 layered chlorenchyma (Fig. B)

Than 2–3 layered Type-D sclerenchymatous with III PVB embedded

Ground tissue: Round to oval-shaped parenchyma(Fig. D), vascular bundles arranged in ring form

Vascular system: All present

I IVB: Type I, sclerenchymatous small cap present (Fig. G)

Size: 15.43 × 16.08 ± 3.16 µm

II IVB: Type P (Fig. F)

Size: 11.78 × 19.28 ± 3.26 µm

III PVB: Type Q (Fig. E)

Size: 3.13 × 3.13 ± 1.25 µm

Kranz arc: Half circle (Fig. E)

Chloroplast type: Centripetal

Shape of Kranz cell: Round to oval

Mestome: Single layered

Bar size: A = 50 µm, B = 10 µm, C-G = 5 µm

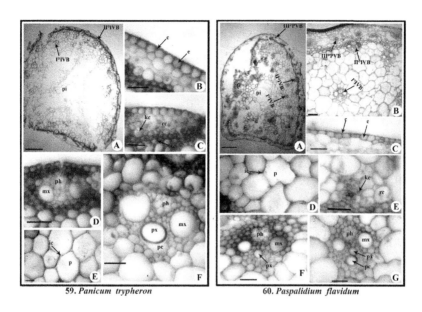

59. *Panicum trypheron*   60. *Paspalidium flavidum*

61. *Paspalidium geminatum* (Fig. 2.33 (61) A–E)

Round to oval in transaction with smooth surface (Fig. A)

Size: 1080.64 × 1677.42 µm, hairs/prickles absent

Hollow internode

Epidermis: Round shaped (Fig. C)

Hypodermis: 3–4 layered parenchymatous cells

Few sclerenchymatous cells interrupted

With III PVB embedded

Type-D sclerenchyma present

Ground tissue: Round to oval-shaped parenchymatous (Fig. B)

Air cavity presents (Fig. A)

Vascular bundles arranged in ring form

Vascular system: All present

I IVB: Type B, sclerenchymatous cap present (Fig. E)

Size: 18.21 × 18.39 ± 4.15 µm

II IVB: Type M (Fig. D)

Size: 10.78 × 11.56 ± 2.36 µm

III PVB: Type Q

Size: 4.27 × 4.03 ± 1.82 µm

Bar size: A = 50 µm, B–E = 5 µm

DIFFERENTIATING FEATURES:

Rectangular-shaped epidermal cell … *P. geminatum*

Round-shaped epidermal cell … *P. flavidum*

62. *Paspalum scrobiculatum* (Fig. 2.33 (62) A–E)

Oval in transaction with undulating surface (Fig. A)

Size: 275 × 362.5 ± 13.54 µm, hairs/prickles absent

Hollow internode

Epidermis: Rectangular-shaped IV (Fig. C)

Hypodermis: 3–4 layered parenchymatous cells (Fig. A)

Than 2–3 layered sclerenchymatous cells

With III PVB embedded

Type-D sclerenchyma present

Ground tissue: Round to oval-shaped parenchymatous (Fig. B)

Vascular bundles arranged in ring form

Vascular system: All present

I IVB: Type B, sclerenchymatous cap absent (Fig. E)

Size: 18.75 × 15.63 ± 3.27 µm

II IVB: Type P (Fig. D)

Size: 17.08 × 21.04 ± 3.65 µm

III PVB: Type Q

Size: 6.34 × 8.34 ± 2.56 µm

Bar size: A = 50 µm, B–E = 5 µm

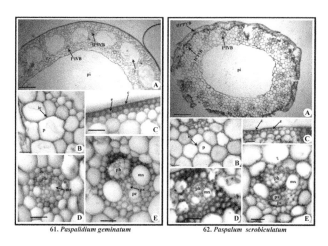

61. *Paspalidium geminatum*   62. *Paspalum scrobiculatum*

63. *Pennisetum setosum* (Fig. 2.34 (63) A–F)

Oval in transaction with smooth surface (Fig. A)

Size: 742.23 × 1151.13 ± 92.36 µm, hairs/prickles absent

Solid internode with parechymatous pith

Epidermis: Rectangular shaped II (Fig. F)

Hypodermis: 3–4 layered of parenchymatous cells (Fig. A)

6–7 layered sclerenchymatous cells

With III PVB embedded

Type-B sclerenchyma present

Ground tissue: Round to oval-shaped parenchymatous (Fig. D)

Vascular bundles arranged in ring form

Vascular system: All present

I IVB: Type F and B (Fig. C)

Sclerenchymatous cells around VB scatterly arranged

Size: 31.43 × 27.85 ± 8.62 µm

II IVB: Type K (Fig. E)

Size: 13.46 × 17.69 ± 4.28 µm

III PVB: Type Q (Fig. B)

Size: 8.12 × 12.5 ± 3.28 µm

Bar size: A = 50 µm, B–F = 5 µm

Setaria

64. *Setaria glauca* 65. *Setaria tomentosa* 66. *Setaria verticillata*

64. *Setaria glauca* (Fig. 2.34 (64) A–G)

Oval in transaction with smooth surface (Fig. A)

Size: 777.79 × 1100 ± 104.26 μm, hairs/prickles absent

Solid internode with parechymatous pith

Epidermis: Barrel-shaped IV (Fig. C)

Hypodermis: 1 layered parenchymatous (Fig. B)

Than 3–4 layered sclerenchymatous

Type-E sclerenchyma present

Ground tissue: Round to oval-shaped parenchymatous (Fig. D)

Vascular bundles arranged in ring form

Vascular system: All present

I IVB: Type H, sclerenchymatous cells around VB (Fig. G)

Size: 22.5 × 29.37 ± 6.37 μm

II IVB: Type J (Fig. F)

Size: 21.36 × 26.36 ± 5.89 μm

III PVB: Type Q (Fig. E)

Size: 15 × 17 ± 4.72 μm

Bar size: A–B = 50 μm, C–G = 5 μm

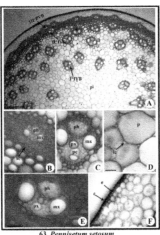

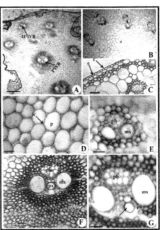

63. *Pennisetum setosum*   64. *Setaria glauca*

65. *Setaria tomentosa* (Fig. 2.35 (65) A–F)

Triangular in transaction with smooth surface (Fig. A)

Size: 263.64 × 290.90 ± 14.52 μm, hairs/prickles absent

Solid internode with parechymatous pith

Epidermis: Barrel shaped III

Hypodermis: 1 layered parenchymatous cells (Fig. C)

Than 2–3 layered sclerenchymatous

With III PVB embedded

Type-D sclerenchyma present

Ground tissue: Round to oval-shaped parenchymatous

Vascular bundles arranged in ring form

Vascular system: All present

I IVB: Type I, sclerenchymatous cap absent (Fig. E)

Size: 20.38 × 13.07 ± 5.28 μm

II IVB: Type P (Fig. D)

Size: 12.67 × 15 ± 3.52 μm

III PVB: Type Q (Fig. F)

Size: 5.23 × 6.36 ± 2.45 μm

Bar size: A = 50 μm, B = 10 μm, C–F = 5 μm

66. *Setaria verticillata* (Fig. 2.35 (66) A–F)

Oblong in transaction with slightly undulating surface (Fig. A)

Size: 241.67 × 325 ± 21.68 μm, hairs/prickles absent

Solid internode with parechymatous pith

Epidermis: Barrel shaped IV (Fig. B)

Hypodermis: 3–4 layered parenchymatous

Interrupted by sclerenchymatous girder

With III PVB embedded

Type-D sclerenchyma present (Fig. C)

Ground tissue: Round to oval-shaped irregular parenchymatous

Vascular bundles arranged in ring form

Vascular system: All present

I IVB: Type B, sclerenchymatous cap present (Fig. F)

Size: 16.43 × 18.21 ± 4.27 μm

II IVB: Type O (Fig. E)

Size: 9.26 × 10.27 ± 3.79 μm

III PVB: Type Q

Size: 6.11 × 6.67 ± 2.74 μm

Bar size: A = 50 μm, B–F = 5 μm

DIFFERENTIATING FEATURES:

Sclerenchyma in form of girder … *S. verticillata*
Sclerenchyma in form of unicylinder
Sclerenchyma type E present … *S. glauca*
Sclerenchyma type D present … *S. tomentosa*

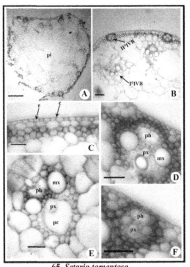

65. *Setaria tomentosa*

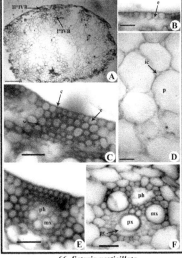

66. *Setaria verticillata*

67. *Isachne globosa* (Fig. 2.36 (67) A–F)

Cordate in transaction with slightly undulating surface (Fig. A)

Size: 268.18 × 431.82 ± 16.29 μm, hairs/prickles absent

Hollow internode. Kranz anatomy present (Fig. B)

Epidermis: Barrel shaped IV (Fig. C)

Hypodermis: 3–4 layered chlorenchymatous cells (Fig. B)

Interrupted by sclerenchymatous with III PVB embedded

Type-E sclerenchyma present

Ground tissue: Round to oval-shaped parenchymatous

Vascular bundles arranged in ring form

Vascular system: All present

I IVB: Type E (Fig. F)

Size: 20.84 × 25.42 ± 6.42 μm

II IVB: Type J (Fig. F)

Size: 11.25 × 13.75 ± 4.82 μm

III PVB: Type Q (Fig. D)

Size: 8.37 × 8.75 ± 2.38 μm

Kranz arc: Half circle (Fig. D)

Chloroplast type: Centrifugal

Shape of Kranz cell: Round to oval

Mestome: Single layered

Bar size: A = 50 μm, B = 10 μm, C–F = 5 μm

Aristida

68. *Aristida adscensionis*

69. *Aristida funiculata*

68. *Aristida adscensionis* (Fig. 2.36 (68) A–E)

Oblong in transaction with slightly undulating surface (Fig. A)

Size: 225.92 × 229.63 ± 15.38 μm, hairs/prickles absent

Hollow internode. Kranz anatomy is present

Epidermis: Barrel shaped IV

Hypodermis: 2–3 layered chlorenchymatous (Fig. B)

Middle layer cell big in size than sclerenchymatouscells present (Fig. B)

With III PVB embedded

Type-D sclerenchyma present

Ground tissue: Round to oval-shaped parenchymatous (Fig. C)

Vascular bundles arranged in ring form

Vascular system: All present

I IVB: Type E, sclerenchymatous cap present (Fig. E)

Size: 13.57 × 12.86 ± 4.89 μm

II IVB: Type O

Size: 8.28 × 9.14 ± 2.19 μm

III PVB: Type Q (Fig. D)

Size: 3.28 × 3.81 ± 1.34 μm

Kranz arc: Half circle (Fig. D)

Chloroplast type: Centrifugal

Shape of Kranz cell: Round to oval

Mestome: Single layered

Bar size: A = 50 μm, B–E = 5 μm

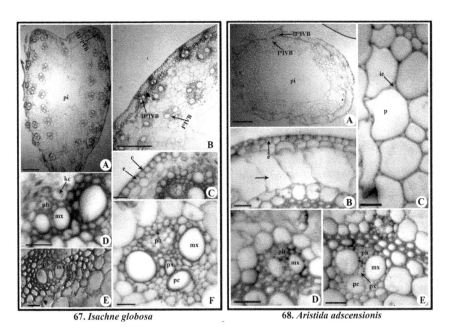

67. Isachne globosa   *68. Aristida adscensionis*

69. *Aristida funiculata* (Fig. 2.37 (69) A–F)

Round in transaction with slightly undulating surface (Fig. A)

Size: 196.37 × 198.32 μm, hairs/prickles absent

Solid internode with parechymatous pith. Kranz anatomy present (Fig. D)

Epidermis: Barrel-shaped III

Hypodermis: 3–4 layered chlorenchyma (Fig. B)

Interrupted by sclerenchymatous with III PVB embedded

Type-E sclerenchyma present

Ground tissue: Round to oval-shaped parenchymatous (Fig. C)

Vascular bundles arranged in ring form

Vascular system: All present

I IVB: Type B, sclerenchymatous samll cap present (Fig. E)

Size: 11.25 × 15 ± 3.16 μm

II IVB: Type P (Fig. F)

Size: 5.71 × 8.57 ± 2.46 μm

III PVB: Type Q (Fig. D)

Size: 3.44 × 6.25 ± 1.37 μm

Kranz arc: Horseshoe (Fig. D)

Chloroplast type: Centrifugal

Shape of Kranz cell: Round to oval (Fig. D)

Mestome: Single layered

Bar size: A = 50 μm, B = 10 μm, C–F = 5 μm

DIFFERENTIATING FEATURES:

Half circle-shaped Kranz arc, sclerenchyma type D present … *A. adscensionis*

Horseshoe-shaped Kranz arc, sclerenchyma type E present … *A. funiculata*

70. *Perotis indica* (Fig. 2.37 (70) A–F)

Round to oval in transaction with smooth surface (Fig. A)

Size: 98.54 × 47.81 ± 9.24 μm, hairs/prickles absent

Solid internode with parechymatous pith. Kranz anatomy is present (Fig. B)

Epidermis: Round shaped (Fig. F)

Hypodermis: 3–4 layered chlorenchymatous interrupted by sclerenchymatous (Fig.B)

With III PVB embedded

Type-E sclerenchyma present

Ground tissue: Round to oval-shaped parenchymatous (Fig. E)

Vascular bundles arranged in ring form

Vascular system: All present

I IVB: Type D, sclerenchymatous cap absent (Fig. C)

Size: 12.5 × 16.67 ± 3.74 μm

II IVB: Type M

Size: 10.36 × 11.73 ± 2.46 μm

III PVB: Type Q (Fig. D)

Size: 7.92 × 10.42 ± 1.53 μm

Kranz arc: Horseshoe (Fig. D)

Chloroplast type: Centrifugal

Shape of Kranz cell: Round to oval (Fig. D)

Mestome: Absent

Bar size: A = 50 μm, B = 10 μm, C–F = 5 μm

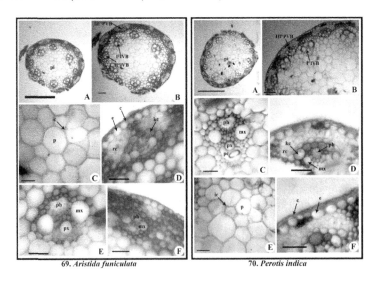

69. *Aristida funiculata*   70. *Perotis indica*

Chloris

71. *Chloris barbata*

72. *Choris montana*

73. *Chloris virgata*

71. *Chloris barbata* (Fig. 2.38 (71) A–G)

Round in transaction with slightly undulating surface (Fig. A)

Size: 129.54 × 118.18 ± 32.64 µm, hairs/prickles absent

Solid internode with parechymatous pith. Kranz anatomy is present (Fig. E).

Epidermis: Barrel shaped V (Fig. C)

Hypodermis: 3–4 layered chlorenchymatous

Interrupted by sclerenchymatous, Type-B sclerenchyma

Ground tissue: Round to oval-shaped parenchymatous (Fig. D)

Vascular bundles arranged in ring form

Vascular system: All present

I IVB: Type I, sclerenchymatous small cap present (Fig. F)

Size: 25 × 10.71 ± 6.63 µm

II IVB: Type M

Size: 12.78 × 17.22 ± 3.46 µm

III PVB: Type Q (Fig. G)

Size: 3.67 × 2.33 ± 1.63 µm

Kranz arc: Horseshoe (Fig. G)

Chloroplast type: Centripetal (Fig. G)

Shape of Kranz cell: Trapezoid (Fig. E)

Mestome: Single layered

Bar size: A = 50 µm, B = 10 µm, C–G = 5 µm

72. *Choris montana* (Fig. 2.38 (72) A–F)

Round in transaction with smooth surface (Fig. A)

Size: 171.44 × 145.24 ± 17.37 µm, hairs/prickles absent

Solid internode with parechymatous pith

Epidermis: Square shaped I (Fig. C)

Hypodermis: 3–4 layered chlorenchymatous interrupted by sclerenchymatous (Fig. B)

With III PVB embedded

Type-D sclerenchyma present

Ground tissue: Round to oval-shaped parenchymatous

Vascular bundles arranged in ring form

Vascular system: All present

I IVB: Type I, sclerenchymatous small cap present (Fig. F)

Size: 20 × 15.45 ± 4.17 μm

II IVB: Type M

Size: 9.25 × 10.15 ± 3.89 μm

III PVB: Type Q (Fig. E)

Size: 6.67 × 7.34 ± 2.34μm

Kranz arc: Horseshoe (Fig. E)

Chloroplast type: Centripetal

Shape of Kranz cell: Round to oval (Fig. E)

Mestome: Single layered

Bar size: A = 50 μm, B = 10 μm, C–F = 5 μm

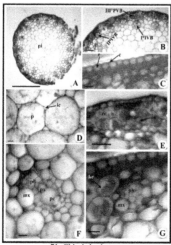 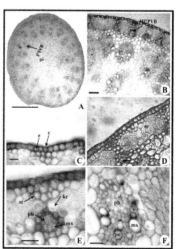

71. *Chloris barbata*   72. *Chloris montana*

73. *Chloris virgata* (Fig. 2.39 (73) A–G)

Oblong in transaction with smooth surface (Fig. A)

Size: 130 × 157.5 ± 26.27 μm, hairs/prickles absent

Solid internode with parechymatous pith. Kranz anatomy is present

Epidermis: Square shaped III (Fig. D)

Hypodermis: 3–4 layered chlorenchymatous interrupted by Type-D sclerenchymatous (Fig.B)

Ground tissue: Round to oval-shaped parenchymatous (Fig. E)

Vascular bundles arranged in ring form

Vascular system: All present

I IVB: Type B, scatterly arranged (Fig. G)

Size: 13.72 × 14.26 ± 2.36 μm

II IVB: Type M

Size: 7.63 × 8.68 ± 2.16 μm

III PVB: Type Q (Fig. F)

Size: 5 × 4.37 ± 1.25 μm

Kranz arc: Horseshoe (Fig. F)

Chloroplast type: Centrifugal

Shape of Kranz cell: Round to oval

Mestome: Single layered

Bar size: A = 50 μm, B = 10 μm, C–G = 5 μm

DIFFERENTIATING FEATURES:

Half circle-shaped Kranz arc … *C. barbata*

Straight-shaped Kranz arc

Centripetal arrangement of chloroplast … *C. montana*

Centrifugal arrangement of chloroplast … *C. virgata*

74. *Cynodon dactylon* (Fig. 2.39 (74) A–F)

Triangular in transaction with smooth surface (Fig. A)

Size: 277.27 × 265.09 ± 41.26 μm, hairs/prickles absent

Hollow internode. Kranz anatomy is present (Fig. C)

Epidermis: Elongated shaped II (Fig. C)

Hypodermis: 4–5 layered chlorenchymatous interrupted by sclerenchymatous (Fig. B)

With III PVB embedded

Type-E sclerenchyma present

Ground tissue: Round to oval-shaped parenchymatous (Fig. B)

Vascular bundles arranged in ring form

Vascular system: All present

I IVB: Type C, sclerenchymatous cap present (Fig. F)

Size: 14.58 × 13.75 ± 2.18 µm

II IVB: Type P (Fig. D)

Size: 9.09 × 12.73 ± 3.17 µm

III PVB: Type Q (Fig. C)

Size: 6.34 × 6.82 ± 2.37 µm

Kranz arc: Half circle (Fig. E)

Chloroplast type: Centrifugal

Shape of Kranz cell: Round to oval (Fig. E)

Mestome: Single layered

Bar size: A = 50 µm, B = 10 µm, C–F = 5 µm

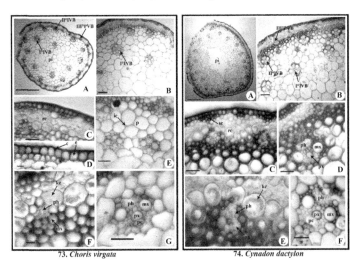

73. *Choris virgata*   74. *Cynadon dactylon*

75. *Melanocenchris jaequemontii* (Fig. 2.40 (75) A–F)

Oblong in transaction with smooth surface (Fig. A)

Size: 115.79 × 126.32 ± 36.26 μm, hairs/prickles absent

Solid internode with parechymatous pith

Epidermis: Rectangular-shaped II (Fig. D)

Hypodermis: 4–5 layered parenchymatous cells (Fig. B)

Than 2–3 layered sclerenchymatous

With III PVB embedded

Type-B sclerenchyma present

Ground tissue: Round to oval-shaped prenchymatous (Fig. B)

Starch grains present (Fig. C)

Vascular bundles arranged in ring form

Vascular system: III PVB absent

I IVB: Type I, sclerenchymatous cap absent (Fig. F)

Size: 10 × 10.59 ± 3.16 μm

II IVB: Type K (Fig. E)

Size: 7.81 × 10.31 μm

III PVB: Size

Bar size: A = 50 μm, B = 10 μm, C–F = 5 μm

76. *Oropetium villosulum* (Fig. 2.34 (76) A–F)

Round in transaction with slightly undulating surface (Fig. A)

Size: 116 × 136 ± 17.28 μm, hairs/prickles absent

Solid internode with parechymatous pith. Kranz anatomy is present

Epidermis: Square shaped III (Fig. D)

Hypodermis: 3–4 layered chlorenchymatous (Fig. B)

Interrupted by sclerenchymatous with III PVB embedded

Type-D sclerenchyma present

Ground tissue: Round to oval-shaped parenchymatous (Fig. C)

Vascular bundles arranged in ring form

Vascular system: All present

I IVB: Type H (Fig. F)

Size: 13 × 12 ± 4.17 µm

II IVB: Type M

Size: 7.78 × 10 ± 2.19 µm

III PVB: Type Q (Fig. E)

Size: 3.70 × 5.18 ± 1.27 µm

Kranz arc: Half circle (Fig. E)

Chloroplast type: Centrifugal

Shape of Kranz cell: Round to oval (Fig. E)

Mestome: Single layered

Bar size: A = 50 µm, B = 10 µm, C–F = 5 µm

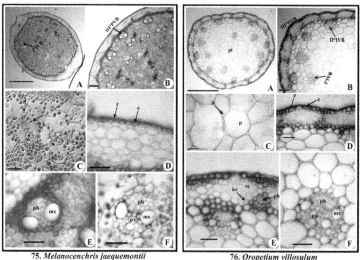

75. *Melanocenchris jaequemontii*   76. *Oropetium villosulum*

77. *Schoenefeldia gracilis* (Fig. 2.41 (77) A–F)

Round to oblong in transaction with smooth surface (Fig. A)

Size: 47.62 × 73.81 ± 11.43 µm, hairs/prickles absent

Solid internode with parechymatous pith

Epidermis: Barrel shaped III (Fig. B)

Hypodermis: 3–4 layered parenchymatous (Fig. D)

Than 3–4 layers Type-E sclerenchymatous

With III PVB embedded

Ground tissue: Round to oval-shaped parenchymatous (Fig. C)

Starch grains present

Vascular bundles arranged in ring form

Vascular system: III PVB absent

I IVB: Type C, sclerenchymatous cap present (Fig. E)

Size: $14.28 \times 15 \pm 4.27$ µm

II IVB: Type P (Fig. F)

Size: $6.04 \times 6.25 \pm 2.17$ µm

III PVB:

Bar size: A = 50 µm, B–F = 5 µm

Tetrapogon

78. *Tetrapogon tenellus* 79. *Tetrapogon villosus*

78. *Tetrapogon tenellus* (Fig. 2.41 (78) A–F)

Round with flat distal end in transaction with slightly undulating surface (Fig. A)

Size: $239.28 \times 271.43 \pm 32.46$ µm, hairs/prickles absent

Hollow internode. Kranz anatomy is present

Epidermis: Barrel shaped IV (Fig. C)

Hypodermis: 3–4 layered chlorenchymatous

Interrupted by sclerenchymatous with III PVB embedded

Type-E sclerenchyma present (Fig. C)

Ground tissue: Round to oval-shaped parenchymatous (Fig. F)

Vascular bundles arranged in ring form

Vascular system: All present

I IVB: Type I, sclerenchymaotus small cap present (Fig. E) scatterly arranged

Size: $19.55 \times 25.45 \pm 5.28$ µm

II IVB: Type P (Fig. D)

Size: $13.5 \times 14.5 \pm 3.17$ µm

III PVB: Type Q (Fig. C)

Size: 7.86 × 11.78 ± 2.16 μm

Kranz arc: Half circle (Fig. C)

Chloroplast type: Centrifugal

Shape of Kranz cell: Round to oval (Fig. C)

Mestome: Single layered

Bar size: A = 50 μm, B = 10 μm, C–F = 5 μm

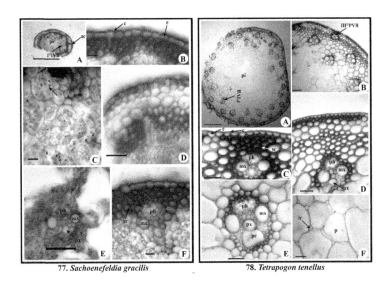

77. *Sachoenefeldia gracilis*   78. *Tetrapogon tenellus*

79. *Tetrapogon villosus* (Fig. 2.42 (79) A–F)

Oblong in transaction with smooth surface (Fig. A)

Size: 131.82 × 168.18 ± 27.46 μm, hairs/prickles absent

Hollow internode. Kranz anatomy is present

Epidermis: Square shaped III (Fig. D)

Hypodermis: 3–4 layered chlorenchymatous interrupted by Type-B sclerenchymatous

With III PVB embedded

Ground tissue: Round to oval-shaped parenchymatous (Fig. C)

Vascular bundles arranged in ring form

Vascular system: All present

I IVB: Type B, sclerenchymaotus cap present (Fig. F)

Size: $10.36 \times 6.43 \pm 2.17$ µm

II IVB: Type P (Fig. D)

Size: $8.64 \times 11.36 \pm 3.12$ µm

III PVB: Type Q (Fig. E)

Size: $4.62 \times 6.15 \pm 2.15$ µm

Kranz arc: Half circle (Fig. E)

Chloroplast type: Centrifugal

Shape of Kranz cell: Round to oval (Fig. E)

Mestome: Single layered

Bar size: A = 50 µm, B–F = 5 µm

DIFFERENTIATING FEATURES:

Round-shaped in transaction, barrel shaped epidermal cells ... *T. tenellus*
Oblong-shaped in transaction, square shaped epidermal cells ... *T. villosus*

80. *Acrachne racemosa* (Fig. 2.42 (80) A–E)

Oval in transaction with slightly undulating surface (Fig. A)

Size: $86.96 \times 93.48 \pm 13.70$ µm, hairs/prickles absent

Hollow internode. Kranz anatomy is present

Epidermis: Barrel shaped IV (Fig. B)

Hypodermis: 3–4 layered chlorenchymatous

Interrupted by sclerenchymatous with III PVB embedded

Type-E sclerenchyma present

Ground tissue: Round to oval-shaped parenchymatous (Fig. C)

Vascular bundles arranged in ring form

Vascular system: All present

I IVB: Type B, sclerenchymaotus cap present (Fig. E)

Size: $11.67 \times 12.92 \pm 1.38$ µm

II IVB: Type P

Size: 8.75 × 9.58 ± 2.61 μm

III PVB: Type Q (Fig. D)

Size: 5.83 × 5 ± 2.15 μm

Kranz arc: Half circle

Chloroplast type: Centrifugal

Shape of Kranz cell: Round to oval

Mestome: Single layered

Bar size: A = 50 μm, B–E = 5 μm

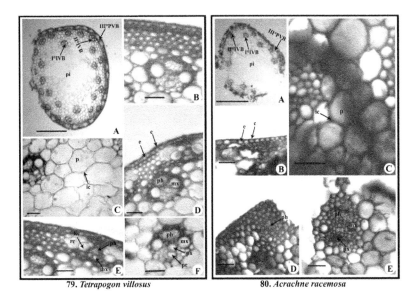

79. *Tetrapogon villosus*   80. *Acrachne racemosa*

Dactyloctenium

81. *Dactyloctenium aegyptium*

82. *Dactyloctenium scindicus*

81. *Dactyloctenium aegyptium* (Fig. 2.43 (81) A–F)

Oblong in transaction with slightly undulating surface (Fig. A)

Size: 294.45 × 433.34 ± 18.64 μm, hairs/prickles absent

Solid internode with parechymatous pith. Kranz anatomy is present (Fig. F)

Epidermis: Barrel shaped IV (Fig. C)

Hypodermis: 2–3 layered chlorenchymatous interrupted by Type-E sclerenchyma cell (Fig.B)

With III PVB embedded

Ground tissue: Round to oval-shaped parenchymatous

Vascular bundles arranged in ring form

Vascular system: All present

I IVB: Type C, sclerenchymaotus cap present (Fig. E)

Size: $18.64 \times 17.5 \pm 3.26$ μm

II IVB: Type P (Fig. D)

Size: $10.52 \times 14.83 \pm 3.17$ μm

III PVB: Type Q (Fig. F)

Size: $5.77 \times 6.15 \pm 2.62$ μm

Kranz arc: Half circle (Fig. F)

Chloroplast type: Centrifugal

Shape of Kranz cell: Round to oval (Fig. F)

Mestome: Single layered

Bar size: A = 50 μm, B = 10 μm, C–E = 5 μm

82. *Dactyloctenium scindicus* (Fig. 2.43 (82) A–F)

Round in transaction with slightly undulating surface (Fig. A)

Size: $242.31 \times 226.92 \pm 38.57$ μm, hairs/prickles absent

Hollow internode. Kranz anatomy is present (Fig. D)

Epidermis: Barrel shaped IV (Fig. C)

Hypodermis: 2–3 layered chlorenchymatous

Interrupted by Type-E sclerenchymatous with III PVB embedded

Ground tissue: Round to oval-shaped parenchyma (Fig. B). Vascular bundles arranged in ring form

Vascular system: All present

I IVB: Type C, sclerenchymaotus cap present (Fig. E, F)

Size: $17.39 \times 22.61 \pm 4.27$ μm

II IVB: Type P

Size: 4.23 × 8.46 ± 2.37 µm

III PVB: Type Q (Fig. D)

Size: 2.18 × 2.61 ± 0.26 µm

Kranz arc: Straight (Fig. D)

Chloroplast type: Centrifugal

Shape of Kranz cell: Square (Fig. D)

Mestome: Single layered

Bar size: A = 50 µm, B–F = 5 µm

DIFFERENTIATING FEATURES:

Half circle-shaped Kranz arc, round to oval-shaped Kranz cell … *D. aegyptium*
Straight-shaped Kranz arc, square to trapezoid-shaped Kranz cell … *D. sindicus*

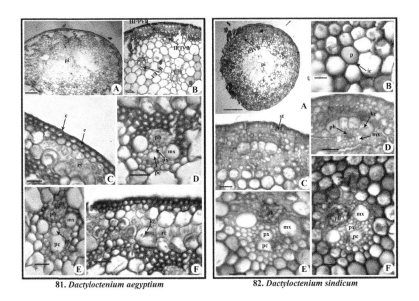

81. Dactyloctenium aegyptium   *82. Dactyloctenium sindicum*

83. *Desmostachya bipinnata* (Fig. 2.44 (83) A–E)

Round in transaction with slightly undulating surface (Fig. A)

Size: 326.67 × 240 ± 21.80 µm, hairs/prickles absent

Hollow internode. Kranz anatomy is present

Epidermis: Round shaped (Fig. D)

Hypodermis: 2–3 layered chlorenchymatous

Interrupted by sclerenchymatous with III PVB embedded

Type-D sclerenchyma present

Ground tissue: Round to oval-shaped parenchymatous (Fig. B)

Air cavities present (Fig. C)

Vascular bundles arranged in ring form

Vascular system: III PVB absent

I IVB: Type B, sclerenchymaotus cap present (Fig. E)

Size: $11.67 \times 11.67 \pm 1.84$ µm

II IVB: Type P

Size: $10 \times 6.67 \pm 2.16$ µm

III PVB: Size

Kranz arc: Half circle

Chloroplast type: Centrifugal

Shape of Kranz cell: Round to oval

Mestome: Single layered

Bar size: A = 50 µm, B–E = 5 µm

84. *Dinebra retroflexa* (Fig. 2.44 (84) A–E)

Round in transaction with smooth surface (Fig. A)

Size: $423.07 \times 346.15 \pm 36.18$ µm, hairs/prickles absent

Solid internode with parechymatous pith. Kranz anatomy is present

Epidermis: Barrel shaped IV (Fig. B)

Hypodermis: 3–4 layered chlorenchymatous interrupted by sclerenchymatous

With III PVB embedded

Type-E sclerenchyma present (Fig. B)

Ground tissue: Round to oval-shaped parenchymatous (Fig. C)

Vascular bundles arranged in ring form

Vascular system: All present

I IVB: Type F, sclerenchymaotus cap absent (Fig. E)

Size: 17.5 × 11.67 ± 2.87 μm

II IVB: Type M

Size: 5.42 × 6.67 ± 1.73 μm

III PVB: Type Q (Fig. D)

Size: 3.17 × 3.17 ± 1.25 μm

Kranz arc: Half circle (Fig. D)

Chloroplast type: Centrifugal

Shape of Kranz cell: Round to oval (Fig. D)

Mestome: Single layered

Bar size: A = 50 μm, B–E = 5 μm

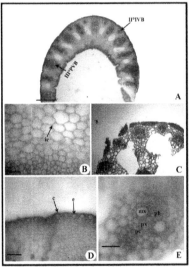 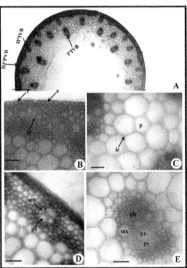

83. *Desmostachya bipinnata*   84. *Dinebra retroflexa*

85. *Eleusine indica* (Fig. 2.45 (85) A–F)

Round to oblong in transaction with smooth surface (Fig. A)

Size: 78.94 × 71.05 ± 6.26 μm, hairs/prickles absent

Hollow internode. Kranz anatomy is present

Epidermis: Barrel shaped II (Fig. B)

Hypodermis: 3–4 layered chlorenchymatous

Interrupted by sclerenchymatous

With III PVB embedded

Type-E sclerenchyma present (Fig. C)

Ground tissue: Round to oval-shaped parenchymatous (Fig. D)

Vascular bundles arranged in ring form

Vascular system: All present

I IVB: Type F (Fig. F)

Size: $8.34 \times 7.5 \pm 2.17$ µm

II IVB: Type P (Fig. E)

Size: $8.24 \times 7.94 \pm 3.17$ µm

III PVB: Type Q

Size: $5.17 \times 4.83 \pm 1.56$ µm

Kranz arc: Half circle

Chloroplast type: Centrifugal

Shape of Kranz cell: Round to oval

Mestome: Single layered

Bar size: A = 50 µm, B–F = 5 µm

Eragrostiella

86. *Eragrostiella bachyphylla* 87. *Eragrostiella bifaria*

86. *Eragrostiella bachyphylla* (Fig. 2.45 (86) A–E)

Oblong in transaction with smooth surface (Fig. A)

Size: $121.25 \times 142.5 \pm 10.27$ µm, hairs/prickles absent

Hollow internode. Kranz anatomy is present

Epidermis: Barrel shaped V (Fig. C)

Hypodermis: 2–3 layered chlorenchymatous interrupted by 6–7 layered sclerenchymatous cells

With III PVB embedded

Type-E sclerenchyma present (Fig. E)

Ground tissue: Round to oval-shaped parenchymatous (Fig. B)

Vascular bundles arranged in ring form

Vascular system: All present

I IVB: Type C, sclerenchymatous cap present (Fig. E)

Size: 9.52 × 13.57 ± 1.56 μm

II IVB: Type P (Fig. D)

Size: 8.89 × 11.12 ± 2.18 μm

III PVB: Type Q   Size: 4.27 × 4.18 ± 1.78 μm

Kranz arc: Half circle

Chloroplast type: Centrifugal

Shape of Kranz cell: Square

Mestome: Single layered

Bar size: A = 50 μm, B–E = 5 μm

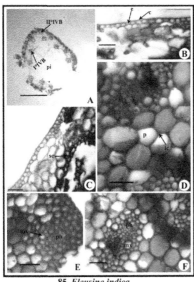

85. *Eleusine indica*

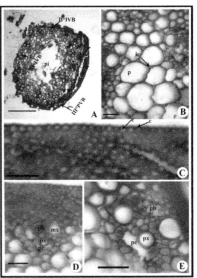

86. *Eragrostiella bachyphylla*

87. *Eragrostiella bifaria* (Fig. 2.46 (87) A–F)

Round in transaction with slightly undulating surface (Fig. A)

Size: 120 × 112.5 ± 34.72 μm, hairs/prickles absent

Hollow internode. Kranz anatomy is present (Fig. C)

Epidermis: Barrel shaped V (Fig. D)

Hypodermis: 4–5 layered chlorenchymatous interrupted by 6–7 layered sclerenchymatous cells

With III PVB embedded, Type-B sclerenchyma present

Ground tissue: Round to oval-shaped parenchyma (Fig. B). Vascular bundles arranged in ring form

Vascular system: All present

I IVB: Type C, sclerenchymatous cap present (Fig. E, F)

Size: $14.85 \times 13.64 \pm 2.78$ µm

II IVB: Type PSize: $7.69 \times 13.46 \pm 1.77$ µm

III PVB: Type Q (Fig. C)

Size: $4.17 \times 4.86 \pm 1.07$ µm

Kranz arc: Half circle (Fig. C)

Chloroplast type: Centrifugal

Shape of Kranz cell: Square (Fig. C)

Mestome: Single layered

Bar size: A = 50 µm, B–F = 5 µm

DIFFERENTIATING FEATURES:

Oblong in transaction, sclerenchyma type E present … *E. bachyaphylla*
Round in transaction, sclerenchyma type B present … *E. bifaria*

Eragrostis

88. *Eragrostis cilianensis*

89. *Eragrostis ciliaris*

90. *Eragrostis japonica*

91. *Eragrostis nutans*

92. *Eragrostis pilosa*

93. *Eragrostis tenella*

94. *Eragrostis tremula*

95. *Eragrostis unioloides*

96. *Eragrostis viscosa*

88. *Eragrostis cilianensis* (Fig. 2.46 (88) A–F)

Oblong in transaction with smooth surface (Fig. A)

Size: 138.46 × 203.85 ± 25.78 µm, hairs/prickles absent

Hollow internode. Kranz anatomy is present

Epidermis: Round shaped (Fig. C)

Hypodermis: 2–3 layered chlorenchymatous interrupted by 2–3 layered sclerenchymatous cells

With III PVB embedded, Type-B sclerenchyma present (Fig. B)

Ground tissue: Round to oval-shaped parenchymatous. Vascular bundles arranged in ring form

Vascular system: All present

I IVB: Type F, sclerenchymatous cap present

Size: 13.75 × 14.58 ± 4.28 µm

II IVB: Type P

Size: 10.84 × 13.34 ± 1.54 µm

III PVB: Type Q

Size: 4.75 × 5.5 ± 1.69 µm

Kranz arc: Half circle

Chloroplast type: Centrifugal

Shape of Kranz cell: Square

Mestome: Single layered

Bar size: A = 50 µm, B–F = 5 µm

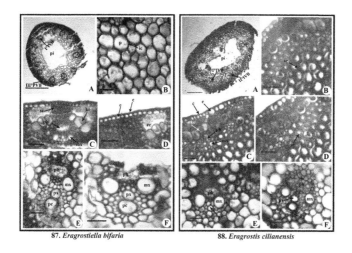

87. *Eragrostiella bifaria*   88. *Eragrostis cilianensis*

89. *Eragrostis ciliaris* (Fig. 2.47 (89) A–F)

Round in transaction with smooth surface (Fig. A)

Size: 168 × 118.18 ± 11.75 µm, hairs/prickles absent

Hollow internode. Kranz anatomy is present (Fig. C)

Epidermis: Barrel shaped IV (Fig. B)

Hypodermis: 2–3 layered chlorenchymatous interrupted by 2–3 layered sclerenchymatous cells

With III PVB embedded, Type-B sclerenchyma present (Fig. B)

Ground tissue: Round to oval-shaped parenchymatous

Vascular bundles arranged in ring form

Vascular system: All present

I IVB: Type B, sclerenchymatous cap present

Size: 11.12 × 11.81 ± 3.32 µm

II IVB: Type P

Size: 6.76 × 6.32 ± 2.17 µm

III PVB: Type Q

Size: 5.6 × 5.4 ± 2.18 µm

Kranz arc: Straight

Chloroplast type: Centripetal

Shape of Kranz cell: Square

Mestome: Single layered

Bar size: A = 50 µm, B–F = 5 µm

90. *Eragrostis japonica* (Fig. 2.47 (90) A–F)

Horseshoe in transaction with smooth surface (Fig. A)

Size: 160.87 × 201.35 ± 48.63 µm, hairs/prickles absent

Hollow internode

Epidermis: Barrel shaped IV (Fig. B)

Hypodermis: 3–4 layered sclerenchymatous (Fig. B)

With III PVB embedded

Type-D sclerenchyma present (Fig. E)

Ground tissue: Round to oval-shaped parenchymatous (Fig. C)

Vascular bundles arranged in ring form

Vascular system: All present

I IVB: Type I, sclerenchymatous cap present (Fig. F)

Size: 17.31 × 13.08 ± 3.87 µm

II IVB: Type P (Fig. D)

Size: 7.69 × 6.35 ± 2.89 µm

III PVB: Type Q (Fig. B)

Size: 3.39 × 3.57 ± 1.66 µm

Bar size: A = 50 µm, B–F = 5 µm

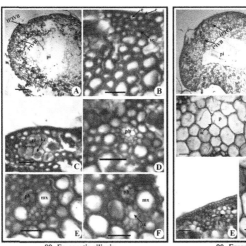

89. *Eragrostis ciliaris*   90. *Eragrostis japonica*

91. *Eragrostis nutans* (Fig. 2.48 (91) A–F)

Horseshoe in transaction with smooth surface (Fig. A)

Size: $150 \times 127.27 \pm 24.18$ µm, hairs/prickles absent

Holow internode

Epidermis: Barrel shaped IV (Fig. C)

Hypodermis: 3–4 layered sclerenchymatous

With III PVB embedded

Type-E sclerenchyma present

Ground tissue: Round to oval-shaped parenchymatous (Fig. B)

Vascular bundles arranged in ring form

Vascular system: All present

I IVB: Type F, sclerenchymatous cap present (Fig. F)

Size: $16.82 \times 11.36 \pm 3.65$ µm

II IVB: Type P (Fig. E)

Size: $11.92 \times 13.46 \pm 2.18$ µm

III PVB: Type Q (Fig. D)

Size: $3.48 \times 4.06 \pm 1.86$ µm

Bar size: A = 50 µm, B–F = 5 µm

92. *Eragrostis pilosa* (Fig. 2.48 (92) A–F)

Horseshoe with less concave distal end in transaction (Fig. A)

Size: $241.28 \times 222.24 \pm 38.20$ µm, hairs/prickles absent

Hollow internode. Kranz anatomy is present.

Epidermis: Barrel shaped IV (Fig. B)

Hypodermis: 3–4 layered chlorenchymatous interrupted by 2–3 layered sclerenchymatous cells (Fig. B)

With III PVB embedded, Type-D sclerenchyma present

Ground tissue: Round to oval-shaped parenchymatous (Fig. C)

Vascular bundles arranged in ring form

Vascular system: All present

I IVB: Type B, sclerenchymatous cap present (Fig. F)

Size: 15.42 × 21.67 ± 3.81 µm

II IVB: Type P (Fig. D, E)

Size: 14.76 × 18.09 ± 2.37 µm

III PVB: Type Q (Fig. B)

Size: 7.5 × 6.5 ± 1.39 µm

Kranz arc: Half circle

Chloroplast type: Centrifugal

Shape of Kranz cell: Round to oval

Mestome: Single layered

Bar size: A = 50 µm, B–F = 5 µm

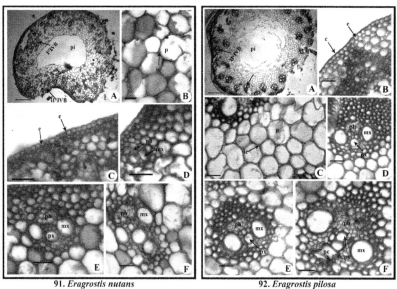

91. *Eragrostis nutans*   92. *Eragrostis pilosa*

93. *Eragrostis tenella* (Fig. 2.49 (93) A–F)

Round in transaction with smooth surface (Fig. A)

Size: 163.64 × 209.09 ± 25.98 µm, hairs/prickles absent

Hollow internode

Epidermis: Barrel shaped V (Fig. D)

Hypodermis: 7–8 layered sclerenchymatous

With III PVB embedded

Type-D sclerenchyma present

Ground tissue: Round to oval-shaped parenchymatous (Fig. B)

Vascular bundles arranged in ring form

Vascular system: All present

I IVB: Type A, sclerenchymatous cap present (Fig. F)

Size: 11.78 × 13.04 ± 3.17µm

II IVB: Type P

Size: 8.32 × 8.96 ± 2.41 µm

III PVB: Type Q

Size: 4.32 × 4.86 ± 1.53 µm

Bar size: A = 50 µm, B–F = 5 µm

94. *Eragrostis tremula* (Fig. 2.49 (94) A–E)

Round in transaction with smooth surface (Fig. A)

Size: 175 × 193.75 ± 21.29 µm, hairs/prickles absent

Hollow internode

Epidermis: Barrel shaped V (Fig. C)

Hypodermis: Parenchymatous cells interrupted by sclerenchymatous cells

With III PVB embedded

Type-B sclerenchyma present

Ground tissue: Round to oval-shaped parenchymatous (Fig. E)

Vascular bundles arranged in ring form

Vascular system: All present

I IVB: Type I, sclerenchymatous cap present (Fig. E)

Size: 11.43 × 12.5 ± 3.17 µm

II IVB: Type M

Size: 7.93 × 7.86 ± 2.18 µm

III PVB: Type Q (Fig. D)

Size: 4.45 × 4.45 ± 0.83 µm

Bar size: A = 50 µm, B–E = 5 µm

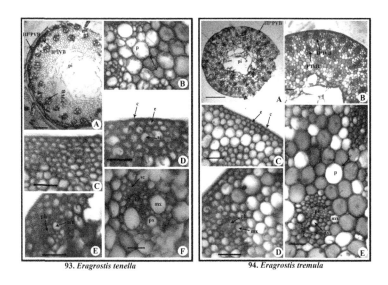

93. *Eragrostis tenella*   94. *Eragrostis tremula*

95. *Eragrostis unioloides* (Fig. 2.50 (95) A–F)

Round in transaction with smooth surface (Fig. A)

Size: 168.75 × 159.38 ± 14.75 μm, hairs/prickles absent

Hollow internode. Kranz anatomy is present (Fig. C)

Epidermis: Barrel shaped V

Hypodermis: 3–4 layered chlorenchymatous (Fig. B)

Than 3–4 layered sclerenchymatous cells

With III PVB embedded, Type-E sclerenchyma present (Fig. B)

Ground tissue: Round to oval-shaped parenchymatous

Vascular bundles arranged in ring form

Vascular system: All present

I IVB: Type B, sclerenchymatous cap present (Fig. F)

Size: 17.08 × 17.92 ± 4.18 μm

II IVB: Type P

Size: 5.86 × 5.52 ± 2.18 μm

III PVB: Type Q (Fig. E)

Size: 5 × 6.56 ± 1.34 μm

Kranz arc: Half circle (Fig. E)

Chloroplast type: Centrifugal

Shape of Kranz cell: Square to trapezoid (Fig. E)

Mestome: Single layered

Bar size: A = 50 µm, B–F = 5 µm

96. *Eragrostis viscosa* (Fig. 2.50 (96) A–F)

Round in transaction with slightly undulating surface (Fig. A)

Size: 150 × 145 ± 13.65 µm, hairs/prickles absent

Solid internode with parechymatous pith. Kranz anatomy is present

Epidermis: Barrel shaped V (Fig. B)

Hypodermis: 2–3 layered chlorenchymatous (Fig. B)

Than 4–5 layered sclerenchymatous cells

With III PVB embedded

Type-E sclerenchyma present

Ground tissue: Round to oval-shaped parenchymatous

Vascular bundles arranged in ring form

Vascular system: All present

I IVB: Type E, sclerenchymatous cap present (Fig. F)

Size: 11.56 × 9.69 ± 1.48 µm

II IVB: Type N (Fig. D)

Size: 8.82 × 8.82 ± 2.68 µm

III PVB: Type (Fig. E)

Size: 5.43 × 5.92 ± 1.54 µm

Kranz arc: Straight (Fig. E)

Chloroplast type: Centrifugal

Shape of Kranz cell: Square to trapezoid (Fig. E)

Mestome: Single layered

Bar size: A = 50 µm, B–F = 5 µm

DIFFERENTIATING FEATURES:

Round-shaped epidermal cells … *E. cilianensis*
Barrel-shaped epidermal cells

Sclerenchyma in form of girder ... *E. tremula*
Sclerenchyma in form of unicylinder
Sclerenchyma type E present
Absence of Kranz arc ... *E. nutans*
Presence of Kranz arc
Half circle-shaped Kranz arc ... *E. unioloides*
Straight-shaped Kranz arc
Centripetal arrangement of chloroplast ... *E. ciliaris*
Centrifugal arrangement of chloroplast ... *E. viscosa*
Sclerenchyma type D present
Presence of Kranz arc ... *E. pilosa*
Absence of Kranz arc
Horseshoe-shaped in transaction ... *E. japonica*
Round-shaped in transaction ... *E. tenella*

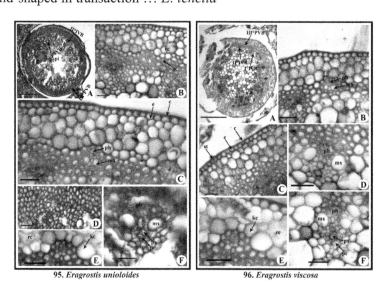

95. *Eragrostis unioloides* 96. *Eragrostis viscosa*

Sporobolus

97. *Sporobolus coromandelianus* 98. *Sporobolus diander* 99. *Sporobolus indicus*

97. *Sporobolus coromandelianus* (Fig. 2.51 (97) A–E)

Round to oval in transaction with slightly undulating surface (Fig. A)

Size: 97.72 × 118.18 ± 24.53 μm, hairs/prickles absent

Hollow internode. Kranz anatomy is present

Epidermis: Barrel-shaped V (Fig. C)

Hypodermis: 2–3 layered chlorenchymatous

Than 4–5 layered sclerenchymatous cells

With III PVB embedded, Type-D sclerenchyma present

Ground tissue: Round to oval-shaped parenchymatous,

Vascular bundles arranged in ring form

Vascular system: All present

I IVB: Type B, sclerenchymatous cap present (Fig. E)

Size: $12.5 \times 13.34 \pm 3.17$ µm

II IVB: Type P (Fig. D)

Size: $9.45 \times 11.38 \pm 2.68$ µm

III PVB: Type Q

Size: $7.04 \times 7.27 \pm 1.13$ µm

Kranz arc: Half circle (Fig. C)

Chloroplast type: Centrifugal

Shape of Kranz cell: Square to trapezoid (Fig. C)

Mestome: Single layered

Bar size: A = 50 µm, B–E = 5 µm

98. *Sporobolus diander* (Fig. 2.51 (98) A–F)

Round in transaction with smooth surface (Fig. A)

Size: $117.5 \times 120 \pm 25.17$ µm, hairs/prickles absent

Solid internode with parechymatous pith. Kranz anatomy absent

Epidermis: Barrel shaped V (Fig. D)

Hypodermis: 2–3 layered chlorenchymatous

Than 2–3 layered sclerenchymatous cells

With III PVB embedded

Type-D sclerenchyma present

Ground tissue: Round to oval-shaped parenchymatous

Vascular bundles arranged in ring form

Vascular system: All present

I IVB: Type I, sclerenchymatous small cap present (Fig. F)

Size: 22.5 × 22.3 ± 4.96 µm

II IVB: Type M (Fig. B)

Size: 17.65 × 20.17 ± 5.28 µm

III PVB: Size

Bar size: A = 50 µm, B–F = 5 µm

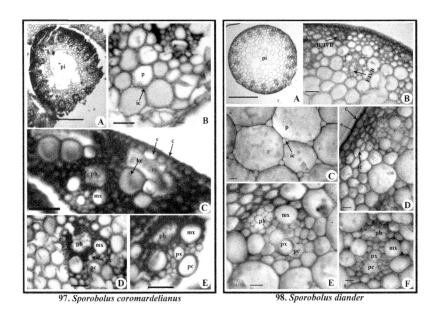

97. *Sporobolus coromardelianus*   98. *Sporobolus diander*

99. *Sporobolus indicus* (Fig. 2.52 (99) A–F)

Round in transaction with smooth surface (Fig. A)

Size: 216.67 × 220.84 ± 25.18 µm, hairs/prickles absent

Solid internode with parechymatous pith. Kranz anatomy is present

Epidermis: Square shaped III (Fig. D)

Hypodermis: 2–3 layered chlorenchymatous (Fig. B)

Interrupted by 5–6 layered Type-E sclerenchymatous cells

With III PVB embedded

Ground tissue: Round to oval-shaped parenchymatous (Fig. D)

Vascular bundles arranged in ring form

Vascular system: All present

I IVB: Type F, sclerenchymatous cap present (Fig. H)

Size: 20 × 14.45 ± 3.17 µm

II IVB: Type O (Fig. F)

Size: 11.25 × 13.34 ± 3.15 µm

III PVB: Type P (Fig. E)

 Size: 3.5 × 5.67 ± 1.78 µm

Kranz arc: Half circle

Chloroplast type: Centrifugal

Shape of Kranz cell: Round to oval

Mestome: Single layered

Bar size: A = 50 aµm, B = 10 µm, C–F = 5 µm

DIFFERENTIATING FEATURES:

Half circle-shaped Kranz arc … *S. coromardelianus*
Straight-shaped Kranz arc
Square-shaped epidermal cell … *S. indicus*
Barrel-shaped epidermal cell … *S. diander*

100. *Tragus biflorus* (Fig. 2.52 (100) A–F)

Round with notch in transaction with slightly undulating surface (Fig. A)

Size: 115.96 × 119.5 ± 9.54 µm, hairs/prickles absent

Solid internode with parechymatous pith. Kranz anatomy is present

Epidermis: Barrel shaped III (Fig. D)

Hypodermis: 3–4 layered chlorenchymatous interrupted by 2–3 layered sclerenchymatous cells (Fig. B)

with III PVB embedded,

Type-D sclerenchyma present (Fig. E)

Ground tissue: Round to oval-shaped parenchymatous (Fig. C)

Vascular bundles arranged in ring form

Vascular system: All present

I IVB: Type E, sclerenchymatous cap absent (Fig. F)

Size: 26 × 27 ± 6.28 μm

II IVB: Type O

Size: 14 × 16 ± 5.17 μm

III PVB: Type Q (Fig. E)

Size: 5.28 × 7.69 ± 2.17 μm

Kranz arc: Half circle (Fig. E)

Chloroplast type: Centripetal

Shape of Kranz cell: Round to ova

Mestome: Single layered

Bar size: A = 50 μm, B = 10 μm, C–F = 5 μm

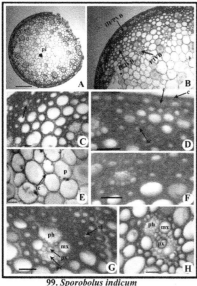

99. *Sporobolus indicum*

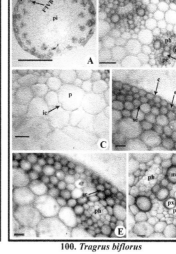

100. *Tragrus biflorus*

2.3   DISCUSSION AND IDENTIFICATION KEY

Distribution of the vascular bundles in the culm varied. Vascular bundles were distributed among the cortical and hypodermal region. I IVB and II IVB were distributed in cortical region and III PVB were generally present in hypodermal region. Vascular bundle distribution is a strength property

TABLE 2.1 Quantitative Anatomical Features of Grass Culm (Anatomy).

Sr No.	Name of plant	Cross section area (µm)	Number of vascular bundles			Size of vascular bundles (µm)			Size of metaxylem (µm)			No. of kranz cells
			I°	II°	III°	I°	II°	III°	I°	II°	III°	
Tribe: Maydeae												
1.	Chionachne koenigii	295×275±24.87	31±3	19±6	28±4	14.68×12.5±3.87	7×9.67±1.73	4.5×3±0.84	3.75×3.43±1.31	4×3.34±1.62	2.25×1.5±0.08	Absent
2	Coix lachryma-jobi	923.07×646.15±31.49	68±5	36±3	27±5	17.78×15.56±2.32	11.07×10.36±3.72	7.34×6.27±1.47	2.78×2.23±0.36	2.85×3.21±0.71	2.04×2.94±0.81	Absent
Tribe: Andropogoneae												
3	Andropogon pumilus	161.17×91.67±18.37	9±2	8±1	10±2	13.22×16.37±3.28	12.34×15.34±2.98	3.62×4.82±.73	3.75×4.25±1.29	2.67×3.67±0.83	1.83×2.21±0.35	absent
4	Apluda mutica	358.25×433.38±28.73	32±8	17±3	15±2	35.71×31.43±6.42	15×18.21±3.72	7.08×14.17±1.38	3.57×13.37±4.28	6.08×4.64±1.35	4.58×3.75±0.72	Absent
5	Arthraxon lanceolatus	116.67×138.09±17.82	14±4	10±2	8±3	14×11.34±4.83	11.07×12.86±2.63	6.67×7.08±1.72	3.67×3±0.13	3.21×3.21±1.62	1.67×1.67±0.08	Absent
6	Bothriochloa pertusa	15.56×13.89±3.78	18±5	10±1	-	11×12.67±3.29	7.5×9.32±2.85	-	4×3.67±1.83	2.72×2.27±0.36	-	Absent
7	Capillipedium hugelii	152.94×138.24±18.63	19±5	14±3	12±4	16×17±6.37	11.36×13.18±2.84	5.67×7.34±1.35	5.5×4.5±0.92	4.54×4.09±1.26	2×1.84±0.74	Absent
8	Chrysopogon fulvus	277.78×283.34±35.22	17±4	12±1	16±3	28.89×21.67±5.18	15.47×20.16±6.18	6.30×10±2.73	7.78×6.12±2.62	4.27×4.92±1.63	2.61×2.61±0.74	Absent
9	Cymbopogon martini	759×900±48.22	42±7	24±3	18±4	24.5×27.5±3.73	14.09×14.09±4.62	3.86×7.95±1.37	7.5×7±2.76	5.45×4.55±2.83	2.5×2.27±0.84	Absent
10	Dichanthium annulatum	253.72×241.67±27.56	18±3	16±3	20±3	21×24.5±7.29	12×16.5±3.81	7.5×11±1.38	5.5×4.5±2.83	5.5×2.75±1.38	3×2.75±0.95	Absent
11	Dichanthium caricosum	527.2×613.23±61.78	24±4	19±5	20±4	22×23.5±4.83	13.18×11.82±1.92	7.1×9.13±2.73	6.5×7.5±1.83	3.64×3.18±0.94	2.97×2.36±0.72	Absent
12	Hackelochloa granularis	218.18×254.54±31.37	22±7	20±6	-	16×23±4.73	8.08×3.85±1.36	-	5.5×4±1.49	3.08×3.85±1.38	-	Absent
13	Heteropogon contortus var. gemuinus sub var. typicus	241.67×316.67±28.82	36±8	21±6	23±4	16.82×17.73±4.39	12.92×16.25±3.82	9.17×9.58±3.78	5.45×4.54±0.63	5×3.75±1.38	2.92×2.5±0.96	Absent

TABLE 2.1 (Continued)

Sr No.	Name of plant	Cross section area (µm)	Number of vascular bundles			Size of vascular bundles (µm)			Size of metaxylem (µm)			No. of kranz cells
			I°	II°	III°	I°	II°	III°	I°	II°	III°	
14	Heteropogon contortus var. genuinus sub var. hispidissimus	256.25×384.37±28.84	23±6	17±2	13±4	15×18.75±4.25	10.36×14.93±4.72	7.5×9.87±4.83	5×5.52±1.38	3.43×3.75±0.36	2.64×2.71±0.92	Absent
15	Heteropogon ritchiei	260×335±35.59	33±5	14±2	11±5	19.61×18.07±3.28	10.84×13.34±4.18	9.23×10±3.75	5.77×4.62±1.37	3.34×3.34±0.76	3.08×2.69±1.74	Absent
16	Heteropogon triliceus	1155.56×1444.45±47.28	38±9	30±6	21±3	25×27.27±5.18	16.67×16.78±5.38	10.5×14±3.28	6.81×7.33±2.47	7.86×5.36±3.48	5×4±1.38	Absent
17	Imperata cylindrica	183×118.27±34.25	29±4	15±2	14±3	22.5×20±4.27	11.12×13.89±2.78	6×5.71±1.39	6×5.5±2.81	3.34×3.34±1.36	2×1.71±0.93	Absent
18	Ischaemum indicus	280×420±21.38	26±3	24±5	20±3	27.5×23±6.28	9.6×11.6±2.74	6.26×7.93±1.83	7×4.75±1.36	4.4×4±1.97	2.04×1.95±0.63	6±2
19	Ischaemum molle	216×224±34.87	18±2	13±2	12±3	15.5×16±3.8	8.6×12.4±3.19	4.38×6.15±1.83	3.5×3.5±1.36	2.4×1.8±0.83	1.98×1.85±0.73	6±1
20	Ischaemum pilosum	250±331.82±41.73	20±4	26±4	19±3	16.36×14.55±4.78	12.73×13.82±3.28	7.27×6.36±2.73	4.54×4.54±1.63	2.73×2.27±0.62	2.03×1.96±0.72	12±6
21	Ischaemum rugosum	330.43×321.74±15.12	19±3	11±1	12±2	24.63×22.37±4.62	9.55×8.64±2.16	8.89×10±1.63	6.28×4.26±2.94	4.55×3.18±1.64	4.45×3.34±0.73	7±3
22	Iseilema laxum	333.34×416.67±24.38	23±3	12±2	15±4	35.45×24.54±7.19	16.25×18.18±4.17	10.91×13.18±2.74	8.18×5.68±3.27	4.17×3.75±1.38	3.18×3.86±1.75	Absent
23	Ophiuros exaltatus	480×545±38.17	34±6	26±3	79±9	22.78×18.34±5.28	17.5×17±3.97	3.82×4.12±1.82	6.12×4.67±1.36	5×3.5±2.74	1.76×0.88±0.63	7±2
24	Rottboellia exaltata	367.28×227.18±24.37	37±5	20±4	25±5	24.37×21.25±3.17	8.27×10.17±3.19	7.83×8.69±3.71	5×5±1.42	2.12×2.69±0.27	2.17×2.61±1.22	Absent
25	Saccharum spontanum	305×355±21.79	28±4	20±1	10±2	17.5×25±5.72	15.03×15.04±3.16	7.69×4.62±1.92	6.63×4.72±1.25	5.38×4.54±1.87	1.35×1.54±0.17	Absent
26	Sehima ischaemoides	203.57×214.28±18.36	17±2	32±5	21±7	17.39×25.22±4.28	8.26×7.93±2.71	5.19×7.41±2.71	7.17×7.39±3.46	3.04×3.27±1.63	2.96×2.59±0.84	Absent
27	Sehima nervosum	257.41×259.26±24.63	23±6	14±4	12±3	21.74×20.22±6.26	17.08×21.25±3.71	3.94×5.76±1.52	6.74×5±2.17	4.79×4.17±2.78	1.82×1.97±0.41	Absent
28	Sehima sulcatum	127.27×102.27±11.72	14±5	11±2	10±4	11.67×13.94±2.78	6.34×8.34±2.19	4.27×4.07±1.18	3.48×3.34±1.86	2.59×2.83±0.93	1.84×1.84±0.g72	Absent
29	Sorghum halepense	1765×1877.5±32.78	21±6	24±2	30±3	19.13×25.22±5.27	17.27×21.82±4.87	4.55×4.31±1.64	8.04×7.83±2.78	6.82×6.36±2.71	4.71×5.17±2.93	Absent
30	Thelepogn elegans	500×550±16.62	24±3	18±2	15±4	16.67×21.43±3.17	7.67×10±2.73	6.82×5.93±2.23	7.38×5.71±1.83	3.34×3.57±1.73	2.83×2.64±0.38	Absent

TABLE 2.1 (Continued)

Sr No.	Name of plant	Cross section area (µm)	Number of vascular bundles			Size of vascular bundles (µm)			Size of meta×ylem (µm)			No. of kranz cells
			I°	II°	III°	I°	II°	III°	I°	II°	III°	
31	*Themeda cymbaria*	269.23×369.23±31.73	26±7	18±4	20±2	20×19.58±2.73	8.28×5.84±2.74	6.86×10.29±1.37	7.29×5.84±2.37	3.17×2.68±1.98	3.14×3.14±1.64	Absent
32	*Themeda laxa*	275×318.75±25.38	32±4	19±2	-	25×24.17±8.78	9.63×11.36±3.18	-	8.56×6.12±2.18	3.63×2.95±1.28	-	Absent
33	*Themeda triandra*	305×362±11.28	36±5	25±3	24±6	19.77×27.5±7.21	13.46×16.15±4.26	8.27×8.92±2.81	9.5×10.5±3.27	6.92×5.19±2.72	2.05×2.48±0.83	Absent
34	*Themeda quadrivalvis*	390.91×363.64±25.81	32±7	24±2	22±5	24.71×24.71±6.17	12.95×16.36±3.19	9.5×10.5±3.47	7.65×6.17±2.91	5.45×4.55±2.37	2.5×3±1.02	Absent
35	*Triplopogon ramosissimus*	1050.14×1108.16±5.28	25±7	16±4	28±3	58.34×91.08 ±2.73	50.34×51.63 ±0.83	33.45×29.35±0.21	10.21×11.16±1.73	9.29×9.18 ±0.40	5.04×5.94±0.63	Absent
36	*Vetivaria zizanioides*	1133.34×1483.34±34.16	60±7	25±4	23±3	36.12×27.78±8.15	13.89×16.67±4.18	8.34×8.34±4.28	16.67×13.89±55.18	5.56×8.34±2.81	2.78×2.78±1.27	Absent
Tribe: Paniceae												
37	*Alloteropsis cimicina*	159.13×153.13±16.28	16±4	9±2	12±1	10.56×15±2.91	13.34×16.67±3.18	4.5×6.5±1.24	4.73×5.56±1.71	6.12×5.27±2.56	1.5×2±0.95	Absent
38	*Brachiaria distachya*	141.17×170.59±32.34	12±2	7±1	8±3	15.45×21.36±2.15	8.57×13.39±3.71	5.88×7.35±1.62	5.45×6.14±1.36	3.57×4.11±1.26	2.94×2.94±1.02	Absent
39	*Brachiaria eruciformis*	112.5×137.5±22.08	15±6	11±2	9±2	14.09×17.27±2.17	8×10±1.27	3.34×4.17±1.34	4.54×5.90±2.04	3.34×3.34±1.09	2.08×1.67±0.46	Absent
40	*Brachiaria ramose*	340.91×572.73±32.72	60±7	22±2	27±5	26.67×22.78±5.16	11.25×11.56±2.18	7.69×11.54±3.18	6.39×5.56±2.10	3.46×3.46±1.72	2.97×2.81±1.04	Absent
41	*Brachiaria reptans*	167.86×164.28±21.91	23±4	12±3	21±2	15.24×17.14±3.16	10.68×9.88±3.16	7.69×5±1.26	4.76×4.76±1.76	3.09×3.08±0.64	2.5×2.95±1.05	9±2
42	*Cenchrus biflorus*	165×480±14.86	53±3	28±1	34±5	23.18×18.18±7.16	11.5×14.5±2.18	5.56×7.78±1.72	6.82×5.91±2.17	3.5×3.5±1.86	1.67×2.08±0.13	Absent
43	*Cenchrus ciliaris*	600×591.67±47.61	48±9	20±4	24±2	36.67×28±6.17	18.63×15.91±4.76	15.2×13.6±5.15	6.34×7.34±2.17	5.45×4.09±1.72	3.6×2.8±1.62	Absent
44	*Cenchrus setigerus*	333.34×311.12±16.24	23±4	19±6	17±3	28.18×32.73±4.16	10.88×13.82±3.28	4.68×4.37±1.62	6.36×7.27±2.06	4.12×3.53±2.7	2.5×2.81±0.61	7±3
45	*Digitaria ciliaris*	320×305±21.63	24±5	15±3	30±2	21.54×23.85±6.18	9.47×11.58±3.71	6.15×5.72±2.16	6.92×6.92±1.52	2.63×4.74±1.53	2.21×3.52±0.35	Absent
46	*Digitaria granularis*	258.34×294.26±13.24	13±3	12±4	-	43.75×40±6.16	16.67×25.34±4.14	-	9.63×11.13±3.71	6×7.34±1.72	-	Absent
47	*Digitaria longiflora*	409.09×463.64±72.22	8±2	7±3	7±2	36×30±9.26	16.24×21.29±3.28	3×3.67±1.62	9×8±2.16	2.5×2.67±0.53	1×1±0.03	Absent

TABLE 2.1 *(Continued)*

Sr No.	Name of plant	Cross section area (μm)	Number of vascular bundles			Size of vascular bundles (μm)			Size of metaxylem (μm)			No. of kranz cells
			I°	II°	III°	I°	II°	III°	I°	II°	III°	
48	Digitaria stricta	250×308.34±24.25	21±5	17±2	18±4	9.56×11.36±3.16	5×5.67±1.25	2×4±0.34	4.09×3.64±1.25	2×2±0.83	1×1±0.04	Absent
49	Echinochloa colonum	1123.08×1184.62±82.63	28±6	20±4	24±3	24.17×31.67±7.25	20.56×25.56±6.24	14×11.5±4.26	11.67×20.84±5.72	8.89×6.67±3.72	5.5×5.5±2.62	Absent
50	Echinochloa crusgalli	1584.61×1638.46±69.18	83±25	41±11	40±7	14.64×14.29±3.12	0.09×9.55±3.17	4.09×3.64±2.13	4.28×2.5±1.52	3.64×2.73±1.85	1.36×1.14±0.46	Absent
51	Echinochloa stagnina	1050×1163±62.58	24±9	14±5	16±3	29.17×25±7.14	23.34×25±5.12	8.34×12.5±3.62	6.67×7.08±2.15	6.25×6.25±2.51	4.17×4.17±1.88	Absent
52	Eremopogon foveolatus	1210×1560±102.13	24±5	16±5	20±3	26.12×16.24±6.15	21.92×16.15±7.36	7×14±3.16	5.77×5±2.61	5.01×4.89±2.61	4.38×3.50±2.17	Absent
53	Eriochloa procera	261.90×304.76±36.41	20±3	15±4	17±3	12.72×16.82±4.52	7.73×11.36±3.17	6.07×8.03±2.76	5.45×4.09±2.75	2.27×4.09±1.35	2.68×2.5±0.72	Absent
54	Oplismenus burmannii	173.34×193.34±36.29	14±3	15±3	12±1	14.34×13.34±5.26	12.34×11.34±4.26	4×5±2.16	5×3.5±2.16	3.34×2.79±1.16	2.25×1.5±1.51	Absent
55	Oplismenus compositus	214.28×257.14±37.34	15±2	23±5	-	12.5×14.75±3.71	5.22×6.52±2.53	-	6×6±1.73	2.61×2.06±0.83	-	Absent
56	Panicum antidotale	480×1146.67±62.25	48±10	23±6	64±12	19.55×20.91±7.21	10×12.61±4.14	6.67×10±3.61	5.91×5.45±2.53	4.35×3.48±1.72	3.34×3.34±1.62	Absent
57	Panicum maximum	533.34×438.09±36.26	36±8	22±5	28±3	22.5×23.5±7.25	8.69×16.95±2.47	5.16×4.17±2.13	6×6.5±2.14	4.35×5.22±1.46	3.02×2.96±1.93	Absent
58	Panicum miliaceum	315.38×323.07±25.16	32±6	24±3	22±4	19.68×23.22±5.27	17.23×19.45±6.16	6.07×9.28±2.72	7.74×7.39±2.68	6.67×6.12±2.17	3.21×2.86±1.28	Absent
59	Panicum trypheron	316.67×308.34±51.56	20±6	14±4	24±6	15.94×19.69±6.17	8.34×12.5±3.17	6.39×10±2.15	5.94×5.94±2.17	4.17×4.17±1.54	3.62×3.05±1.13	11±4
60	Paspalidium flavidum	361.90×280.95±47.25	32±6	24±6	14±3	15.43×16.08±3.16	11.78×19.28±3.26	3.13×3.13±1.25	4.78×4.35±1.53	3.57×4.64±1.26	2.08×2.75±0.7vb4	6±2
61	Paspalidium geminatum	1080.64×1677.42	28±4	24±3	12±3	18.21×18.39±4.15	10.78×11.56±2.36	4.27×4.03±1.82	6.07×5±2.64	2.18×2.5±0.73	1.64×1.82±0.28	Absent
62	Paspalum scrobiculatum	275×362.5±13.54	17±3	15±4	16±4	18.75×15.63±3.27	17.08×21.04±3.65	6.34×8.34±2.56	6.25×5±2.63	6.25×4.68±2.38	3×2.67±1.36	Absent
63	Pennisetum setosum	742.23×1151.13±92.36	67±8	19±3	18±4	31.43×27.85±8.62	13.46×17.69±4.28	8.12×12.5±3.28	9.28×8.57±3.45	5.19×4.61±2.78	3.75×3.75±1.62	Absent

TABLE 2.1 (Continued)

Sr No.	Name of plant	Cross section area (µm)	Number of vascular bundles			Size of vascular bundles (µm)			Size of metaxylem (µm)			No. of kranz cells
			I°	II°	III°	I°	II°	III°	I°	II°	III°	
64	Setaria glauca	777.79×1100±104.26	26±3	14±2	12±2	22.5×29.37±6.37	21.36×26.36±5.89	15×17±4.72	9.09×5±2.73	8.75×8.13±2.67	6.5×5.75±2.63	Absent
65	Setaria tomentosa	263.64×290.90±14.52	16±3	11±2	13±2	20.38×13.07±5.28	12.67×15±3.52	5.23×6.36±2.45	5.77×5±2.74	3.67×4.34±1.67	1.59×1.82±0.51	Absent
66	Setaria verticillata	241.67×325±21.68	14±3	16±3	12±3	16.43×18.21±4.27	9.26×10.27±3.79	6.11×6.67±2.74	4.64×5.36±2.47	3.47×3.97±1.29	2.23×2.23±0.96	Absent
	Group : Pooideae											
	Tribe: Isachneae											
67	Isachne globosa	268.18×431.82±16.29	29±7	18±3	20±5	20.84×25.42±6.42	11.25×13.75±4.82	8.37×8.75±2.38	8.34×7.5±2.43	5.63×4.38±2.67	2.75×2.98±0.65	6±2
	Tribe: Aristideae											
68	Aristida adscensionis	225.92×229.63±15.38	17±2	15±3	12±2	13.57×12.86±4.89	8.28×9.14±2.19	3.28×3.81±1.34	5.71×4.28±2.78	3.71×2.86±1.03	1.85×1.84±0.06	6±2
69	Aristida funiculata	196.37×198.32±21.38	9±1	10±3	10±2	11.25×15±3.16	5.71×8.57±2.46	3.44×6.25±1.37	5.31×4.69±2.67	2.32×2.14±0.84	1.88×1.88±0.52	6±2
	Tribe: Perotideae											
70	Perotis indica	98.54×47.81±9.24	10±2	11±2	13±3	12.5×16.67±3.74	10.36×11.73±2.46	7.92×10.42±1.53	5×4.58±2.46	3.46×3.06±1.34	2.5×2.08±0.85	8±3
	Tribe: Chlorideae											
71	Chloris barbata	129.54×118.18±32.64	17±3	19±2	24±3	25×10.71±6.63	12.78×17.22±3.46	3.67×2.33±1.63	7.86×10±2.63	5×6.11±2.39	1.67×1.67±0.25	7±2
72	Chloris montana	171.44×145.24±17.37	30±6	14±3	26±3	20×15.45±4.17	9.25×10.15±3.89	6.67×7.34±2.43	5×3.18±	3.36×3.89±1.37	1.67×1.34±0.93	6±2
73	Chloris virgata	130×157.5±26.27	12±3	8±2	10±3	13.72×14.26±2.36	7.63×8.68±2.16	5×4.37±1.25	5×2.63±1.25	2.5×2.5±0.68	1.46×1.71±0.48	6±2
74	Cynadon dactylon	277.27×265.09±41.26	27±4	16±3	48±5	14.58×13.75±2.18	9.09×12.73±3.17	6.34×6.82±2.37	5×4.17±2.37	4.09×4.09±1.38	1.53×1.38±0.92	7±2
75	Melanocenchris jaequemontii	115.79×126.32±36.26	19±3	11±2	-	10×10.59±3.16	7.81×10.31	-	4.06×3.13±1.36	3.53×2.94±1.37	-	Absent
76	Oropetium villosulum	116×136±17.28	10±2	8±2	19±4	13×12±4.17	7.78×10±2.19	3.70×5.18±1.27	3×2.67±1.32	2.23×2.23±0.95	0.74×1.12±0.39	10±3
77	Sachoenefeldia gracilis	47.62×73.81±11.43	11±2	14±3	-	14.28×15±4.27	6.04×6.25±2.17	-	6.04×6.25±2.67	5×5±1.94±2.18	-	Absent
78	Tetrapogon tenellus	239.28×271.43±32.46	32±8	14±4	22±3	19.55×25.45±5.28	13.15×14.5±3.17	7.86×11.78±2.16	5.91×5±2.48	4×5.5±1.83	3.21×3.21±0.82	5±1
79	Tetrapogon villosus	131.82×168.18±27.46	22±3	12±3	26±4	10.36×6.43±2.17	8.64×11.36±3.12	4.62×6.15±2.15	3.18×2.73±1.28	2.86×1.43±0.62	1.15×1.15±0.38	4±1

TABLE 2.1 *(Continued)*

Sr No.	Name of plant	Cross section area (μm)	Number of vascular bundles			Size of vascular bundles (μm)			Size of metaxylem (μm)			No. of kranz cells
			I°	II°	III°	I°	II°	III°	I°	II°	III°	
Tribe: Eragrosteae												
80	Acrachne Racemosa	86.96×93.48±13.70	18±3	10±1	11±2	11.67×12.92±1.38	8.75×9.58±2.61	5.83×5±2.15	2.5×2.5±0.93	1.25×1.25±0.46	0.84×0.84±0.04	Absent
81	Dactyloctenium aegyptium	294.45×433.34±18.64	28±6	24±3	30±5	18.64×17.5±3.26	10.52×14.83±3.17	5.77×6.15±2.62	4.54×4.54±1.73	4.14×3.62±2.18	2.31×1.92±0.92	10±2
82	Dactyloctenium sindicus	242.31×226.92±38.57	12±1	14±2	24±2	17.39×22.61±4.27	4.23×8.46±2.37	2.18×2.61±0.26	4.56×5.65±1.86	1.35×1.73±0.43	1.17×1.05±0.04	6±2
83	Desmostachya bipinnata	326.67×240±21.80	22±5	20±2	-	11.67×11.67±1.84	10×6.67±2.16	-	4.27×4.25±2.46	2.92×3.75±1.35	-	Absent
84	Dinebra retroflexa	423.07×346.15±36.18	26±4	20±3	30±2	17.5×11.67±2.87	5.42×6.67±1.73	3.17×3.17±1.25	3.75×2.92±1.02	1.67×1.67±0.29	1.27×1.25±0.76	Absent
85	Eleusine indica	78.94×71.05±6.26	15±2	10±2	18±3	8.34×7.5±2.17	8.24×7.94±3.17	5.17×4.83±1.56	2.08×2.5±0.36	1.76×1.91±0.18	1.55×1.38±0.27	7±2
86	Eragrostiella bachyphylla	121.25×142.5±10.27	12±1	10±2	18±2	9.52×13.57±1.56	8.89×11.12±2.18	4.27×4.18±1.78	3.09×3.57±1.87	1.85×2.59±0.53	1.28×1.45±0.15	6±2
87	Eragrostiella bifaria	120×112.5±34.72	13±4	14±2	14±2	14.85×13.64±2.78	7.69×13.46±1.77	4.17×4.86±1.07	4.09×4.24±1.67	1.92×1.73±0.63	1.38×1.25±0.32	6±1
88	Eragrostis cilianensis	138.46×203.85±25.78	20±2	16±2	32±3	13.75×14.58±4.28	10.84×13.34±1.54	4.75×5.5±1.69	3.96×3.75±1.42	3.34×2.92±1.65	1.75×1.75±0.21	5±2
89	Eragrostis ciliaris	168±118.18±11.75	24±5	30±2	31±4	11.12×11.81±3.32	6.76×6.32±2.17	5.6×5.4±2.18	4.17×3.47±2.21	3.53×3.24±1.54	2.4×2.8±0.54	4±1
90	Eragrostis japonica	160.87×201.35±48.63	30±2	27±4	44±6	17.31×13.08±3.87	7.69×6.35±2.89	3.39×3.57±1.66	6.73×4.62±1.89	2.5×2.12±0.83	1.61×1.07±0.58	5±2
91	Eragrostis nutans	150×127.27±24.18	26±2	28±2	29±2	16.82×11.36±3.65	11.92×13.46±2.18	3.48×4.06±1.86	4.32×3.41±1.29	4.23×3.46±1.37	1.52×1.06±0.78	5±1
92	Eragrostis pilosa	241.28×222.24±38.20	30±5	21±2	24±1	15.42×21.67±3.81	14.76×18.09±2.37	7.5×6.5±1.39	6.25×6.25±1.36	4.76×4.29±2.61	1.5×1.75±0.83	7±2
93	Eragrostis tenella	163.64×209.09±25.98	22±4	18±2	14±2	11.78×13.04±3.17	8.32×8.96±2.41	4.32×4.86±1.53	4.29×3.93±0.93	1.62×1.22±0.48	1.08×1.17±0.48	Absent
94	Eragrostis tremula	175×193.75±21.29	25±3	23±3	27±5	11.43×12.5±3.17	7.93×7.86±2.18	4.45×4.45±0.83	2.5×2.5±0.64	1.97×1.94±0.54	1.25×1.39±0.28	Absent
95	Eragrostis unioloides	168.75×159.38±14.75	16±3	10±3	19±3	17.08×17.92±4.18	5.86×5.52±2.18	5×6.56±1.34	4.58×5±1.45	1.72×1.55±0.53	1.25×1.56±0.48	7±2
96	Eragrostis viscosa	150×145±13.65	17±2	15±3	16±2	11.56×9.69±1.48	8.82×8.82±2.68	5.43×5.92±1.54	3.44×2.5±1.87	3.24×3.08±1.03	1.43×1.71±0.04	5±1

TABLE 2.1 *(Continued)*

Sr No.	Name of plant	Cross section area (µm)	Number of vascular bundles			Size of vascular bundles (µm)			Size of metaxylem (µm)			No. of kranz cells
			I°	II°	III°	I°	II°	III°	I°	II°	III°	
Tribe: Sporoboleae												
97	*Sporobolus coromandelianus*	97.72×118.18±24.53	15±1	12±2	14±2	12.5×13.34±3.17	9.45×11.38±2.68	7.04×7.27±1.13	3.34×3.34±1.79	2.78×3.89±1.86	2.27×2.27±0.81	6±2
98	*Sporobolus diander*	117.5×120±25.17	16±3	18±2	-	22.5×22.3±4.96	17.65×20.17±5.28	-	10×10±3.18	5.29×6.47±2.18	-	Absent
99	*Sporobolus indicus*	216.67×220.84±25.18	17±4	15±2	24±3	20×14.45±3.17	11.25×13.34±3.15	3.5×5.67±1.78	4.45×4.17±1.59	2.92×3.75±0.83	1.34×1.17±0.07	Absent
Tribe: Zoysieae												
100	*Tragrus biflorus*	115.96×119.5±9.54	20±4	9±2	18±3	26×27±6.28	14×16±5.17	5.28×7.69±2.17	10×8±3.97	6×7±2.16	2.31×2.31±0.61	Absent

of the internode (Sulthani, 1989). Increased number of vascular bundle's might be accompanied by increment in the greater number and density of sclerenchyma and conducting cells thereby increasing the strength properties of the internodes (Mohmod et al., 1992). Number of vascular bundles vary from species to species. *Coix-lachryma jobi* had maximum number of I IVB (68) and II IVB (36) while *Ophiuros exaltatus* showed maximum of III PVB (79) among the studied species.

Liese and Grosser (1972) noted considerable variation in the appearance of the vascular bundles within a single culm species of bamboo. Grosser and Liese (1971) describes four types of vascular bundles in different bamboo species. Hypodermal vascular bundles are smaller in size reported in many monocotyledons species (Grosser and Liese, 1971; Hossain 2011; Joarder et al., 2010). The center vascular bundles, that is, which were near to the pith, were larger in sizes and outer ones were smaller in size. A wide range of vascular bundles area has been reported in cogon grass ecotypes (Homeed et al., 2009). According to Liese (1985), in bamboos smaller vascular bundles are denser in distribution than that of bigger ones resulting in the higher density and mechanical strength for the outer zone than both inner and middle. Among the 100 studied species *Triplopogon ramosissimus* showed maximum size of all vascular bundles (58.34 × 91.08 ± 2.73 (μm) I IVB, 50.34 × 51.63 ± 0.83 (μm) II IVB, 33.45 × 29.35 ± 0.21 (μm) III°PVB). Quantitative data shows a wide standard error in the size of the vascular bundles which is an indication of existence of differences between plants with in a sample. High variation in measurements of anatomical traits has been reported in different ecotypes of cogon grass (Hameed et al., 2009). Polygenic controlled rice and wheat exhibits high variation grown under different environmental conditions (Joarder, 1980; Sima, 2010). Hossain (2010) noted strong environmental effects on the expression of anatomical traits and each estimate is attached with high standard error in wild sugarcane.

Peripheral vascular bundles being completely encircled by thick walled sclerenchymatous, bundle sheath, indicates an anatomical adaptation for tolerating drought and harsh environment (Matumura and Nokajima, 1988). Species like *Panicum miliaceum, Panicum trypheron, Paspalum scorbiculatum, Setaria sp., Melanocenchris jaquimontii, Chionachnae koenigii* showed peripheral vascular bundles were covered by sclerenchymatous cells while species like *Pennisetum setosum, Dinebra, Eragrostiella, Chrysopogon, Cymbopogon, Dicanthium* showed peripheral vascular bundles totally embedded within sclerenchymatous tissue.

Sometimes few vascular bundles did not develop into full vascular bundles (i.e., with xylem and phloem), they had only phloem cap and more tissues which were surrounded by fibrous cell. This is an adaptation for efficient phloem transport to undergo and rhizome for reserve (Sima et al., 2015).

In cross section, vessel diameter increased significantly from outer to inner. Vascular bundles an observation which is in accordance with noticed by that authors Liese (1985), Hishman et al. (2006), and Wang et al. (2011). The metaxylem vessels wide in the inner and middle zones, a feature attributed to the fact that these zones are mainly functioned for water and nutrient transportation while in outer/peripheral zone incomplete developed vascular vessel was observed (Huong x y, 2015). Vessel diameter of I IVB are more than the other vascular bundles while III PVB showed less diameter. Among the studied species, metaxylem of I IVB of *Vetiveria zinzanoides* (16.67 × 13.89 ± 5.18 (μm)) showed maximum size. *Triplopogon ramosissimus* showed maximum vessel diameter (9.29 × 9.19 ± 0.40 (μm)) of II IVB among the studied species. Likewise, *Setaria glauca* showed high vessel diameter (6.5 × 5.75 ± 2.63) of III PVB.

In ground tissue, cell size increases from the outer to the inner layers (Grosser, 1970; Parameswaran and Liese, 1981) a similar feature noticed in all the studied species.

Presence of Schizo-lysigeny to lysogenous arenchyma in culm is normal feature observed in *Cynodon dactylon, Eremochloa ophiuroides, Hemarthria altissima,* and *Miscanthus sacchariflorus* (Sangster, 1985; Seago et al., 2005; Jung et al., 2008; Yang et al., 2011). In the present study, *Cynodon dactylon,* did not show air cavities but the pith region was hollow. *Vetivaria zinzanoides, Paspalidium geminatum* shows air cavities in cortical region. *Aristida adscensionis* did not show any cavity but parenchyma cells below hypodermis region are large in very big size and elongated. These consecutively formed air spaces (Kawai et al., 1998) which obviously stored and transported oxygen to organs in hypoxic environments (Justin and Armstong, 1987; Vartapetian and Jackson, 1997; Colmer, 2003; Armstrong and Armstrong, 2006).

Kranz anatomy is observed only in species like *Eragrostis, Dinebra, Cenchrus setigerus, Dactyloctenium sps., Tragus biflors, Perotis indica, Tetrapogon sp.,* etc. Kranz layer shows prominently distinct rounded cells in *Arundinella* (Sanchez et al., 1989) a similar feature observed in the present study in species like *Chloris montana, Tetrapogon sps., Tragus biflorus, Paspalidium flavidum, Sporobolus dianer, Aristida adscenionis, Ischanae globosa,* etc. The cells are thickened and contained specialized distinct large chloroplasts embedded in distinctly larger chlorenchyma cells. They are not part of vascular tissue and therefore cannot be considered as degenerate intercalary vascular bundles (Brown, 1977). Instead, they are seen to be long, isolated Kranz cell columns not physically connected to the vascular bundles but presumably functionally linked at regular intervals by vascular strands. The occurrence of Kranz anatomy in both the culms and leaves is noteworthy because grass species with Kranz leaf anatomy do not necessarily exhibit Kranz structure in the culm as well (Sanchez et al., 1989). Kranz arc, radial cholrenchyma, culm outline and chloroplast shape and position, are some of the culm anatomical features suggested to be useful for inferring phylogenies (Siqueiros and Herrera, 1996). Four different types of Kranz arc were observed, that is, straight, half circle, horse shoe shaped, circular in the studied species. *Eragrostis viscosa, Eragrostis ciliaris, Tetraoogon tenellus, Tragus biflorus* showed straight type of Kranz arc, *Sporobolus indicus, Chloris montana, Dactyloctenium sindicus, Tetrapogon villosus, Dactyloctenium aegyptium, Cenchrus setigerus* showed half circle-shaped Kranz arc, while species like *Aristida funiculata, Perotis indica* showed horse shoe-shaped Kranz arc and only *Ophiuros exaltatus* showed circular-shaped Kranz arc. Two types of chloroplast arrangements were present: centripetal and centrifugal. Species like *Eragrostis viscosa, Sporobolus sp., Tetrapogon sp., Dactyloctenium sp., Eragrostiella bachyaphylla, Sporobolus coromandelianus, Dinebra* showed centrifugal type of chloroplast while species like *Eragrostis ciliaris, Chloris sps., Tragus biflorus* showed centripetal type of choloroplast.

Based on the characteristic features, a dichotomous key for the studied species has been prepared.

1. Presence of Kranz cells..**2**
1. Absence of Kranz cells...**39**
2. Elongated/ round-shaped epidermal cell....................................**3**
2. Barrel/ rectangular/ square-shaped epidermal cell....................**11**
3. Square to trapezoid Kranz cell shape...................***Eragrostis cilianensis***
3. Round to oval Kranz cell shape **4**
4. Sclrenchyma in form of girder.. **5**
4. Sclrenchyma in form of unicylinder **6**
5. Kranz arc shape is half circle...............................***Ishcaemum pilosum***
5. Kranz arc shape is horse shoe ....***Perotis indica***
6. Circular-shaped Kranz arc ....***Ophiuros exaltatus***
6. Half circle-shaped Kranz arc **7**
7. Type D sclerenchyma....................................***Desmostachya bipinnata***
7. Type E sclerenchyma **8**
8. Elongated-shaped epidermal cell...**9**

27. Elliptical shaped of I IVB, cross section size 78.94
    Schizo-lysigeny 71.05 ± 6.25......................................*Eleusine indica*
27. Oval shaped of I IVB, Cross section
    size 239.28 × 271 ± 32.46....................................*Tetrapogon tenellus*
28. Hollow internode ....*Tetrapogon villosus*
28. Solid internode...*Dactyloctenium aegyptium*
29. Square-shaped epidermal cells..**30**
29. Barrel-shaped epidermal cells..**32**
30. Type E sclerenchyma present................................*Sporobolus indicus*
30. Type D sclerenchyma present **31**
31. Epidermal cells covered with thick cuticle
    including tangential wall....................................*Oropetium villosulum*
31. Epidermal cell covered with thick cuticle
    penetrating between epidermal cell ....*Chloris virgata*
32. Type E sclerenchyma present..**33**
32. Type D sclerenchyma present **35**
32. Cordate shape in cross section ....*Isachne globosa*
32. Round/oblong shape in cross section..**34**
34. Diamond-shaped vascular bundle...........................*Acrachne racemosa*
34. Elliptical-shaped vascular bundle ....*Dinebra retroflexa*
35. Hollow pith region **36**
35. Parenchymatous pith region...**37**
36. Type E I IVBs present..................................*Aristida adscensionis*
36 Type B I IVBs present................................... *Eragrostris pilosa*
37. Round-shaped in cross section............................ *Sporobolus diander*
37. Oblong to oval shaped in cross section....................................**38**
38. Elliptical in transaction ....*Ischaemum indicus*
38. Heart shaped in transaction.................................*Ischaemum molle*
39. Sclerenchyma present in form of girder.................................**40**
39. Sclerenchyma present in form of cylinder.............................**50**
40. Type C sclerenchyma present .... *Arthraxon lanceolatus*
40. Type D/E sclerenchyma present..**41**
41. Type E sclerenchyma present..**42**
41. Type D sclerenchyma present **43**
42. Barrel-shaped epidermal cell ....*Oplismenus burmanii*
42. Square-shaped epidermal cells.................... *Triplopogon ramosissimus*
43. Elongated/ round-shaped epidermal cell....................................**44**
43. Barrel-shaped epidermal cells..**45**
44. Elongated-shaped epidermal cells................................ *Apluda mutica*
44. Round-shaped epidermal cells .... *Cenchrus ciliaris*

61. Oblong shaped in cross section **62**
61. Round to oval shaped in cross section **63**
62. Angular parenchymatous ground tissue .... *Hackelochloa granularis*
62. Round-shaped parenchymatous ground tissue...........*Cenchrus biflorus*
63. Barrel-shaped- V shaped epidermal cells, cross section
size in1155.56 × 1444.45 ± 47.28 µm ....*Heteropogon triticeus*
63. Barrel-shaped- IV shaped epidermal cells, cross
section size in between 185–280 × 195–230 µm....................................**64**
64. Inverted triangular shaped I IVB ....*Sehima ischaemoides*
64. Triangular diamond shaped I IVB.............................*Sehima nervosum*
65. Type D sclerenchyma present **66**
65. Type E sclerenchyma present..**86**
66. Round-Shaped epidermal cell.......................... *Paspalidium geminatum*
66. Barrel/rectangular/square/elongated-shaped epidermal cell...............**67**
67. Square-shaped epidermal cell **68**
67. Barrel/rectangular/elongated-shaped epidermal cell..........................**70**
68. Triangle diamond-shaped vascular bundle......................*Isilema laxum*
68. Diamond-shaped vascular bundle..**69**
69. Type B I IVB...*Vetivaria zinzanoides*
69. Type I I IVB .... *Themeda laxa*
70. Elongated-shaped epidermal cell **71**
70. Barrel/rectangular-shaped epidermal cells......................................**74**
71. Hollow pith region **72**
71. Parenchymatous pith region...**73**
72. Type- B sclerenchyma present, number of III
PVBs is 22 ± 5 .... *Themeda quadrivalvis*
72. Type- D sclerenchyma present, number of III
PVBs is 64 ± 12 ....*Panicum antidotale*
73. Angular parenchymatous ground tissue...................*Rottboellia exaltata*
73. Round to oval-shaped parenchymatous
ground tissue ....*Thelepogon elegans*
74. Barrel-shaped epidermal cells..**75**
74. Rectangular-shaped epidermal cells...**81**
75. Parenchymatous pith region...**76**
75. Hollow pith region **77**
76. Irregular-shaped simple parenchymatous
ground tissue ....*Setaria tomentosa*
76. Round shaped with thick walled
parenchymatous ground tissue...........................*Capillidpedium hugelii*
77. Horseshoe shaped in cross section.......................... *Eragrostis japonica*

2.4   CLUSTER ANALYSIS

The software used displays a single tree among the possible ones (Fig. 2.53). The *dendrogram* based on the anatomical characters of the cluster.

I.   Epidermal cell shape (5 criteria):
        Barrel shaped (1), rectangle shaped (2), square shaped (3), elongated shaped (4), round shaped (5).

II.   Presence of sclerenchyma in different forms (3 criteria):
        Unicylindrical (1), bicylindrical (2) in form of girder (3)

III.   Type of sclerenchyma cell (5 criteria):
        Type-A (1), Type-B (2), Type-C (3), Type-D (4), Type-E (5)

IV.   Presence or absence of Kranz anatomy (2 criteria):
        Present (1), absent (2)

V.   Shape of Kranz are (4 criteria):
        Straight (1), half circle (2), horse shoe (3), circular (4)

VI.   Kranz cell shape (3 criteria):
        Round to oval (1), square to trapezoid (2), absence (3)

VII.   Chloroplast type (2 criteria):
        Centripetal (1), centrifugal (2)

VIII.   Number of Kranz cells (9 criteria):
        0 (1), 12 (2), 7 (3), 9 (4), 11 (5), 6 (6), 10 (7), 4 (8), 5 (9)

Two major clusters with the similarity at 57.38% could be obtained under one major cluster two subcluster are present, that is, cluster 1 and 2 while under second major cluster, eight subclusters are present, that is, cluster 3,4,5,6,7,8,9, and 10. The observations per clusters are presented in Table 2.2 and *dendrogram* tree was shown in Fig. 2.53.

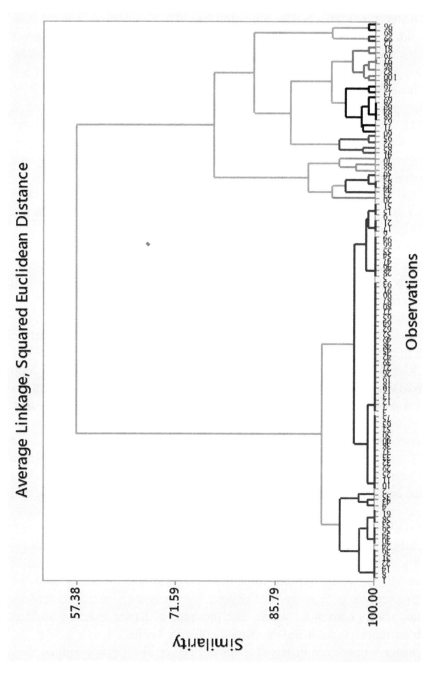

FIGURE 2.53 A dendrogram showing clustering of grass species on the basis of culm anatomy.

Cluster 1 has 16 grass species and has 2.013 maximum values from centroid. It shows around 95% similarity with other clusters. Cluster 3 contains species like *Chrypopogan fulvus, fsiloma laxum, Apluda mutica, Thelepogan,* etc. and show common features like absence of Kranz anatomy.

Cluster shows maximum of species and are 52 and shows around 97% similarity with other clusters. Cluster 2 shows maximum value from centroid is 1.610. The species like *coix luchryma-jobi, Dicanthium cimiciana, Andropogan plumis, Hackelochloa, Fanicum maximum, Saccharum spontonum,* etc belong to cluster 2 and share two common features like absence of Kranz anatomy, either barrel-shaped or rectangular-shaped epidermal cells.

Cluster 3 is a simplified cluster, that is, it has only a single species in cluster. It has *Ischaemum pilosum* and shows similarity around 94%. It has maximum value from centroid which is 0.00.

Clusters 4 and 5 have four species; cluster 4 has 1.58% maximum value from centroid while cluster 5 has 1.479. Cluster 4 has *ophiorus exaltate, cenchrus seligerus, Desmostachya bipinnata,* and *cynodon dactylon* and shows around 96.88% similarity with other clusters.

Cluster 5 has *Brachiaria reptans, Eleusine indica, Eragroslis pilosa, and Eragrostis unioloides* and shows around 96% similarity with other clusters. It has maximum value from centroid is 1.479.

Clusters 8 and 10 have two species. Cluster 8 has 1.050 values from centroid while cluster 10 has 0.500 distances from the centroid. Cluster 8 has *Chloris montang* and *sporobolus indicus* and shows almost 99% similarity with other clusters. Cluster 10 has *Eragraslis ciliaris* and *Eragraslis viscosa* and also shows around 99% similarity with other clusters.

Cluster 6 contains three species. Cluster 6 has 1.453 maximum value from centroid and has aboud 95% similarities with other clusters. This cluster contains species like *Panium trypheron, Eragrostis cilinensis,* and *Perotis indica.* These species show few common features like presence of Kranz cell, centrifugal type of chloroplast.

Cluster 7 contains nine species. It has maximum value from centroid is 1.718 and it shows around 95% similarity with other clusters. It contains species like *Pasalidium flovidum, Chloris barbata, Ischanae globose, Dinebra retroflexa, Sporobolus diander, Choris virgata, Aristida funiculate,* etc. and shows common features like presence of Kranz anatomy and has either barrel/rectangular/square-shaped epidermal cells.

Cluster 9 has seven different species. Species like *Tetrapogon tenellus, Tetrapogon villorus, Tragus biflorus, Eragrostiella bachyaphylla, Sporobolus coromanelalianus, Dactyloctenium aegyptium,* etc. show common features like presence of Kranz anatomy and barrel-shaped epidermal cell. This cluster has maximum value from centroid which is 1.641 and shows around 97.28% similarity with other clusters.

Cluster analysis carried out of anatomical features of culm was done to find the relationship between different groups of species. Square Euclidean distance showed good resolution of the genera based on qualitative characters. Dendrogram clearly showed two major groups/clusters. This dendrogram showed similarity with the dichotomous key which was prepared by using anatomical features. Two major clusters are formed out of which one major cluster contained 68 species and other major cluster showed 57.38% similarity between them. Cluster 1 and 2 form one major while cluster 3, 4, 5, 6, 7, 8, 9, and 10 form other major group. Cluster 1 and 2 showed around 93% similarity between them and they shared common character like absence of Kranz anatomy, while 3–10 showed around 76% similarity between them and they shared common character like prescence of Kranz anatomy.

Cluster 1 and 2 combine and form one major group. Out of that cluster 1 had 16 species while cluster 2 had 52 species. Cluster 1 showed around 95.5% similarity with other clusters of group. Species of cluster showed characters like either square/elongated/round-shaped epidermal cells, or sclerenchyma present either in form of girder or in form of unicylindrical ring. Likewise, cluster 2 contains 52 species and showed around 97% similarity with other cluster of group. These species showed either barrel or rectangular-shaped epidermal cells.

TABLE 2.2 Distribution of Grass Species into Clusters on the Basis of Their Culm Anatomical Features.

Cluster Number	Number of observation (grass species) per cluster	Maximum value from centroid	Average distance from centroid
1	16	2.013	1.0155
2	52	1.610	0.798
3	1	0.000	0.000
4	4	1.581	0.926
5	4	1.479	1.239
6	3	1.453	1.223
7	9	1.718	1.002
8	2	1.050	0.500
9	7	1.641	0.956
10	2	0.500	0.500

From the second major group, cluster 3 was a simplifolius cluster and showed around 97.3% similarity with other clusters of group. Cluster 3 and 4 were close together and both these clusters close to cluster 6. All these three clusters shared few common features and these were closer to each other than the other clusters. All the species of clusters 3, 4, and 6 show prescence of Kranz anatomy, centrifugal type of chloroplast. Cluster 8 and 10 were close together; both the clusters contained two species. These two clusters unite and form one subunit. Cluster 10 contains *Eragrostis viscosa* and *Eragrostis ciliaris*. Both were very close to each other. The difference between these two was only of type of chloroplast, that is, they shared more common characters. Likewise, *Sporobolus indicus* and *Chloris montana* were forming cluster 8, they also share more common characters. These four species showed close relation with each other apart from other species.

Tetrapogon tenellus, Tragus biflorus, Dactyloctenium aegyptium, Dactyloctenium sindicum, Eragrostiella bachyphylla, Sporobolus coromardelianus form cluster 9 which showed more close relation with *Paspalidium flavidum, Chloris barbata, Orpetium villosulum, Ischanae globosa, Dinebra retroflexa, Aristida* spp., etc. were forming cluster 7. All the species of cluster 9 showed barrel-shaped epidermal cells while species of cluster 7 showed either rectangular/square/barrel-shaped epidermal cells. Based on cladogram study, we can conclude that all the studied species shared 57.38% similar characters. Cluster 1 and 2 forming one group, showing few similar features while cluster 3–10 which were forming other group showing some different characteristic features than the first group. *Ischaemum pilosum* showed unique characteristic features and separating it from all other species and forming single species from one cluster. Most of species of cluster 1 and 2 belong to group Panicoideae which showed around 93% similarity and cluster 3–10 belonged to group Pooideae which showed around 76% similarity with others.]

KEYWORDS

- **culm**
- **internode**
- **anatomy**
- **Kranz anatomy**

REFERENCES

Agrasar, Z. E. R. D. and Rodríguez, M. F. Cauline Anatomy of Native Woody Bamboos in Argentina and Neighboring Areas: Epidermis. *Bot. J. Linn. Soc.* **2002**, *138* (1), 45–55.

Arber A. *The Gramineae*. Cambridge University Press: Cambridge, 1934.

Armstrong, J. M. A Cytological Study of the Genus *Poa* L. *Can. J. Res.* **1937**, *15* (6), 281–297.

Armstrong, J.; Jones, R. E.; Armstrong, W. Rhizome Phyllosphere Oxygenation in *Phragmites* and Other Species in Relation to Redox Potential, Convective Gas Flow, Submergence and Aeration Pathways. *New Phytol.* **2006**, *172* (4), 719–731.

Bews, J. W. *World'S Grasses: Their Differentiation, Distribution, Economic and Ecology.* Longmans: Green and Co. Ltd.: London, 1929; pp 408.

Brown W. V. The Kranz Syndrome and its Subtypes in Grass Systematics. *Mem. Torrey Bot. Club.* **1977**, *23* (3), 1–97.

Canfield R. H. Stem Structure of Grasses on the *Jornada* Experimental Range. *Bot. Gazette.* **1934**, *95* (4), 636–648.

Cenci, C. A.; Grando, S.; Ceccarelli, S. Culm Anatomy in Barley (Hordeumvulgare). *Can. J. Bot.* **1984**, *62* (10), 2023–2027.

Colmer, T. D. Long-Distance Transport of Gases in Plants: a Perspective on Internal Aeration and Radial Oxygen Loss from Roots. *Plant Cell Environ.* **2003**, *26* (1), 17–36.

Gasser, M.; Vegetti, A. C.; Tivano, J. A. Anatomía de las estructurasfoliares y caulinare senvástagosre productivos de *Eleusine indica* (L.) Gaertner y E. tristachya (Lam.) Lam. In *Resúmenes VI Congreso Latino americano de Botánica, Mar del Plata,* 1994; pp 39.

Grosser, D.; Liese, W. On the Anatomy of Asian Bamboos, with Special Reference to their Vascular Bundles. *Wood Sci. Technol.* **1971**, *5* (4), 290–312.

Hameed, M.; Ashraf, M.; Na, N. Anatomical Adaptations to Salinity in Cogon Grass [*Imperata cylindrica* (L.) Raeuschel] from the Salt Range, Pakistan. *Plant Soil.* **2009**, *322* (1–2), 229–238.

Hisham, H. N.; Othman, S.; Rokiah, H.; Latif, M. A.; Ani, S.; Tamizi, M. M. Characterization of Bamboo *Gigantochloa scortechinii* at Different Ages. *J. Trop. For. Sci.* **2006**, *18*, 236–242.

Hitchcock, A. S. *A Text-Book of Grasses*. The Macmillan Company: New York, 1914.

Hossain, M. A. Culm Internode and Leaf Anatomy of Two Economic Grass of Bangladesh [Napier and Kas]. M.S. Thesis, Rajshahi University, Bangladesh 2010.

Hossain, M. A. *Culm Internode and Leaf Anatomy of Two Economic Grass of Bangladesh* [Napier and Kas]. M. S. Thesis, Rajshahi University, Bangladesh 2010.

Huang, X. Y.; Qi, J. Q.; Xie, J. L.; Hao, J. F.; Qin, B. D.; Chen, S. M. Variation in Anatomical Characteristics of Bamboo, Bambusarigida. *Sains Malays.* **2015**, *44* (1), 17–23.

Joarder, N. Comparative Anatomy of Basal Internodes of Some Lodging and Non-Lodging Rice Varieties. M. Phil Thesis, Rajshahi University, 1980.

Joarder, N.; Roy, A. K.; Sima, S. N.; Parvin K. Leaf Blade and Midrib Anatomy of Two Sugarcane Cultivars of Bangladesh. *J. Bio Sci.* **2010**, *18*, 66–73.

Jung, M. J.; Veldkamp, J. F.; Kuoh, C. S. Notes on *Eragrostis* Wolf (Poaceae) for the Flora of Taiwan. *Taiwania* **2008**, *53* (1), 96–102.

Justin, S. H. F. W.; Armstrong, W. The Anatomical Characteristics of Roots and Plant Response to Soil Flooding. *New Phytol.* **1987**, *106* (3), 465–495.

Kawai, M.; Umeda, M.; Chimiya, H. Stimulation of Adenylate Kinase in Rice Seedlings under Submergence Stress. *J. Plant Physiol.* **1998**, *152* (4–5), 533–539.

Liese; Grosser. Noted Considerable Variation in the Appearance of the Vascular Bundles Within a Single Culm Species of Bamboo. *Holzforsch* **1972,** *26* (6), 202–211.

Liese, W. Anatomy of Bamboo. In *Bamboo Research in Asia*, Proceedings of a Workshop Held in Singapore. International Development Research Centre Singapore: Ottawa, 1980; pp 161–164.

Liese, W. Anatomy and Properties of Bamboo. In *International Bamboo Workshop* (INBAR). October 6–14. China, 1985; pp 196–208.

Liese, W. *The Anatomy of Bamboo Culms.* Technical Report: International Network for Bamboo and Rattan, Beijing, 1998; pp 207.

Londoño, X.; Camayo, G. C.; Riaño, N. M.; López, Y. Characterization of the Anatomy of *Guadua angustifolia* (Poaceae: Bambusoideae) Culms. *Bamboo Sci. Cult. J. Am. Bamboo Soc.* **2002,** *16* (1), 18–31.

Matumura, M.; Nakajima, N. Comparative Ecology of Intraspecific Variation of the Chigaya, Imperata Cylindrica Var. koenigii (Alono-along). III. Annual Growth of the 3rd Year Communities Originated from the Seedlings. *J. Jap. Soc. Grassl.* **1988,** *34*, 77–84.

Metcalfe, C. R. *Anatomy of the Monocotyledons*, Vol. 1, *Gramineae;* Clarendon Press: Oxford, 1960; p 731.

Mohmod, A. L.; Mustafa, M. T. Variation in Anatomical Properties of Three Malaysian Bamboos From Natural Stands. *J. Trop. For. Sci.*1992, *5* (1), 90–96.

Parameswaran, N.; Liese, W. Torus-Like Structures in Interfibre Pits of *Prunus* and *Pyrus. IAWA J.* **1981,** *2* (2–3), 89–93.

Ramos, J. C.; Tivano, J. C.; Vegetti, A. C. Estudioanatomico de vastagos reproductivosen Bromusauleticus Trin.ex Ness (Poaceae). *Gayana. Botánica.* **2002,** *59* (2), 51–60.

Sanchez, E.; Arriaga, M. O.; Ellis, R. P. Kranz Distinctive Cells in the Culm of *Arundineua* (Arundinelleae; Panicoideae; Poaceae). *Bothalia* **1989,** *19* (1), 45–52.

Sangster, A. G. Silicon Distribution and Anatomy of the Grass Rhizome, with Special Reference to *Miscanthus sacchariflorus* (Maxim.) Hackel. *Ann. Bot.* **1985,** *55* (5), 621–634.

Seago, J. L.; Marsh, L. C.; Stevens, K. J.; Soukup, A.; Votrubova, O. A Re-examination of the Root Cortex in Wetland Flowering Plants with Respect to Aerenchyma. *Ann. Bot.* **2005,** *96*, 565–579.

Sekar, T.; Balasubramanian, A. Culm Anatomy of *Guadua* and its Systematic Position. *BIC, Indian Bull.* **1994,** *4*, 6–9.

Sima, S. N. Studies on Culm Anatomy and Yield Related Traits in Some Genotypes of Wheat (Triticum aestivum L.). Ph. D Thesis, University of Rajshahi, Bangladesh, 2010.

Sima, S. N.; Roy, A. K.; Joarder N. Erect Culm Internodal Anatomy and Properties of Sun Ecotype of *Imperata cylindrica* (L.) P. Beauv. *Bangladesh J. Bot.* **2015,** *44* (1), 67–72.

Siqueiros Delgado, M. E.; Herrera Arrieta, Y. Taxonomic Value of Culm Anatomical Characters in the Species of *Bouteloua lagasca* (Poaceae: Eragrostideae). *Phytologia* **1996,** *81* (2), 124–141.

Siqueiros Delgado, M. E.; Herrera Arrieta, Y. Taxonomic Value of Culm Anatomical Characters in the Species of *Bouteloua lagasca* (Poaceae: Eragrostideae). *Phytologia* **1996,** *81* (2), 124–141.

Siqueiros-Delgado, M. Anatomía del tallo de *Bouteloua* y génerosrelacionados (Gramineae: Chloridoideae: Boutelouinae). *Acta botánica Mexicana* **2007,** *78*, 39–54.

Sulthoni, A. *Bamboo:* In *Physical Properties, Testing Methods and Means of Preservation*, Proceedings of a Workshop on Design and Manufacture of Bamboo and Rattan Furniture; Bassiii, A. V., Davies, W. G., Eds.; March 3–14 1989; pp 1–15.

Vartapetian, B. B.; Jackson, M. B. Plant Adaptations to Anaerobic Stress. *Ann. Bot.* **1997,** 79, 3–20.

Wang, S. G.; Pu, X. L.; Ding, Y. L.; Wan, X. C.; Lin, S. Y. Anatomical and Chemical Properties of *Fargesia yunnanensis*. *J. Trop. For. Sci.* **2011,** *23* (1), 73–81.

Yang, C.; Zhang, X.; Zhou, C.; Seago, J. L. Root and Stem Anatomy and Histochemistry of Four Grasses from the *Jianghan* Floodplain along the Yangtze River, China. *Flora Morphol. Distrib. Funct. Ecol. Plants.* **2011,** *206* (7), 653–661.

Yao, X.; Yi, T.; Ma, N.; Wang, Y.; Li, Y. *Bamboo Culm Anatomy of China,* Vol. 187; Science Press: Beijing, 2002.

CHAPTER 3

Caryopsis Anatomy

ABSTRACT

Structurally caryopsis of the different grass species shows a great variation. General anatomy of grass caryopsis reveals that it can be divided into three distinct parts/regions: (1) the caryopsis coat which includes the pericarp, the seed coat, and the nucellus; (2) the endosperm; and (3) the embryo. The grass embryo is highly specialized and appears as an oval depression on the flat side of the caryopsis next to the lemma. On the opposite side of the grain of the embryo is the hilum; a line marking the point of attachment of the seed to the pericarp. The scutellum is a haustorial organ, equivalent to the single cotyledon, and functions in enzyme secretion and absorption of nutrients from the endosperm. Endosperm occurs as a large elliptical structure and is usually solid and starchy.

Characteristic features like course of vascularization, presence or absence of epiblast, presence or absence of cleft and arrangements of embryonic leaves in cross section have been adapted from Reeder (1957). Other than these, pericarp, angle of embryo with respect to anterior-posterior position of caryopses, coleoptiles, plumule, mesocotyl, coleorhiza, types of epiblast, scutellum, angle of vascularization, endosperm, type of starch grains, types of aleurone layer were also taken into consideration for the study. Other significant diagnostic features were observed for the first time in caryopses anatomy. This includes: (1) Embryo placed at different angles with respect to anterior-posterior position of caryopsis. Based on this feature, angle of embryo with respect to caryopses has been measured and it could be classified into three categories: less than 130°, between 130° and 160° and between 160° and 190.° (2) Previous studies report only presence or absence of epiblast in the different species studied according to which in group Panicoideae all the members showed absence of epiblast except *Chionachnae koengii*, while in group Pooideae all members showed presence of epiblast except *Ischanae globosa* from tribe Ischaneae, *Aristida adscensionsis, Aristida funiculata* from Aristideae and *Melanocenchris* and *Oropetium* from tribe Chlorideae. Five different types of epiblast could be identified and categorized. This was found to be an important diagnostic feature for identification of grass species. (3) Different authors have categorized the type of scutellum into two main categories: sickle shaped and V shaped but in the present study based on the shape and further on the angle which forms the different shapes, scutellum could be broadly divided into three main categories: V shaped, U shaped, and Δ shaped. Further 18 different types of V shaped scutellum, 5 different types of U shaped scutellum, and 6 different types of Δ shaped scutellum were observed and categorized. Based on above characters, identification key was prepared which help to identify caryopses on the basis of anatomical features. (4) Angle of vascularization was also studied in detail and the grass species could be divided into four different categories.

Quantitative features like cross section size of caryopses, seed coat thickness, aleurone cell size, occupied % of endosperm, thickness of endosperm, occupied % of embryo, thickness of embryo, number of starch grains per endospermic cell were also observed. Species belonging to the same genera could be identified distinctly. On the basis of embryo anatomy caryopses has been divided into six categories by Reeder (1957). According to this the grass species could be categorized into the following: All Panicoideae members belonged to True panicoids except *Chionachnae koengii* which belonged to Oryzoid-olyroid group. Most of the members of Pooideae group belonged to Chloridoid-Eragrostoid group except *Chloris montana, Chloris virgata, Sachoenefeldia gracilis* belong to Bambusoid and *Aristida sp., Melanocenchris jaquemontii, Orepetium villosum* belonged to Arundinoi-Danthonioid group. A dichotomous key has been prepared with the help of these characteristic features and dendrogram has been prepared by using cluster analysis.

3.1 INTRODUCTION

Caryopsis is a basic dispersal unit in the grasses. It simply consists of embryo and endodermis surrounded by several different structural layers that are contributed by the flower of the parent plant. Reeder (1957) has used unique characters like coleoptiles, epiblast, colelrhiza, and scutellum as a basis to classify grass taxa. The growth of the endosperm causes expansion, modification, and compaction of their enveloping layers. This study may contribute to dormancy in several species (Thornton, 1966; Rost, 1971), and is also known to contain nutritionally important materials in certain food grains. Seed coat has anatomical and agricultural importance, since it may contribute to dormancy in several weed species (Thornton, 1966a; 1966b; Rost 1971). These weed species known to contain nutritionally important materials (e.g. nicotinic acid and Vitamin B_1) in certain food grains (Hinton, 1948; Hinton and Shaw, 1954).

Few representative studies have been conducted on barley (Mann and Harlan, 1916), corn (Kiesselbach, 1949; Wolf et al., 1952), Indian millets (Narayanaswami, 1953; 1955a; 1955b; 1955c; 1956), Johanson grass (Harrington and Crocker, 1923), sorghum (Sanders, 1955), sugarcane (Artschwager et al., 1929), and other Poaceae (Guerin, 1899).

The coat is continuous around the entire caryopsis except at the point where it is connected to the axis of the inflorescence. The outermost layers are covered by the thick cuticle layer. The pericarp, derived from the ovary of the flower, may be a thin membrane, or it may be composed of one or several cell layers. General structure of caryopsis coat consisted of the adnate layers of the pericarp, seed coat, and nucellus that surround the endosperm and embryonic axis (Izaguirre de Artucio and La-guardia, 1987).

The embryo of Poaceae presents characteristics that are common to other poales, such as lateral disposition, the presence of coleoptiles, meso-cotyl formation, and the complete reduction of the primary root (Tillich, 2007). Structures of the Poaceae embryo that are characteristics used in the taxonomy of this group include the presence or absence of the sceutellum slit, the epiblast, vascularization, and the overlap of the plumule margins (Reeder, 1957).

Eichemberg and Scatena (2013) studied morphology and anatomy of the diasporas and seedlings of *Paspalum.* They observed that the caryopsis involves the seed that presents the differentiated embryo and dispored later-ally, an elliptical hilum in all the studied species.

Wleaherwax (1930) studied the endosperm of *Zea* and *Coix* and summa-rized that the embryo of *Zea* and *Coix* is embedded in one side of the endo-sperm but surrounded by the latter except at the base. Endosperm shows at maturity a higher degree of differentiation than is ordinarily attributed to it. Rost (1973) studied caryopsis coat in mature caryopses of the *Setaria lutescens.* Rost et al. (1990) studied caryopsis anatomy of the *Briza maxima.* They observed that seed coat cuticle extends all around the caryopsis, except in the placental pad region.

3.2   ANATOMICAL FEATURES

Both cross and longitudinal sections were taken serially and observed. Observations for the characteristic features were noted from the sections passing through the central portion of the embryo located in the caryopsis.

General anatomy of grass caryopsis reveals that it can be distinguished into three distinct parts/regions (Diagram 1): (1) the caryopsis coat that includes the pericarp, the seed coat, and the nucellus; (2) the endosperm; and (3) the embryo.

Few caryopses were free from the pericarp, for example *Sporobolus* and *Eleusine*, while few caryopses were permanently enclosed the lemma and palea, for example members of tribe Paniceae, *Aristida,* and *Stipa* (which were members of group Pooideae). In the tribe Andropogoneae members, glumes permanently enclose the caryopsis along with a pedicel and a section of the rachis. The grain may enlarge during development and ripening exceeding the size of the glumes, lemma, and palea.

1. ***Caryopsis coat:*** There are several layers of cells in caryopsis coat and it can be differentiated into pericarp, integuments, and nucellus. Pericarp is generally defined as the fruit wall developed from the ovary wall. Seed coat is the protective covering of the seed formed from integuments. In grasses, the pericarp and the remains of the integuments of the single seed are completely fused.

During development, the expansion of the caryopsis causes these cells to become crushed and form a complex caryopsis coat. Pericarp layer can be further distinguished into five different layers: the outer epidermis, the hypodermis, a zone of thin walled cells, cross cells, and tube cells.

The outer epidermis and the hypodermis together form the exocarp. The cells of the exocarp are elongated in a direction parallel to the longitudinal axis of the caryopsis, and they become compressed and their walls thicken considerably so that when the caryopsis is ripe, cell lumina cannot easily be distinguished in them.

Cross cells are found below the parenchymatous layer and they have thick walls with pits that are elongated transversely to the cell. The longitudinal axis of these cells is at right angles to that of the exocarp cells. The tube cells constitute the inner epidermis of the pericarp and they occur on the inside of the cross walls. There are large intercellular spaces between the tube cells, the walls of which are pitted and thinner than those of cross walls. The longitudinal axis of these cells is parallel to that of exocarp cells.

Seed coat that is united with the pericarp is crushed in the mature caryopsis and hence difficult to discern cells in this zone. A fully mature seed coat consists only of the inner integument, the outer being completely crushed.

Nucellus is discernible as one to two layers of thin walled cells between aleurone layer and the seed coat. In a complete mature seed, these nucellar cells are crushed and appear as thin glassy zone that is bright and colorless.

Depending upon the number of distinguishable layers in the pericarp, it can be categorized into four different types.

1. Type – A: (Fig. 3.1 A):

    Pericarp is differentiated into five layers:

    First: Epidermis that is made up of square/rectangular flat or broad cells.

    Second: Subepidermal/hypodermis that is- made up of rectangular/square cells.

    Third: Thin walled cells that are crushed.

    Fourth: Cross cells

    Fifth: Nucellus that is made up of rectangular cells.

2. Type – B: (Fig. 3.1 B):

    Pericarp is differentiated into four layers:

    First: Epidermis that is made up of square/rectangular flat or broad cells.

    Second: Subepidermal/hypodermis that is made up of rectangular/square cells.

    Third: Cross cells

    Fourth: Nucellus that is made up of rectangular cells.

3. Type – C: (Fig. 3.1 C):

    Pericarp is differentiated into three layers:

    First: Epidermis that is made up of square/rectangular flat or broad cells.

    Second: Cross cells

    Third: Nucellus that is made up of rectangular cells.

    Other layers like hypodermis and crushed parenchyma may be crushed during development.

4. Type – D: (Fig. 3.1 D):

    Pericarp is differentiated into three layers:

    First: Epidermis that is made up of square/rectangular flat or broad cells.

    Second: Crushed parenchyma cells

    Third: Nucellus that is made up of rectangular but flat or broad layer of cross cells that not develop from beginning and hypodermis may be crushed during development.

Embryo: The grass embryo is highly specialized and appears as an oval depression on the flat side of the caryopsis next to lemma. On the opposite side of the grain to the embryo is the hilum; a line (e.g. *Lolium perenne, Avena fatua*) or dot (*Holcus lanatus, Briza media*) marking the point of attachment of the seed to the pericarp. The embryo includes primordia of one, two, or more foliage leaves and the primary root, and primordia of several adventitious roots. The scutellum is a haustorial organ, equivalent to the single cotyledon, and functions in enzyme secretion and absorption of nutrients from the endosperm. Also in the embryo is the coleoptiles (part of the single cotyledon, an open sheath with a pore at tip, homolog of first foliage leaf), the epiblast (an outgrowth of the coleorhiza), and the coleorhiza (the first part of the plant to emerge on germination, producing a tuft of anchoring hairs, protecting the radicle; the primary root emerges through it); all are peculiar to the grass embryo. The embryonic shoot is known as the plumule, and the embryonic primary root as the radicle. The mesocotyl is a vascular trace in the scutellum extending down into the coleorhiza and up into the coleoptile. Endosperm occurs as a large elliptical structure in the seed formed from the fusion of two polar nuclei of the embryo sac and a sperm nucleus.

The endosperm provides nutrition for the embryo and developing seedling, and is usually solid and starchy (liquid in some including *Koeleria, Trisetum*, and all Aveneae). Variation in embryo structure is related to the presence or absence of the mesocotyl, the epiblast, the scutellum cleft (i.e. whether or not the scutellum is fused to the coleorhiza), and whether the first leaf is rolled or folded, and allows an evolutionary sequence of embryo types to be recognized (Renvoize, 2002). It is the new immature plant, developing from the zygote. Embryo is placed dorsally and is surcirculared by caryopsis coat on its dorsal surface and crushed cells that composed depletion layer on its ventral surface from where the scutellum comes in contact with the endosperm.

It contains two major parts: (i) embryo axis and (ii) scutellum. Embryo axis represents coleorhiza, radicle, mesocotyl, plumule, and coleoptiles (Fig. 3.2 A and B).

Embryo is placed at different angles with respect to anterior–posterior position of caryopsis. Taking this as a characteristic feature of variation caryopsis is categorized as:

A: less than 130°: (Fig. 3.1 E): embryo placed almost nearly to the right angle of the anterior–posterior axis of the caryopsis.

B: 130°–160°: (Fig. 3.1 F)

C: 160°–190°: (Fig. 3.1 G)

With an increase in the angle, position of embryo becomes more parallel to the anterior–posterior axis and with a decrease in the angle; the position of embryo is almost at right angle and placed perpendicular to the axis.

Embryo axis is an imaginary line drawn longitudinally through the embryo, comprising of the following parts:

1. Root axis: It comprises of coleorhiza and radicle.

- **Coleorhiza** is a cone-shaped protective sheath covering the radicle, at the terminal end of the embryonic axis. On the side nearest the scutellum, it extends to the level of the subapical meristem of the radicle. On opposite side, it extends as a thin sheet 4–5 cells thick to the top of the radicle. Coleorhiza shows outer and inner epidermis. Epidermal cells are square shaped (Fig. 3.2 C, arrowhead) and in between two epidermises, loosely arranged columnar to isodiametric cells are present, which are thin walled (Fig. 3.2 C).

  Coleorhiza shows two variations. Sometimes, it extends toward the dorsal surface and forms the epiblast. Epiblast is produced laterally from the embryo away from the scutellum and forming a small extension of the coleorhiza, while in others coleorhiza is directly connected to the mesocotyl with an absence of epiblast.

  Epiblast, distinguished in many caryopses, is a small projection of tissue of the embryonic axis opposite the scutellum. Based on the length and sharpness of the tip, different types of epiblast are present among the studied caryopses (Fig. 3.2 H–L).

  Type I: Epiblast as a small outgrowth form colerhiza with blunt end (Fig. 3.2 H).

  Type II: Epiblast in form of small outgrowth with sharp pointed end (Fig. 3.2 I).

  Type III: Epiblast long reaching upto half length of plumule, thin with a pointed (like a needle) end (Fig. 3.2 J).

  Type IV: Epiblast long, thick and pointed (blunt) end (Fig. 3.2 K).

  Type V: Epiblast long, nearly reaching upto entire length of plumule and with pointed end (Fig. 3.2 L).

- **Radicle:** It has a closed apical organization typical of grasses, surcir-culared by a densely stained layer and terminates at a root cap. Cells of the root cap at the tip of the radicle are closely adherent to the cells of coleorhiza (Fig. 3.2 D, arrowhead). It has a thick birefringent cell wall on its lateral surfaces. Cells are isodiametric to penta-hexagonal and compactly arranged. Cells are arranged in radial tiers one above the other (Fig. 3.2 D). Differentiation into a distinct epidermis and other layers is not prominent.

2. *Shoot axis:* It comprises coleoptile and plumule.

- **Coleoptile** is a cone-shaped sheath-like protective covering over the upper end of the embryonic axis. It is two to three layers thick and is a flattened cone surcircularing the first leaves and shoot apex. Coleoptile shows outer epidermis and inner epidermis. Outer epidermal cells are rectangular, long, and flat. Inner epidermal cells are broad, rectangular cells. Between two epidermis, 3–4 layers are present, and cells are square to rectangular shaped cells (Fig. 3.2 F).
- **Plumule** is present at the upper end of the axis under the coleoptiles, which is compactly arranged a terminal bud. Plumule has embryonic leaf. Generally, three leaves are there, namely, first leaf, second leaf, and third leaf. Third leaf is the innermost leaf and it is very small. Cells of plumule are similar to that of coleoptiles, that is, squarish to rectangular, but they are smaller in size compared to the coleoptiles cells, and cells are arranged in row (Fig. 3.2 G). In plumule, no such differentiation in epidermis and any other layers are seen. All cells are rectangular to square shaped, arranged one above the other and compactly arranged.
- **The anatomical features of coleorhiza, radicle, coleoptiles, and plumule are similar in all the studied species, and hence are not described individually in the description of the embryo axis of each species.**

3. *Mesocotyl:* Mesocotyl is an undifferentiated zone of compact internodes and meristematic initials present between the radicle and plumule. It extends downward to the radicle. The mesocotyl cells are procambial cells from where radicle and plumule originate. It also vascularizes rhizome and embryos. Mesocotyl is differentiated into the following layers: an outer and inner epidermis, and in between them few cells are present. Outer epidermal cells are square to rectangular cells. Inner epidermal cells are square shaped compactly arranged one above the other. In between radially arranged rectangular, flat radial cells are present, which are arranged one above the other and are compactly arranged. But in few species, they are loosely arranged (Fig. 3.2 E).

- **Anatomical features of mesocotyl are similar for all studied species so that it will not be described again in description of plants.**

4. Scutellum is considered as the single cotyledon of grasses. It is shield like and attached to the embryo axis functioning in absorbing food stored in the endosperm by secreting enzymes and transporting the food back to the embryo via vascular strands. It partially surcirculars the embryo axis and is continuous with coleorhiza. In between scutellum and coleorhiza, a small gap is present that is known as scutellum slit or cleft. Cleft may or may not be present in all caryopsis. The outermost layer of the scutellum consists of short, uniformly columnar cells known as scutellar epithelium. Scutellum extends to the apical portion of the embryonic axis.

When viewed in cross-section, scutellum is "V" or "U" or "Δ" shaped.

(I) "V" shaped: It is further classified into different categories: V1–V9: (Fig. 3.3 A–R). The angle of the arms of "V" is measured as shown below:

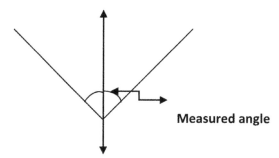

Measured angle

Further, the depth of the scutellum has been categorized as narrow and short on the basis of the distance (measurement) of anterior–posterior axis as shown below:

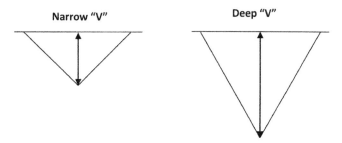

Narrow "V"   Deep "V"

Similarly, the base of the scutellum may be circular or tapering as shown below:

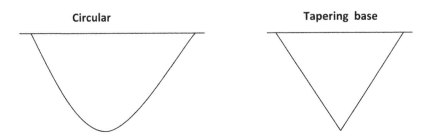

Microphotographs of the different types of scutellum based on the above-mentioned characteristic features have been represented in Figure 3.3.

V1: Deep "V" shaped with flat anterior side and having circular base. Based on the angle of arms of "V", this is again divided into two categories:

V1-1: Angle is 63.2 ± 4.72 (Fig. 3.3 A).

V1-2: Angle is 75.6 ± 5.83 (Fig. 3.3 B).

V2: Deep "V" shaped with flat anterior side and having pointed base, at an angle of 52.1 ± 4.27 (Fig. 3.3 C).

V3: Deep "V" shaped with convex anterior side and having circular base. Based on the angle of arms of "V", this is again divided into six categories:

V3-1: Arms of "V" at an angle of 89.4 ± 6.48 (Fig. 3.3 D).

V3-2: Arms of "V" at an angle of 68.9 ± 3.46 (Fig. 3.3 E).

V3-3: Arms of "V" at an angle of 63.9 ± 5.37 (Fig. 3.3 F).

V3-4: Arms of "V" tapering having angle 53.4 ± 5.42 (Fig. 3.3 G).

V3-5: Arms of "V" broad arm end with angle 82.23 ± 3.46 (Fig. 3.3 H).

V3-6: Having angle 43.2 ± 7.38 (Fig. 3.3 I).

V4: Deep "V" shaped with convex anterior side and having pointed base. Based on angle of arms of "V", type V4 can be categorized into:

V4-1: Angle is 27.2 ± 4.39 (Fig. 3.3 J).

V4-2: Angle is 45.3 ± 6.78 (Fig. 3.3 K).

V5: Deep "V" shaped with concave anterior side and having circular base with angle 52.9 ± 3.78 (Fig. 3.3 L).

V6: Narrow "V" shaped with flat anterior side and having circular base with angle 61.2 ± 6.24 (Fig. 3.3 M).

V7: Narrow "V" shaped with convex anterior side and having circular base with angle 69.5 ± 7.46 (Fig. 3.3 O).

V8: Narrow "V" shaped with convex anterior side and having pointed base.

Based on angle of arms of "V", this is again divided into three categories:

V8-1: Angle is 55.6 ± 9.67 (Fig. 3.3 P).

V8-2: Angle is 69.9 ± 2.47 (Fig. 3.3 Q).

V8-3: Angle is 63.7 ± 3.21 (Fig. 3.3 R).

V9: Narrow "V" shaped with concave anterior side and having circular base with angle 64.9 ± 2.58 (Fig. 3.3 N).

(II) "U" shaped: It is divided into two categories as shown below (Fig. 3.4 A–E):

Type U1: Arms of "U" overarches   Type U2: Arms of "U" extend
cover the embryo   only upto embryo surface

Type U1: Arms of "U" overarch and cover the embryo (Fig. 3.4 A and B). On the basis of the angle of arms and shape, it is further divided into two categories:

U1-1: Deep "U" with concave anterior side having an angle 41.9 ± 5.73 (Fig. 3.4 A). The arms of extend out giving a horseshoe-shaped form to the embryo.

U1-2: Deep "U" with convex circular anterior side and having an angle 65.7 ± 8.31 (Fig. 3.4 B).

Type U2: The arms of "U" extend only upto the embryo surface, not extending further (Fig. 3.4 C–E). On the basis of the angle of arms and shape, it is divided into three categories:

U2-1: Narrow "U" with flat anterior side having an angle 59.3 ± 4.84 (Fig. 3.4 C).

U2-2: Deep "U" with concave anterior side having an angle 55.3 ± 2.39 (Fig. 3.4 D).

U2-3: Narrow "U" with convex anterior side having an angle 62.1 ± 4.83 (Fig. 3.4 E).

(III):"Δ" shaped: It is divided into seven categories (Fig. 3.4, F–L).

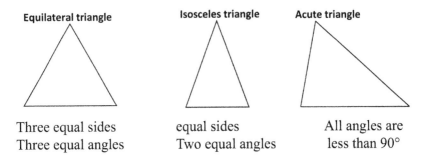

Equilateral triangle	Isosceles triangle	Acute triangle
Three equal sides Three equal angles	equal sides Two equal angles	All angles are less than 90°

Δ1: Isosceles with flat base having angle 35.8 ± 3.74 (Fig. 3.4 F).
Δ2: Isosceles with flat base having angle 52.4 ± 5.51 (Fig. 3.4 G).
Δ3: Isosceles with circular base having angle 38.2 ± 5.37 (Fig. 3.4 H).
Δ4: Acute triangle with circular base having angle (Fig. 3.4 I).
Δ5: Equilateral with circular base having angle 58.9 ± 2.93 (Fig. 3.4 J).
Δ6: Equilateral with flat base having angle 58.7 ± 4.73 (Fig. 3.4 K).

Scutellum cells are circular to oval in shape but sizes vary from species to species. The scutellum extends downward to blend with the ventral portion of the coleorhiza.

Vascularization: Vascular strand of the embryo extends lengthwise through the embryo, and splits at one point with one strand going into the scutellum and the other going to the coleoptiles. Vascular bundle connecting the embryonal axis with the scutellum traves from the mesocotyl. In grass caryopses, most striking feature is the curvature of the embryo, which is so strong that the inner faces of plumule and the radicle are contiguous. It is evidently due to this condition that no trace of an epiblast is seen. According to Schlickum (1896), the coleoptiles of the grasses differ essentially from the leaves following it in the absence of stomata and the presence of only two vascular bundles. The vascular system that comes from spikelet is diverted into florets through rachilla.

Reeder (1957) analyzed the structure of caryopsis in detail, especially structure of embryo that helps in grass systematics. Characteristic features described by Reeder have been considered to differentiate and characterize caryopses of grasses in the present study. Reeder concentrated on the following four characteristic features of embryo that are having taxonomic significance:

1. The course of the vascular system that may be

(a) Panicoid (denoted as P): The trace to the scutellum and embry-onic leaves diverge at approximately the same point.

(b) Festucoid (denoted as F): The trace to the scutellum and embry-onic leaves are separated by a more or less elongated internode.

2. The presence or absence of epiblast.

Epiblast as described earlier is an erect flap, long or short, with an absence of vascular tissue and located on the opposite side of the embryonic axis from the scutellum. Depending on its presence or absence, it has been denoted as '+' when present and '-' when absent.

3. The presence or absence of a cleft.

Cleft means the space present between scutellum and coleorhiza, that is, lower part of scutellum whether free from the coleorhiza or fused to it (if distinct cleft present than denoted as "P" and if absent than denoted as "F").

4. In the cross-section of the embryonic leaf, the vasculature has been observed. The vasculature has many bundles and overlapping margins or few bundles and margins that merely meet (if overlapping margins present than denoted as "P" and if margins merely meet than denoted as "F").

On the basis of above four characters, nine different formulae of grass caryopsis have been listed. These nine different formulae categorized into six different groups. Out of which oryzoid–olyroid group have three formulae and true festucoids have two forumulae, while other groups have single formula. Diagrammatic representation of different types has been described in Diagram 2.

Characteristic features of six groups are given as below:

(1) True festucoids: The species having embryo with the formula "F+FF". These are characterized by festucoid vascularization, presence of epiblast, no cleft between the scutellum and coleorhiza, and in cross-section the primary leaf is seen to have relatively few bundles and margins that do not overlap. Few species having the formula "F-FF" are also true festucoid, but they showed absence of epiblast.

(2) True panicoids: The genera in which the embryo has the formula "P-PP". The embryos are characterized by having panicoid vascularization, no epiblast, a distinct cleft between the scutellum and coleorhizae, and in transverse section the primary leaf with its numerous vascular bundles has overlapping margins.

(3) **Chloridoid–Eragrostoid type**: The embryo has the formula **"P+PF"**. These embryos are basically panicoid in that they are characterized by that type of vascularization and lower part of the scutellum free from the coleorhiza. They resemble festucoids in that they have an epiblast, and in transverse section the margins of the leaves do not ordinarily overlap.

(4) **Bambusoid type**: The embryo has the formula **"P+PP"**. These are basically panicoid but differing principally in the presence of a more or less developed epiblast.

(5) **Oryzoid–Olyroid type**: Embryos in these combination of genera have formulae **"F-PP"**, **"F+PP"**, and **"F+FP"**. All have a festucoid type of vascularization and panicoid embryonic leaf. In most, genera show the presence of epiblast, and there is usually a cleft between the coleorhiza and scutellum.

(6) **Arundinoid–Danthonioid type**: The genera in which the embryo has the formula **"P-PF"**. They are basically panicoid and differ only in that the embryonic leaf has a few vascular bundles and margins that merely meet rather than numerous bundles and overlapping margins.

Two types of vascularizations are present (shown in longitudinal section):

1. Panicoid (Fig. 3.4 L)
2. Festucoid (Fig. 3.4 M)

In the cross-section, embryonic leaf shows two different types of arrangements:

1. P-type: overlapping margins with many bundles (Fig. 3.5 A)
2. F-type: margins merely meet with few bundles (Fig. 3.5 B)

Angle of vascularization: Vascular bundle that comes from the radicle and goes into the plumule. It is classified into different classes:

Type A: less than 130° (Fig. 3.5 C)
Type B: 130°–150° (Fig. 3.5 D)
Type C: 150°–170° (Fig. 3.5 E)
Type D: greater than and equal to170° (Fig. 3.5 F)

Endosperm: Endosperm is the major reserve food storage of caryopsis. It is a source of feed for embryo when it germinates and cells may be multi-nucleate. It has polygon, that is, isodiametric storage cells that contain starch grains, the type of which varies in different species. Starch grains may be simple, elliptic to circular, hexagonal, pentagonal or rectangular or may be a mixture of simple and compound grains with few granules.

Generally, the endospermal cells that are present toward the pericarp are smaller, flattened next to the aleurone layer, and cells that are present toward the center are bigger in size (Fig. 3.6 H). Cells that are interior contain mostly starch while those toward the periphery contain mostly protein. Seeds were contain storage starch, that is, starch grains designed for long-term storage of energy produced in the amyloplast of seed, roots, tubers, corms, fruits, and rhizomes. And the morphology is genetically controlled (Reichert, 1913).

ISCN (2011) described the international code for starch nomenclature. On the basis of which, in present study, four different types of starch grains were observed:

1. Simple grains (i.e. grains form single) with elliptic to circular shape (i.e. appearing as a circle in which all radii are of equal length) (Fig. 3.6 A).
2. Simple grains with polygon shape (Fig. 3.6 B).
3. Simple starch grains having circular and polygon (i.e having more than four sides) shape grains (Fig. 3.6 C).
4. Mixture of starch grains, that is, simple and compound grains (grains that form in aggregates or grains that have more than one center of formation within or single amyloplast, Fahn, 1990) both are present together (Fig. 3.6 D and E).

Aleurone layer: It is a single-layered structure present after pericarp. It is uniseriate and continuously circled the entire endosperm. The alerurone layer is lacking at the scutellum adaxial apex above its ventral scale downward to the place where the caryopsis comes in contact with the basal scutellum.

Aleurone cells are of different shapes:

- Vertically elongated rectangular shape (Fig. 3.6 G)
- Both types, that is, square and rectangular shapes of cells present together (Fig. 3.6 H).
- Tangentially elongated rectangular shape (Fig. 3.6 I)
- Square shape (Fig. 3.6 J)

All the quantitative features are mentioned in Table 3.1.

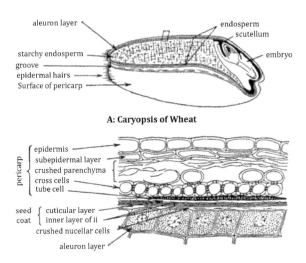

A: Caryopsis of Wheat

B: Pericarp of Wheat caryopsis

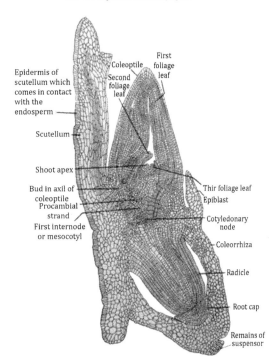

C: Longitudinal section of Wheat caryopsis

DIAGRAM 1:   General anatomy of grass caryopsis.
[Easu (1953) Plant anatomy]

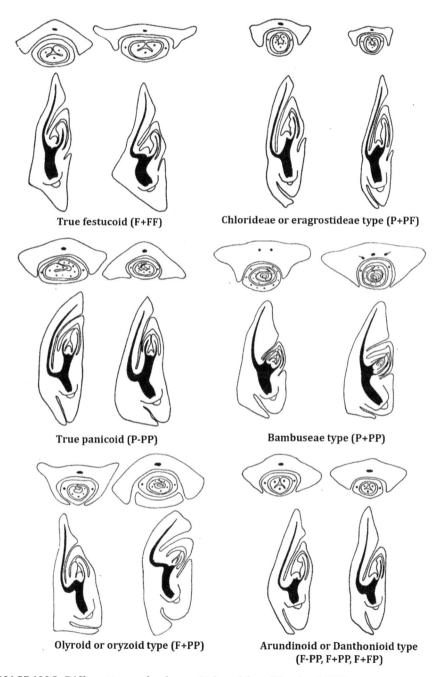

True festucoid (F+FF)

Chlorideae or eragrostideae type (P+PF)

True panicoid (P-PP)

Bambuseae type (P+PP)

Olyroid or oryzoid type (F+PP)

Arundinoid or Danthonioid type
(F-PP, F+PP, F+FP)

DIAGRAM 2: Different types of embryos. (Adapted from Rheeder (1957)).
Moprhometric analysis/Histoarchitechtural analysis of caryopsis.

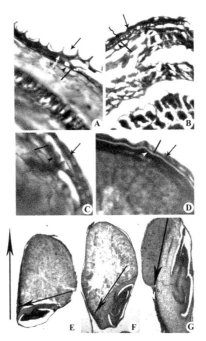

FIGURE 3.1   Variations of pericarp in caryopsis and angles of embryo.

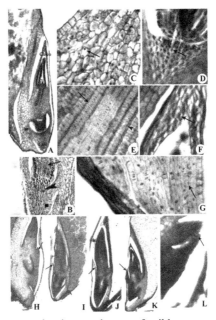

FIGURE 3.2   Different parts of embryo and types of epiblast.

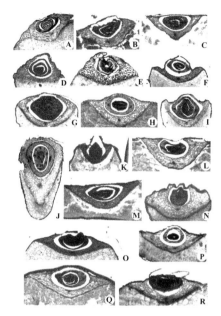

FIGURE 3.3   Variations in "V" shaped scutellum.

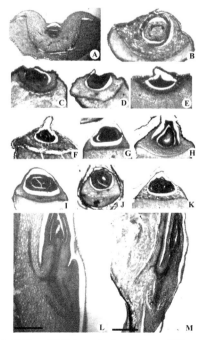

FIGURE 3.4   Variations in "U" and "Δ" shaped of scutellum, different types of vascularization.

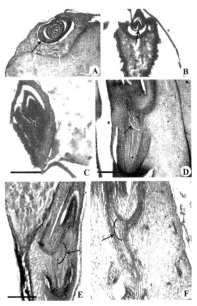

FIGURE 3.5 Different arrangements of embryonic leaf and different angles of vascularization.

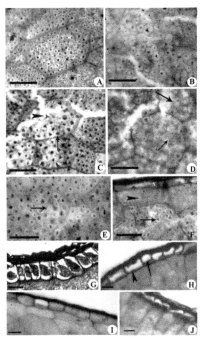

FIGURE 3.6 Variations in types of starch grain and aleurone layer.

Morphometric/Histoarchitechtural Analysis of Caryopsis Anatomy

1. *Chionachne koenigii* (Fig. 3.7(1) A–G)

Shape: Kidney shape in transaction and having size of 1.39 (*L*) × 3.34 (*B*) mm² (Fig. 3.7 C).

Pericarp: Type B (Fig. 3.7 G; epidermis (arrowhead), nucellus (arrowhead)).

Embryo: Type C, Embryo is placed at an angle of 172.3°. Type I epiblast (Fig. 3.7 A and B).

Scutellum: "U" shape: Type U1-1, having an angle of 41.9°. Scutellum cleft/slit is present. Oval-shaped scutellum cells. Depletion layer is present (Fig. 3.7 C and E).

Vascularization: Type of vascularization: Festucoid (Fig. 3.7 A)

Angle of vascularization: 165.6°, Type C.

Pattern of embryonic leaf: P-type

Endosperm: Polygon-shaped endospermic cells.

Starch grain: Some cells have simple circular shaped starch grains; some cells have simple circular shaped and compound starch grains that are present together (Fig. 3.7 F).

Number of starch grains per endosperm cell: 63 ± 4; **Starch grain size:** 2.09 ± 0.73 (µm)

Aleurone layer: Vertically elongated rectangular shaped aleurone cells (Fig. 3.7 D).

Caryopsis type and formula: Oryzoid–Olyroid type having formula "F+PP".

Bar size: A and C = 125 µm; B = 85.5 µm; D–G = 12.5 µm

2. *Coix lachryma-jobi* (Fig. 3.7(2) A–H)

Shape: Oblong to circular in transaction with wavy anterior and posterior end and having size of 2.57 (*L*) × 4.38 (*B*) mm² (Fig. 3.7 D and E).

Pericarp: Type B (Fig. 3.7 G; epidermis (arrowhead), hypodermis (arrow)).

Embryo: Type C, Embryo is placed at an angle of 163.9°. Epiblast is absent (Fig. 3.7 D).

Scutellum: "U" shape: Type U1-1, having an angle of 35.82°. Scutellum cleft/slit is present (Fig. 3.7 A). Oval-shaped scutellum cells. Depletion layer is present (Fig. 3.7 D and H).

Vascularization: Type of vascularization: Panicoid (Fig. 3.7 B)

Angle of vascularization: 177.9°, Type D.

Pattern of embryonic leaf: P-type

Endosperm: Polygon-shaped endospermic cells.

Starch grain: Simple circular and polygon-shaped starch grains (Fig. 3.7 C).

No. of starch grains per endosperm cell: 87 ± 7; **Starch grain size:** 5 ± 2.54 (μm)

Aleurone layer: Vertically elongated rectangular shaped aleurone cells (Fig. 3.7 F).

Caryopsis type and formula: True panicoids type and having formula "P-PP".

Bar size: A, B, D, E =125 μm; C, E–G = 12.5 μm

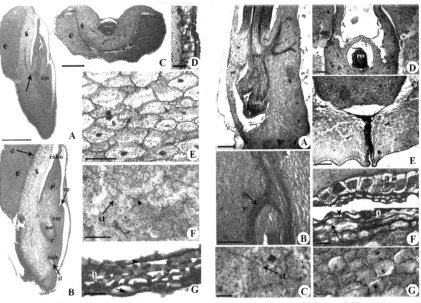

1. *Chionachne koenigii*   2. *Coix lachryma-jobi*

3. *Andropogon pumilus* (Fig. 3.8(3) A−E)

Shape: Rectangular to oval shape in transaction and having size of 0.27 (*L*) × 0.45 (*B*) mm² (Fig. 3.8 B).

Pericarp: Type C.

Embryo: Type C, Embryo is placed at an angle of 167.7°. Epiblast is absent.

Scutellum: "V" shape: Type V3-4, having an angle of 52.67°. Scutellum cleft/slit is present. Oval-shaped scutellum cells. Depletion layer is present (Fig. 3.8 D).

Vascularization: Type of vascularization: Panicoid

Angle of vascularization: 167.5°, Type C.

Pattern of embryonic leaf: P-type

Endosperm: Polygon-shaped endospermic cells.

Starch grain: Simple polygon-shaped starch grains (Fig. 3.8 E).

No. of starch grains per endosperm cell: 48 ± 5; **Starch grain size:** 7.29 ± 3.72 (µm)

Aleurone layer: Tangentially elongated rectangular shaped alerurone cells (Fig. 3.8 C).

Caryopsis type and formula: True panicoids type and having formula "P-PP".

Bar size: A and B =125 µm; C−E = 12.5 µm

4. *Apluda mutica* (Fig. 3.8(4) A−G)

Shape: Circular to oval shaped in transaction and having size of 0.54 (*L*) × 0.39 (*B*) mm².

Pericarp: Type C (Fig. 3.8 G; epidermis (arrowhead), hypodermis (arrow)).

Embryo: Type B, Embryo is placed at an angle of 154.4°. Epiblast is absent (Fig. 3.8 A).

Scutellum: "V" shaped: Type V3-2, having an angle of 66.93°. Scutellum cleft/slit is present. Oval-shaped scutellum cells. Depletion layer is present (Fig. 3.8 D).

Vascularization: Type of vascularization: Panicoid (Fig. 3.8 C)

Angle of vascularization: 141.3°, Type B.

Pattern of embryonic leaf: P-type

Endosperm: Polygon-shaped endospermic cells.

Starch grain: Simple circular and polygon-shaped starch grains (Fig. 3.8 F, arrow).

No. of starch grains per endosperm cell: 43 ± 3; **Starch grain size:** 5 ± 1.50 (μm)

Aleurone layer: Tangentially elongated rectangular shaped aleurone cells (Fig. 3.8 F).

Caryopsis type and formula: True panicoids type and having formula "P-PP".

Bar size: A and B =125 μm; C = 85.5 μm; D and E=12.5 μm

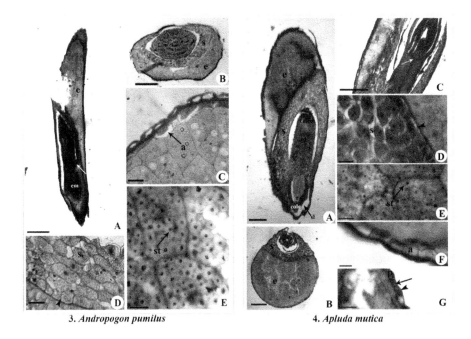

3. *Andropogon pumilus*   4. *Apluda mutica*

5. *Arthraxon lanceolatus* (Fig. 3.9(5) A−F)

Shape: Kidney shape in transaction and having size of 0.38 (*L*) × 0.31 (*B*) mm² (Fig. 3.9 B).

Pericarp: Type C (Fig. 3.9 F; epidermis (arrowhead), hypodermis (arrow)).

Embryo: Type C, Embryo is placed at an angle of 170.5°. Epiblast is absent (Fig. 3.9 A).

Scutellum: "V" shaped: Type V8-3, having an angle of 63.1°. Scutellum cleft/slit is present. Oval-shaped scutellum cells. Depletion layer is present (Fig. 3.9 C).

Vascularization: Type of vascularization: Panicoid (Fig. 3.9 A)

Angle of vascularization: 168.5°, Type C.

Pattern of embryonic leaf: P-type

Endosperm: Polygon-shaped endospermic cells.

Starch grain: Simple polygon-shaped starch grains (Fig. 3.9 E).

No. of starch grains per endosperm cell: 38 ± 6; **Starch grain size:** 4.55 ± 0.64 (µm)

Aleurone layer: Tangentially elongated rectangular shaped aleurone cells (Fig. 3.9 D).

Caryopsis type and formula: True panicoids type and having formula "P-PP".

Bar size: A and B = 125 µm; C–F = 12.5 µm

6. *Bothriochloa pertusa* (Fig. 3.9(6) A–G)

Shape: Oval shape in cross-section, convex anterior side and having size of 0.35 (*L*) × 0.67 (*B*) mm² (Fig. 3.9 C).

Pericarp: Type C (Fig. 3.9 G; cross cells (arrowhead), nucellus (arrow)).

Embryo: Type C, -Embryo is placed at an angle of 172.3°. Epiblast is absent (Fig. 3.9 B).

Scutellum: "V" shaped: Type V8-1-, having an angle of 52.5°. Scutellum cleft/slit is present, and it is small, present above the coleorhiza, and covers colerorhiza. Oval-shaped scutellum cells. Depletion layer is present (Fig. 3.9 D).

Vascularization: Type of vascularization: Panicoid (Fig. 3.9 B)

Angle of vascularization: 156.5°with Type C.

Pattern of embryonic leaf: P-type

Endosperm: Polygon-shaped endospermic cells.

Starch grain: Some cells have simple circular shaped starch grains, and some cells have simple circular shaped and compound starch grains that are present together (Fig. 3.9 E; simple (arrow), compound (arrowhead)).

No. of starch grains per endosperm cell: 49 ± 11; **Starch grain size:** 3.41 ± 0.42 (µm)

Aleurone layer: Tangentially elongated rectangular shaped aleurone cells.

Caryopsis type and formula: True panicoids type and having formula "P-PP".

Bar size: A–C =125 µm; D–G = 12.5 µm

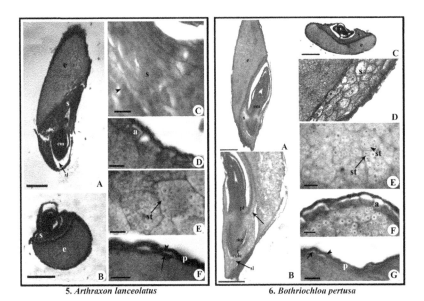

5. *Arthraxon lanceolatus*   6. *Bothriochloa pertusa*

7. *Capillipedium hugelii* (Fig. 3.10(7) A–F)

Shape: Oval to oblong in cross-section, convex dorsal surface and having size of 0.42 (L) × 0.70 (B) mm² (Fig. 3.10 C).

Pericarp: Type C (Fig. 3.10 F).

Embryo: Type B, Embryo is placed at an angle of 172.3° (Fig. 3.10 A). Epiblast is absent.

Scutellum: "U" shape: Type U2-3, having an angle of 64.5°. Scutellum cleft/slit is present (Fig. 3.10 B), and it is present above the coleorhiza and covers colerorhiza. Oval-shaped scutellum cells (Fig. 3.10 E). Depletion layer is present (Fig. 3.10 E, arrowhead).

Vascularization: Type of vascularization: Panicoid

Angle of vascularization: 167.0° with Type C.

Pattern of embryonic leaf: P-type

Endosperm: Polygon-shaped endospermic cells.

Starch grain: Simple circular and polygon-shaped starch grains that are present together (Fig. 3.10 E, arrow).

No. of starch grains per endosperm cell: 29 ± 7; **Starch grain size:** 3.75 ± 1.64 (μm)

Aleurone layer: Tangentially elongated rectangular shaped aleurone cells (Fig. 3.10 D).

Caryopsis type and formula: True panicoids type and having formula "P-PP".

Bar size: A–C = 125 μm; B = 85.5 μm; D–F = 12.5 μm

8. *Chrysopogon fulvus* (Fig. 3.10(8) A–F)

Shape: Inverted triangular shaped in transaction and having size of 1.14 (L) $\times$ 0.59 (B) mm² (Fig. 3.10 C).

Pericarp: Type D (Fig. 3.10 E).

Embryo: Type C, Embryo is placed at an angle of 173.1° (Fig. 3.10 A). Epiblast is absent.

Scutellum: "V" shaped: Type V4-1 (Fig. 3.10 C), having an angle of 26.8°. Scutellum cleft/slit is present. Oval-shaped scutellum (Fig. 3.10 F). Depletion layer is present (Fig. 3.10 F, arrowhead).

Vascularization: Type of vascularization: Panicoid

Angle of vascularization: 149.9°, Type B.

Pattern of embryonic leaf: P-type

Endosperm: Polygon-shaped endospermic cells.

Starch grain: Simple circular (arrow) and polygon-shaped starch grains (Fig. 3.10 D).

No. of starch grains per endosperm cell: 15 ± 4; **Starch grain size:** 6.82 ± 2.49 (μm)

Aleurone layer: Tangentially elongated rectangular shaped aleurone cells.

Caryopsis type and formula: True panicoids type and having formula "P-PP".

Bar size: A–C = 125 μm; D–F = 12.5 μm

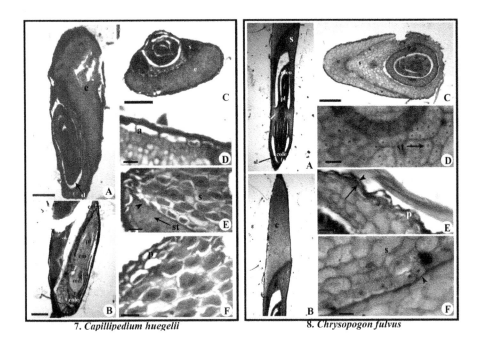

7. *Capillipedium huegelii*   8. *Chrysopogon fulvus*

9. *Cymbopogon martini* (Fig. 3.11(9) A–F)

Shape: Oval shape in cross-section, convex dorsal surface, and concave ventral surface having size of 0.36 (L) × 0.69 (B) mm² (Fig. 3.11 B).

Pericarp: Type D (Fig. 3.11 C).

Embryo: Type C, Embryo is placed at an angle of 172.6° (Fig. 3.11 A). Epiblast is absent.

Scutellum: "V" shape: Type V3-3 (Fig. 3.11 B), having an angle of 61.7°. Scutellum cleft/slit is present, andis seen toward scutellum side. Oval-shaped scutellum cells (Fig. 3.11 E), and depletion layer is also present (Fig. 3.11 E, arrowhead).

Vascularization: Type of vascularization: Panicoid

Angle of vascularization: Type C with 164.2°.

Pattern of embryonic leaf: P-type

Endosperm: Polygon-shaped endospermic cells.

Starch grain: Simple circular starch grains (Fig. 3.11 F).

No. of starch grains per endosperm cell: 40 ± 6; **Starch grain size:** 9.38 ± 3.72 (μm)

Aleurone layer: Tangentially elongated rectangular shaped aleurone cells (Fig. 3.11 D).

Caryopsis type and formula: True panicoids type and having formula "P-PP".

Bar size– A and B =125 μm; C–F = 12.5 μm

Dichanthium

10. *Dichanthium annulatum*

11. *Dichanthium caricosum*

10. *Dicanthium annulatum* (Fig. 3.11(10) A–F)

Shape: Oblong shape in transaction and having size of 0.47 (*L*) × 0.71 (*B*) mm² (Fig. 3.11 C).

Pericarp: Type B (Fig. 3.11 E; epidermis (arrowhead), nucellus (arrow)).

Embryo: Type B, Embryo is placed at an angle of 159.4° (Fig. 3.11 A). Epiblast is absent.

Scutellum: "V" shape: Type V3-3 (Fig. 3.11 C), having an angle of 58.4°. Scutellum cleft/slit is present. Oval-shaped scutellum cells. Depletion layer is present.

Vascularization: Type of vascularization: Panicoid (Fig. 3.11 B)

Angle of vascularization: 177.5°, Type D.

Pattern of embryonic leaf: P-type

Endosperm: Polygon-shaped endospermic cells.

Starch grain: Some cells have simple circular shaped starch grains, and some cells have simple circular shaped and compound starch grains that are present together (Fig. 3.11 F).

No. of starch grains per endosperm cell: 29 ± 8; **Starch grain size:** 5.56 ± 1.48 (μm)

Aleurone layer: Vertically elongated rectangular shaped aleurone cells (Fig. 3.11 D).

Caryopsis type and formula: True panicoids type and having formula "P-PP".

Bar size: A–C =125 μm; D–F = 12.5 μm

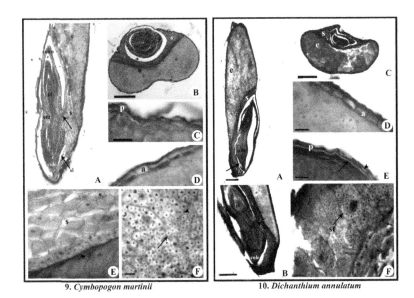

9. *Cymbopogon martinii*   10. *Dichanthium annulatum*

11. *Dichanthium caricosum* (Fig. 3.12(11) A−F)

Shape: Oval shape in transection, convex dorsal surface having size of 0.47 (*L*) × 0.93 (*B*) mm² (Fig. 3.12 B).

Pericarp: Type D (Fig. 3.12 D).

Embryo: Type C, Embryo is placed at an angle of 161.42° (Fig. 3.12 A). Epiblast is absent.

Scutellum: "U" shaped: Type U2-3 (Fig. 3.12 B), having an angle of 67.1°. Scutellum cleft/slit is present (Fig. 3.12 A). Oval-shaped scutellum cells (Fig. 3.12 E), and depletion layer is also present (Fig. 3.12 E, arrowhead).

Vascularization: Type of vascularization: Panicoid

Angle of vascularization: 169.9°, Type C.

Pattern of embryonic leaf: P-type

Endosperm: Polygon-shaped endospermic cells.

Starch grain: Simple polygon starch grains (Fig. 3.12 F).

No. of starch grains per endosperm cell: 22 ± 5; **Starch grain size:** 5 ± 2.84 (μm)

Aleurone layer: Tangentially elongated rectangular shaped aleurone cells (Fig. 3.12 C).

Caryopsis type and formula: True panicoids type and having formula "P-PP".

Bar size: A and B = 125 µm; C–F = 12.5 µm

DIFFERENTIATING FEATURES:

Angle of vascularization, Type D, vertically elongated aleurone cells, embryo type- B ... *D. annulatum*

Angle of vascularization, Type C, tangential elongated aleurone cells, embryo type- C ... *D. caricosum*

12. *Hackelochloa granularis* (Fig. 3.12(12) A–E)

Shape: Oblong to circular shaped in transaction and having size of 0.19 (*L*) × 0.72 (*B*) mm (Fig. 3.12 B).

Pericarp: Type C (Fig. 3.12 E).

Embryo: Type B, Embryo is placed at an angle of 158.6° (Fig. 3.12 A). Epiblast is absent.

Scutellum: "V" shape: Type Δ5 (Fig. 3.12 B), having an angle of 41.9°. Scutellum cleft/slit is present. Oval-shaped scutellum cells (Fig. 3.12 D). Depletion layer is present (Fig. 3.12 D, arrowhead).

Vascularization: Type of vascularization: Panicoid

Angle of vascularization: 152.5°, Type C.

Pattern of embryonic leaf: P-type

Endosperm: Polygon-shaped endospermic cells.

Starch grain: Simple circular and polygon-shaped starch grains (Fig. 3.12 C).

No. of starch grains per endosperm cell: 11 ± 4; **Starch grain size:** 4.17 ± 1.33 (µm)

Aleurone layer: Tangentially elongated rectangular shaped alerurone cells (Fig. 3.12 E, arrow).

Caryopsis type and formula: True panicoids type and having formula "P-PP".

Bar size: A and B = 125 μm; C–E = 12.5 μm

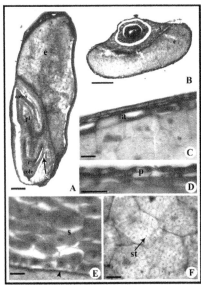

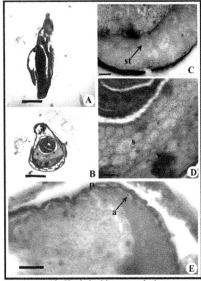

11. *Dicanthium caricpsum*   12. *Hackelochloa granularis*

Heteropogon

13. *Heteropogon contortus var. genuinus sub var. typicus*

14. *Heteropogon contortus var. genuinus sub var. hispidissimus*

15. *Heteropogon ritcheii*

16. *Heteropogon triticeus*

13. *Heteropogon contortus* **var.** *genuinus* **sub var.** *typicus* **(Fig. 3.13(13) A–G)**

Shape: Crock (Matka) shape in transaction with convex anterior side and having size of 0.54 (*L*) × 0.60 (*B*) mm² (Fig. 3.13 C).

Pericarp: Type B (Fig. 3.13 E; epidermis (arrowhead), nucellus (arrow)).

Embryo: Type C, Embryo is placed at an angle of 178.8° (Fig. 3.13 A). Epiblast is absent.

Scutellum: "V" shape: Type V3-6 (Fig. 3.13 C), having an angle of 43.7°. Scutellum cleft/slit is present. Oval-shaped scutellum cells (Fig. 3.13 F). Depletion layer is present (Fig. 3.13 F, arrowhead).

Vascularization: Type of vascularization: Panicoid

Angle of vascularization: 177.9°, Type D.

Pattern of embryonic leaf: P-type

Endosperm: Polygon-shaped endospermic cells (Fig. 3.13 B).

Starch grain: Simple circular to polygon shaped starch grains (Fig. 3.13 G).

No. of starch grains per endosperm cell: 25 ± 3; **Starch grain size:** 4.69 ± 1.68 (μm)

Aleurone layer: Tangentially elongated rectangular shaped aleurone cells (Fig. 3.13 D).

Caryopsis type and formula: True panicoids type and having formula "P-PP".

Bar size: A–C = 125 μm; D–G = 12.5 μm

14. *Heteropogon contortus* var. *genuinus* sub var. *typicus* (Fig. 3.13(14) A–F)

Shape: Crock (Matka) shaped in transaction with convex anterior side and having size of 0.75 (*L*) × 0.83 (*B*) mm² (Fig. 3.13 B).

Pericarp: Type D (Fig. 3.13 F; epidermis (arrowhead), nucellus (arrow)).

Embryo: Type C, Embryo is placed at an angle of 175.7° (Fig. 3.13 A). Epiblast is absent.

Scutellum: "U" shape: Type U1-1 (Fig. 3.13 B), having an angle of 41.9°. Scutellum cleft/slit is present. Oval-shaped scutellum cells (Fig. 3.13 C). Depletion layer is present (Fig. 3.13 C, arrowhead).

Vascularization: Type of vascularization: Panicoid

Angle of vascularization: 178.4°, Type D.

Pattern of embryonic leaf: P-type

Endosperm: Polygon-shaped endospermic cells.

Starch grain: Simple circular (arrow) to polygon shaped (arrowhead) starch grains (Fig. 3.13 E).

No. of starch grains per endosperm cell: 21 ± 5; **Starch grain size:** 6.94 ± 2.38 (μm)

Aleurone layer: Tangentially elongated rectangular shaped aleurone cells (Fig. 3.13 D).

Caryopsis type and formula: True panicoids type and having formula "P-PP".

Bar size: A and B = 125 μm; C–F = 12.5 μm

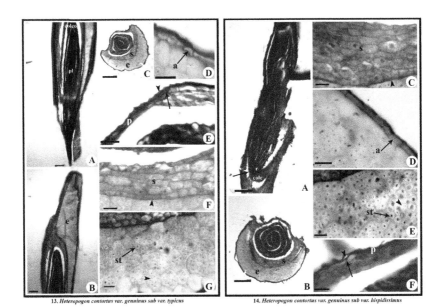

13. *Heteropogon contortus var. genuinus sub var. typicus*　　　14. *Heteropogon contortus var. genuinus sub var. hispidissimus*

15. *Heteropogon ritchiei* (Fig. 3.14(15) A–G)

Shape: Crock shaped in transection and concave from ventral surface, having size of $0.52(L) \times 0.51(B)$ mm^2 (Fig. 3.14 C).

Pericarp: Type D (Fig. 3.14 G).

Embryo: Type C, Embryo is placed at an angle of 176.9° (Fig. 3.14 A). Epiblast is absent.

Scutellum: "U" shape: Type U1-2 (Fig. 3.14 C), having an angle of 57.7°. Scutellum cleft/slit is present toward scutellum side; it is becoming broad toward the proximal end. Oval-shaped, compactly arranged scutellum cells (Fig. 3.14 E).

Vascularization: Type of vascularization: Panicoid

Angle of vascularization: 177.8°, Type D; **Pattern of embryonic leaf:** P-type

Endosperm: Polygon-shaped endospermic cells (Fig. 3.14 B).

Starch grain: Simple circular and polygon-shaped starch grains (Fig. 3.14 D).

No. of starch grains per endosperm cell: 19 ± 6; **Starch grain size:** 5.68 ± 2.13 (μm)

Aleurone layer: Tangentially elongated rectangular shaped aleurone cells (Fig. 3.14 F).

Caryopsis type and formula: True panicoids type and having formula "P-PP".

Bar size: A–C = 125 μm; D–G = 12.5 μm

16. *Heteropogon triticeus* (Fig. 3.14(16) A–H)

Shape: Crock shaped in transection, concave from ventral surface, having size of 1.25 (L) × 1.23 (B) mm^2 (Fig. 3.14 B).

Pericarp: Type A (Fig. 3.14 G; cross cells (arrowhead), epidermal cell (arrow)).

Embryo: Type C, Embryo is placed at an angle of 169.9° (Fig. 3.14 A). Epiblast is absent.

Scutellum: "U" shape: Type U1-1 (Fig. 3.14 B), having an angle of 57.7°. Scutellum cleft/slit is present and it is seen toward scutellum side; it is becoming broad toward the proximal end. Oval-shaped, compactly arranged scutellum cells (Fig. 3.14 H) and depletion layer is present.

Vascularization: Type of vascularization: Panicoid (Fig. 3.14 D)

Angle of vascularization: 146.4°, Type B. **Pattern of embryonic leaf:** P-type

Endosperm: Polygon-shaped endospermic cells (Fig. 3.14 C).

Starch grain: Some cells have simple polygon-shaped, and some have simple circular shaped and compound starch grains that are present together (Fig. 3.14 E).

No. of starch grains per endosperm cell: 20 ± 5; **Starch grain size:** 8.75 ± 3.48 (μm)

Aleurone layer: Tangentially elongated rectangular shaped aleurone cells (Fig. 3.14 F).

Caryopsis type and formula: True panicoids type and having formula "P-PP".

Bar size: A–D = 125 μm; E–H = 12.5 μm

DIFFERENTIATING FEATURES:

Angle of vascularization, Type B ... *H. triticeus*
Angle of vascularization, Type D
Scutellum Type- V3-6 ... *H. contortus* var. *genuinus* sub var. *typicus*
Scutellum Type- U1-1 ... *H. contortus var. genuinus sub var. hispidissimus*
Scutellum Type- U1-2 ... *H. ritchiei*

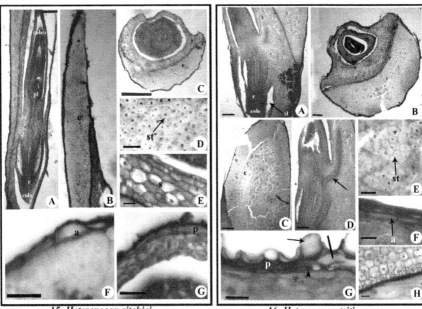

15. *Heteropoogn ritchiei*   16. *Heteropogon triticeus*

17. *Imperata cylindrica* (Fig. 3.15(17) A–F)

Shape: Oblong shape in transaction with convex anterior side and having size of 0.19 (*L*) × 0.24 (*B*) mm² (Fig. 3.15 C).

Pericarp: Type D (Fig. 3.15 F; epidermis (arrowhead), nucellus (arrow)).

Embryo: Type C, Embryo is placed at an angle of 163.7° (Fig. 3.15 B). Epiblast is absent.

Scutellum: "Δ" shape: Type Δ5 (Fig. 3.15 C), having an angle of 58.9°. Scutellum cleft/slit is present. Oval-shaped scutellum cells (Fig. 3.15 E). Depletion layer is present (Fig. 3.15 E, arrowhead).

Vascularization: Type of vascularization: Panicoid

Angle of vascularization: 152.3°, Type B.

Pattern of embryonic leaf: P-type

Endosperm: Polygon-shaped endospermic cells.

Starch grain: Simple polygon-shaped starch grains (Fig. 3.15 D).

No. of starch grains per endosperm cell: 12 ± 2; **Starch grain size:** 3.47 ± 1.26 (μm)

Aleurone layer: Vertically elongated rectangular shaped aleurone cells.

Caryopsis type and formula: True panicoids type and having formula "P-PP".

Bar size: A and C = 125 μm; B = 85.5 μm; D–F = 12.5 μm

Ischaemum

18. *Ischaemum indicus*

19. *Ischaemum molle*

20. *Ischaemum pilosum*

21. *Ischaemum rugosum*

18. *Ischaemum indicus* (Fig. 3.15(18) A–F)

Shape: Oval to rectangular shaped in transaction and having size of 0.41 (*L*) × 0.71 (*B*) mm^2 (Fig. 3.15 C).

Pericarp: Type C (Fig. 3.15 F; epidermis (arrowhead), nucellus (arrow)).

Embryo: Type C, Embryo is placed at an angle of 171.2° (Fig. 3.15 A). Epiblast is absent.

Scutellum: "V" shape: Type V3-2 (Fig. 3.15 C), having an angle of 72.64°. Scutellum cleft/slit is present. Oval-shaped scutellum cells (Fig. 3.15 D). Depletion layer is present (Fig. 3.15 D, arrowhead).

Vascularization: Type of vascularization: Panicoid (Fig. 3.15 B)

Angle of vascularization: 167.3°, Type C.

Pattern of embryonic leaf: P-type

Endosperm: Polygon-shaped endospermic cells.

Starch grain: Simple polygon-shaped starch grains.

No. of starch grains per endosperm cell: 18 ± 3; **Starch grain size:** 4.17 ± 1.67 (μm)

Aleurone layer: Tangentially elongated rectangular and square shaped aleurone cells (Fig. 3.15 E).

Caryopsis type and formula: True panicoids type and having formula "P-PP".

Bar size: A–C = 125 μm; D–F = 12.5 μm

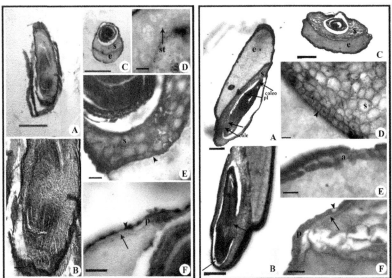

17. *Imperata cylindrica*   18. *Ischaemum indicum*

19. *Ischaemum molle* (Fig. 3.16(19) A–E)

Shape: Oval shaped in transection, concave from dorsal surface and having size of 0.67 (*L*) × 0.76 (*B*) mm² (Fig. 3.16 B).

Pericarp: Type A (Fig. 3.16 C; cross cells (arrowhead), epidermis (arrow)).

Embryo: Type 3.16 C, Embryo is placed at an angle of 173.4° (Fig. 3.16 A). Epiblast is absent.

Scutellum: "V" shape: Type V3-1 (Fig. 3.16 B), having an angle of 89.4°. Scutellum cleft/slit is present and it covers coleorhiza. Oval-shaped, compactly arranged scutellum cells (Fig. 3.16 D), and also show depletion layer (Fig. 3.16 D, arrowhead).

Vascularization: Type of vascularization: Panicoid

Angle of vascularization: 170.34°, Type D.

Pattern of embryonic leaf: P-type

Endosperm: Polygon-shaped endospermic cells.

Starch grain: Simple circular and polygon-shaped starch grains (Fig. 3.16 E, arrow).

No. of starch grains per endosperm cell: 17 ± 5; **Starch grain size:** 2.78 ± 0.63 (μm)

Aleurone layer: Tangentially elongated rectangular shaped aleurone cells (Fig. 3.16 E).

Caryopsis type and formula: True panicoids type and having formula "P-PP".

Bar size: A and B = 125 μm; C–E = 12.5 μm

20. *Ischaemum pilosum* (Fig. 3.16(20) A–F)

Shape: Oblong to circular shaped in transection, convex from dorsal surface and having size of 0.91 (*L*) × 1.03 (*B*) mm² (Fig. 3.16 C).

Pericarp: Type B (Fig. 3.16 D; nucellus (arrow)).

Embryo: Type C, Embryo is placed at an angle of 172.1° (Fig. 3.16 A). Epiblast is absent.

Scutellum: "Δ" shape: Type Δ3 (Fig. 3.16 C), having an angle of 40.4°. Scutellum cleft/slit is present, and it is present toward scutellum. Oval to circular shaped, compactly arranged scutellum cells (Fig. 3.16 F), and also show depletion layer (Fig. 3.16 F, arrowhead).

Vascularization: Type of vascularization: Panicoid

Angle of vascularization: 144.8°, Type B.

Pattern of embryonic leaf: P-type

Endosperm: Polygon-shaped endospermic cells.

Starch grain: Some cells have simple circular shaped starch grains (arrow), and some cells have simple pentagonal shaped (arrowhead) and compound starch grains that are present together (Fig. 3.16 E).

No. of starch grains per endosperm cell: 24 ± 4; **Starch grain size:** 4.76 ± 2.57 (μm)

Aleurone layer: Tangentially elongated rectangular shaped aleurone cells (Fig. 3.16 D).

Caryopsis type and formula: True panicoids type and having formula "P-PP".

Bar size: A–C = 125 μm; D–F = 12.5 μm

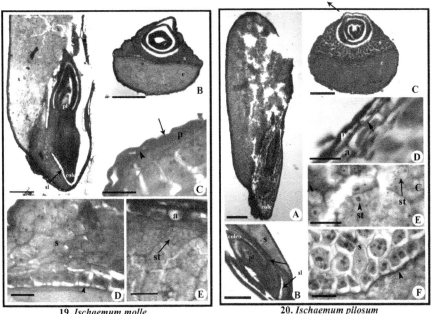

19. *Ischaemum molle*   20. *Ischaemum pilosum*

21. *Ischaemum rugosum* (Fig. 3.17(21) A−F)

Shape: Oblong shape in transection, convex from dorsal surface, having size of 0.91(*L*) × 1.34 (*B*) mm² (Fig. 3.17 C).

Pericarp: Type B (Fig. 3.17 D; epidermis (arrow), nucellus (arrowhead)).

Embryo: Type C, Embryo is placed at an angle of 174.5° 6 (Fig. 3.17 A). Epiblast is absent.

Scutellum: "V" shape: Type V7 (Fig. 3.17 C), having an angle of 72.8°. Scutellum cleft/slit is present, and it is present toward scutellum (Fig. 3.17 B). Oval-shaped, compactly arranged scutellum cells (Fig. 3.17 F), and also show depletion layer (Fig. 3.17 F, arrowhead).

Vascularization: Type of vascularization: Panicoid

Angle of vascularization: 173.8°, Type B. **Pattern of embryonic leaf:** P-type

Endosperm: Polygon-shaped endospermic cells.

Starch grain: Simple polygon-shaped starch grains (Fig. 3.17 E, arrow).

No. of starch grains per endosperm cell: 28 ± 3; **Starch grain size:** 2.21 ± 0.62 (μm)

Aleurone layer: Both shaped i.e. tangentially elongated and square-shaped aleurone cells (Fig. 3.17 E).

Caryopsis type and formula: True panicoids type and having formula "P-PP".

Bar size: A–C = 125 μm; D–F = 12.5 μm

DIFFERENTIATING FEATURES:

Tangentially elongated aleurone layer
Scutellum type – V3-1 … *I. molle*
Scutellum type- Δ3 … *I. pilosum*
Both square and rectangular shaped aleurone cells
Scutellum type – V3-2 … *I. indicus*
Scutellum type – V7 … *I. rugosum*
OR
Angle of vascularization Type C … *I. indicus*
Angle of vascularization Type D … *I. molle*
Angle of vascularization Type C
Tangentially elongated aleurone layer … *I. pilosum*
Both square and rectangular shaped aleurone cells … *I. rugosum*

22. *Iseilema laxum* (Fig. 3.17(22) A–F)

Shape: Oblong shape in transection and having size of 0.38 (*L*) × 0.79 (*B*) mm² (Fig. 3.17 C).

Pericarp: Type B (Fig. 3.17 D; cross cells (arrowhead)).

Embryo: Type C, Embryo is placed at an angle of 171.8° (Fig. 3.17 A). Epiblast is absent.

Scutellum: "V" shape: Type V7 (Fig. 3.17 C), having an angle of 71.2°. Scutellum cleft/slit is present, and it is broad, and covers the coleorhiza.

Oval shape, compactly arranged scutellum cells (Fig. 3.17 E), and also shows depletion layer (Fig. 3.17 E, arrowhead).

Vascularization: Type of vascularization: Panicoid (Fig. 3.17 B).

Angle of vascularization: 166.4°, Type C. **Pattern of embryonic leaf:** P-type

Endosperm: Polygon-shaped endospermic cells.

Starch grain: Simple circular shaped starch grains present together (Fig. 3.17 E, arrow).

No. of starch grains per endosperm cell: 27 ± 6; **Starch grain size:** 2.21 ± 0.23 (μm)

Aleurone layer: Tangentially elongated rectangular shaped aleurone cells (Fig. 3.17 F).

Caryopsis type and formula: True panicoids type and having formula "P-PP".

Bar size: A–C = 125 μm; D–F = 12.5 μm

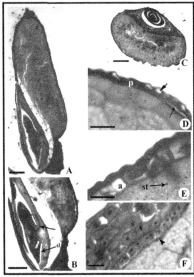
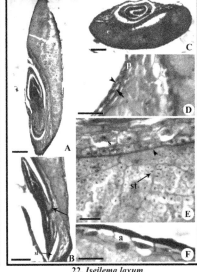

21. Ischaemum rugosum   *22. Iseilema laxum*

23. *Ophiuros exaltatus* (Fig. 3.18(23) A−F)

Shape: Circular shaped with flat anterior side in cross-section and having size of 0.69 × 0.83 mm² (Fig. 3.18 B).

Pericarp: Type C (Fig. 3.18 F; cross cells (arrowhead)).

Embryo: Type C, Embryo is placed at an angle of 168.7° (Fig. 3.18 A). Epiblast is absent.

Scutellum: "V" shape: Type V1-2, having an angle of 74.3°. Scutellum cleft/slit is present and it is very thin, just like slit. Oval-shaped, compactly arranged scutellum cells (Fig. 3.18 D), and also shows depletion layer (Fig. 3.18 D, arrowhead).

Vascularization: Type of vascularization: Panicoid (Fig. 3.18 A)

Angle of vascularization: 167.3°, Type C.

Pattern of embryonic leaf: P-type

Endosperm: Polygon-shaped endospermic cells.

Starch grain: Simple circular and polygon-shaped starch grains (Fig. 3.18 C).

No. of starch grains per endosperm cell: 69 ± 8; **Starch grain size:** 3.85 ± 1.27 (µm)

Aleurone layer: Square and tangentially rectangular, both type of aleurone cells are present (Fig. 3.18 E).

Caryopsis type and formula: True panicoids type and having formula "P-PP".

Bar size: A and B = 125 µm; C–F = 12.5 µm

24. *Rottboellia exaltata* (Fig. 3.18(24) A−G)

Shape: Circular shaped with flat anterior side in transection and having size of 2.00 (*L*) × 2.35 (*B*) mm^2 (Fig. 3.18 D).

Pericarp: Type C (Fig. 3.18 E, cross cells (arrowhead)).

Embryo: Type C, Embryo is placed at an angle of 173.3° (Fig. 3.18 A and B). Epiblast is absent.

Scutellum: "V" shape: Type V1-1 (Fig. 3.18 D), having an angle of 63.2°. Scutellum cleft/slit is present, and it is very thin, just like slit. Oval-shaped, compactly arranged scutellum cells (Fig. 3.18 C), and also shows depletion layer.

Vascularization: Type of vascularization: Panicoid

Angle of vascularization: 165.3°, Type C.

Pattern of embryonic leaf: P-type

Endosperm: Polygon-shaped endospermic cells.

Starch grain: Simple circular and polygon-shaped starch (Fig. 3.18 F).

No. of starch grains per endosperm cell: 67 ± 5; **Starch grain size:** 2.5 ± 1.04 (μm)

Aleurone layer: Vertically elongated rectangular shaped aleurone cells (Fig. 3.18 G).

Caryopsis type and formula: True panicoids type and having formula "P-PP".

Bar size: A, B, D = 125 μm; C, E–G = 12.5 μm

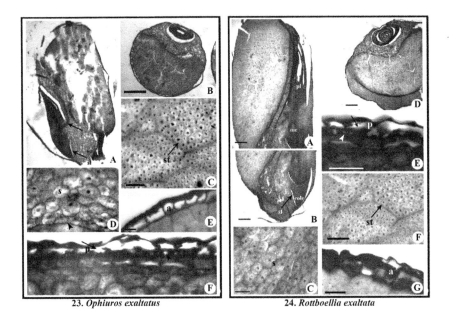

23. *Ophiuros exaltatus*   24. *Rottboellia exaltata*

25. *Saccharum spontanum* (Fig. 3.19(25) A–F)

Shape: Rectangular shape with circular corners and flat anterior side in transection and having size of 2.00 (L) × 2.35 (B) mm^2 (Fig. 3.19 B).

Pericarp: Type B (Fig. 3.19 C; hypodermis (arrow)).

Embryo: Type C, Embryo is placed at an angle of 171.4° (Fig. 3.19 A). Epiblast is absent.

Scutellum: Δ shape: Type $\Delta 1$ (Fig. 3.19 B), having an angle of 35.8°. Scutellum cleft/slit is present, and it covers the coleorhiza. Oval-shaped, compactly arranged scutellum cells (Fig. 3.19 E), and also shows depletion layer (Fig. 3.19 F, arrowhead).

Vascularization: Type of vascularization: Panicoid

Angle of vascularization: 168.9°, Type C.

Pattern of embryonic leaf: P-type

Endosperm: Polygon-shaped endospermic cells.

Starch grain: Simple circular and polygon-shaped starch grains (Fig. 3.19 D).

No. of starch grains per endosperm cell: 36 ± 7; **Starch grain size:** 2.68 ± 0.38 (μm)

Aleurone layer: Tangentially elongated rectangular shaped aleurone cells (Fig. 3.19 C, arrow).

Caryopsis type and formula: True panicoids type and having formula "P-PP".

Bar size: A, B = 125 μm; C–F = 12.5 μm

Sehima

26. *Sehima ischaemoides* 27. *Sehima nervosum*

28. *Sehima sulcatum*

26. *Sehima ischaemoides* (Fig. 3.19(26) A–F)

Shape: Lanceolate shape in transaction with convex anterior side and concave posterior side and having size of 0.30 (*L*) × 0.97 (*B*) mm² (Fig. 3.19 B).

Pericarp: Type B (Fig. 3.19 F; epidermis (arrow), cross cells (arrowhead)).

Embryo: Type C, Embryo is placed at an angle of 169.9° (Fig. 3.19 A). Epiblast is absent.

Scutellum: "V" shape: Type V7 (Fig. 3.19 B), having an angle of 77.31°. Scutellum cleft/slit is present. Oval-shaped scutellum cells (Fig. 3.19 D). Depletion layer is present (Fig. 3.19 D, arrowhead).

Vascularization: Type of vascularization: Panicoid

Angle of vascularization: 157.1°, Type C.

Pattern of embryonic leaf: P-type

Endosperm: Polygon-shaped endospermic cells.

Starch grain: Simple polygon shaped and compounds in which both starch grains are present together (Fig. 3.19 E).

No. of starch grains per endosperm cell: 91 ± 5; **Starch grain size:** 3.13 ± 1.63 (μm)

Aleurone layer: Tangentially elongated rectangular shaped aleurone cells (Fig. 3.19 C).

Caryopsis type and formula: True panicoids type and having formula "P-PP".

Bar size: A, B = 125 μm; C–F =12.5 μm

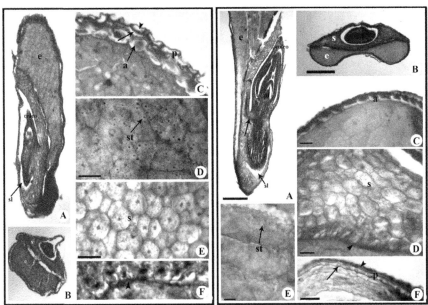

25. *Saccharum spontaneum*　　　26. *Sehima ischaemoides*

27. *Sehima nervosum* (Fig. 3.20(27) A−F)

Shape: Triangle shaped with circular corners and flat anterior side in transection and having size of 0.47 (*L*) × 0.97 (*B*) mm (Fig. 3.20 B).

Pericarp: Type A (Fig. 3.20 D; cross cells (arrowhead), nucellus (arrow)).

Embryo: Type C, Embryo is placed at an angle of 174.2° (Fig. 3.20 A). Epiblast is absent.

Scutellum: Δ shape: Type Δ1 (Fig. 3.20 B), having an angle of 38.8°. Scutellum cleft/slit covers the coleorhiza (Fig. 3.20 E). Oval-shaped, compactly arranged scutellum cells show depletion layer (Fig. 3.20 E, arrowhead).

Vascularization: Type of vascularization: Panicoid

Angle of vascularization: 177.8°, Type D. **Pattern of embryonic leaf:** P-type

Endosperm: Polygon-shaped endospermic cells.

Starch grain: Some cells have simple circular shaped starch grains, and some cells have simple circular shaped and compound starch grains that are present together (Fig. 3.20 F).

No. of starch grains per endosperm cell: 83 ± 7; **Starch grain size:** 3.13 ± 0.74 (μm)

Aleurone layer: Tangentially rectangular aleurone cells (Fig. 3.20 C).

Caryopsis type and formula: True panicoids type and having formula "P-PP".

Bar size: A, B = 125 μm; C–F = 12.5 μm

28. *Sehima sulcatum* (Fig. 3.20(28) A–F)

Shape: Kidney shaped, convex anterior, two lobed structure on posterior side in transection and having size of 1.87 $(L) \times 1.69 (B)$ mm^2 (Fig. 3.20 C).

Pericarp: Type A (Fig. 3.20 F; cross cells (arrowhead), nucellus (arrow)).

Embryo: Type C, Embryo is placed at an angle of 177.8° (Fig. 3.20 A). Epiblast is absent.

Scutellum: V shape: Type V3-1 (Fig. 3.20 C), having an angle of 83.3°. Scutellum cleft/slit is present, and it covers the coleorhiza. Oval shaped, compactly arranged scutellum cells (Fig. 3.20 D), and also shows depletion layer (Fig. 3.20 D, arrowhead).

Vascularization: Type of vascularization: Panicoid

Angle of vascularization: 171.3°, Type D. **Pattern of embryonic leaf:** P-type

Endosperm: Polygon-shaped endospermic cells.

Starch grain: Simple circular shaped starch grains (Fig. 3.20 B).

No. of starch grains per endosperm cell: 78 ± 6; **Starch grain size:** 2.5 ± 0.68 (μm)

Aleurone layer: Tangentially elongated rectangular shaped aleurone cells (Fig. 3.20 E).

Caryopsis type and formula: True panicoids type and having formula "P-PP".

Bar size: A, C = 125 μm; B, D–F = 12.5 μm

DIFFERENTIATING FEATURES:

Angle of vascularization Type C … *S. ischaemoides*
Angle of vascularization Type D
Scutellum type Δ1 … *S. nervosum*
Scutellum type V3-1 … *S. sulcatum*

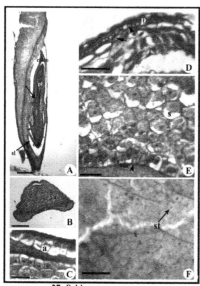

27. *Sehima nervosum*

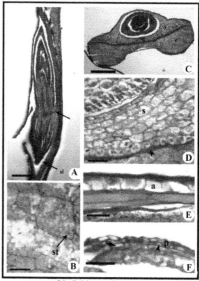

28. *Sehima sulcatum*

29. *Sorghum halepense* (Fig. 3.21(29) A–F)

Shape: Oval shaped in transection and having size of 1.87 (*L*) × 1.69 (*B*) mm² (Fig. 3.21 C).

Pericarp: Type A.

Embryo: Type C, Embryo is placed at an angle of 170.1° (Fig. 3.21 B). Epiblast is absent.

Scutellum: V shaped: Type V1-1 (Fig. 3.21 C), having an angle of 66.7°. Scutellum cleft/slit is present, and it covers the coleorhiza. Oval shaped, compactly arranged scutellum cells (Fig. 3.21 D), and also shows depletion layer (Fig. 3.21 D, arrowhead).

Vascularization: Type of vascularization: Panicoid

Angle of vascularization: 157.5°, Type C.

Pattern of embryonic leaf: P-type

Endosperm: Polygon-shaped endospermic cells.

Starch grain: Simple circular and polygon-shaped starch grains are present (Fig. 3.21 E).

No. of starch grains per endosperm cell: 47 ± 5; **Starch grain size:** 5.21 ± 2.17 (µm)

Aleurone layer: Tangentially elongated rectangular shaped aleurone cells (Fig. 3.21 F).

Caryopsis type and formula: True panicoids type and having formula "P-PP".

Bar size: A–C = 125 µm; D–F = 12.5 µm

30. *Thelepogon elegans* (Fig. 3.21(30) A-G)

Shape: Square shaped in transaction with circular base and having size of 1.19 (*L*) × 1.53 (*B*) mm² (Fig. 3.21 C).

Pericarp: Type B (Fig. 3.21 F; cross cells (arrowhead)).

Embryo: Type C, Embryo is placed at an angle of 167.7° (Fig. 3.21 A). Epiblast is absent.

Scutellum: V shape: Type V1-1 (Fig. 3.21 C), having an angle of 60.9°. Scutellum cleft/slit is present, and it covers the coleorhiza. Oval shaped, compactly arranged scutellum cells (Fig. 3.21 D), and also shows depletion layer (Fig. 3.21 D, arrowhead).

Vascularization: Type of vascularization: Panicoid (Fig. 3.21 B)

Angle of vascularization: 145.2°, Type B.

Pattern of embryonic leaf: P-type

Endosperm: Polygon-shaped endospermic cells.

Starch grain: Some cells have simple circular shaped starch grains, and some cells have simple circular shaped and compound starch grains that are present together (Fig. 3.21 G).

No. of starch grains per endosperm cell: 45 ± 3; **Starch grain size:** 3.47 ± 1.28 (μm)

Aleurone layer: Tangentially elongated rectangular shaped aleurone cells (Fig. 3.21 E).

Caryopsis type and formula: True panicoids type and having formula "P-PP".

Bar size: A–C = 125 μm; D–H = 12.5 μm

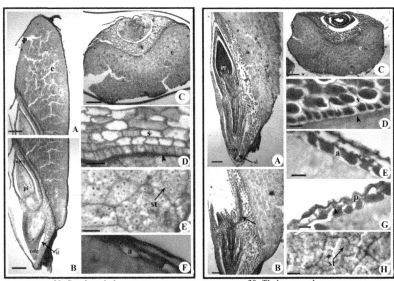

29. *Sorghum halepense*   30. *Thelepogon elegans*

Themeda

31. *Themeda cymbaria* 32. *Themeda laxa*

33. *Themeda triandra* 34. *Themeda quadrivalvis*

31. *Themeda cymbaria* (Fig. 3.22(31) A-G)

Shape: Oval shaped in transaction with dorsal convex surface and having size of 0.63 (*L*) × 0.38 (*B*) mm² (Fig. 3.22 B).

Pericarp: Type A (Fig. 3.22 D; cross cells (arrowhead), nucellus (arrow)).

Embryo: Type C, Embryo is placed at an angle of 171.4° (Fig. 3.22 A). Epiblast is absent.

Scutellum: U shape: Type U1-2 (Fig. 3.22 B), having an angle of 65.7°. Scutellum cleft/slit is present, and it is small slit. Oval shaped, compactly arranged scutellum cells (Fig. 3.22 F), and also shows depletion layer.

Vascularization: Type of vascularization: Panicoid

Angle of vascularization: 163.3°, Type C.

Pattern of embryonic leaf: P-type

Endosperm: Polygon-shaped endospermic cells.

Starch grain: Simple circular and polygon-shaped starch grains are present (Fig. 3.22 G).

No. of starch grains per endosperm cell: 41 ± 6; **Starch grain size:** 5.21 ± 3.09 (μm)

Aleurone layer: Tangentially elongated rectangular shaped flat aleurone cells (Fig. 3.22 E, arrow).

Caryopsis type and formula: True panicoids type and having formula "P-PP".

Bar size: A, B = 125 μm; C–G = 12.5 μm

32. *Themeda laxa* (Fig. 3.22(32) A−F)

Shape: Kidney shaped in transaction with dorsal convex surface and having size of 0.36 (L) × 0.64 (B) mm² (Fig. 3.22 A).

Pericarp: Type B (Fig. 3.22 F).

Embryo: Type C, Embryo is placed at an angle of 173.2° (Fig. 3.22 A). Epiblast is absent.

Scutellum: U shape: Type U1-2 (Fig. 3.22 B), having an angle of 57.3°. Scutellum cleft/slit is present, and it covers the coleorhiza. Oval shaped, compactly arranged scutellum cells (Fig. 3.22 E), and also shows depletion layer (Fig. 3.22 E, arrowhead).

Vascularization: Type of vascularization: Panicoid

Angle of vascularization: 168.7°, Type C.

Pattern of embryonic leaf: P-type

Endosperm: Polygon-shaped endospermic cells.

Starch grain: Simple circular and polygon-shaped starch grains are present (Fig. 3.22 D).

No. of starch grains per endosperm cell: 38 ± 9; **Starch grain size:** 4.76 ± 1.48 (μm)

Aleurone layer: Tangentially elongated rectangular shaped aleurone cells (Fig. 3.22 C).

Caryopsis type and formula: True panicoids type and having formula "P-PP".

Bar size: A, B = 125 μm; C–F = 12.5 μm

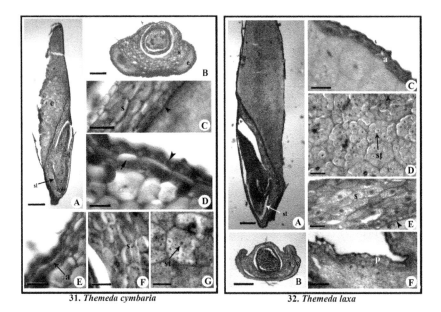

31. *Themeda cymbaria*   32. *Themeda laxa*

33. *Themeda triandra* (Fig. 3.23(33) A–F)

Shape: Cork shaped in transaction with convex anterior and posterior side and having size of 0.65 (L) × 0.73 (B) mm^2 (Fig. 3.23 B).

Pericarp: Type A (Fig. 3.23 E).

Embryo: Type C, Embryo is placed at an angle of 172.7° (Fig. 3.23 A). Epiblast is absent.

Scutellum: "V" shape: Type V3-3 (Fig. 3.23 A), having an angle of 62.17°. Scutellum cleft/slit is present. Oval-shaped scutellum cells (Fig. 3.23 D). Depletion layer is present (Fig. 3.23 D, arrowhead).

Vascularization: Type of vascularization: Panicoid

Angle of vascularization: 164.4°, Type C; **Pattern of embryonic leaf:** P-type

Endosperm: Polygon-shaped endospermic cells.

Starch grain: Simple circular and polygon-shaped starch grains (Fig. 3.23 C).

No. of starch grains per endosperm cell: 44 ± 3; **Starch grain size:** 4.69 ± 2.16 (µm)

Aleurone layer: Tangentially elongated rectangular shaped aleurone cells (Fig. 3.23 F).

Caryopsis type and formula: True panicoids type and having formula "P-PP".

Bar size: A, B = 125 µm; C–F = 12.5 µm

34. *Themeda quadrivalvis* (Fig. 3.23(34) A–G)

Shape: Crock (Matka) shaped in transaction with dorsal convex surface and having size of 0.85 (*L*) × 0.98 (*B*) mm^2 (Fig. 3.23 B).

Pericarp: Type A (Fig. 3.23 C, cross cells (arrowhead), D nucellus (arrow).

Embryo: Type C, Embryo is placed at an angle of 173.1°. Epiblast is absent.

Scutellum: V shape: Type V3-3, having an angle of 61.3°. Scutellum cleft covers the coleorhiza. Oval-shaped, compactly arranged scutellum cells (Fig. 3.23 F), and shows depletion layer (Fig. 3.23 G, arrowhead).

Vascularization: Type of vascularization: Panicoid

Angle of vascularization: 168.4°, Type C.; **Pattern of embryonic leaf:** P-type

Endosperm: Polygon-shaped endospermic cells.

Starch grain: Simple circular and polygon-shaped starch grains are present (Fig. 3.23 G, arrow).

No. of starch grains per endosperm cell: 40 ± 5; **Starch grain size:** 3.41 ± 1.39 (µm)

Aleurone layer: Tangentially elongated rectangular shaped aleurone cells (Fig. 3.23 E).

Caryopsis type and formula: True panicoids type and having formula "P-PP".

Bar size: A, B = 125 µm; C–G = 12.5 µm

DIFFERENTIATING FEATURES:

Perricarp type B … *T. laxa*
Perricarp type A
Scutellum type U1-2 … *T. cymbaria*
Scutellum type V3-4
Simple polygonal-shaped starch grains, thickness of endosperm: 546.31 ± 7.82 μm … *T. triandra*
Simple circular shaped starch grains, thickness of endosperm: 877.3 ± 7.63 μm … *T. qurqdrivalvis*

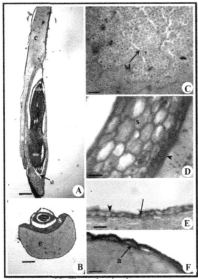

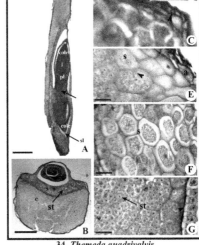

33. *Themeda triandra*   34. *Themeda quadrivalvis*

35. *Triplopogon ramosissimus* (Fig. 3.24(35) A−F)

Shape: Rectangular to oval shaped in transaction with dorsal convex surface and having size of 0.48 (*L*) × 0.95 (*B*) mm² (Fig. 3.24 B).

Pericarp: Type A (Fig. 3.24 D; epidermis (arrowhead), hypodermis (arrow), cross cells (arrow)).

Embryo: Type C, Embryo is placed at an angle of 172.1° (Fig. 3.24 A). Epiblast is absent.

Scutellum: U shape: Type U2-2 (Fig. 3.24 B), having an angle of 61.7°. Scutellum cleft/slit is present, and it covers the coleorhiza. Oval-shaped, compactly arranged scutellum cells (Fig. 3.24 C), and also shows depletion layer (Fig. 3.24 C, arrowhead).

Vascularization: Type of vascularization: Panicoid

Angle of vascularization: 166.5°, Type C.

Pattern of embryonic leaf: P-type

Endosperm: Polygon-shaped endospermic cells.

Starch grain: Simple circular and polygon-shaped starch grains present (Fig. 3.24 C).

No. of starch grains per endosperm cell: 37 ± 8; **Starch grain size:** 2.5 ± 1.08 (μm)

Aleurone layer: Tangentially elongated rectangular shaped aleurone cells (Fig. 3.24 F).

Caryopsis type and formula: True panicoids type and having formula "P-PP".

Bar size: A, B = 125 μm; C–F = 12.5 μm

36. *Vetivaria zizanioides* (Fig. 3.24(36) A–F)

Shape: Triangular shaped with circular base in transaction with dorsal convex surface and having size of 0.96 (*L*) × 0.60 (*B*) mm^2 (Fig. 3.24 B).

Pericarp: Type C (Fig. 3.24 E; epidermis (arrowhead), nucellus (arrow)).

Embryo: Type C, Embryo is placed at an angle of 171.1° (Fig. 3.24 A). Epiblast is absent.

Scutellum: Δ shape: Type Δ5 (Fig. 3.24 B), having an angle of 56.9°. Scutellum cleft/slit is present, and it covers the coleorhiza. Oval-shaped, compactly arranged scutellum cells (Fig. 3.24 C), and also shows depletion layer (Fig. 3.24 C, arrowhead).

Vascularization: Type of vascularization: Panicoid (Fig. 3.24 A, arrow)

Angle of vascularization: 167.9°, Type C.

Pattern of embryonic leaf: P-type

Endosperm: Polygon-shaped endospermic cells.

Starch grain: Simple circular and polygon-shaped starch grains present (Fig. 3.24 D).

No. of starch grains per endosperm cell: 32 ± 7; **Starch grain size:** 10 ± 3.86 (μm)

Aleurone layer: Tangentially elongated rectangular shaped aleurone cells (Fig. 3.24 F).

Caryopsis type and formula: True panicoids type and having formula "P-PP".

Bar size: A, B = 125 μm; C–F = 12.5 μm

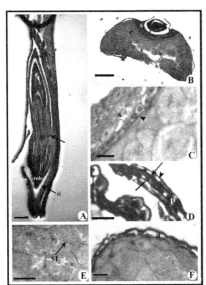

35. *Triplopogon ramosissimus*

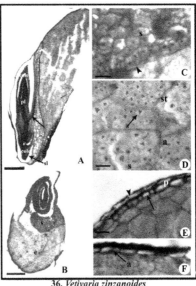

36. *Vetivaria zinzanoides*

37. *Alloteropis cimicina* (Fig. 3.25(37) A–F)

Shape: Triangular shaped with circular base in transaction with dorsal convex surface and having size of 0.34 (*L*) × 3.45 (*B*) mm² (Fig. 3.25 B).

Pericarp: Type C (Fig. 3.25 F).

Embryo: Type C, Embryo is placed at an angle of 166.7° (Fig. 3.25 A). Epiblast is absent.

Scutellum: V shape: Type V6 (Fig. 3.25 B), having an angle of 61.2°. Scutellum cleft/slit is present in the form of small slit. Oval-shaped, compactly arranged scutellum cells (Fig. 3.25 D), and also shows depletion layer (Fig. 3.25 D, arrowhead).

Vascularization: Type of vascularization: Panicoid (Fig. 3.25 A, arrow)

Angle of vascularization: 141.7°, Type B.

Pattern of embryonic leaf: P-type

Endosperm: Polygon-shaped endospermic cells.

Starch grain: Simple circular and polygon-shaped starch grains present (Fig. 3.25 C).

No. of starch grains per endosperm cell: 54 ± 5; **Starch grain size:** 5 ± 1.68 (μm)

Aleurone layer: Tangentially elongated rectangular and square-shaped aleurone cells (Fig. 3.25 E).

Caryopsis type and formula: True panicoids type and having formula "P-PP".

Bar size: A, B = 125 μm; C–F = 12.5 μm

Brachiaria

38. *Brachiaria distachya* 39. *Brachiaria eruciformis*

40. *Brachiaria ramose* 41. *Brachiaria reptans*

38. *Brachiaria distachya* (Fig. 3.25(38) A–F)

Shape: Oval to rectangular shaped with circular base in transaction with dorsal convex surface and having size of 0.52 (L) $\times 1.50$ (B) mm^2 (Fig. 3.25 B).

Pericarp: Type C (Fig. 3.25 E; cross cells (arrowhead), nucellus (arrow)).

Embryo: Type C, Embryo is placed at an angle of 166.8° (Fig. 3.25 A). Epiblast is absent.

Scutellum: V shape: Type V3-5 (Fig. 3.25 B), having an angle of 82.33°. Scutellum cleft/slit is present in the form of small slit. Oval-shaped, compactly arranged scutellum cells (Fig. 3.25 F), and also shows depletion layer (Fig. 3.25 F, arrowhead).

Vascularization: Type of vascularization: Panicoid (Fig. 3.25 A, arrow)

Angle of vascularization: 154.0°, Type C.

Pattern of embryonic leaf: P-type

Endosperm: Polygon-shaped endospermic cells.

Starch grain: Simple circular and polygon starch grains are present (Fig. 3.25 D).

No. of starch grains per endosperm cell: 29 ± 11; **Starch grain size:** 3.13 ± 1.93 (μm)

Aleurone layer: Tangentially elongated rectangular shaped aleurone cells (Fig. 3.25 C).

Caryopsis type and formula: True panicoids type and having formula "P-PP".

Bar size: A, B = 125 μm; C–F = 12.5 μm

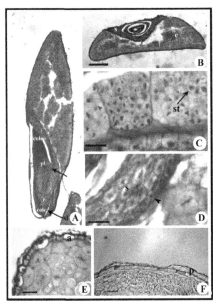

37. *Alloperopsis cimicina*

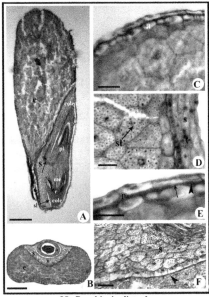

38. *Brachiaria distachya*

39. *Brachiaria eruciformis* (Fig. 3.26(39) A–F)

Shape: Oval shaped with flat base and convex anterior side in transaction and having size of 0.36 (*L*) × 0.67 (*B*) mm^2 (Fig. 3.26 B).

Pericarp: Type C (Fig. 3.26 E; cross cells (arrowhead)).

Embryo: Type C, Embryo is placed at an angle of 170.3° (Fig. 3.26 A). Epiblast is absent.

Scutellum: V shape: Type V3-6 (Fig. 3.26 B), having an angle of 46.06°. Scutellum cleft/slit is present, and it covers the coleorhiza. Oval-shaped, compactly arranged scutellum cells (Fig. 3.26 D).

Vascularization: Type of vascularization: Panicoid

Angle of vascularization: 168.8°, Type C.

Pattern of embryonic leaf: P-type

Endosperm: Polygon-shaped endospermic cells.

Starch grain: Simple circular and polygon-shaped starch grains are present (Fig. 3.26 C).

No. of starch grains per endosperm cell: 26 ± 9; **Starch grain size:** 2.88 ± 0.36 (μm)

Aleurone layer: Tangentially elongated rectangular shaped aleurone cells (Fig. 3.26 F).

Caryopsis type and formula: True panicoids type and having formula "P-PP".

Bar size: A, B = 125 μm; C–F = 12.5 μm

40. *Brachiaria ramosa* (Fig. 3.26(40) A–F)

Shape: Oval shaped with flat base and convex anterior side in transaction and having size of 0.62 (L) × 1.35 (B) mm² (Fig. 3.26 B).

Pericarp: Type C (Fig. 3.26 D; nucellus (arrow), cross cells (arrowhead)).

Embryo: Type C, Embryo is placed at an angle of 160.3° (Fig. 3.26 A). Epiblast is absent.

Scutellum: V shape: Type V3-5 (Fig. 3.26 B), having an angle of 78.37°. Scutellum cleft/slit is present. Oval-shaped, compactly arranged scutellum cells (Fig. 3.26 C), and depletion layer is also present (Fig. 3.26 C, arrowhead).

Vascularization: Type of vascularization: Panicoid (Fig. 3.26 A, arrowhead)

Angle of vascularization: 152.38 °, Type C.

Pattern of embryonic leaf: P-type

Endosperm: Polygon-shaped endospermic cells.

Starch grain: Simple circular and polygon starch grains are present (Fig. 3.26 F).

No. of starch grains per endosperm cell: 23 ± 8; **Starch grain size:** 4.17 ± 1.74 (μm)

Aleurone layer: Tangentially elongated rectangular shaped small aleurone cells (Fig. 3.26 D, arrow).

Caryopsis type and formula: True panicoids type and having formula "P-PP".

Bar size: A, B = 125 μm; C–F = 12.5 μm

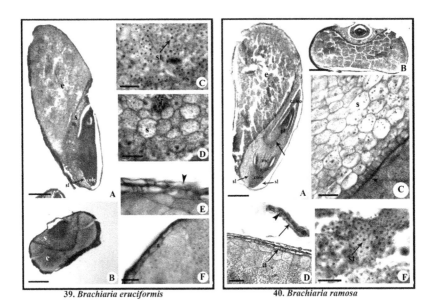

39. *Brachiaria eruciformis*   40. *Brachiaria ramosa*

41. *Brachiaria reptans* (Fig. 3.27(41) A–F)

Shape: Oval to dumble shaped with convex anterior side in T.S., having size of 0.43 (*L*) × 1.42 (*B*) mm² (Fig. 3.27 B).

Pericarp: Type C (Fig. 3.27 E; epidermis (arrowhead), cross cells (arrow)).

Embryo: Type C, Embryo is placed at an angle of 173.2 ° (Fig. 3.27 A). Epiblast is absent.

Scutellum: V shape: Type V3-3 (Fig. 3.27 B), an angle of 61.24°. Scutellum cleft present in the center of the proximal end. Oval-shaped, compactly arranged scutellum cells and depletion layer are present (Fig. 3.27 C).

Vascularization: Type of vascularization: Panicoid (Fig. 3.27 A, arrow)

Angle of vascularization: 156.2°, Type C. **Pattern of embryonic leaf:** P-type

Endosperm: Polygon-shaped endospermic cells.

Starch grain: Simple circular and polygon-shaped starch grains are present (Fig. 3.27 E).

No. of starch grains per endosperm cell: 21 ± 7; **Starch grain size:** 2.88 ± 0.34 (μm)

Aleurone layer: Tangentially elongated rectangular small aleurone cells (Fig. 3.27 D).

Caryopsis type and formula: True panicoids type and having formula "P-PP".

Bar size: A, B = 125 μm; C–F = 12.5 μm

DIFFERENTIATING FEATURES:

Scutellum type V3-6 ... *B. eruciformis*
Scutellum type V3-3 ... *B. reptans*
Scutellum type V3-5
Endosperm thickness: 335.8 ± 11.21 μm ... *B. distachya*
Endosperm is very thick, almost three times more (897.71 ± 23.63 μm) ...
B.ramosa

Cenchrus

42. *Cenchrus biflorus* 43. *Cenchrus ciliaris* 44. *Cenchrus setigerus*

42. *Cenchrus biflorus* (Fig. 3.27(42) A–F)

Shape: Semicircle shaped with flat anterior side and circular posterior side in transaction and having size of 0.82 (*L*) × 1.34 (*B*) mm² (Fig. 3.27 B).

Pericarp: Type C (Fig. 3.27 F; epidermis (arrowhead)).

Embryo: Type C, Embryo is placed at an angle of 172.7 ° (Fig. 3.27 A). Epiblast is absent.

Scutellum: V shape: Type V2 (Fig. 3.27 B), an angle of 53.9°. Scutellum cleft is present. Oval-shaped, compactly arranged scutellum cells (Fig. 3.27 C), and depletion layer is also present (Fig. 3.27 C, arrowhead).

Vascularization: Type of vascularization: Panicoid (Fig. 3.27 A)

Angle of vascularization: 164.1°, Type C. **Pattern of embryonic leaf:** P-type

Endosperm: Polygon-shaped endospermic cells.

Starch grain: Simple polygon-shaped starch grains present (Fig. 3.27 E).

No. of starch grains per endosperm cell: 45 ± 8; **Starch grain size:** 5.68 ± 2.27 (μm)

Aleurone layer: Tangentially elongated rectangular, small squarish shaped aleurone cells (Fig. 3.27 D).

Caryopsis type and formula: True panicoids type and having formula "P-PP".

Bar size: A, B = 125 μm; C–F = 12.5 μm

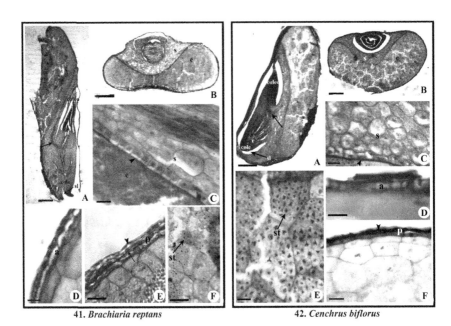

41. *Brachiaria reptans*   42. *Cenchrus biflorus*

43. *Cenchrus ciliaris* (Fig. 3.28(43) A–E)

Shape: Oblong to oval shaped with convex anterior side and circular posterior side in transaction, and having size of 0.52 (*L*) × 0.64 (*B*) mm² (Fig. 3.28 B).

Pericarp: Type C (Fig. 3.28 C).

Embryo: Type C, Embryo is placed at an angle of 172.9 ° (Fig. 3.28 A). Epiblast is absent.

Scutellum: V shape: Type V8-2, having an angle of 71.3°. Scutellum cleft/ slit is present. Oval-shaped, slightly flat, compactly arranged scutellum cells (Fig. 3.28 E), and depletion layer is also present (Fig. 3.28 E, arrowhead).

Vascularization: Type of vascularization: Panicoid

Angle of vascularization: 167.8°, Type C. **Pattern of embryonic leaf:** P-type

Endosperm: Polygon-shaped endospermic cells.

Starch grain: Simple circular and polygon-shaped starch grains present (Fig. 3.28 E, arrow).

No. of starch grains per endosperm cell: 39 ± 6; **Starch grain size:** 5 ± 1.83 (μm)

Aleurone layer: Tangentially elongated rectangular small aleurone cells (Fig. 3.28 D).

Caryopsis type and formula: True panicoids type and having formula "P-PP".

Bar size: A, B = 125 μm; C–E = 12.5 μm

44. *Cenchrus setigerus* (Fig. 3.28(44) A–G)

Shape: Kidney shaped with flat anterior side and circular posterior side in transaction, and having size of 0.83 (L) × 1.39 (B) mm² (Fig. 3.28 B).

Pericarp: Type C (Fig. 3.28 E; epidermis (arrow)).

Embryo: Type C, Embryo is placed at an angle of 166.2 ° (Fig. 3.28 A). Epiblast is absent.

Scutellum: V shape: Type V6 (Fig. 3.28 B), having an angle of 58.96°. Scutellum cleft/slit is present, and it is covering the coleorhiza. Oval-shaped, compactly arranged scutellum cells (Fig. 3.28 G), depletion layer is also present (Fig. 3.28 G, arrowhead).

Vascularization: Type of vascularization: Panicoid (Fig. 3.28 B)

 Angle of vascularization: 146.3°, Type B. **Pattern of embryonic leaf:** P-type

Endosperm: Polygon-shaped endospermic cells.

Starch grain: Simple circular and polygon-shaped starch grains present (Fig. 3.28 F).

No. of starch grains per endosperm cell: 41 ± 4; **Starch grain size:** 6.25 ± 2.57 (μm)

Aleurone layer: Tangentially elongated rectangular small aleurone cells (Fig. 3.28 D).

Caryopsis type and formula: True panicoids type and having formula "P-PP".

Bar size: A, C = 125 μm; B = 85.5 μm; D–G = 12.5 μm

DIFFERENTIATING FEATURES:

Both square and rectangular shaped aleurone cells … *C. biflorus*
Tangentially elongated aleurone cells
Angle of vascularization Type C … *C. ciliaris*
Angle of vascularization Type B … *C. setigerus*

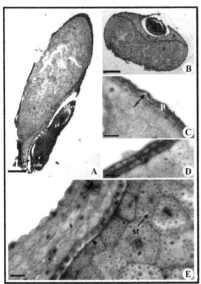
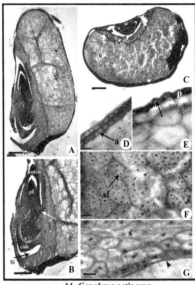

43. *Cenchrus ciliaris*   44. *Cenchrus setigerus*

Digitaria

45. *Digitaria ciliaris*

46. *Digitaria granularis*

47. *Digitaria longiflora*

48. *Digitaria stircta*

45. *Digitaria ciliaris* (Fig. 3.29(45) A–F)

Shape: Oblong to oval shaped with flat anterior side and circular posterior side in transaction, and having size of 0.62 (L) × 0.43 (B) mm^2 (Fig. 3.29 A).

Pericarp: Type D (Fig. 3.29 F; epidermis (arrow), nucellus (arrowhead)).

Embryo: Type B, Embryo is placed at an angle of 157.76 ° (Fig. 3.29 A). Epiblast is absent.

Scutellum: V shape: Type V7 (Fig. 3.29 B), having an angle of 67.5°. Scutellum cleft/slit is present. Oval-shaped, compactly arranged scutellum cells (Fig. 3.29 C), and depletion layer is also present (Fig. 3.29 C, arrowhead).

Vascularization: Type of vascularization: Panicoid (Fig. 3.29 A, arrow)

Angle of vascularization: 145.45°, Type B.

Pattern of embryonic leaf: P-type

Endosperm: Polygon-shaped endospermic cells.

Starch grain: Simple circular and polygon-shaped starch grains are present (Fig. 3.29 D).

No. of starch grains per endosperm cell: 30 ± 9; **Starch grain size:** 3.85 ± 1.63 (µm)

Aleurone layer: Tangentially elongated rectangular shaped aleurone cells (Fig. 3.29 E).

Caryopsis type and formula: True panicoids type and having formula "P-PP".

Bar size: A, B = 125 µm; C–F = 12.5 µm

46. *Digitaria granularis* (Fig. 3.29(46) A–F)

Shape: Oblong shape with convex anterior side and flat posterior side in transaction, and having size of $0.49\ (L) \times 0.64(B)$ mm² (Fig. 3.29 B).

Pericarp: Type C (Fig. 3.29 E; epidermis cells (arrowhead), nucellus (arrow)).

Embryo: Type C, Embryo is placed at an angle of 148.2° (Fig. 3.29 A). Epiblast is absent.

Scutellum: V shape: Type V4-2, having an angle of 51.44° (Fig. 3.29 B). Scutellum cleft/slit is present, and it is present in the center of the proximal end. Oval-shaped, compactly arranged scutellum cells (Fig. 3.29 F), and depletion layer is also present (Fig. 3.29 F, arrowhead).

Vascularization: Type of vascularization: Panicoid

Angle of vascularization: 142.5°; Type C.

Pattern of embryonic leaf: P-type

Endosperm: Polygon-shaped endospermic cells.

Starch grain: Simple circular and polygon starch grains are present (Fig. 3.29 C).

No. of starch grains per endosperm cell: 38 ± 7; **Starch grain size:** 2.68 $\pm$ 0.74 (μm)

Aleurone layer: Tangentially elongated rectangular shaped aleurone cells (Fig. 3.29 D, arrow).

Caryopsis type and formula: True panicoids type and having formula "P-PP".

Bar size: A, B = 125 μm; C–F = 12.5 μm

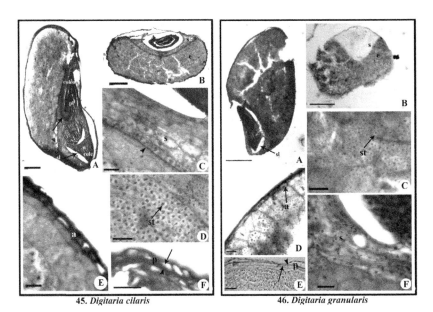

45. *Digitaria cilaris*   46. *Digitaria granularis*

47. *Digitaria longiflora* (Fig. 3.30(47) A–E)

Shape: Oblong to oval shaped with convex anterior side and concave posterior side in transaction, and having size of 0.70 $(L) \times 1.13$ (B) mm² (Fig. 3.30 B).

Pericarp: Type C (Fig. 3.30 E; epidermis (arrowhead), cross cells (arrow)).

Embryo: Type C, Embryo is placed at an angle of 163.4° (Fig. 3.30 A). Epiblast is absent.

Scutellum: V shape: Type U1-2 (Fig. 3.30 B), an angle of 59.72°. Scutellum cleft is present. Oval-shaped, compactly arranged scutellum cells (Fig. 3.30 F); depletion layer is present (Fig. 3.30 F, arrowhead).

Vascularization: Type of vascularization: Panicoid

Angle of vascularization: 146.1°, Type B; **Pattern of embryonic leaf:** P-type

Endosperm: Polygon-shaped endospermic cells.

Starch grain: Simple polygon-shaped starch grains are present (Fig. 3.30 D).

No. of starch grains per endosperm cell: 44 ± 5; **Starch grain size:** 8.34 ± 3.04 (μm)

Aleurone layer: Tangentially elongated rectangular shaped aleurone cells (Fig. 3.30 C, arrow).

Caryopsis type and formula: True panicoids type and having formula "P-PP".

Bar size: A, B = 125 μm; C–F = 12.5 μm

48. *Digitaria stricta* (Fig. 3.30(48) A–E)

Shape: Oblong shape with convex anterior side and flat posterior side in transaction, and having size 0.47 (*L*) × 0.78 (*B*) mm² (Fig. 3.30 B).

Pericarp: Type D (Fig. 3.30 C; epidermis cells (arrowhead)).

Embryo: Type C, Embryo is placed at an angle of 162.1° (Fig. 3.30 A). Epiblast is absent.

Scutellum: V shape: Type V8-3, having an angle of 62.4° (Fig. 3.30 B). Scutellum cleft/slit is present. Oval-shaped, compactly arranged scutellum cells (Fig. 3.30 D).

Vascularization: Type of vascularization: Panicoid

Angle of vascularization: 143.4°, Type B. **Pattern of embryonic leaf:** P-type

Endosperm: Polygon-shaped endospermic cells.

Starch grain: Simple circular and polygon-shaped starch grains are present (Fig. 3.30 E).

No. of starch grains per endosperm cell: 43 ± 8; **Starch grain size:** 6.95 ± 2.56 (μm)

Aleurone layer: Tangentially elongated rectangular shaped aleurone cells (Fig. 3.30 C, arrow).

Caryopsis type and formula: True panicoids type and having formula "P-PP".

Bar size: A, B = 125 µm; C–E = 12.5 µm

DIFFERENTIATING FEATURES:

Angle of vascularization Type C … *D. granularis*
Angle of vascularization Type B
Embryo type B … *D. ciliaris*
Embryo type C
Scutellum type U1-2 … *D. longiflora*
Scutellum type V8-3 … *D. stricta*

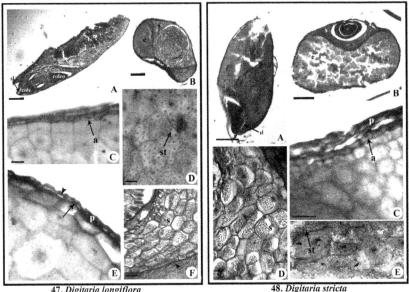

47. *Digitaria longiflora*   48. *Digitaria stricta*

Echinichloa

49. *Echinichloa colona* 50. *Echinochloa crusgalli* 51. *Echionochloa stagnina*

49. *Echinochloa colona* (Fig. 3.31(49) A-F)

Shape: Oblong shape with convex anterior side and concave posterior side in transaction, and having size of 0.88 (*L*) × 1.42 (*B*) mm^2 (Fig. 3.31 B).

Pericarp: Type B (Fig. 3.31 E; epidermis (arrowhead), cross cells (arrow)).

Embryo: Type C, Embryo is placed at an angle of 163.8 ° (Fig. 3.31 A). Epiblast is absent.

Scutellum: V shape: Type V7 (Fig. 3.31 B), having an angle of 65.74°. Scutellum cleft/slit is present, and it covers the whole coleorhiza. Oval-shaped, compactly arranged scutellum cells (Fig. 3.31 C), and depletion layer is also present (Fig. 3.31 C, arrowhead).

Vascularization: Type of vascularization: Panicoid

Angle of vascularization: 131.3°, Type B.

Pattern of embryonic leaf: P-type

Endosperm: Polygon-shaped endospermic cells.

Starch grain: Simple circular and polygon starch grains present (Fig. 3.31 F).

No. of starch grains per endosperm cell: 48 ± 12; **Starch grain size:** 3.75 ± 1.64 (μm)

Aleurone layer: Tangentially elongated rectangular shaped aleurone cells (Fig. 3.31 D).

Caryopsis type and formula: True panicoids type and having formula "P-PP".

Bar size: A, B = 125 μm; C–F = 12.5 μm

50. *Echinochloa crusgalli* (Fig. 3.31(50) A–F)

Shape: Oblong shape with convex anterior side and concave posterior side in transaction, and having size of 0.93 (*L*) × 1.67 (*B*) mm² (Fig. 3.31 B).

Pericarp: Type B (Fig. 3.31 E; epidermis cells (arrow), cross cells (arrowhead)).

Embryo: Type C, Embryo is placed at an angle of 163.4 ° (Fig. 3.31 A). Epiblast is absent.

Scutellum: V shape: Type V8-2 (Fig. 3.31 B), having an angle of 69.9°. Scutellum cleft/slit is present, and it is present in the form of small slit. Oval-shaped, compactly arranged scutellum cells (Fig. 3.31 D), and deple-tion layer is also present (Fig. 3.31 E, arrowhead).

Vascularization: Type of vascularization: Panicoid (Fig. 3.31 A, arrow)

Angle of vascularization: 143.5°, Type B.

Pattern of embryonic leaf: P-type

Endosperm: Polygon-shaped endospermic cells.

Starch grain: Simple circular and polygon starch grains are present (Fig. 3.31 D).

No. of starch grains per endosperm cell: 47 ± 6; **Starch grain size:** 5 ± 1.88 (μm)

Aleurone layer: Tangentially elongated rectangular shaped aleurone cells (Fig. 3.31 C).

Caryopsis type and formula: True panicoids type and having formula "P-PP".

Bar size: A, B = 125 μm; C–E = 12.5 μm

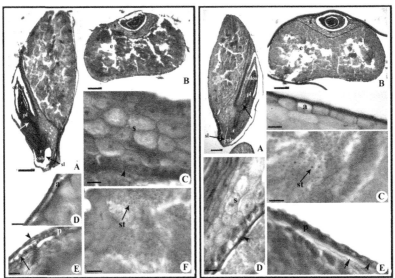

49. *Echinochloa colona*   50. *Echinochloa crusgalli*

51. *Echinochloa stagnina* (Fig. 3.32(51) A−E)

Shape: Oblong to oval shaped with flat anterior side and circular posterior side in transaction, and having size of 1.25 (L) × 2.33 (B) mm^2 (Fig. 3.32 B).

Pericarp: Type B (Fig. 3.32 D; epidermis (arrow), cross cells (arrowhead)).

Embryo: Type C, Embryo is placed at an angle of 163.2° (Fig. 3.32 A). Epiblast is absent.

Scutellum: V shape: Type V8-1 (Fig. 3.32 B), having an angle of 45.89°. Scutellum cleft/slit is present, and it is present at the side of coleorhiza.

Oval-shaped, compactly arranged scutellum cells (Fig. 3.32 E); depletion layer is also present (Fig. 3.32 E, arrowhead).

Vascularization: Type of vascularization: Panicoid

Angle of vascularization: 143.6°, Type B.

Pattern of embryonic leaf: P-type

Endosperm: Polygon-shaped endospermic cells.

Starch grain: Simple circular and polygon starch grains are present (Fig. 3.32 D).

No. of starch grains per endosperm cell: 41 ± 9; **Starch grain size:** 5.56 ± 3.02 (µm)

Aleurone layer: Tangentially elongated rectangular and square shaped mixed aleurone cells (Fig. 3.32 C).

Caryopsis type and formula: True panicoids type and having formula "P-PP".

Bar size: A, B = 125 µm; C–F = 12.5 µm

DIFFERENTIATING FEATURES:

Both square and rectangular shaped aleurone cells … *E. stagnina*
Tangentially elongated aleurone cells
Scutellum type V7 … *E. colona*
Scutellum type V8-2 … *E. crusgalli*

52. *Eremopogon foveolatus* (Fig. 3.32(52) A–F)

Shape: Oblong to oval shaped with flat anterior side and circular posterior side in transaction, and having size of 0.5 (*L*) × 1.54 (*B*) mm² (Fig. 3.32 B).

Pericarp: Type C (Fig. 3.32 D; epidermis cells (arrow), cross cells (arrowhead)).

Embryo: Type C, Embryo is placed at an angle of 167.9 ° (Fig. 3.32 A). Epiblast is absent.

Scutellum: V shape: Type V5 (Fig. 3.32 B), having an angle of 52.9°. Scutellum cleft/slit is present, and it is present at the side of coleorhiza. Oval-shaped, compactly arranged scutellum cells (Fig. 3.32 E).

Vascularization: Type of vascularization: Panicoid

Angle of vascularization: 155.0°, Type C. **Pattern of embryonic leaf:** P-type

Endosperm: Polygon-shaped endospermic cells.

Starch grain: Simple circular and polygon starch grains are present (Fig. 3.32 F).

No. of starch grains per endosperm cell: 35 ± 7; **Starch grain size:** 7.14 ± 2.95 (µm)

Aleurone layer: Tangentially elongated rectangular shaped aleurone cells (Fig. 3.32 C).

Caryopsis type and formula: True panicoids type and having formula "P-PP".

Bar size: A, B = 125 µm; C–F = 12.5 µm

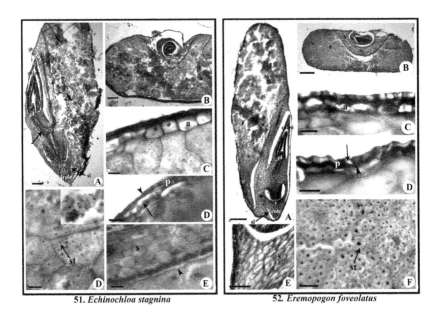

51. *Echinochloa stagnina*   52. *Eremopogon foveolatus*

53. *Eriochloa procera* (Fig. 3.33(53) A–F)

Shape: Oblong shape in transaction with concave anterior side and flat posterior side, and having size 0.64 (*L*) × 1.83 (*B*) mm² (Fig. 3.33 B).

Pericarp: Type A (Fig. 3.33 F; epidermis (arrowhead), cross cells (arrow)).

Embryo: Type C, Embryo is placed at an angle of 160.2° (Fig. 3.33 A). Epiblast is absent.

Scutellum: "V" shape: Type V5 (Fig. 3.33 B), having an angle of 52.9°. Scutellum cleft/slit is present. Oval-shaped scutellum cells (Fig. 3.33 D). Depletion layer is present.

Vascularization: Type of vascularization: Panicoid

Angle of vascularization: 154.1°, Type C.

Pattern of embryonic leaf: P-type

Endosperm: Polygon-shaped endospermic cells.

Starch grain: Simple circular and polygon-shaped starch grains (Fig. 3.33 E).

No. of starch grains per endosperm cell: 77 ± 6; **Starch grain size:** 2.78 ± 1.17 (µm)

Aleurone layer: Tangentially elongated rectangular shaped aleurone cells (Fig. 3.33 C).

Caryopsis type and formula: True panicoids type and having formula "P-PP".

Bar size: A, B = 125 µm; C–F = 12.5 µm

Oplismenus

54. *Oplismenus burmannii* 55. *Oplismenus composites*

54. *Oplismenus burmannii* (Fig. 3.33(54) A–F)

Shape: Oblong to oval shaped with flat anterior side and circular posterior side in transaction, and having size of 0.39 (*L*) × 0.55 (*B*) mm² (Fig. 3.33 B).

Pericarp: Type C (Fig. 3.33 C; epidermis cells (arrow), cross cells (arrowhead)).

Embryo: Type B, Embryo is placed at an angle of 154.4 ° (Fig. 3.33 A). Epiblast is absent.

Scutellum: V shape: Type U2-2 (Fig. 3.33 B), having an angle of 55.3°. Scutellum cleft/slit is present, and it is present at the side of coleorhiza. Oval-shaped, compactly arranged scutellum cells (Fig. 3.33 E), and depletion layer is also present (Fig. 3.33 E, arrowhead).

Vascularization: Type of vascularization: Panicoid (Fig. 3.33 A, arrow)

Angle of vascularization: 150.2°, Type C.

Pattern of embryonic leaf: P-type

Endosperm: Polygon-shaped endospermic cells.

Starch grain: Simple polygon-shaped starch grains are present (Fig. 3.33 D).

No. of starch grains per endosperm cell: 37 ± 10; **Starch grain size:** 4.35 ± 2.05 (µm)

Aleurone layer: Tangentially elongated rectangular shaped aleurone cells (Fig. 3.33 F).

Caryopsis type and formula: True panicoids type and having formula "P-PP".

Bar size: A, B = 125 µm; C–F = 12.5 µm

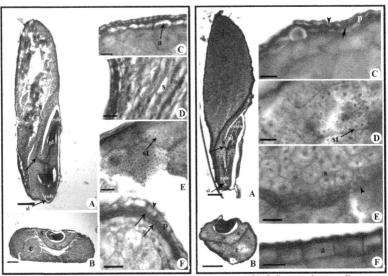

53. *Oplismenus burmannii*　　　54. *Oplismenus burmannii*

55. *Oplismenus composites* (Fig. 3.34(55) A–F)

Shape: Rectangular shaped with curved angles, and wavy anterior side and convex posterior side in transaction, and having size of $0.52 \, (L) \times 0.94 \, (B)$ mm² (Fig. 3.34 B).

Pericarp: Type B (Fig. 3.34 D; epidermis (arrowhead), nucellus (arrow)).

Embryo: Type C, Embryo is placed at an angle of $164.9°$ (Fig. 3.34 A). Epiblast is absent.

Scutellum: V shape: Type V9 (Fig. 3.34 B), having an angle of 64.9°. Scutellum cleft/slit is present, and it is present at the side of the coleorhiza. Oval-shaped, compactly arranged scutellum cells (Fig. 3.34 C).

Vascularization: Type of vascularization: Panicoid (Fig. 3.34 A, arrow)

Angle of vascularization: 154.7°, Type C. **Pattern of embryonic leaf:** P-type

Endosperm: Polygon-shaped endospermic cells.

Starch grain: Simple circular and polygon-shaped starch grains are present (Fig. 3.34 F).

No. of starch grains per endosperm cell: 59 ± 7; **Starch grain size:** 5 ± 1.38 (μm)

Aleurone layer: Tangentially elongated rectangular shaped aleurone cells (Fig. 3.34 E).

Caryopsis type and formula: True panicoids type and having formula "P-PP".

Bar size: A, B = 125 μm; C–F = 12.5 μm

DIFFERENTIATING FEATURES:

Pericarp type C, scutellum type U2-2 … *O. burmanii*
Pericarp type B, scutellum type V9 … *O. composites*

Panicum

56. *Panicum antidotale* 57. *Panicum maximum*

58. *Panicum miliaceum* 59. *Panicum trypheron*

56. *Panicum antidotale* (Fig. 3.34(56) A–F)

Shape: Oblong to rectangular shaped with convex anterior side and flat posterior side in transaction, and having size of $0.58(L) \times 0.89 (B)$ mm^2 (Fig. 3.34 B).

Pericarp: Type B (Fig. 3.34 F; epidermis cells (arrowhead), cross cells (arrow)).

Embryo: Type C, Embryo is placed at an angle of 164.4° (Fig. 3.34 A). Epiblast is absent.

Scutellum: V shape: Type V4-2 (Fig. 3.34 B), an angle of 42.79°. Scutellum cleft is present. Oval-shaped, compactly arranged scutellum cells (Fig. 3.34 D); depletion layer is present (Fig. 3.34 D, arrowhead).

Vascularization: Type of vascularization: Panicoid (Fig. 3.34 A, arrow)

Angle of vascularization: 132.2°, Type B. **Pattern of embryonic leaf:** P-type

Endosperm: Polygon-shaped endospermic cells.

Starch grain: Simple circular and polygon starch grains are present (Fig. 3.34 E).

No. of starch grains per endosperm cell: 42 ± 5; **Starch grain size:** 2.78 ± 0.83 (μm)

Aleurone layer: Tangentially elongated rectangular shaped aleurone cells (Fig. 3.34 C).

Caryopsis type and formula: True panicoids type and having formula "P-PP".

Bar size: A, B = 125 μm; C–F = 12.5 μm

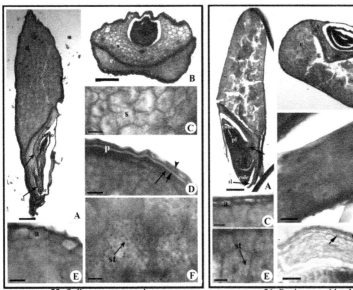

55. *Oplismenus compositus* **56. *Panicum antidotale***

57. *Panicum maximum* (Fig. 3.35(57) A–F)

Shape: Oblong shape with convex anterior side and circular posterior side in transaction, and having size of 0.57 (*L*) × 1.07 (*B*) mm² (Fig. 3.35 B).

Pericarp: Type B (Fig. 3.35 F; epidermis (arrow), cross cells (arrowhead)).

Embryo: Type C, Embryo is placed at an angle of 174.7° (Fig. 3.35 A). Epiblast is absent.

Scutellum: U shape: Type U2-3 (Fig. 3.35 B), having an angle of 64.32°. Scutellum cleft/slit is present, and it is covering the whole coleorhiza. Oval-shaped, compactly arranged scutellum cells (Fig. 3.35 C), and depletion layer is also present (Fig. 3.35 C, arrowhead).

Vascularization: Type of vascularization: Panicoid

Angle of vascularization: 157.5°, Type C.

Pattern of embryonic leaf: P-type

Endosperm: Polygon-shaped endospermic cells.

Starch grain: Simple circular and polygon starch grains present (Fig. 3.35 E).

No. of starch grains per endosperm cell: 33 ± 7; **Starch grain size:** 4.17 ± 1.66 (μm)

Aleurone layer: Tangentially elongated rectangular shaped aleurone cells (Fig. 3.35 D).

Caryopsis type and formula: True panicoids type and having formula "P-PP".

Bar size: A, B = 125 μm; C–F = 12.5 μm

58. *Panicum miliaceum* (Fig. 3.35(58) A–G)

Shape: Oblong shape with flat anterior side and concave posterior side in transaction, and having size of 1.07 (*L*) × 1.82 (*B*) mm² (Fig. 3.35 B).

Pericarp: Type B (Fig. 3.35 E; epidermis cells (arrow), cross cells (arrowhead)).

Embryo: Type C, Embryo is placed at an angle of 165.7° (Fig. 3.35 A). Epiblast is absent.

Scutellum: V shape: Type V6 (Fig. 3.35 B), having an angle of 55.03°. Scutellum cleft/slit is present, and it is present at the side of the coleorhiza.

Oval-shaped, compactly arranged scutellum cells (Fig. 3.35 F), and depletion layer is also present (Fig. 3.35 F, arrowhead).

Vascularization: Type of vascularization: Panicoid (Fig. 3.35 D)

Angle of vascularization: 155.3°, Type C.

Pattern of embryonic leaf: P-type

Endosperm: Polygon-shaped endospermic cells.

Starch grain: Simple circular and polygon starch grains are present (Fig. 3.35 G).

No. of starch grains per endosperm cell: 58 ± 5; **Starch grain size:** 4.17 ± 0.83 (μm)

Aleurone layer: Tangentially elongated rectangular shaped flat aleurone cells (Fig. 3.35 C).

Caryopsis type and formula: True panicoids type and having formula "P-PP".

Bar size: A, B = 125 μm; D = 85.5 μm; C, E–G = 12.5 μm

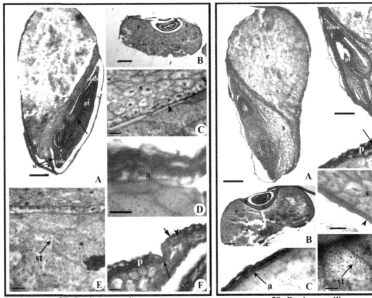

57. *Panicum maximum*   **58. *Panicum miliaceum***

59. *Panicum trypheron* (Fig. 3.36(59) A–F)

Shape: Oblong shaped with flat anterior side and concave posterior side in transaction, and having size of 0.68 (*L*) × 1.49 (*B*) mm^2 (Fig. 3.36 B).

Pericarp: Type B (Fig. 3.36 F; epidermis (arrowhead), nucellus (arrow)).

Embryo: Type B, Embryo is placed at an angle of 157.7° (Fig. 3.36 A). Epiblast is absent.

Scutellum: V shape: Type V2 (Fig. 3.36 B), having an angle of 52.1°. Scutellum cleft/slit is present, and it is covering the whole coleorhiza. Oval-shaped, compactly arranged scutellum cells (Fig. 3.36 E), and depletion layer is also present (Fig. 3.36 E, arrowhead).

Vascularization: Type of vascularization: Panicoid (Fig. 3.36 A, arrow)

Angle of vascularization: 131.1°, Type B. **Pattern of embryonic leaf:** P-type

Endosperm: Polygon-shaped endospermic cells.

Starch grain: Simple circular and polygon starch grains present (Fig. 3.36 C).

No. of starch grains per endosperm cell: 47 ± 10; **Starch grain size:** 2.78 ± 1.43 (μm)

Aleurone layer: Tangentially elongated rectangular and square aleurone cells (Fig. 3.36 D).

Caryopsis type and formula: True panicoids type and having formula "P-PP".

Bar size: A, B = 125 μm; C–F = 12.5 μm

DIFFERENTIATING FEATURES:

Embryo type B ... *P. trypheron*
Embryo type C
Angle of vascularization type C ... *P. miliaceum*
Angle of vascularization type B
Scutellum type V4-2 ... *P. antidotalae*
Scutellum type U2-3 ... *P. maximum*

Paspalidium

60. *Paspalidium flavidum* 61. *Paspalidium geminatum*

60. *Paspalidium flavidum* (Fig. 3.36(60) A−F)

Shape: Oval shape with convex anterior side and circular posterior side in transaction, and having size of 0.67 (*L*) × 1.62 (*B*) mm² (Fig. 3.36 B).

Pericarp: Type A (Fig. 3.36 F; cross cells (arrow), nucellus (arrowhead), subepidermis (line)).

Embryo: Type B, Embryo is placed at an angle of 147.2° (Fig. 3.36 A). Epiblast is absent.

Scutellum: U shape: Type U2-3 (Fig. 3.36 B), having an angle of 59.36°. Scutellum cleft/slit is present, and it is present at side of coleorhiza. Oval-shaped, compactly arranged scutellum cells (Fig. 3.36 D), and depletion layer is also present (Fig. 3.36 D, arrowhead).

Vascularization: Type of vascularization: Panicoid (Fig. 3.36 A)

Angle of vascularization: 156.4°, Type C. **Pattern of embryonic leaf:** P-type

Endosperm: Polygon-shaped endospermic cells.

Starch grain: Simple circular and polygon starch grains are present (Fig. 3.36 E).

No. of starch grains per endosperm cell: 71 ± 12; **Starch grain size:** 5.36 ± 2.37 (μm)

Aleurone layer: Tangentially elongated rectangular shaped aleurone cells (Fig. 3.36 C).

Caryopsis type and formula: True panicoids type and having formula "P-PP".

Bar size: A, B = 125 μm; C−F = 12.5 μm

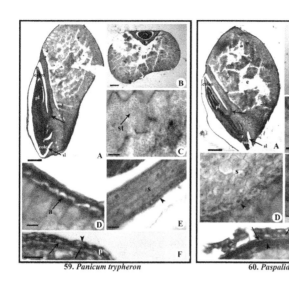

59. *Panicum trypheron*   60. *Paspalidium flavidum*

61. *Paspalidium geminatum* (Fig. 3.37(61) A–E)

Shape: Kidney shape in transaction, and having size of 0.44 (*L*) × 1.11 (*B*) mm² (Fig. 3.37 B).

Pericarp: Type A (Fig. 3.37 D, epidermis cells (arrowhead), cross cells (arrow)).

Embryo: Type C, Embryo is placed at an angle of 162.5° (Fig. 3.37 A). Epiblast is absent.

Scutellum: "V" shape: Type V1-2 (Fig. 3.37 B), having an angle of 69.34°. Scutellum cleft/slit is present. Oval-shaped scutellum cells (Fig. 3.37 C). Depletion layer is present (Fig. 3.37 C, arrowhead).

Vascularization: Type of vascularization: Panicoid

Angle of vascularization: 167.2°, Type C.

Pattern of embryonic leaf: P-type

Endosperm: Polygon-shaped endospermic cells.

Starch grain: Simple polygon-shaped starch grains (Fig. 3.37 E).

No. of starch grains per endosperm cell: 60 ± 9; **Starch grain size:** 3.75 ± 1.36 (μm)

Aleurone layer: Tangentially elongated rectangular shaped aleurone cells (Fig. 3.37 E, arrow).

Caryopsis type and formula: True panicoids type and having formula "P-PP".

Bar size: A, B = 125 µm; C–E = 12.5 µm

DIFFERENTIATING FEATURES:

Embryo type B, Scutellum type U2-3 … *P. flavidum*
Embryo type C, Scutellum type V1-2 … *P. geminatum*

62. *Paspalum scrobiculatum* (Fig. 3.37(62) A–F)

Shape: Oblong to oval shape with flat anterior side and circular posterior side in transaction, and having size of 1.08 (L) × 2.25 (B) mm^2 (Fig. 3.37 B).

Pericarp: Type D (Fig. 3.37 E; epidermis cells (arrowhead), nucellus (arrow)).

Embryo: Type B, Embryo is placed at an angle of 152.2° (Fig. 3.37 A). Epiblast is absent.

Scutellum: V shape: Type V2 (Fig. 3.37 B), having an angle of 48.18°. Scutellum cleft/slit is present, and it is present in the center of the proximal end. Oval-shaped, compactly arranged scutellum cells (Fig. 3.37 D), and depletion layer is also present (Fig. 3.37 D, arrowhead).

Vascularization: Type of vascularization: Panicoid

Angle of vascularization: 135.5°, Type B.

Pattern of embryonic leaf: P-type

Endosperm: Polygon-shaped endospermic cells.

Starch grain: Simple circular shaped starch grains are present (Fig. 3.37 F).

No. of starch grains per endosperm cell: 49 ± 7; **Starch grain size:** 5 ± 0.97 (µm)

Aleurone layer: Tangentially elongated rectangular shaped small aleurone cells (Fig. 3.37 C).

Caryopsis type and formula: True panicoids type and having formula "P-PP".

Bar size: A, B = 125 µm; C–F = 12.5 µm

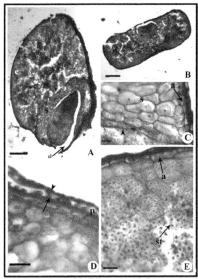

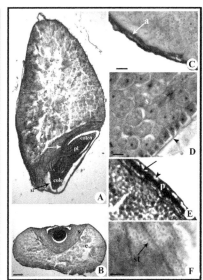

61. Paspalidium geminatum   *62. Paspalum scrobiculatum*

63. *Pennisetum setosum* (Fig. 3.38(63) A−F)

Shape: Oval shape with convex anterior side and flat posterior side in transaction, and having size of 0.52 (*L*) × 0.59 (*B*) mm² (Fig. 3.38 B).

Pericarp: Type C (Fig. 3.38 F, epidermis (arrowhead), cross cells (arrow)).

Embryo: Type C, Embryo placed at an angle of 164.8 ° (Fig. 3.38 A). Epiblast is absent.

Scutellum: V shape: Type V7 (Fig. 3.38 B), having an angle of 79.83°. Scutellum cleft/slit is present. Oval to circular shaped, compactly arranged scutellum cells (Fig. 3.38 D), and depletion layer is also present (Fig. 3.38 D, arrowhead).

Vascularization: Type of vascularization: Panicoid

Angle of vascularization: 160.2°, Type C.

Pattern of embryonic leaf: P-type

Endosperm: Polygon-shaped endospermic cells.

Starch grain: Simple circular and polygon starch grains present (Fig. 3.38 D).

No. of starch grains per endosperm cell: 16 ± 4; **Starch grain size:** 10 ± 4.26 (μm)

Aleurone layer: Tangentially elongated rectangular and square-shaped aleurone cells (Fig. 3.38 E).

Caryopsis type and formula: True panicoids type and having formula "P-PP".

Bar size: A, B = 125 μm; C–F = 12.5 μm

Setaria

64. *Setaria glauca* 65. *Setaria tomentosa* 66. *Setaria verticillata*

64. *Setaria gluca* (Fig. 3.38(64) A–E)

Shape: Oblong shape with convex anterior side and concave posterior side in transaction, and having size of 0.53 (*L*) × 1.08 (*B*) mm² (Fig. 3.38 B).

Pericarp: Type C.

Embryo: Type B, Embryo is placed at an angle of 155.7° (Fig. 3.38 A). Epiblast is absent.

Scutellum: V shape: Type V8-3 (Fig. 3.38 B), having an angle of 63.7°. Scutellum cleft/slit is present, and it is present at the side of the coleorhiza. Oval-shaped, compactly arranged scutellum cells (Fig. 3.38 D), and depletion layer is also present (Fig. 3.38 D, arrowhead).

Vascularization: Type of vascularization: Panicoid

Angle of vascularization: 144.1°, Type B.

Pattern of embryonic leaf: P-type

Endosperm: Polygon-shaped endospermic cells.

Starch grain: Simple circular and polygon starch grains present (Fig. 3.38 C).

No. of starch grains per endosperm cell: 73 ± 3; **Starch grain size:** 4.17 ± 1.00 (μm)

Aleurone layer: Tangentially elongated rectangular shaped aleurone cells (Fig. 3.38 E).

Caryopsis type and formula: True panicoids type and having formula "P-PP".

Bar size: A, B = 125 μm; C–F = 12.5 μm

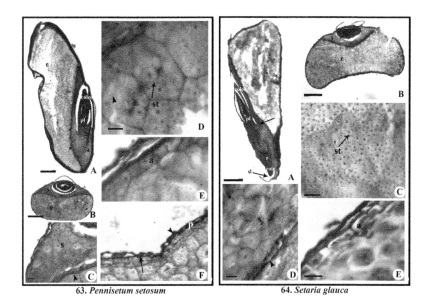

63. *Pennisetum setosum*   64. *Setaria glauca*

65. *Setaria tomentosa* (Fig. 3.39(65) A–F)

Shape: Kidney shape with convex anterior side and concave posterior side in transaction, and having size 0.91 (*L*) × 0.66 (*B*) mm² (Fig. 3.39 B).

Pericarp: Type D (Fig. 3.39 F; nucellus (arrowhead), epidermis (arrowhead)).

Embryo: Type C, Embryo is placed at an angle of 167.1° (Fig. 3.39 A). Epiblast is absent.

Scutellum: V shape: Type V3-4 (Fig. 3.39 B), having an angle of 54.37°. Scutellum cleft/slit is present, and it is present at the side of the coleorhiza. Oval-shaped, compactly arranged scutellum cells (Fig. 3.39 C), and depletion layer is also present (Fig. 3.39 C, arrowhead).

Vascularization: Type of vascularization: Panicoid (Fig. 3.39 A, arrow)

Angle of vascularization: 142.1°, Type B. **Pattern of embryonic leaf:** P-type

Endosperm: Polygon-shaped endospermic cells.

Starch grain: Simple circular and polygon starch grains present (Fig. 3.39 D).

No. of starch grains per endosperm cell: 58 ± 6; **Starch grain size:** 3.57 ± 0.89 (μm)

Aleurone layer: Tangentially elongated rectangular shaped aleurone cells (Fig. 3.39 E).

Caryopsis type and formula: True panicoids type and having formula "P-PP".

Bar size: A, B = 125 μm; C–F = 12.5 μm

66. *Setaria verticillata* (Fig. 3.39(66) A–F)

Shape: Rectangular shaped with convex anterior side and concave posterior side in transaction, and having size of 1.06 (*L*) × 1.24 (*B*) mm² (Fig. 3.39 B).

Pericarp: Type B (Fig. 3.39 F; epidermis cells (arrowhead), cross cells (arrow), nucellus (line)).

Embryo: Type B, Embryo is placed at an angle of 141.8 ° (Fig. 3.39 A). Epiblast is absent.

Scutellum: U shape: Type U2-3 (Fig. 3.39 B), having an angle of 66.38°. Scutellum cleft/slit is present, and it is covering the whole colerorhiza. Oval-shaped, compactly arranged scutellum cells (Fig. 3.39 D), and depletion layer is also present (Fig. 3.39 D, arrowhead).

Vascularization: Type of vascularization: Panicoid (Fig. 3.39 A, arrow)

Angle of vascularization: 135.8°, Type B. **Pattern of embryonic leaf:** P-type

Endosperm: Polygon-shaped endospermic cells.

Starch grain: Simple circular and polygon starch grains are present (Fig. 3.39 E).

No. of starch grains per endosperm cell: 74 ± 5; **Starch grain size:** 2.27 ± 1.09 (μm)

Aleurone layer: Tangentially elongated rectangular shaped aleurone cells (Fig. 3.39 C).

Caryopsis type and formula: True panicoids type and having formula "P-PP".

Bar size: A, B = 125 μm; C–F = 12.5 μm

DIFFERENTIATING FEATURES:

Embryo type C ... *S. tomentosa*
Embryo type B

Scutellum type V8-3 … *S. glauca*
Scutellum type U2-3 … *S. verticillata*

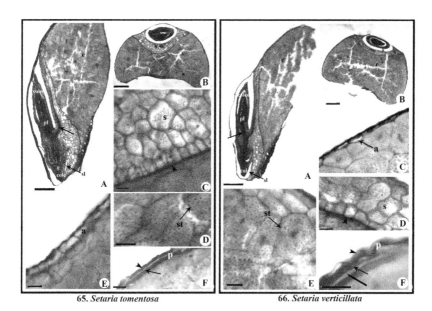

65. *Setaria tomentosa*   66. *Setaria verticillata*

67. *Isachne globosa* (Fig. 3.40(61) A–F)

Shape: Oval or circular shaped with convex anterior side and pointed posterior side in transaction, and having size of 0.68 (*L*) × 0.91 (*B*) mm^2 (Fig. 3.40 B).

Pericarp: Type C (Fig. 3.40 F; epidermis (arrowhead), hypodermis (arrow), nucellus (line)).

Embryo: Type A, Embryo is placed at an angle of 108.8° (Fig. 3.40 A). Epiblast is absent.

Scutellum: V shape: Type V7 (Fig. 3.40 B), having an angle of 71.63°. Scutellum cleft/slit is present, and it is covering the whole coleorhiza. Oval-shaped, compactly arranged scutellum cells (Fig. 3.40 D).

Vascularization: Type of vascularization: Panicoid (Fig. 3.40 A, arrow)

Angle of vascularization: 99.0°, Type A.

Pattern of embryonic leaf: P-type

Endosperm: Polygon-shaped endospermic cells.

Starch grain: Simple circular and polygon-shaped starch grains are present.

No. of starch grains per endosperm cell: 46 ± 11; **Starch grain size:** 1.97 ± 0.62 (µm)

Aleurone layer: Tangentially elongated rectangular and square-shaped mixed aleurone cells (Fig. 3.40 E).

Caryopsis type and formula: True panicoids type and having formula "P-PP".

Bar size: A, B = 125 µm; C–F = 12.5 µm

Aristida

68. *Aristida adscensionis*

69. *Aristida funiculata*

68. *Aristida adscensioinis* (Fig. 3.40(69) A–E)

Shape: Circular shaped in transaction with convex anterior and posterior side, and having size of 0.43 (L) × 0.35 (B) mm^2 (Fig. 3.40 B).

Pericarp: Type B (Fig. 3.40 D).

Embryo: Type B, Embryo is placed at an angle of 153.73° (Fig. 3.40 A). Epiblast is absent.

Scutellum: "Δ" shape: Type Δ4 (Fig. 3.40 B). Scutellum cleft/slit is present. Oval-shaped scutellum cells (Fig. 3.40 E). Depletion layer is present (Fig. 3.40 E, arrowhead).

Vascularization: Type of vascularization: Panicoid

Angle of vascularization: 149.98°, Type B.

Pattern of embryonic leaf: F-type

Endosperm: Polygon-shaped endospermic cells.

Starch grain: Simple circular and polygon-shaped starch grains (Fig. 3.40 C).

No. of starch grains per endosperm cell: 43 ± 6; **Starch grain size:** 2.63 ± 0.56 (µm)

Aleurone layer: Tangentially elongated rectangular shaped aleurone cells (Fig. 3.40 C, arrow).

Caryopsis type and formula: Arundionoia–Danthonioid type, having formula "P-PF".

Bar size: A, B = 125 µm; C–E = 12.5 µm

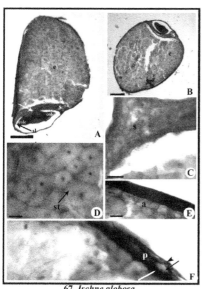

67. *Ischne globosa*

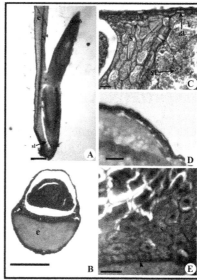

68. *Aristida adscensionis*

69. *Aristida funiculata* (Fig. 3.41(1) A–E)

Shape: Oblong shape in transaction with convex anterior and posterior side, and having size of 0.44 (*L*) × 0.42 (*B*) mm² (Fig. 3.41 B).

Pericarp: Type B (Fig. 3.41 C; epidermis (arrowhead), hypodermis (arrow)).

Embryo: Type B, Embryo is placed at an angle of 152.3° (Fig. 3.41 A). Epiblast is absent.

Scutellum: "Δ" shape: Type Δ4 (Fig. 3.41 B). Scutellum cleft/slit is present. Oval-shaped scutellum cells; Depletion layer is present.

Vascularization: Type of vascularization: Panicoid

Angle of vascularization: 148.38°, Type B.

Pattern of embryonic leaf: F-type

Endosperm: Polygon-shaped endospermic cells.

Starch grain: Simple circular and polygon-shaped starch grains (Fig. 3.41 D).

No. of starch grains per endosperm cell: 38 ± 5; **Starch grain size:** 3.41 ± 1.29 (μm)

Aleurone layer: Tangentially elongated rectangular shaped aleurone cells (Fig. 3.41 E).

Caryopsis type and formula: Arundionoia–Danthonioid type and having formula "P-PF".

Bar size: A, B = 125 μm; C–E = 12.5 μm

DIFFERENTIATING FEATURES:

Circular shaped in transection … *A. adscensionis*
Oblong shaped in transection … *A. funiculata*

70. *Perotis indica* (Fig. 3.41(70) A–F)

Shape: Circular to oblong shape in transaction with convex anterior and flat posterior side, and having size of 0.26 (*L*) × 0.28 (*B*) mm² (Fig. 3.41 C).

Pericarp: Type C (Fig. 3.41 E; epidermis (arrowhead), hypodermis (arrow)).

Embryo: Type C, Embryo is placed at an angle of 167.0° (Fig. 3.41 A). Type IV epiblast.

Scutellum: "Δ" shape: Type Δ3 (Fig. 3.41 C), having an angle of 41.9°. Scutellum cleft/slit is present. Oval-shaped scutellum cells (Fig. 3.41 F). Depletion layer is present (Fig. 3.41 F, arrowhead).

Vascularization: Type of vascularization: Panicoid

Angle of vascularization: 163.7°, Type C.

Pattern of embryonic leaf: F-type

Endosperm: Polygon-shaped endospermic cells.

Starch grain: Simple polygon-shaped starch grains.

No. of starch grains per endosperm cell: 31 ± 7; **Starch grain size:** 2.5 ± 1.32 (μm)

Aleurone layer: Tangentially elongated rectangular shaped alerurone cells (Fig. 3.41 D).

Caryopsis type and formula: Chloridoid–Eragrostoid type and having formula "P+PF".

Bar size: A–C = 125 μm; D–F = 12.5 μm

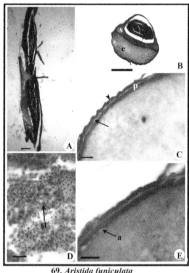

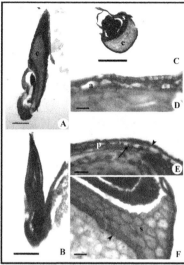

69. *Aristida funiculata*   **70.** *Perotis indica*

Chloris

71. *Chloris barbata* 72. *Choris montana* 73. *Chloris virgata*

71. *Chloris barbata* (Fig. 3.42(71) A−F)

Shape: Triangular shaped with convex anterior side and flat posterior side in transaction, and having size of 0.38 (*L*) × 0.46 (*B*) mm² (Fig. 3.42 B).

Pericarp: Type C (Fig. 3.42 F; epidermis (arrowhead), nucellus (arrow), cross cells (arrow)).

Embryo: Type C Embryo is placed at an angle of 176.0° (Fig. 3.42 A). Type II, Epiblast is present.

Scutellum: Δ shape: Type–Δ6 (Fig. 3.42 B), having an angle of 59.53°. Scutellum cleft is present, and it is at the side of the coleorhiza. Oval-shaped, compactly arranged scutellum cells (Fig. 3.42 D), depletion layer is also present (Fig. 3.42 D, arrowhead).

Vascularization: Type of vascularization: Panicoid

Angle of vascularization: 166.1°, Type C.

Pattern of embryonic leaf: F-type

Endosperm: Polygon-shaped endospermic cells.

Starch grain: Simple polygon-shaped starch grains present (Fig. 3.42 E).

No. of starch grains per endosperm cell: 33 ± 6; **Starch grain size:** 7.03 ± 3.14 (μm)

Aleurone layer: Tangentially elongated rectangular shaped aleurone cells (Fig. 3.42 C).

Caryopsis type and formula: Chloridoid–Eragrostoid type, having formula "P+PF".

Bar size: A, B = 125 μm; C–F = 12.5 μm

72. *Chloris Montana* (Fig. 3.42(72) A–F)

Shape: Oval shape with convex anterior side and concave posterior side in transaction, and having size of 0.49 (*L*) × 0.69 (*B*) mm² (Fig. 3.42 B).

Pericarp: Type C (Fig. 3.42 F; epidermis (arrowhead), nucellus (arrow), cross cells (line)).

Embryo: Type C, Embryo is placed at an angle of 170.7° (Fig. 3.42 A). Type II, Epiblast is present.

Scutellum: Δ shape: Type Δ3 (Fig. 3.42 B), having an angle of 35.64°. Scutellum cleft is present; it is present at the side of the coleorhiza. Oval-shaped, compactly arranged scutellum cells (Fig. 3.42 C); depletion layer is also present (Fig. 3.42 C, arrowhead).

Vascularization: Type of vascularization: Panicoid

Angle of vascularization: 152.5°, Type C.

Pattern of embryonic leaf: P-type

Endosperm: Polygon-shaped endospermic cells.

Starch grain: Simple circular and polygon starch grains present (Fig. 3.42 E).

No. of starch grains per endosperm cell: 49 ± 7; **Starch grain size:** 4.86 ± 2.04 (μm)

Aleurone layer: Tangentially elongated rectangular shaped aleurone cells (Fig. 3.42 D).

Caryopsis type and formula: Bambusoid type and having formula "P+PP".

Bar size: A, B = 125 μm; C–F = 12.5 μm

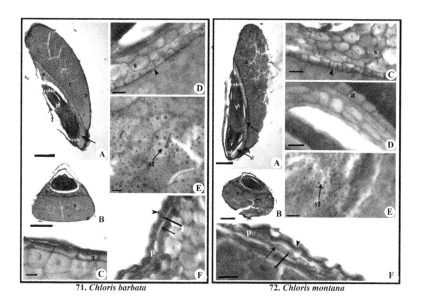

71. *Chloris barbata*   72. *Chloris montana*

73. *Chloris virgata* (Fig. 3.43(73) A−F)

Shape: Circular to oval shape with convex anterior side and circular posterior side in transaction, and having size of 0.62 (*L*) × 0.52 (*B*) mm² (Fig. 3.43 B).

Pericarp: Type B (Fig. 3.43 F; epidermis (arrowhead), nucellus (arrow), cross cells (arrow)).

Embryo: Type C, Embryo is placed at an angle of 172.3°(Fig. 3.43 A). Type III, Epiblast is present.

Scutellum: Δ shape: Type Δ2 (Fig. 3.43 B), having an angle of 48.92°. Scutellum slit is present, and it is present at the side of the coleorhiza. Oval-shaped, compactly arranged scutellum cells (Fig. 3.43 D); depletion layer is also present (Fig. 3.43 D, arrowhead).

Vascularization: Type of vascularization: Panicoid

Angle of vascularization: 154.7°, Type C. **Pattern of embryonic leaf:** P-type

Endosperm: Polygon-shaped endospermic cells.

Starch grain: Simple circular and polygon starch grains are present (Fig. 3.43 E).

No. of starch grains per endosperm cell: 63 ± 12; **Starch grain size:** 6.25 ± 1.83 (μm)

Aleurone layer: Tangentially elongated rectangular shaped aleurone cells (Fig. 3.43 C).

Caryopsis type and formula: Bambusoid type and having formula "P+PP".

Bar size: A, B = 125 µm; C–F = 12.5 µm

DIFFERENTIATING FEATURES:

Caryopsis type, Chloridoid–Eragrostoid … *C. barbata*
Caryopsis type, Bambusoid
Epiblast type II … *C. montana*
Epiblast type III … *C. virgata*

74. *Cynodon dactylon* (Fig. 3.43(74) A–E)

Shape: Oblong to oval shape with flat anterior side and circular posterior side in transaction, and having size of 0.58 (*L*) × 0.52 (*B*) mm² (Fig. 3.43 B).

Pericarp: Type A (Fig. 3.43 E; epidermis cells (arrow), crushed parenchyma (arrowhead), nucellus (line)).

Embryo: Type 3.43 B, Embryo is placed at an angle of 144.2° (Fig. 3.43 A). Type II, Epiblast is present.

Scutellum: Δ shape: Type Δ4 (Fig. 3.43 B). Scutellum cleft/slit is present, and it covers the whole coleorhiza. Oval-shaped, compactly arranged scutellum cells (Fig. 3.43 C), and depletion layer is also present (Fig. 3.43 C, arrowhead).

Vascularization: Type of vascularization: Panicoid

Angle of vascularization: 132.9°, Type B. **Pattern of embryonic leaf:** F-type

Endosperm: Polygon-shaped endospermic cells.

Starch grain: Simple circular and polygon starch grains present (Fig. 3.43 D).

No. of starch grains per endosperm cell: 58 ± 9; **Starch grain size:** 1.97 ± 0.22 (µm)

Aleurone layer: Tangentially elongated rectangular shaped aleurone cells (Fig. 3.43 E).

Caryopsis type and formula: Chloridoid–Eragrostoid type, having formula "P+PF".

Bar size: A, B = 125 µm; C–E = 12.5 µm

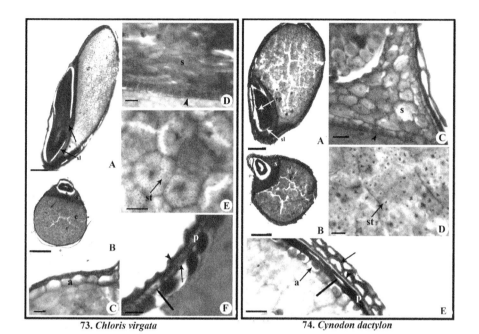

73. *Chloris virgata*   74. *Cynodon dactylon*

75. *Melanocenchris jaequemontii* (Fig. 3.44(75) A−F)

Shape: Oblong to oval shape in transaction with convex anterior side, and having size of 0.20 (*L*) × 0.32 (*B*) mm^2 (Fig. 3.44 B).

Pericarp: Type D (Fig. 3.44 F; epidermis (arrowhead), cross cells (arrow)).

Embryo: Type C, Embryo is placed at an angle of 165.4° (Fig. 3.44 A). Epiblast is absent.

Scutellum: "U" shape: Type U2-3 (Fig. 3.44 B), having an angle of 66.24°. Scutellum cleft/slit is present. Oval-shaped scutellum cells (Fig. 3.44 E); depletion layer is present (Fig. 3.44 E, arrowhead).

Vascularization: Type of vascularization: Panicoid

Angle of vascularization: 163.3°, Type – C.

Pattern of embryonic leaf: F-type

Endosperm: Polygon-shaped endospermic cells.

Starch grain: Some cells have simple circular shaped starch grains, and some cells have simple circular shaped and compound starch grains that are present together (Fig. 3.44 D).

No. of starch grains per endosperm cell: 58 ± 9; **Starch grain size:** 1.97 ± 0.22 (μm)

Aleurone layer: Tangentially elongated rectangular shaped aleurone cells (Fig. 3.44 C).

Caryopsis type and formula: Arundinoid–Danthonioid type and having formula "P-PF".

Bar size: A, B = 125 μm; C–F = 12.5 μm

76. *Oropetium villosulum* (Fig. 3.44(76) A–F)

Shape: Oblong to oval shape in transaction, and having size of 0.20 (*L*) × 0.18 (*B*) mm² (Fig. 3.44 A).

Pericarp: Type C (Fig. 3.44 E; epidermis (arrowhead), cross cells (arrow)).

Embryo: Type C, Embryo is placed at an angle of 175.5° (Fig. 3.44 A). Epiblast is absent.

Scutellum: "U" shape: Type U2-3 (Fig. 3.44 B), having an angle of 41.9°. Scutellum cleft/slit is present. Oval-shaped scutellum cells (Fig. 3.44 D); depletion layer is present (Fig. 3.44 D).

Vascularization: Type of vascularization: Panicoid

Angle of vascularization: 172.5°, Type D.

Pattern of embryonic leaf: F-type

Endosperm: Polygon-shaped endospermic cells.

Starch grain: Simple circular and polygon-shaped starch grains (Fig. 3.44 F).

No. of starch grains per endosperm cell: 51 ± 8; **Starch grain size:** 5.56 ± 1.37 (μm)

Aleurone layer: Tangentially elongated rectangular shaped aleurone cells (Fig. 3.44 C).

Caryopsis type and formula: Arundinoid–Danthonioid type, having formula "P-PF".

Bar size: A, B = 125 μm; C–F = 12.5 μm

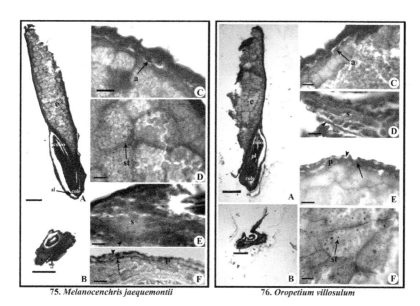

75. *Melanocenchris jaequemontii*   76. *Oropetium villosulum*

77. *Schoenefeldia gracilis* (Fig. 3.45(77) A–F)

Shape: Oval to circular shape with convex anterior side and circular posterior side in transaction, and having size of 0.68 (*L*) × 0.41 (*B*) mm² (Fig. 3.45 C).

Pericarp: Type A (Fig. 3.45 F; epidermis (arrowhead), nucellus (arrow), crushed parenchyma (line)).

Embryo: Type C, Embryo is placed at an angle of 174.7 ° (Fig. 3.45 A). Type IV, Epiblast is present (Fig. 3.45 B).

Scutellum: Δ shape: Type Δ3 (Fig. 3.45 C), having an angle of 42.61°. Scutellum cleft/slit is present; it is covering the whole coleorhiza. Oval-shaped, compactly arranged scutellum cells (Fig. 3.45 D); depletion layer is also present (Fig. 3.45 D, arrowhead).

Vascularization: Type of vascularization: Panicoid

Angle of vascularization: 170.6°, Type D.

Pattern of embryonic leaf: P-type

Endosperm: Polygon-shaped endospermic cells.

Starch grain: Compound polygon-shaped starch grains present (Fig. 3.45 E).

No. of starch grains per endosperm cell: 39 ± 8; **Starch grain size:** 5 ± 2.36 (μm)

Aleurone layer: Tangentially elongated rectangular shaped aleurone cells (Fig. 3.45 F).

Caryopsis type and formula: Bambusoid type and having formula "P+PP".

Bar size: AC = 125 μm; D–F = 12.5 μm

Tetrapogon

78. *Tetrapogon tenellus* 79. *Tetrapogon villosus*

78. *Tetrapogon tenellus* (Fig. 3.45(78) A–F)

Shape: Rectangular shape with convex anterior side and concave posterior side in transaction, and having size of 0.49 (*L*) × 0.47 (*B*) mm² (Fig. 3.45 C).

Pericarp: Type C

Embryo: Type C, Embryo is placed at an angle of 172.3° (Fig. 3.45 A). Type II, Epiblast is present (Fig. 3.45 B).

Scutellum: Δ shaped: Type – Δ3 (Fig. 3.45 C), having an angle of 32.26°. Scutellum cleft/slit is present; it is present at the side of the coleorhiza. Oval-shaped, compactly arranged scutellum cells (Fig. 3.45 D); depletion layer is also present (Fig. 3.45 D, arrowhead).

Vascularization: Type of vascularization: Panicoid

Angle of vascularization: 158.1°, Type C.

Pattern of embryonic leaf: F-type

Endosperm: Polygon-shaped endospermic cells.

Starch grain: Simple circular and polygon starch grains present (Fig. 3.45 E).

No. of starch grains per endosperm cell: 13 ± 5; **Starch grain size:** 5.29 ± 1.89 (μm)

Aleurone layer: Tangentially elongated rectangular shaped aleurone cells (Fig. 3.45 F).

Caryopsis type and formula: Chloridoid–Eragrostoid type, having formula "P+PF".

Bar size: A–C = 125 μm; D–F = 12.5 μm

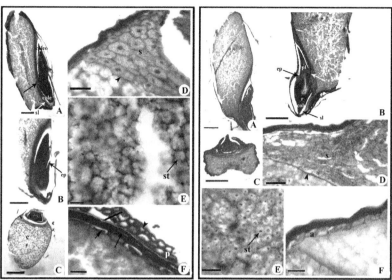

77. *Schoenefeldia gracilis*   78. *Tetrapogon tenellus*

79. *Tetrapogon villosus* (Fig. 3.46(79) A–F)

Shape: Rectangular shape with curved angles and having convex anterior side and concave posterior side in transaction, and having size of 0.44 (*L*) × 0.79 (*B*) mm² (Fig. 3.46 C).

Pericarp: Type C

Embryo: Type C, Embryo is placed at an angle of 169.4° (Fig. 3.46 A). Type III, Epiblast is present.

Scutellum: V shape: Type – V7 (Fig. 3.46 C), having an angle of 69.5°. Scutellum cleft/slit is present; it is covering the whole coleorhiza. Oval-shaped, compactly arranged scutellum cells (Fig. 3.46 D); depletion layer is also present (Fig. 3.46 D, arrowhead).

Vascularization: Type of vascularization: Panicoid (Fig. 3.46 B)

Angle of vascularization: 156.7°, Type C.

Pattern of embryonic leaf: F-type

Endosperm: Polygon-shaped endospermic cells.

Starch grain: Simple circular and polygon starch grains present (Fig. 3.46 F).

No. of starch grains per endosperm cell: 16 ± 4; **Starch grain size:** 7.03 ± 3.16 (μm)

Aleurone layer: Tangentially elongated rectangular shaped aleurone cells (Fig. 3.46 E).

Caryopsis type and formula: Chloridoid–Eragrostoid type, having formula "P+PF".

Bar size: AC = 125 μm; D–F = 12.5 μm

DIFFERENTIATING FEATURES:

Epiblast type II, Scutellum type Δ3 … *T. tenellus*
Epiblast type III, Scutellum type V7 … *T. villosus*

80. *Acrachne racemosa* (Fig. 3.46(80) A–G)

Shape: Kidney shape in transaction with flat anterior side and concave posterior side, and having size of 0.27 (*L*) × 0.51 (*B*) mm² (Fig. 3.46 D).

Pericarp: Type D (Fig. 3.46 G; epidermis (arrowhead), nucellus (arrow)).

Embryo: Type C, Embryo is placed at an angle of 175.1° (Fig. 3.46 A). Type V, Epiblast (3.46 Fig. B).

Scutellum: "U" shape: Type U2-1(Fig. 3.46 D), having an angle of 62.18°. Scutellum cleft/slit is present. Oval-shaped scutellum cells (Fig. 3.46 F). Depletion layer is present (Fig. 3.46 F, arrowhead).

Vascularization: Type of vascularization: Panicoid

Angle of vascularization: 151.0°, Type C.

Pattern of embryonic leaf: F-type

Endosperm: Polygon-shaped endospermic cells.

Starch grain: Simple circular and polygon starch grains present (Fig. 3.46 E).

No. of starch grains per endosperm cell: 17 ± 5; **Starch grain size:** 5.47 ± 1.53 (μm)

Aleurone layer: Tangentially elongated rectangular and square aleurone cells (Fig. 3.46 C).

Caryopsis type and formula: Chloridoid–Eragrostoid type, having formula "P+PF".

Bar size: A, B, D = 125 μm; C, E–G = 12.5 μm

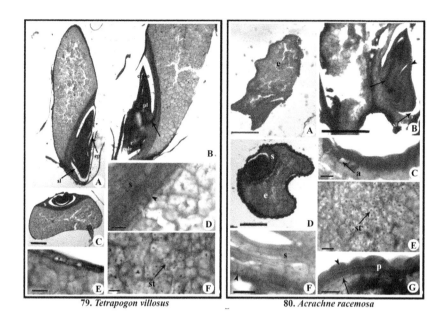

79. *Tetrapogon villosus*   80. *Acrachne racemosa*

Dactyloctenium

81. *Dactyloctenium aegyptium*

82. *Dactyloctenium scindicus*

81. *Dactyloctenium aegyptium* (Fig. 3.47(81) A−G)

Shape: Rectangular shape with curved margins and convex anterior side in transaction, and having size of 0.75 (L) × 0.58 (B) mm² (Fig. 3.47 B).

Pericarp: Type B (Fig. 3.47 G; epidermis (arrow), nucellus (arrowhead)).

Embryo: Type B, Embryo is placed at an angle of 141.3° (Fig. 3.47 A). Type V epiblast (Fig. 3.47 C, arrowhead).

Scutellum: "V" shape: Type V4-2 (Fig. 3.47 B), having an angle of 44.76°. Scutellum cleft/slit is present. Oval-shaped scutellum cells (Fig. 3.47 D); depletion layer is present (Fig. 3.47 D, arrowhead).

Vascularization: Type of vascularization: Panicoid (Fig. 3.47 B, arrow)

Angle of Vascularization: 123.6°, Type – A. **Pattern of embryonic leaf:** F-type

Endosperm: Polygon-shaped endospermic cells.

Starch grain: Some cells have simple circular shaped starch grains, and some cells have simple circular shaped and compound starch grains that are present together (Fig. 3.47 E).

No. of starch grains per endosperm cell: 53 ± 7; **Starch grain size:** 3.13 ± 1.05 (μm)

Aleurone layer: Tangentially elongated rectangular shaped aleurone cells (Fig. 3.47 F).

Caryopsis type and formula: Chloridoid–Eragrostoid type, having formula "P+PF".

Bar size: A, B = 125 μm; C–E = 12.5 μm

82. *Dactyloctenium scindicus*(Fig. 3.47(82) A–F)

Shape: Rectangular shape with convex anterior side and concave posterior side in transaction, and having size of 0.81 (*L*) × 0.58 (*B*) mm² (Fig. 3.47 C).

Pericarp: Type D (Fig. 3.47 E; nucellus (arrow))

Embryo: Type C, Embryo is placed an angle of 160.8° (Fig. 3.47 A). Type V, Epiblast is present (Fig. 3.47 B).

Scutellum: Δ shape: Type Δ5 (Fig. 3.47 C), having an angle of 55.78°. Scutellum slit is present; it is present at the side of the coleorhiza. Oval-shaped, compactly arranged scutellum cells (Fig. 3.47 D); depletion layer is also present (Fig. 3.47 D, arrowhead).

Vascularization: Type of vascularization: Panicoid

Angle of vascularization: 159.75°, Type C. **Pattern of embryonic leaf:** F-type

Endosperm: Polygon-shaped endospermic cells.

Starch grain: Simple circular and polygon-shaped (arrowhead) starch grains (Fig. 3.47 F).

No. of starch grains per endosperm cell: 52 ± 5; **Starch grain size:** 3.34 ± 0.97 (μm)

Aleurone layer: Tangentially elongated rectangular shaped aleurone cells.

Caryopsis type and formula: Chloridoid–Eragrostoid type, having formula "P+PF".

Bar size: A, C = 125 μm; B, D–F = 12.5 μm

DIFFERENTIATING FEATURES:

Angle of vascularization type A, embryo type B, Scutellum type- V4-2 … *D. aegyptium*
Angle of vascularization type C, embryo type B, Scutellum type Δ5 … *D. sindicus*

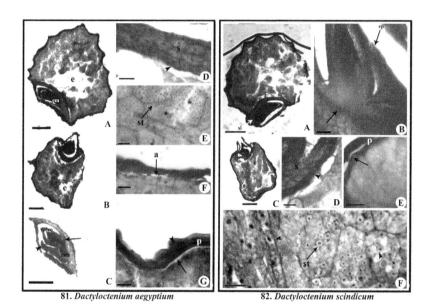

81. *Dactyloctenium aegyptium*   82. *Dactyloctenium scindicum*

83. *Desmostachya bipinnata* (Fig. 3.48(83) A−G)

Shape: Square shaped with curved margins in transaction, and having size of 0.37 (*L*) × 0.26 (*B*) mm (Fig. 3.48 D).

Pericarp: Type B (Fig. 3.48 G; nucellus (arrow), cross cells (arrowhead)).

Embryo: Type C, Embryo is placed at an angle of 174.8° (Fig. 3.48 A). Type II epiblast (Fig. 3.48 B, arrow).

Scutellum: "U" shape: Type U2-3 (Fig. 3.48 D), having an angle of 65.83°. Scutellum cleft/slit is present. Oval-shaped scutellum cells (Fig. 3.48 F). Depletion layer is present (Fig. 3.48 F, arrowhead).

Vascularization: Type of vascularization: Panicoid

Angle of vascularization: 168.5°, Type C.

Pattern of embryonic leaf: F-type

Endosperm: Polygon-shaped endospermic cells.

Starch grain: Simple polygon-shaped starch grains (Fig. 3.48 E).

No. of starch grains per endosperm cell: 15 ± 3; **Starch grain size:** 2.88 ± 1.13 (µm)

Aleurone layer: Vertically elongated rectangular shaped aleurone cells (Fig. 3.48 C).

Caryopsis type and formula: Chloridoid–Eragrostoid type, having formula "P+PF".

Bar size: A, D = 125 µm; B, C, E–G = 12.5 µm

84. *Dinebra retroflexa* (Fig. 3.48(84) A–F)

Shape: Oblong shape in transaction and having size 0.42 (*L*) × 0.47 (*B*) mm² (Fig. 3.48 B).

Pericarp: Type B (Fig. 3.48 F; cross cells (arrowhead), nucellus (arrow)).

Embryo: Type A, Embryo is placed an angle of 126.37° (Fig. 3.48 A). Type I epiblast (Fig. 3.48 A).

Scutellum: "V" shape: Type V8-2 (Fig. 3.48 B), having an angle of 66.81°. Scutellum cleft/slit is present. Oval-shaped scutellum cells (Fig. 3.48 C). Depletion layer is present (Fig. 3.48 C, arrowhead).

Vascularization: Type of vascularization: Panicoid (Fig. 3.48 A, arrow)

Angle of vascularization: 127.63°, Type A.

Pattern of embryonic leaf: F-type

Endosperm: Polygon-shaped endospermic cells.

Starch grain: Some cells have simple circular shaped starch grains, and some cells have simple circular shaped and compound starch grains that are present together (Fig. 3.48 D).

No. of starch grains per endosperm cell: 44 ± 9; **Starch grain size:** 4.69 ± 1.74 (µm)

Aleurone layer: Tangentially elongated aleurone cells (Fig. 3.48 E).

Caryopsis type and formula: Oryzoid–Olyroid type and having formula "F+PP".

Bar size: A, B = 125 µm; C–F = 12.5 µm

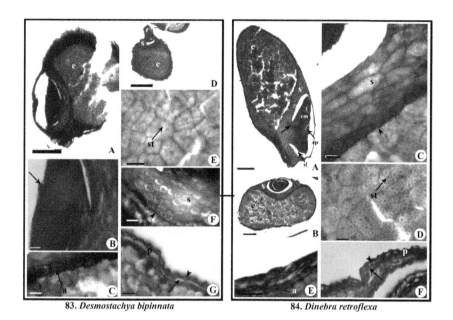

83. *Desmostachya bipinnata*   **84. *Dinebra retroflexa***

85. *Eleusine indica* (Fig. 3.49(85) A–F)

Shape: Kidney shape in transaction with concave posterior end, and having size of 0.46 (*L*) × 0.60 (*B*) mm² (Fig. 3.49 B).

Pericarp: Type D (Fig. 3.49 E; cross cells (arrow), crushed parenchyma (arrowhead)).

Embryo: Type B, Embryo is placed at an angle of 132.6° (Fig. 3.49 A). Type V epiblast (Fig. 3.49 C, arrow).

Scutellum: "U" shape: Type U2-1 (Fig. 3.49 B), having an angle of 55.36°. Scutellum cleft/slit is present. Oval-shaped scutellum cells. Depletion layer is present.

Vascularization: Type of vascularization: Panicoid

Angle of vascularization: 104.8°, Type A.

Pattern of embryonic leaf: F-type

Endosperm: Polygon-shaped endospermic cells.

Starch grain: Simple circular and polygon starch grains are present (Fig. 3.49 F).

No. of starch grains per endosperm cell: 32 ± 8; **Starch grain size:** 4.17 ± 2.04 (µm)

Aleurone layer: Tangentially elongated rectangular and square aleurone cells (Fig. 3.49 D).

Caryopsis type and formula: Chloridoid–Eragrostoid type, having formula "P+PF".

Bar size: A–C = 125 μm; D–F = 12.5 μm

Eragrostiella

86. *Eragrostiella bachyphylla*

87. *Eragrostiella bifaria*

86. *Eragrostiella bachyphylla* (Fig. 3.49(86) A−F)

Shape: Circular to oval shape in transaction, and having size of 0.32 (*L*) × 0.35 (*B*) mm² (Fig. 3.49 B).

Pericarp: Type D (Fig. 3.49 D; cross cells (arrowhead), nucellus (arrow)).

Embryo: Type B, Embryo is placed at an angle of 143.9° (Fig. 3.49 A). Type V epiblast (Fig. 3.49 A).

Scutellum: "U" shape: Type U2-3 (Fig. 3.49 B), having an angle of 58.74°. Scutellum cleft/slit is present. Oval-shaped scutellum cells (Fig. 3.49 C). Depletion layer is present (Fig. 3.49 C, arrowhead).

Vascularization: Type of vascularization: Panicoid

Angle of vascularization: 154.4°, Type C.

Pattern of embryonic leaf: F-type

Endosperm: Polygon-shaped endospermic cells.

Starch grain: Simple circular and polygon starch grains are present (Fig. 3.49 E).

No. of starch grains per endosperm cell: 45 ± 6; **Starch grain size:** 8.34 ± 3.77 (μm)

Aleurone layer: Tangentially elongated rectangular shaped aleurone cells (Fig. 3.49 F).

Caryopsis type and formula: Chloridoid–Eragrostoid type, having formula "P+PF".

Bar size: A, B = 125 μm; C–F = 12.5 μm

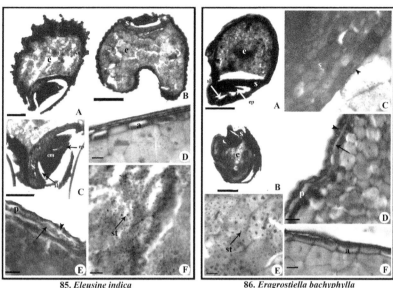

85. *Eleusine indica*   86. *Eragrostiella bachyphylla*

87. *Eragrostiella bifaria* (Fig. 3.50(87) A–F)

Shape: Rectangular to oval shape in transaction, and having size of 0.27 (*L*) × 0.18 (*B*) mm² (Fig. 3.50 B).

Pericarp: Type D (Fig. 3.50 F; cross cells (arrow), epidermis (arrowhead)).

Embryo: Type B, Embryo is placed at an angle of 146.1° (Fig. 3.50 A). Type V epiblast (Fig. 3.50 A).

Scutellum: "U" shape: Type U2-3 (Fig. 3.50 B), having an angle of 60.53°. Scutellum cleft/slit is present. Oval-shaped scutellum cells (Fig. 3.50 E). Depletion layer is present (Fig. 3.50 E, arrowhead).

Vascularization: Type of vascularization: Panicoid

Angle of vascularization: 152.6°, Type C.

Pattern of embryonic leaf: F-type

Endosperm: Polygon-shaped endospermic cells.

Starch grain: Simple circular and polygon shaped starch grains are present (Fig. 3.50 D).

No. of starch grains per endosperm cell: 21 ± 8; **Starch grain size:** 5 ± 2.48 (μm)

Aleurone layer: Tangentially elongated rectangular shaped aleurone cells (Fig. 3.50 C).

Caryopsis type and formula: Chloridoid–Eragrostoid type and having formula "P+PF".

Bar size: A, B = 125 μm; C–F = 12.5 μm

DIFFERENTIATING FEATURES:

Number of starch grains per endospermic cell are 45 ± 6 ... *E. bachyaphylla*
Number of starch grains per endospermic cell are 21 ± 8 ... *E. bifaria*

Eragrostis

88. *Eragrostis cilianensis*

89. *Eragrostis ciliaris*

90. *Eragrostis japonica*

91. *Eragrostis nutans*

92. *Eragrostis pilosa*

93. *Eragrostis tenella*

94. *Eragrostis tremula*

95. *Eragrostis unioloides*

96. *Eragrostis viscosa*

88. *Eragrostis cilianensis* (Fig. 3.50(88) A–F)

Shape: Oblong to oval shaped in transaction with convex anterior side and concave posterior side, and having size of 0.52 (*L*) × 0.53 (*B*) mm² (Fig. 3.50 B).

Pericarp: Type B (Fig. 3.50 E; epidermis cells (arrowhead), crushed parenchyma (arrow)).

Embryo: Type B, Embryo is placed at an angle of 146.2° (Fig. 3.50 A). Type V epiblast (Fig. 3.50 A).

Scutellum: "U" shape: Type U2-3 (Fig. 3.50 B), having an angle of 68.43°. Scutellum cleft/slit is present. Oval-shaped scutellum cells (Fig. 3.50 D). Depletion layer is present (Fig. 3.50 D, arrowhead).

Vascularization: Type of vascularization: Panicoid

Angle of vascularization: 155.2°, Type – C. **Pattern of embryonic leaf:** F-type

Endosperm: Polygon-shaped endospermic cells.

Starch grain: Simple polygon-shaped starch grains present (Fig. 3.50 C).

No. of starch grains per endosperm cell: 28 ± 9; **Starch grain size:** 2.94 ± 0.73 (μm)

Aleurone layer: Tangentially elongated rectangular shaped aleurone cells (Fig. 3.50 F).

Caryopsis type and formula: Chloridoid–Eragrostoid type, having formula "P+PF".

Bar size: A, B = 125 μm; C–F = 12.5 μm

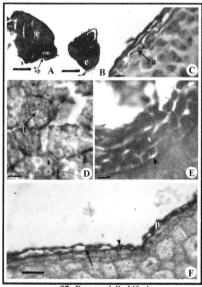

87. *Eragrostiella bifaria*

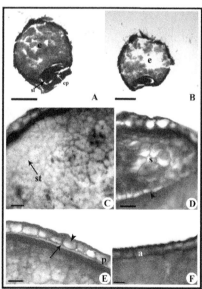

88. *Eragrostis cilianensis*

89. *Eragrostis ciliaris* (Fig. 3.51(89) A−F)

Shape: Circular to oval shape in transaction and having size of 0.18 (*L*) × 0.18 (*B*) mm² (Fig. 3.51 B).

Pericarp: Type B (Fig. 3.51 E; epidermis cells (arrowhead), crushed parenchyma (arrow)).

Embryo: Type B, Embryo is placed at an angle of 141.6° (Fig. 3.51 A). Type V epiblast (Fig. 3.51 A).

Scutellum: "U" shape: Type U2-3 (Fig. 3.51 B), having an angle of 61.9°. Scutellum cleft/slit is present. Oval-shaped scutellum cells (Fig. 3.51 F). Depletion layer is present (Fig. 3.51 F, arrowhead).

Vascularization: Type of vascularization: Panicoid

Angle of vascularization: 145.0°, Type B.

Pattern of embryonic leaf: F-type

Endosperm: Polygon-shaped endospermic cells.

Starch grain: Simple polygon-shaped starch grains present (Fig. 3.51 C).

No. of starch grains per endosperm cell: 23 ± 9; **Starch grain size:** 5.56 ± 1.63 (μm)

Aleurone layer: Tangentially elongated rectangular shaped aleurone cells (Fig. 3.51 D).

Caryopsis type and formula: Chloridoid–Eragrostoid type and having formula "P+PF".

Bar size: A, B = 125 μm; C–F = 12.5 μm

90. *Eragrostis japonica* (Fig. 3.51(90) A–G)

Shape: Oblong to oval shape in transaction, and having size of 0.24 (*L*) × 0.15 (*B*) mm² (Fig. 3.51 C).

Pericarp: Type B (Fig. 3.51 G; epidermis cells (arrowhead), nucellus (arrow)).

Embryo: Type B, Embryo is placed at an angle of 142.4° (Fig. 3.51 A). Type II epiblast.

Scutellum: "U" shape: Type U2-3 (Fig. 3.51 C), having an angle of 62.7°. Scutellum cleft/slit is present. Oval-shaped scutellum cells (Fig. 3.51 E). Depletion layer is present (Fig. 3.51 E, arrowhead).

Vascularization: Type of vascularization: Panicoid

Angle of vascularization: 148.8°, Type B.

Pattern of embryonic leaf: F-type

Endosperm: Polygon-shaped endospermic cells.

Starch grain: Simple polygon-shaped starch grains are present (Fig. 3.51 E, arrow).

No. of starch grains per endosperm cell: 34 ± 7; **Starch grain size:** 1.67 ± 0.08 (μm)

Aleurone layer: Tangentially elongated rectangular shaped aleurone cells (Fig. 3.51 F).

Caryopsis type and formula: Chloridoid–Eragrostoid type and having formula "P+PF".

Bar size: A, C = 125 μm; B, D–G = 12.5 μm

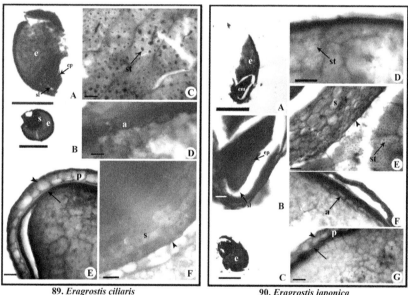

89. *Eragrostis ciliaris*   90. *Eragrostis japonica*

91. *Eragrostis nutans* (Fig. 3.52(91) A–G)

Shape: Oval to oblong shape in transaction, and having size of 0.32 (*L*) × 0.19 (*B*) mm² (Fig. 3.52 C).

Pericarp: Type B (Fig. 3.52 G; epidermis cells (arrowhead), nucellus (arrow)).

Embryo: Type B, Embryo is placed at an angle of 149.5° (Fig. 3.52 A). Type II epiblast (Fig. 3.52 B).

Scutellum: "U" shape: Type U2-3 (Fig. 3.52 C), having an angle of 57.26°. Scutellum cleft/slit is present. Oval-shaped scutellum cells (Fig. 3.52 F). Depletion layer is present (Fig. 3.52 F, arrowhead).

Vascularization: Type of vascularization: Panicoid

Angle of vascularization: 152.4°, Type C.

Pattern of embryonic leaf: F-type

Endosperm: Polygon-shaped endospermic cells.

Starch grain: Simple polygon-shaped starch grains are present (Fig. 3.52 D).

No. of starch grains per endosperm cell: 16 ± 5; **Starch grain size:** 3.12 ± 0.86 (μm)

Aleurone layer: Tangentially elongated rectangular shaped aleurone cells (Fig. 3.52 E).

Caryopsis type and formula: Chloridoid–Eragrostoid type and having formula "P+PF".

Bar size: A–C = 125 μm; D–G = 12.5 μm

92. *Eragrostis pilosa* (Fig. 3.52(92) A–E)

Shape: Rectangular shape in transaction, and having size of 0.26 (*L*) × 0.25 (*B*) mm² (Fig. 3.52 B).

Pericarp: Type B (Fig. 3.52 E; epidermis cells (arrowhead), nucellus (arrow))

Embryo: Type C, Embryo is placed at an angle of 161.6° (Fig. 3.52 A). Type II epiblast (Fig. 3.52 A).

Scutellum: "U" shape: Type U2-3 (Fig. 3.52 B), having an angle of 57.84°. Scutellum cleft/slit is present. Oval-shaped scutellum cells (Fig. 3.52 C). Depletion layer is present (Fig. 3.52 C, arrowhead).

Vascularization: Type of vascularization: Panicoid

Angle of vascularization: 157.9°, Type C.

Pattern of embryonic leaf: F-type

Endosperm: Polygon-shaped endospermic cells.

Starch grain: Simple polygon-shaped starch grains present (Fig. 3.52 E).

No. of starch grains per endosperm cell: 13 ± 6; **Starch grain size:** 4.54 ± 1.62 (μm)

Aleurone layer: Tangentially elongated rectangular shaped aleurone cells (Fig. 3.52 D).

Caryopsis type and formula: Chloridoid–Eragrostoid type, having formula "P+PF".

Bar size: A, B = 125 μm; C–E = 12.5 μm

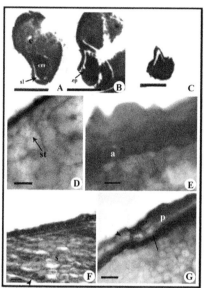
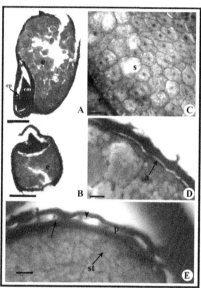

91. Eragrostis nutans   *92. Eragrostis pilosa*

93. *Eragrostis tenella* (Fig. 3.53(93) A–F)

Shape: Oblong shape in transaction, and having size of 0.48 (*L*) × 0.37 (*B*) mm^2 (Fig. 3.53 C).

Pericarp: Type B (Fig. 3.53 F; epidermis cells (arrowhead), nucellus (arrow)).

Embryo: Type C, Embryo is placed an angle of 161.1° (Fig. 3.53 A). Type II epiblast (Fig. 3.53 B).

Scutellum: "U" shape: Type U2-3 (Fig. 3.53 C), having an angle of 59.38°. Scutellum cleft/slit is present. Oval shape scutellum cells (Fig. 3.53 E). Depletion layer is present (Fig. 3.53 E, arrowhead).

Vascularization: Type of vascularization: Panicoid

Angle of vascularization: 150.3°, Type C.

Pattern of embryonic leaf: F-type

Endosperm: Polygon-shaped endospermic cells.

Starch grain: Simple polygon-shaped starch grains are present (Fig. 3.53 D).

No. of starch grains per endosperm cell: 37 ± 12; **Starch grain size:** 3.12 ± 0.71 (μm)

Aleurone layer: Tangentially elongated rectangular shaped aleurone cells (Fig. 3.53 D).

Caryopsis type and formula: Chloridoid–Eragrostoid type and having formula "P+PF".

Bar size: A–C = 125 μm; D–F = 12.5 μm

94. *Eragrostis tremula* (Fig. 3.53(94) A–H)

Shape: Oblong shape in transaction, and having size of $0.41(L) \times 0.35$ (B) mm² (Fig. 3.53 B).

Pericarp: Type B (Fig. 3.53 H; epidermis cells (arrowhead), nucellus (arrow)).

Embryo: Type B, Embryo is placed at an angle of 155.4° (Fig. 3.53 A). Type V epiblast (Fig. 3.53 A).

Scutellum: "U" shape: Type U2-3 (Fig. 3.53 B), having an angle of 66.25°. Scutellum cleft/slit is present. Oval-shaped scutellum cells (Fig. 3.53 G). Depletion layer is present (Fig. 3.53 G, arrowhead).

Vascularization: Type of vascularization: Panicoid

Angle of vascularization: 156.5°, Type C.

Pattern of embryonic leaf: F-type

Endosperm: Polygon-shaped endospermic cells.

Starch grain: Simple polygon-shaped starch grains present (Fig. 3.53 F).

No. of starch grains per endosperm cell: 25 ± 7; **Starch grain size:** 3.12 ± 1.76 (μm)

Aleurone layer: Tangentially elongated rectangular shaped aleurone cells (Fig. 3.53 E).

Caryopsis type and formula: Chloridoid–Eragrostoid type, having formula "P+PF".

It shows slimy glands on embryo portion. Figure 3.53 C and D shows slime cells.

Bar size: A, B = 125 μm; C–H = 12.5 μm

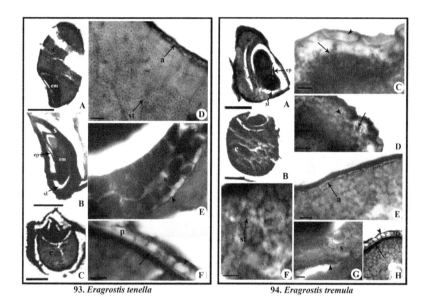

93. *Eragrostis tenella*   94. *Eragrostis tremula*

95. *Eragrostis unioloides* (Fig. 3.54(95) A−F)

Shape: Oblong shape in transaction, and having size of 0.48 (L) × 0.40 (B) mm^2 (Fig. 3.54 B).

Pericarp: Type B (Fig. 3.54 D; cross cells (arrow), epidermis (arrowhead)).

Embryo: Type B, Embryo is placed at an angle of 154.1° (Fig. 3.54 A). Type V epiblast (Fig. 3.54 A).

Scutellum: "U" shape: Type U2-3 (Fig. 3.54 B), having an angle of 66.4°. Scutellum cleft/slit is present. Oval-shaped scutellum cells (Fig. 3.54 C). Depletion layer is present.

Vascularization: Type of vascularization: Panicoid

Angle of vascularization: 147.9°, Type B.

Pattern of embryonic leaf: F-type

Endosperm: Polygon-shaped endospermic cells.

Starch grain: Simple polygon-shaped starch grains present (Fig. 3.54 E).

No. of starch grains per endosperm cell: 19 ± 7; **Starch grain size:** 2.68 ± 0.83 (μm)

Aleurone layer: Tangentially elongated rectangular shaped aleurone cells (Fig. 3.54 F).

Caryopsis type and formula: Chloridoid–Eragrostoid type and having formula "P+PF".

Bar size: A, B = 125 μm, C–F =12.5 μm

96. *Eragrostis viscosa* (Fig. 3.54(96) A–F)

Shape: Kidney shape in transaction and having size of 0.21 (*L*) × 0.25 (*B*) mm^2 (Fig. 3.54 B).

Pericarp: Type B (Fig. 3.54 D; epidermis cells (arrowhead), nucellus (arrow)).

Embryo: Type B, Embryo is placed at an angle of 151.28° (Fig. 3.54 A). Type V epiblast.

Scutellum: "U" shape: Type U2-3 (Fig. 3.54 B), having an angle of 59.28°. Scutellum cleft/slit is present. Oval-shaped scutellum cells (Fig. 3.54 F). Depletion layer is present.

Vascularization: Type of vascularization: Panicoid

Angle of vascularization: 145.63°, Type B.

Pattern of embryonic leaf: F-type

Endosperm: Polygon-shaped endospermic cells.

Starch grain: Simple polygon-shaped starch grains present (Fig. 3.54 E).

No. of starch grains per endosperm cell: 18 ± 3; **Starch grain size:** 1.89 ± 0.51 (μm)

Aleurone layer: Tangentially elongated rectangular shaped aleurone cells (Fig. 3.54 C).

Caryopsis type and formula: Chloridoid–Eragrostoid type and having formula "P+PF".

Bar size: A, B = 125 μm; C–F = 12.5 μm

DIFFERENTIATING FEATURES:

Square-shaped aleurone cells … *E. pilosa*

Tangentially elongated aleurone cells
Epiblast type II
Angle of vascularization type B … *E. japonica*
Angle of vascularization type C
Embryo type B … *E. nutans*
Embryo type C … *E. tenella*
Epiblast type V
Angel of vascularization type C
Absence of slimy glands on embryo … *E. cilianensis*
Presence of slimy glands on embryo … *E. tremula*
Angel of vascularization type B
Kidney shape in transection … *E. viscosa*
Circular to oval or oblong in transection
Circular to oval in transection with size of 0.18 (*L*) × 0.18 (*B*) mm^2 … *E. ciliaris*
Oblong in transection with size of 0.48 (*L*) × 0.40 (*B*) mm^2 … *E. unioloides*

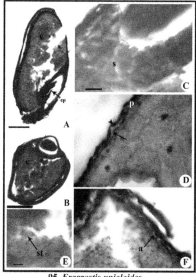

95. *Eragrostis unioloides*

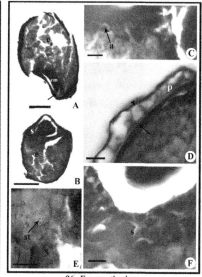

96. *Eragrostis viscosa*

Sporobolus

97. *Sporobolus coromandelianus*

98. *Sporobolus diander*

99. *Sporobolus indicus*

97. *Sporobolus coromandelianus* (Fig. 3.55(97) A−G)

Shape: Rectangular to oblong shape in transaction, and having size of 0.55 (*L*) × 0.35 (*B*) mm² (Fig. 3.55 C).

Pericarp: Type D (Fig. 3.55 D; epidermis cells (arrow), nucellus (arrowhead)).

Embryo: Type B, Embryo is placed at an angle of 147.36° (Fig. 3.55 A). Type V epiblast (Fig. 3.55 B).

Scutellum: "U" shape: Type U2-3 (Fig. 3.55 C), having an angle of 63.8°. Scutellum cleft/slit is present. Oval-shaped scutellum cells (Fig. 3.55 E). Depletion layer is present (Fig. 3.55 E, arrowhead).

Vascularization: Type of vascularization: Panicoid

Angle of vascularization: 138.36°, Type B.

Pattern of embryonic leaf: F-type

Endosperm: Polygon-shaped endospermic cells.

Starch grain: Simple polygon-shaped starch grains present (Fig. 3.55 F).

No. of starch grains per endosperm cell: 15 ± 6; **Starch grain size:** 4.69 ± 1.28 (μm)

Aleurone layer: Tangentially elongated rectangular shaped aleurone cells (Fig. 3.55 G).

Caryopsis type and formula: Chloridoid–Eragrostoid type and having formula "P+PF".

Bar size: AC = 125 μm; D−G = 12.5 μm

98. *Sporobolus diander* (Fig. 3.55(98) A−F)

Shape: Oval shape in transaction and having size of 0.58 (*L*) × 0.43 (*B*) mm² (Fig. 3.55 B).

Pericarp: Type C (Fig. 3.55 E; epidermis cells (arrowhead), nucellus (arrow)).

Embryo: Type B, Embryo is placed at an angle of 145.26° (Fig. 3.55 A). Type IV epiblast (Fig. 3.55 A).

Scutellum: "U" shape: Type U2-3 (Fig. 3.55 B), having an angle of 41.9°. Scutellum cleft/slit is present. Oval-shaped scutellum cells (Fig. 3.55 D). Depletion layer is present (Fig. 3.55 D, arrowhead).

Vascularization: Type of vascularization: Panicoid

Angle of vascularization: 131.77°, Type B.

Pattern of embryonic leaf: F-type

Endosperm: Polygon-shaped endospermic cells.

Starch grain: Simple polygon-shaped starch grains present (Fig. 3.55 F).

No. of starch grains per endosperm cell: 20 ± 4; **Starch grain size:** 2.59 ± 0.17 (μm)

Caryopsis type and formula: Chloridoid–Eragrostoid type, having formula "P+PF".

Bar size: A, B = 125 μm; C–F = 12.5 μm

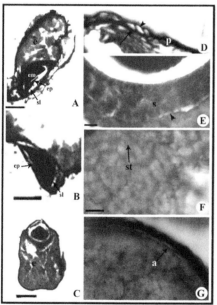

97. *Sporobolus coromandelianus*

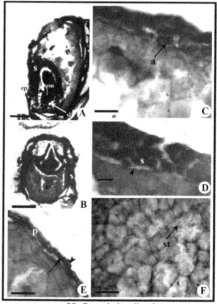

98. *Sporobolus diander*

99. *Sporobolus indicus* (Fig. 3.56(99) A–F)

Shape: Round shaped in transaction, and having size of 0.31 $(L) \times 0.25$ (B) mm^2 (Fig. 3.56 B).

Pericarp: Type C (Fig. 3.56 C; epidermis cells (arrow), nucellus (arrowhead)).

Embryo: Type B, Embryo is placed at an angle of 142.25° (Fig. 3.56 A). Type V epiblast (Fig. 3.56 A).

Scutellum: "U" shape: Type U2-3 (Fig. 3.56 B), having an angle of 60.23°. Scutellum cleft/slit is present. Oval-shaped scutellum cells (Fig. 3.56 E). Depletion layer is present (Fig. 3.56 E, arrowhead).

Vascularization: Type of vascularization: Panicoid

Angle of vascularization: 135.61°, Type B.

Pattern of embryonic leaf: F-type

Endosperm: Polygon-shaped endospermic cells.

Starch grain: Simple polygon-shaped starch grains present (Fig. 3.56 F).

No. of starch grains per endosperm cell: 16 ± 4; **Starch grain size:** 2.5 ± 0.93 (μm)

Aleurone layer: Tangentially elongated rectangular shaped aleurone cells (Fig. 3.56 D).

Caryopsis type and formula: Chloridoid–Eragrostoid type and having formula "P+PF".

Bar size: A, B = 125 μm; C–F = 12.5 μm

DIFFERENTIATING FEATURES:

Square-shaped aleurone cells … *S. coromardelianus*
Tangentially elongated aleurone cells
Round shape in transection with size 0.31 (*L*) × 0.25 (*B*) mm² … *S. indicus*
Oval shape in transection with size 0.58 (*L*) × 0.43 (*B*) mm² … *S. diander*

100. *Tragrus biflorus* (Fig. 3.56(100) A−F)

Shape: Oval shape in transaction, and having size of 0.31 (*L*) × 0.42 (*B*) mm² (Fig. 3.56 B).

Pericarp: Type D (Fig. 3.56 C; epidermis cells (arrow), nucellus (arrowhead))

Embryo: Type C, Embryo is placed at an angle of 163.7° (Fig. 3.56 A). Type III epiblast.

Scutellum: "U" shape: Type U2-3 (Fig. 3.56 B), having an angle of 62.41°. Scutellum cleft/slit is present. Oval-shaped scutellum cells (Fig. 3.56 E). Depletion layer is present (Fig. 3.56 E, arrowhead).

Vascularization: Type of vascularization: Panicoid

Angle of vascularization: 162.9°, Type C.

Pattern of embryonic leaf: F-type

Endosperm: Polygon-shaped endospermic cells.

Starch grain: Simple polygon-shaped starch grains present (Fig. 3.56 F).

No. of starch grains per endosperm cell: 26 ± 6; **Starch grain size:** 2.27 ± 1.03 (µm)

Aleurone layer: Tangentially elongated rectangular shaped aleurone cells (Fig. 3.56 D).

Caryopsis type and formula: Chloridoid–Eragrostoid type and having formula "P+PF".

Bar size: A, B = 125 µm; C–F = 12.5 µm

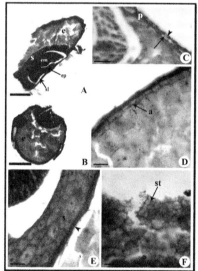

99. *Sporobolus indicus*

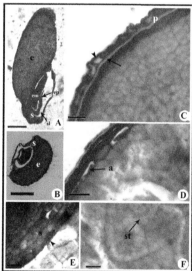

100. *Tragrus biflorus*

3.3 DISCUSSION AND IDENTIFICATION KEY

Anatomically, all the (100 species) grass caryopses studies showed variations in the histoarchitecture on the basis of which a distinct identification could be done. Based on the diagnostic characteristic features, a key to the identification of all the 100 species has been prepared.

Anatomical features of caryopsis in a few grass species have been documented by different authors. Grass species included in the present study have been compared with these species and discussed here.

TABLE 3.1 Quantitative and Qualitative Features of Caryopsis (Anatomy).

Sr No	Name of plant	A	B	C	D		E		F	G
					D1	D2	E1	E2		
Tribe: Maydeae										
1.	Chionachne koenigii	1.39 × 3.34± 0.27	2.32± 0.7	2.86 × 1.98± 1.4	24.76±2.54	459 ± 4.65	75.24±2.78	842± 8.54	165.6±1.05	172.8±4.09
2	Coix lachryma-jobi	2.57 × 4.38± 0.96	1.41± 0.8	2.1 × 1.4± 0.6	26.65±3.67	415±.45	73.37 ±5.84	943±16.3	177.9±0.94	163.9±1.87
Tribe: Andropogoneae										
3	Andropogon pumilus	0.27 × 0.45±0.07	3.74±1.28	4.68 × 9.32±1.68	40.51±1.75	69.18± 4.87	59.49±4.16	121.3±7.45	167.5±2.86	167.7±3.18
4	Apluda mutica	0.54 × 0.39± 0.14	6.25±2.54	7.42 × 9.23± 3.12	32.83±3.18	478.34±9.64	67.16±3.14	345.57±11.76	141.3±1.18	154.5±3.05
5	Arthraxon lanceolatus	0.38 × 0.31±0.12	3.96±0.42	10.7 × 10.2±2.75	48.29±2.96	352.69±7.53	51.70±2.19	327.89±5.46	168.5±3.02	170.5±4.51
6	Bothriochloa pertusa	0.35 × 0.67± 0.09	3.24±1.89	13.6 × 11.7±4.78	55.89±4.18	302.65±13.76	44.09±3.87	404.23±11.78	156.5±3.18	161.2±2.94
7	Capillipedium hugelii	0.42×0.70±0.19	5.29±1.38	6.36×13.79±2.88	57.94±2.67	501.38±16.37	42.05±2.49	319.36±10.67	167.0±2.15	154.5±3.87
8	Chrysopogon fulvus	1.14×0.59±0.73	3.29±0.8	6.43×1.38±2.67	40.17±3.09	490.26±13.68	59.83±5.12	438.98±7.42	149.9±5.08	173.1±2.16
9	Cymbopogon martini	0.36×0.69±0.18	3.74±1.07	6.78×11.86±3.78	37.01±2.06	528.78±16.34	62.98±3.32	486.78±17.78	164.2±1.94	172.6±2.19
10	Dichanthium annulatum	0.47×0.71±0.21	6.47±2.35	3.56×8.45±1.98	37.46±3.18	418.78±7.45	62.51±5.16	308.67±12.80	177.5±4.97	159.4±2.19
11	Dichanthium caricosum	0.47×0.93±0.18	3.79±1.06	8.45×1.59±3.80	42.22±2.93	629.56±14.65	58.44±1.73	447.62±7.24	169.9±1.16	161.42±3.7
12	Hackelochloa granularis	0.19×0.72±0.13	2.86±1.77	3.13×5.93±1.37	59.82±4.53	485.92±3.48	40.18±3.88	286.2±3.84	152.5±2.29	158.6±2.86

TABLE 3.1 (Continued)

Sr No	Name of plant	A	B	C	D		E		F	G
					D1	D2	E1	E2		
13	*Heteropogon contortus var. genuinus sub var. typicus*	0.54×0.60±0.18	3.07±1.74	3.58×4.69±1.73	43.07±6.08	563.43±14.78	56.80±3.06	432.78±17.93	177.9±4.03	178.8±3.64
14	*Heteropogon contortus var. genuinus sub var. hispidissimus*	0.75×0.83±0.24	3.25±1.36	3.72×9.79±0.83	44.02±4.04	528.76±9.38	55.98±4.05	442.87±13.64	178.4±4.69	175.7±4.74
15	*Heteropogon ritchiei*	0.52×0.51±0.27	3.59±0.73	4.12×7.24±2.84	42.16±4.18	223.56±7.23	63.68±8.45	290.67±11.64	177.8±3.16	176.9±2.68
16	*Heteropogon triticeus*	1.25×1.23±0.65	1.2±0.96	7.76×1.28±4.26	45.13±2.59	1778.57±6.83	57.19±4.13	1284.6±21.54	146.4±4.87	169.9±3.57
17	*Imperata cylindrical*	0.19×0.24±0.04	2.99±1.04	2.49×5.22±1.74	41.73±7.28	80.57±14.83	58.27±6.32	80.57±10.37	152.3±3.28	163.7±3.81
18	*Ischaemum indicus*	0.41×0.71±0.23	5.73±1.64	9.23×1.24±3.76	28.41±4.53	328.67±6.34	71.59±3.18	338.34±11.46	167.3±1.17	171.2±2.74
19	*Ischaemum molle*	0.67×0.76±0.14	6.96±0.94	13.56×16.56±5.12	40.5±3.19	468.32±24.37	59.50±6.13	374.78±18.34	170.34±1.04	173.4±3.47
20	*Ischaemum pilosum*	0.91×1.03±0.37	3.79±1.87	6.23×14.21±2.73	40.32±3.75	932.47±7.23	59.68±4.84	451.87±20.45	144.8±3.19	172.1±4.74
21	*Ischaemum rugosum*	0.91×1.34±0.26	5.49±2.17	5.72×16.23±1.58	35.48±4.03	462.67±13.76	64.52±3.68	640.63±8.39	173.8±3.09	174.5±5.12
22	*Iseilema laxum*	0.38×0.79±0.26	5.03±1.39	7.42×19.42±2.16	34.06±2.13	514.27±21.68	65.94±4.63	329.08±17.56	166.4±2.09	171.8±3.83
23	*Ophiuros exaltatus*	0.69×0.83±0.36	6.27±2.96	9.41×13.73±1.76	37.39±3.89	693.76±18.54	62.61±2.67	421.67±15.97	167.3±3.08	168.7±1.02
24	*Rottboellia exaltata*	2.00×2.35±0.73	7.48±3.72	13.75×12.51±3.54	21.57±4.16	1883.8±27.76	78.43±6.07	1220.3±41.36	165.3±1.16	173.3±3.45

TABLE 3.1   (Continued)

Sr No	Name of plant	A	B	C	D		E		F	G
					D1	D2	E1	E2		
25	Saccharum spontanum	0.54 × 0.67± 0.13	1.27±0.35	2.87 × 3.42± 1.27	58.23±2.38	482.67±2.78	41.76±2.53	281.67±13.27	169.8±1.93	171.4±2.67
26	Sehima ischaemoides	0.30×0.97±0.27	4.72±1.38	4.24×8.23±2.49	32.85±2.14	672.46±11.36	67.15±6.74	521.78±8.37	157.1±3.05	169.9±5.42
27	Sehima nervosum	0.47×0.97±0.33	4.97±2.06	11.2×12.5±4.18	31.97±1.82	462.28±7.18	68.03±3.69	526.8±1.92	177.1±2.05	174.2±4.18
28	Sehima sulcatum	1.87×1.69±0.27	7.71±1.08	7.22×11.92±2.43	39.49±5.72	376.3±5.03	60.51±2.19	413.9±1.93	171.3±1.68	177.8±3.61
29	Sorghum halepense	1.87×1.69±0.92	6.47±1.53	9.46×18.16±2.17	30.57±4.13	633.14±4.57	69.43±6.19	427.20±5.13	157.5±3.67	170.1±4.36
30	Thelepogn elegans	1.19×1.53±0.04	11.2±2.32	9.46×18.16±2.17	31.79±3.21	1160.38±7.82	68.21±7.18	787.71±8.32	145.2±1.03	167.7±3.14
31	Themeda cymbaria	0.63 × 0.38±0.16	3.82±1.03	9.23 × 14.57±2.87	44.51±5.41	443.23±7.34	55.49±2.74	390.85±6.23	163.3±1.32	171.4±2.93
32	Themeda laxa	0.36×0.64±0.11	4.73±0.87	5.2×10.45±1.87	34.24±8.34	672.0±15.14	65.76±12.18	500.0±11.72	168.7±4.32	173.2±5.26
33	Themeda triandra	0.65×0.73±0.27	3.73±1.83	5.72×16.43±3.83	41.58±2.73	546.31±7.82	58.42±3.16	441.29±6.73	164.4±3.68	172.7±2.83
34	Themeda quadrivalvis	0.85 × 0.98± 0.08	4.35±1.68	9.56×13.54± 0.21	47.36±4.67	877.3±7.63	52.64±8.15	607.43±5.19	168.5±1.64	173.1±2.43
35	Triplopogon ramosissimus	0.48×0.95±0.23	6.23±2.67	3.48×4.98±2.16	40.45±5.72	451.61±7.36	59.55±6.78	462.28±6.15	166.5±4.13	172.1±3.26
36	Vetivaria zizanioides	0.96×0.60±0.08	5.22±1.84	4.27×6.28±1.94	37.84±5.12	338.67±7.38	62.16±3.14	278.09±10.71	167.90±2.18	171.1±4.73
Tribe: Paniceae										
37	Alloteropsis cimicina	0.34×3.45±7.13	6.47±3.89	4.73×12.19±1.72	45.58±2.73	348.59±8.71	54.35±4.73	387.04±8.19	141.7±3.51	166.7±2.19
38	Brachiaria distachya	0.52 × 1.50± 0.13	3.16±0.8	4.92 × 5.17± 1.09	22.79±5.72	335.8±11.21	77.21±8.13	440.76±7.81	154.0±3.34	166.8±1.14
39	Brachiaria eruciformis	0.36×0.67±0.21	2.90±1.07	2.71× 3.03± 1.98	38.43±5.62	326.67±12.04	61.57±8.92	171.34±7.42	168.8±1.34	170.3±1.21

TABLE 3.1 (Continued)

Sr No	Name of plant	A	B	C	D		E		F	G
					D1	D2	E1	E2		
40	Brachiaria ramosa	0.62 × 1.35± 0.08	1.18± 0.6	2.69×4.23± 0.85	23.75±9.18	897.71±23.63	76.25±8.23	467.61±9.12	152.38±1.19	160.3±3.92
41	Brachiaria reptans	0.43×1.42±0.17	5.48±2.15	7.47×11.91±1.94	26.72±3.19	473.90±13.16	73.28±6.19	462.28±11.10	156.2±4.62	173.2±3.12
42	Cenchrus biflorus	0.82×1.34±0.52	3.22±0.73	3.48×9.95±0.97	38.43±5.62	415.23±5.61	71.54±6.13	322.48±4.74	164.1±3.17	172.7±3.94
43	Cenchrus ciliaris	0.52×0.64±0.27	3.24±1.07	3.73×11.21±1.43	43.55±3.17	245.90±3.91	56.45±2.98	471.81±7.18	167.8±3.98	172.9±1.71
44	Cenchrus setigerus	0.83×1.39±0.32	3.73±1.26	5.22×10.77±2.19	27.85±2.18	709.52±10.34	72.15±7.11	556.38±17.83	146.3±2.10	166.2±3.43
45	Digitaria ciliaris	0.62×0.98±0.43	6.51±2.71	4.28×11.48±1.86	47.26±2.80	516.75±14.69	52.74±3.17	256.34±6.32	145.45±2.62	157.76±1.41
46	Digitaria gramularis	0.49×0.64±0.23	1.99±0.73	3.74×10.71±1.43	51.9±2.93	520.19±7.83	48.15±3.63	153.14±6.27	142.5±1.34	148.2±3.19
47	Digitaria longiflora	0.70×1.13±0.07	8.64±2.83	5.63×15.21±2.14	49.73±4.32	505.56±10.12	50.27±3.83	392.16±11.03	146.1±5.62	163.4±2.17
48	Digitaria stricta	0.47×0.78±0.18	4.89±2.07	3.74×13.68±1.72	53.76±7.14	435.04±7.83	46.24±5.20	342.47±10.12	143.4±2.16	162.1±3.12
49	Echinochloa colonum	0.88×1.42±0.34	6.22±1.09	9.21×14.43±3.46	29.51±1.71	725.71±6.82	70.49±3.46	639.61±7.42	131.3±3.10	163.8±1.31
50	Echinochloa crusgalli	0.93×1.67±0.51	4.25±0.13	8.96×23.39±4.36	23.85±0.81	548.0±4.38	76.15±2.77	447.24±4.62	143.5±1.16	163.4±3.23
51	Echinochloa stagnina	1.25×2.33±0.46	6.79±2.03	14.45×20.5±5.21	21.65±3.16	874.85±6.34	78.35±3.84	971.61±7.82	143.6±3.71	163.2±4.83
52	Eremopogon foveolatus	0.5 × 1.54± 0.48	1.78±0.76	3.71 × 4.21± 1.04	68.42±5.63	521.52±8.12	31.58±2.17	344.78±12.63	155.0±1.15	167.9±2.36
53	Eriochloa procera	0.64 × 1.83± 0.07	4.34±2.34	3.89×3.09± 1.68	42.76±3.55	897.71±9.24	57.24±4.83	467.7±14.13	154.1±2.13	160.2±4.27
54	Oplismenus burmannii	0.39×0.55±0.17	5.22±3.45	4.73×10.21±2.89	54.62±8.72	737.71±18.61	45.38±3.77	302.47±7.81	150.2±1.46	154.4±1.38

TABLE 3.1 *(Continued)*

Sr No	Name of plant	A	B	C	D1	D2	E1	E2	F	G
55	Oplismenus compositus	0.52×0.94±0.07	4.98±2.63	4.24×13.43±1.93	42.44±2.64	396.70±6.45	57.56±6.72	398.86±8.25	154.7±2.41	164.9±3.43
56	Panicum antidotale	0.58×0.89±2.74	4.97±1.54	5.2×8.95±0.82	39.43±5.87	487.43±9.87	60.57±4.62	423.05±16.73	132.2±4.36	164.4±2.27
57	Panicum maximum	0.57×1.07±0.37	2.73±1.68	2.72×7.71±0.93	40.43±2.77	542.09±10.22	59.57±5.84	391.04±7.87	157.5±2.65	174.7±2.93
58	Panicum miliaceum	1.07×1.82±0.86	4.24±2.08	4.48×13.91±1.67	24.54±1.58	989.14±11.69	75.46±3.46	553.71±8.10	155.3±4.63	165.7±2.93
59	Panicum trypheron	0.68×1.49±0.05	3.74±1.07	6.47×9.95±3.34	23.85±2.34	779.42±19.36	76.15±7.66	446.09±7.63	131.1±3.73	157.7±1.9
60	Paspalidium flavidum	0.67 × 1.62 ± 0.18	4.67±1.95	4.34 ×3.32± 1.8	22.89±0.70	927.24±8.31	77.11±3.39	437.93±9.13	156.4±3.75	147.2±4.62
61	Paspalidium geminatum	0.44×1.107±0.28	6.47±2.24	5.98×11.31±2.06	35.32±3.68	601.14±15.02	64.68±7.08	358.85±11.58	167.2±1.75	162.5±1.83
62	Paspalum scrobiculatum	1.08×2.25±0.62	3.32±1.37	2.66×10.29±1.05	43.43±2.46	1333.14±15.9	56.57±3.39	532.19±5.23	135.5±2.12	152.2±3.16
63	Pennisetum setosum	0.52×0.59±0.09	2.98±1.85	3.48×7.96±2.15	55.18±6.93	519.81±5.89	44.81±3.77	173.34±3.77	160.02±3.39	164.8±3.73
64	Setaria glauca	0.53×1.08±0.78	1.49±0.65	3.73×9.21±1.65	22.68±4.72	560.38±16.78	77.32±8.42	237.90±14.83	144.1±4.56	155.7±5.69
65	Setaria tomentosa	0.91×0.66±0.21	2.33±0.56	3.32×12.28±2.69	29.98±3.18	473.14±20.56	70.02±6.22	220.38±8.34	142.1±2.45	167.1±2.23
66	Setaria verticillata	1.06×1.24±0.75	2.85±0.34	2.99×13.61±1.89	24.72±4.56	612.76±13.65	75.28±2.75	333.34±15.78	135.8±1.43	141.8±2.74
	Group: Pooideae									
	Tribe: Isachneae									
67	Isachne globosa	0.68×0.91±0.48	5.47±2.78	6.47×11.21±3.82	68.35±4.57	628.95±17.63	31.65±2.85	177.34±4.65	99.0±1.72	108.8±2.45
	Tribe: Aristideae									
68	Aristida adscensionis	0.43×0.35±0.21	3.48±1.67	3.11×7.72±1.76	75.19±8.62	345.09±6.27	24.81±5.18	142.76±7.85	149.98±2.92	153.73±3.2

TABLE 3.1 (Continued)

Sr No	Name of plant	A	B	C	D		E		F	G
					D1	D2	E1	E2		
69	Aristida funiculata	0.44×0.42±0.17	4.98±1.63	3.48×6.96±2.59	58.47±6.30	296.73±46.21	41.53±8.75	139.21±8.45	148.38±3.73	152.3±4.48
Tribe: Perotideae										
70	Perotis indica	0.26×0.28±0.17	4.25±2.76	3.24×5.98±1.39	64.17±7.63	193.52±9.36	35.83±5.18	157.34±6.38	163.7±3.37	167.0±3.20
Tribe: Chlorideae										
71	Chloris barbata	0.38×0.46±012	2.98±0.83	3.48×7.53±2.97	38.37±2.51	298.28±10.42	61.63±4.78	315.81±16.32	166.1±4.29	176.0±3.93
72	Chloris montana	0.49×0.69±0.08	7.71±2.41	2.24×8.22±1.77	16.63±2.65	439.42±12.76	82.37±6.38	286.28±3.56	152.5±2.987	170.7±2.65
73	Chloris virgata	0.62×0.52±0.28	3.98±0.38	5.98×12.45±2.17	23.53±1.58	451.62±9.37	76.47±5.83	241.90±7.29	154.7±5.27	172.3±4.27
74	Cynadon dactylon	0.58×0.52±0.16	7.96±1.86	4.72×9.21±2.76	51.08±8.74	249.71±5.27	48.92±3.04	314.48±7.57	132.9±3.76	144.2±2.59
75	Melanocenchris jaequemontii	0.20×0.32±0.11	2.98±0.74	2.94×6.82±1.39	46.24±2.67	298.28±8.96	53.76±4.78	278.09±4.26	163.3±3.56	165.4±1.28
76	Oropetium villosulum	0.20×0.18±0.05	2.99±0.84	3.74×6.96±0.65	69.47±6.26	88.76±4.92	30.53±2.18	92.76±8.23	172.5±3.19	175.5±2.94
77	Sachoenefeldia gracilis	0.68×0.41±0.31	4.48±2.17	3.73×5.98±1.78	47.26±2.46	632.57±10.43	52.74±3.83	241.71±5.28	170.6±1.68	174.7±2.29
78	Tetrapogon tenellus	0.49×0.47±0.11	5.98±1.75	5.74×8.96±2.75	38.63±2.63	836.38±21.76	61.37±5.72	360.19±6.46	158.1±4.18	172.3±2.87
79	Tetrapogon villosus	0.44×0.79±0.21	2.49±0.54	3.98×10.95±1.65	44.68±6.29	221.52±8.23	55.32±3.18	314.47±56.28	156.7±3.28	169.4±1.37
Tribe: Eragrosteae										
80	Acrachne Racemosa	0.27×0.51±0.07	3.74±1.72	4.24×8.46±1.68	41.74±2.36	423.43±24.82	58.26±3.27	177.34±6.29	151.0±1.26	175.1±4.79
81	Dactyloctenium aegyptium	0.75×0.58±0.41	3.99±0.84	5.22×10.51±1.94	51.22±4.62	351.58±11.39	48.78±2.62	189.34±4.89	123.6±2.92	141.3±3.17
82	Dactyloctenium sindicus	0.81×0.58±0.38	3.97±1.59	4.24×9.21±2.26	61.48±8.78	384.71±9.41	38.52±1.63	237.71±7.13	159.75±3.09	160.8±2.19

TABLE 3.1 *(Continued)*

Sr No	Name of plant	A	B	C	D1	D2	E1	E2	F	G
83	*Desmostachya bipinnata*	0.37×0.26±0.16	3.24±1.62	4.73×6.22±1.64	54.28±2.83	276.38±6.37	45.72±6.28	205.71±4.72	168.5±4.89	174.8±5.83
84	*Dinebra retroflexa*	0.42×0.47±0.15	3.14±0.93	3.78×8.27±2.61	57.13±6.18	427.73±14.48	48.29±2.73	316.35±18.63	127.63±3.18	126.37±5.37
85	*Eleusine indica*	0.46×0.60±0.21	6.96±2.71	5.72×9.96±2.87	64.28±5.92	665.34±18.58	35.72±2.88	201.52±5.92	104.8±1.37	132.6±3.42
86	*Eragrostiella bachyphylla*	0.32×0.35±0.18	1.74±0.06	2.64×6.62±1.53	59.37±8.92	383.05±6.63	40.63±2.93	128.95±4.82	154.5±0.28	143.9±2.83
87	*Eragrostiella bifaria*	0.27×0.18±0.09	3.73±1.02	4.98×7.72±2.71	63.53±4.61	256.35±4.85	36.47±1.84	144.96±7.37	152.6±3.72	146.1±1.43
88	*Eragrostis ciliaensis*	0.52×0.53±0.28	3.72±1.78	2.49×0.70±1.05	66.73±3.27	427.05±8.65	33.27±4.83	116.95±7.28	155.2±5.75	146.2±4.39
89	*Eragrostis ciliaris*	0.18×0.18±0.05	3.24±1.37	2.49×3.99±1.68	62.74±4.56	120.95±4.83	37.26±2.46	84.57±9.02	145.0±2.34	141.6±2.32
90	*Eragrostis japonica*	0.24×0.15±0.04	4.21±1.94	3.74×5.98±1.03	65.73±4.80	261.86±5.98	34.27±4.68	108.7±5.90	148.8±2.96	142.4±2.67
91	*Eragrostis nutans*	0.32×0.19±0.17	3.23±0.86	3.48×4.97±0.94	59.74±3.74	214.78±7.72	40.26±5.37	112.76±7.53	152.4±4.89	149.5±1.63
92	*Eragrostis pilosa*	0.26×0.25±0.07	5.47±2.81	3.23×6.47±1.28	36.28±2.85	544.38±12.36	63.72±7.34	173.34±6.53	157.9±4.69	161.6±3.04
93	*Eragrostis tenella*	0.48×0.37±0.15	3.47±1.02	2.15×5.82±1.46	48.71±3.48	355.85±7.37	51.29±3.60	157.34±3.457	150.31±3.64	161.1±4.55
94	*Eragrostis tremula*	0.41×0.35±0.41	4.24±1.67	1.99×4.97±0.51	43.18±2.41	394.85±6.29	56.82±2.93	177.34±6.18	156.54±6.28	155.4±7.43
95	*Eragrostis unioloides*	0.48×0.40±0.27	2.98±0.85	1.76×5.22±0.38	53.73±3.87	254.09±5.83	46.27±2.45	133.14±5.94	147.9±4.73	154.1±5.81
96	*Eragrostis viscosa*	0.21×0.25±0.07	2.64±0.36	1.61×6.02±0.83	33.58±1.43	272.46±9.32	66.42±4.27	142.74±4.38	145.63±3.72	151.28±4.61
	Tribe: Sporoboleae									
97	*Sporobolus coromandelianus*	0.55×0.35±0.38	3.11±1.78	2.51×5.97±1.20	52.27±8.75	238.36±9.63	47.73±3.87	137.83±7.84	138.36±4.87	147.36±2.29
98	*Sporobolus diander*	0.58×0.43±0.28	2.96±1.04	2.11×5.38±1.83	48.16±6.73	193.26±17.54	51.84±3.46	115.72±7.41	131.77±1.36	145.26±4.42

TABLE 3.1 *(Continued)*

Sr No	Name of plant	A	B	C	D			E		F	G
					D1	D2	E1	E2			
99	*Sporobolus indicus*	0.31×0.25±0.16	3.03±1.37	2.31×4.28±1.26	50.72±2.68	216.27±11.72	49.28±276	128.27±7.59		135.61±5.28	142.25±6.41
	Tribe: Zoysieae										
100	*Tragrus biflorus*	0.31×0.42±0.17	7.21±2.72	4.97×8.96±1.73	51.74±3.86	443.43±7.81	48.26±1.83	233.90±4.28		162.9±3.84	163.7±5.17

A: cross section size (μM), B: seed coat thickness (μM), C: aleurone cell size (μM), D1: occupied % of endosperm, D2: thickness of endosperm, E1: occupied % of embryo, E2: thickness of embryo (μM), F: angle of vascularization (in °), G: angle of embryo in caryopsis (in °).

Anatomically, grass caryopsis showed three genetically distinct compartments: the Pericarp, the embryo, and a prominent and persistent endosperm for which grass seeds have been domesticated (Hands et al., 2012). Irving (1983) worked on anatomy and histochemistry of *Echinochloa turnerana* (Channel millet). He described the major typical structural characteristics of Gramineae, such as aleurone cells, transfer cells, type of embryo, scutellum, presence of protein and starch grains etc. In the present study, three different species of *Echinochloa* was studied, that is, *Echinochloa colona, Echinochloa crushgalli,* and *Echinochloa stagnina.* All three species showed Type C embryo, "V"-shaped scutellum, simple circular and polygon-shaped starch grains, and Type B pericarp, that is, epidermis, hypodermis, cross cells, and nucellus layers of pericarp.

Narayanswami (1954) worked on the structure and development of the caryopsis in *Paspalum scorbiculatum.* Mature caryopsis consists of the suberized nucellar epidermis forming the perisperm, pericarp, and insertion region of the inner integument. Pericarp consists of epidermis, hypodermis, and nucellus. In the present study, the pericarp consists of epidermis, crushed parenchymatous cells, and nucellus. Narayanswami (1956) described the caryopsis in *Setaria italica* with an absence of epiblast, single-layered aleurone, and cellular endosperm. These same features were observed in the *Setaria gluca, Setaria tomentosa,* and *Setaria verticillata* of the present study. Rost (1973) described the anatomy of the caryopsis coat in mature caryopsis of the yellow foxtail grass (*Setaria lutescens*). He concluded that several layers of cells from parent plant subcircular the developing caryopsis; nucellus, integuments, and pericarp. Pericarp showed the different layers: epidermis, crushed parenchyma, cross cells, and nucellus. Pericarp in *Setaria* species also showed the same anatomical features. Eichemberg and Scatena (2013) studied morphology and anatomy of the diaspores and seedling of *Paspalum* (Poaceae, Poales). They described caryopses in species *Paspalum dilatatum, Paspalum mandiocanum, Paspalum pumilum,* and *Paspalum urvillei* to be albuminous, with a starchy endosperm and one aleurone layer in the peripheral region. Embryos in the four species have a slit scutellum, and epiblast is absent. vascularization of the embryos is of the panicoid type. These features have been noted in the *Paspalum scorbiculatum* included in the present study.

Osman et al. (2012) studied anatomy of 31 different grass species belonging to different tribes and concluded that the unique anatomical structure of the very marked caryopsis, by its general construction or even by each of its components singly, can serve significantly for identifying a plant group. *Aristida adscensionis* included in the present study had 6.03-µm thick seed coat, transversely rectangular shaped aleurone cells and starchy endosperm, as also recorded by Osman et al. (2012), but seed coat thickness was 3.48 µm. Cross-section of *Aristida funiculata* was circled linear with 2-fold pointed ends, but in the present study, cross-section appeared oblong shape with a convex anterior and posterior side. *Dactyloctenium aegyptium* showed characters like thin, rectangular shaped aleurone cells, 8.50-µm seed coat thickness and polygonal in outline, but in the present study it was rectangular shaped with curved margins and convex anterior side in cross-section and tangentially elongated rectangular shaped aleurone cells with seed coat of 3.99-µm thickness. According to Osman, *Eragrostis cilianensis* showed characters like circular outline, rectangular shaped aleurone cells, but in the present study the cross-section was oblong to oval shape and tangentially elongated rectangular shaped aleurone cells. Similarly, *Cenchrus ciliaris* showed anatomical characters like oval shape in outline, not completely compressed pericarp, but the present study showed that cross-section was oblong to oval shape and pericarp showed epidermis, cross cells and nucellus. *Echinochloa colona* showed characters like triangular in cross-section, horizontally rectangular shaped aleurone cells and well-compressed pericarp layers, and in the present study cross-section was oblong shaped with convex anterior and concave posterior side, tangentially elongated rectangular shaped aleurone cells and well-developed pericarp.

Seed coat anatomy is species specific and can serve as an aid in identification and taxonomy (Carlquist et al., 1997). Seed coat of Gramineae plants under investigation appeared with a uniform structure: the whole seed coat thickness showed a great variation among taxa at a specific level and even between genera. Many light microscopy studies have been conducted both to determine the development of the coat and to examine its mature structure. Some representative studies include: Harrington and Crocker (1923) were the first to describe the caryopsis coat; caryopsis coat consists of pericarp, seed coat, and nucellus (Rost et al., 1984). Morrison and Dashnicky (1982) worked on the structure of the covering layers of the *Avena fatua* caryopsis, and he told that seed coat envelopes the entire caryopsis. This seed coat continues through the entire length of the grain. Bechtel and Pomereanz (1977) studied rice caryopsis coat. They reported that surcirculating the rice caryopsis was the pericarp, seed coat, and nucellus layers. Interior to the pericarp was a single crushed layer of cells, the seed coat, and with a thick cuticle located to the inner side of crushed cells.

The anatomy of the seed coat in mature caryopses is remarkably similar in the studied grass species. The coat is continuous circular the entire caryopsis except at the point where it is connected to the axis of the inflorescence. The exterior of the caryopsis is covered by a thick cuticle layer. The pericarp derived from the ovary of the flower, may be a thin membrane, or may be composed of one or several layers. In the studied 100 species, four different types of pericarp were observed. Few members of group Panicoideae showed "A" type of Pericarp (*Heteropogon triticeus, Ischaemum molle, Sehima spp., Themeda spp., Erichloa procera, Paspalidium spp.* etc.), while from group Pooideae *Cynadon datylon, Sachoenefeldia gracilis* showed "A" type of pericarp. Other studied species from different types of pericarp i.e. B/C/D. Both members of tribe Maydeae showed "B" type of pericarp.

Caryopsis showed two major portions: embryo and endosperm. Embryos of grasses are considered to be unique, because they are highly complex, possessing a leaf-like coleoptile, a coleorhiza, up to six leaf primordial and a characteristic prominent outgrowth termed a scutellum that has not been recorded in other families and has been hypothesized to represent a modified cotyledon (Sargant et al., 1905; Reeder, 1953; Shah and Sreekumari, 1980; Negbi, 1984; Rudall et al., 2005). Variation in embryo structure is related to the presence or absence of mesocotyl, the epiblast, the scutellum cleft, that is, whether or not scutellum is fused to the coleorhiza and whether the first leaf is rolled or followed and allows an evolutionary sequence of embryo types to be recognized (Renvoize, 2002; David, 2009). Epiblast was a peculiar organ, which was situated opposite to scutellum. It occurred with much greater frequency than is generally supposed. Few workers studied grass caryopsis (Warming and polter, 1895) but they did not mention epiblast. Scutellum that showed vascular system but epiblast did not showed vascular system. Might be epiblast occur many instances of reduced organs without fibrovascular systems. It was very small structure in comparison with the scutellum, but there are many grasses in which it reaches considerable sizes.

It has been observed that with an increase in the percentage of embryo the percentage of endosperm decreases and vice versa, that is, *Aristida adscensionis* showed lowest embryo percentage (24.81%) and highest endosperm percentage (75.19%), while *Chloris Montana* showed highest embryo percentage (82.37%) and lowest endosperm percentage (16.63%). Within the caryopsis, embryos of the different species were positioned at different angles with respect to the anterioposterior axis of the caryopsis. On the basis of this, embryo was categorized into three different types: Type A (less than 130° angle), Type B (angle between 130° and 160°), and Type C (angle between 160° and 190°). Among the 100 species, *Isachne globosa* and *Dinebra retroflexa* showed "A" type of embryo, while the 28 species were showed "B" type of embryo and 70 species were showed "C" type of embryo.

Among the studied species, few species showed absence of epiblast and few species showed presence of epiblast. In the present study, epiblast has been classified in five different categories. Characteristic features of epiblast varied tribe wise that facilitated in characterizing different groups and tribes. Group Panicoideae members showed the absence of epiblast, except *Chionachne koenigii* that showed Type I epiblast. From group Pooideae, members of tribe Ischaneae and Aristideae showed the absence of epiblast. Apart from this, *Melanocenchris jaquemontii* and *Oropetium villosum* from tribe Chlorideae showed the absence of epiblast. Other species of group Pooideae showed the presence of epiblast. Tribe Sporoboleae members (like *Sporobolus coromardelianus*, *Sporobolus diander*, and *Sporobolus indicus*) showed Type II epiblast. Members of Eragrostideae tribe showed two types of epiblast: Type II and IV. Species like *Acrachne racemosa, Dactyloctenium, Eleusine indica, Eragrostiella, Eragrostis cilianensis, Eragrostis ciliaris, Eragrostis tremula, Eragrostis unioloides, Eragrostis viscosa,* etc., had Type V epiblast, while species like *Eragrostis japonica, Eragrostis nutans, Eragrostis pilosa, Eragrostis tenella, Dinebra retrofelxa,* and *Desmostachya bipinnata* had Type II epiblast. Tribe Chlorideae contains eight species, out of which *Tetrapogon villosus* and *Chloris virgata* had Type III epiblast, *Sachoenefeldia gracilis* and *Perotis indica* had Type IV epiblast, while *Tetrapogon tenellus, Chloris barbata, Chloris montana,* and *Cynadon dactylon* had Type II epiblast.

Negbi (1984) described vascular system of the scutellum. Genus like *Zea* showed the maing connecting the embryonal axis with the scutellum travels from the mesocotyl almost to the scutellar tip where manu of their turn downward like a weeping willow. The vascular system in the scutellum of *Hordeum vaulgare* has a double main bundle, but wild variety showed single main bundle. In many grasses, there were also bundles that supply the posterior scutellar appendices. Four types of angle of vascularization were present among the studied species. Only three species showed "A" type (i.e. angle was less than 130°) of angle of vascularization, viz. *Isachne globosa* (99.0°), *Dactyloctenium aegyptium* (123.6°), and *Eleusine indica* (104.8°). Tribe Sporoboleae includes three species and all three species had "B" type (angle between 130° and 150°) of angle of vascularization. Tribe Eragrostideae showed "A", "B", and "C" type of vascularization. Species like *Eragrostis ciliaris* (145.0°), *Eragrostis japonica* (148.8°), *Eragrostis unioloides* (147.9°), and *Eragrostis viscosa* (145.6°) showed "B" type of angle of vascularization. *Eleusine indica* and *Dactyloctenium aegyptium* showed "A" type of angle of vascularization. Tribe Chlorideae members showed "B"/ "C"/ "D" type of angle of vascularization. Group Panicoideae members showed "B"/ "C"/ "D" type of angle of vascularization. "A" type of vascularization was not observed in any species of Panicoideae group.

Scutellum is the single cotyledon of embryo in the caryopsis. It lies in contact with the endosperm (Kar et al., 2002). In the caryopsis scutellum, there is no starch grains (swift and O'Brien, 1972). Negbi (1984) described four types of scutellum on the basis of the development of scutellum during caryopsis maturation: (i) scutellum Sensu Stricto: In which scutellum maintains the shield-like structure during germination and early seedling growth. This type of scutellum is found in major grass species like *Oropetium, Coix, Sorghum, Triticum, Oryza, Zea, Lolium, Oryzopsis* etc. (ii) The *Avena* kind of scutellum: It has a shield-like shape only in the mature grain. During germination and early seedling development, the tip of the scutellum elongates inside the bulk of the storage endosperm. (iii) The *Zizania* kind of scutellum: This kind of scutellum is that of *Zizania aqatica*. It is also found in *Spartina*. During grain development, the scutellum reaches the distal end of the embryo sac, and sometimes even turns backward the very long tip of this scutellum that is linear and is appressed to the aleurone layer with its adaxial surface, while its abaxial surface is appressed to the hard endosperm. The length of diffusion pathways in the caryopsis of *Zizania* must be similar to that of the *Avena* kind and hence shorter than in the scutellum Sensu Stricto. (iv) The *Melocanna* kind of scutellum: This kind of scutellum is quite dissimilar in structure and function from that of the other three kinds. It was found in genera of bamboos like *Ochlandra* and *Dinochloa*. Caryopses of these species were relatively large and had a thick pericarp. Most of seed cavity is taken up by large scutellum, which is supplied with vascular system. Majorly three types of scutellum shapes were observed in studied species. Group Panicoideae members majorly showed either "U"- or "V"-shaped scutellum; "Δ"-shaped scutellum was observed only in five species of group Panicoideae members and species like *Imperata cylindrica, Ischaemum pilosum, Saccharum spontanum, Sehima nervosum,* and *Vetivaria zinzanoides*. Tribe Maydeae species, *Chionachne koenigii* and *Coix lachryma jobi*, showed "U"-shaped scutellum. Group Pooideae members showed all three types of scutellum. Tribe Chlorideae members majorly showed "Δ"-shaped scutellum, and species were *Chloris barbata, Chloris montana, Chloris virgata, Cynodon dactylon*, and *Schoenefeldia gracilis*, and *Tetrapogon tenellus* had "Δ"-shaped scutellum. *Melanocenchris jaquemontii* and *Oropetium villosulum* had "U"-shaped scutellum. Tribe Aristideae members, like *Aristida adscendionis* and *Aristida funiculate*, showed "Δ"-shaped scutellum.

Next to the nucellar region the aleurone layer is present, which is outermost endosperm tissue. Aleurone cells are thick walled and rectangular in outline, containing closely packed protein grains (Bechtel and Pomeranz, 1977). Aleurone layer is not continuous acircular the entire caryopsis (Harrington and Crocker, 1923; Artschwager et al., 1929). Aleurone layer consists of protein bodies. It is important for both the developing seed and the mature plant. The aleurone tissue accumulates large quantities of oils and lipids, which are useful during seed development. In some species, it also accumulates minerals, which function in seed dormancy. Aleurone layer performs a variety of functions to help maintain proper development of the seed. Aleurone layer releases organic and phosphoric acid in order to keep the pH of the endosperm between 3.5 and 4 (Nonogaki and Bradford, 2007). During seed germination, the plant embryo produces the hormone gibberellin that triggers the aleurone for the hydrolysis of starch, proteases, and storage proteins into endosperm (Taiz and Zeiger, 2002). Species like *Holcus, Saccharum* etc. have an opening called the hilar orifice that is made up of densely packed thick walled, collapsed dark pigmented cells. Four types of aleurone cells were present in studied species: (i) tangentially elongated and rectangular shaped, (ii) vertically elongated and rectangular shaped, (iii) square shaped cells, and (iv) both rectangular and square shaped aleurone cells. Majority of the studied species show tangentially elongated rectangular shaped aleurone cells. Group Pooideae did not showed vertically elongated rectangular shaped aleurone cells, but they showed other three types of aleurone cells. From the tribe Sporoboleae, *Sporobolus coromardelianus* had square-shaped aleurone cells, while other species showed tangentially elongated rectangular shaped aleurone cells. Likewise from tribe Eragrostideae, only one species *Eleusine indica* showed square-shaped scutellum. Two species of tribe Eragrostideae, *Acrachne ramosa* and *Desmostachya bipinnata*, showed both the type of aleurone cells. Other than these species, all species of the tribe Eragrostideae showed tangentially elongated rectangular shaped aleurone cells. All the species of tribe Chlorideae, Perotideae, Aristideae members showed tangentially elongated rectangular shaped aleurone cells. *Ischane globosa* (Tribe Ischaneae) had both the type of aleurone cells. Tribe Maydeae species, like *Chionachne koenigii* and *Coix lachryma jobi*, showed vertically elongated rectangular shaped aleurone cells. Tribe Andropogoneae members showed tangentially elongated rectangular shaped aleurone cells/vertically elongated rectangular shaped/both types of aleurone cells. *Dicanthium annulatum, Imperata cylindrica*, and *Rottbelia exaltata* showed vertically elongated rectangular shaped aleurone cells, while *Ischaemum indicus, Ischaemum rugosum,* and *Ophiuros exaltatus* showed both types of aleurone cells. All other species of tribe Andropogoneae had tangentially elongated rectangular aleurone cells. Tribe Paniceae showed tangentially elongated aleurone cells, square-shaped aleurone cells, and both types of aleurone cells. *Panicum trypheron* had square-shaped aleurone cells; *Alloteropis cimicina, Echinochloa stagnina,* and *Pennisetum setosum* showed both types of aleurone cells. Other species of tribe Paniceae showed tangentially elongated rectangular aleurone cells. Aleurone cells were typically cuboid in shape with much thicker cell walls, because they contain a large amount of protein bodies and oil; they also releases organic acids (Burton and Fincher, 2014).

Majority of monocot seeds were endospermic, for example, wheat, maize, etc. and it occupies greater portion of the seed (Kar et al., 2002). Endosperm constitutes a main portion for storing the food material even with the supply from the mother plant. The endosperm was considered to be starchy, according to Olsen et al. (1999) and Olsen (2001), functioned for reserving a large amount of starch granules that are consumed by the embryo upon germination and development. These components are characterized by their appearance where they appear with light color, unstained or white color as quoted by Zeleznak and Varriano-Marsteon (1982); Irving (1983) and Jane (2004). Choi et al. (2004) observed the starch granules of *Setaria italica*, Panicoideae. According to them, Panicoideae tribe has large, simple granules with a wide range of shapes from spherical/ovoid to regular polyhedrons. In the present study, most of the Panicoideae members showed simple circular or polygon starch grains. Endosperm showed polygon-shaped cells, which contain the different shaped starch grains. *Sehima sulcatum, Pasaplum scorbiculatum* and *Cymbopogon martini* showed only simple circular shaped starch grains, while species like *Dicanthium caricosum, Digitaria stricta, Oplismenous burmanii, Paspalidium geminatum, Perotis indica*, and *Chloris barbata* showed simple pent to hexagonal shaped starch grains. Few species showed mixture of simple and compound starch grains (e.g. *Chionachne koenigii, Bothriochloa pertusa, Dicanthium annulatum, Heteropogon triticeus, Ischaemum pilosum, Sehima ischaemoides, Sehima nervosum, Thelepogon elegans, Melanocenchris jaquemontii, Schoenefeldia gracilis*, and *Dactloctenium aegyptium*). All other species show combination of a simple circular and polygon-shaped starch grains. Shapter et al. (2008) studied on endosperm and starch granule morphology in wild cereal relatives. They worked on 19 Australian native wild grass species. They reported that *Setaria italica* showed compound starch grains but in the present study *Setaria gluca, Setaria tomentosa,* and *Setaria verticillata* showed simple starch grains with circular and polygon shapes. There were a specific numbers of starch granules per endospermic cell and the number showed great variation. From the group Panicoideae, *Coix* showed around 87 starch granules per endospermic cell and *Hackelochloa granularis* showed around 11 starch granules per endospermic cell. Among the group Pooideae, *Chloris virgata* showed 68 starch granules while *Eragrostis pilosa* and *Tetrapogon tenellus* showed 13 starch granules per endospermic cell. When numbers of starch granules are more per endospermic cell, then the size of starch granules are less and vice versa. *Coix* that has more number of starch granules then it showed around 2.08 μm sizes of starch granule and *Vetiveria zinzanoides* showed 10 μm size of starch granule from the group Panicoideae. Among the group Pooideae, *Eragrostiella bachyphylla* had around 8.34 μm size of starch granule, while *Eragrostis japonica* had 1.67 μm size of starch granule.

Rheeder (1957) classified caryopsis into six different groups and nine different formulae on the basis of the type of embryo. Decker (1964); Barker et al. (1999); Gopal and Ram (1985) studied systematic significance of mature embryo in the systematic of bamboos. All species of bamboos showed same basic pattern of embryo organization. The scutellum and coleoptilar bundles were not separated by an internode, lower portion of the scutellum, and the coleorhiza were separated by a cleft and the margins of embryonic leaves overlapped.

In the studied 100 species, five different caryopsis formula (like P-PP, P+PF, P+PP, F+PP, P-PF) and five different caryopsis types (like True festucoids, True panicoids, Chloridoid–Eragrostoid, Bambusoid, Oryzoid–Olyroid, Arundinoid–Danthonioid) could be identified. In general, Group Panicoideae, members showed only "P-PP" caryopsis formula a characteristic features of true panicoids, while group Pooideae members showed Bambusoid, Arundinoid–Danthonioid, or Chloridoid–Eragrostoid caryopsis type. Exceptions were observed among a few. In the group Panicoideae, only *Chionachne koenigii* showed "F+PP" caryopsis formula and Oryzoid–Olyroid caryopsis type. In Tribe Andropogoneae and tribe Paniceae members showed "P-PP" caryopsis formula, that is, True Panicoids. In Group Pooideae only *Ischanae globosa*, the single genera of tribe Ischaneae showed "P-PP" caryopsis formula and the true Panicoids type of caryopsis. Few species of group Pooideae like *Aristida adscensionis* and *Aristida funiculata* of Tribe Aristideae, *Melanocenchris jaquemontii, Oropetium villosum* of the Tribe Chlorideae showed "P-PF" caryopsis type and characteristic Arundinoid–Danthonioid caryopsis. *Chloris montana, Chloris virgata, Sachoenefeldia gracilis* of Tribe Chlorideae had "P+PP" caryopsis formula and type of caryopsis Bambusoid. Other members of group Pooideae had characteristic features of Chloridoid–Eragrostoid type and "P+PF" caryopsis formula.

Key criteria used to describe the characteristic features in the works on grass systematics by Stebbins (1956); Reeder (1957); Brown (1958); Stebbins and Crampton (1961); Arber (2010) included leaf anatomy, epidermal cells, lodicules and embryo, identification of non-kranz and kranz syndromes based predictably on the combination of anatomical and photosynthetic and cytological data such as chromosome size and number. In the classification put forward by the grass phylogeny work group (GPWG, 2001), characters related to fruit and embryo (Reeder, 1957), leaf anatomy (Ellis, 1987), biochemistry of photosynthesis (Brown, 1977; Hattersley et al., 1986), and structure of reproductive units, both flowers and spikelets were some of the major characteristic features considered.

'Genera Graminum' (Clayton and Renvoize, 1986) and Clayton et al. (2002) were responsible for the development of the project 'Grass Base'. The online World Grass Flora (http://www.key.org/data/grasses_db/.html) uses 1081 descriptions, most of them morphological, but some also related to geographic distribution, for describing the members of Poaceae species using the Delta system (Dallwits et al. 1993).

Based on the characteristic features a dichotomous key for the studied species has been prepared.

1. U- and Δ-shaped scutellum **2**
1. V-shaped scutellum **52**
2. Caryopsis formula: F+PP .... ***Chionachne koenigii***
2. Caryopsis formula: P-PP/P+PF/P+PP/P-PF **3**
3. Caryopsis type: Arundinoid–Danthonioid **4**
3. Caryopsis type: True panicoids/Chloridoid–Eragrostoid/Bambusoid.. **7**
4. Δ-shaped scutellum **5**
4. U-shaped scutellum **6**
5. Circular shaped in transection and embryo %
    occupied: 24.81 ± 5.18 .... ***Aristida adscensionis***
5. Oblong shape in transection and embryo %
    occupied: 41.53 ± 8.75 .... ***Aristida funiculata***
6. Angle of vascularization: Type C .... ***Melanocenchris jaequemontii***
6. Angle of vascularization: Type D .... ***Oropetium villosulum***
7. Δ-shaped scutellum **8**

39. Type v epiblast.. *Acrachne racemosa*
40. Embryo Type C **41**
40. Embryo Type B **42**
41. Type C Pericarp.. *Dinebra retroflexa*
41. Type B Pericarp.. *Eragrostis tenella*
42. Type II epiblast.. **43**
42. Type V epiblast.. **44**
43. Angle of vascularization: Type B .... *Eragrostis japonica*
43. Angle of vascularization: Type C .... *Eragrostis nutans*
44. Angle of vascularization: Type C.. **45**
44. Angle of vascularization: Type B **48**
45. Type D Pericarp.. **46**
45. Type B Pericarp... **47**
46. Number of starch grains per endospermic
    cell 21 ± 8...*Eragrostiella bifaria*
46. Number of starch grains per endespermic
    cell 45 ± 6...*Eragrostiella bachyphylla*
47. Absence of slime glands on hilum surface..........*Eragrostis cilianensis*
47. Presence of slime glands on hilum surface .... *Eragrostis tremula*
48. Type C Pericarp **49**
48. Type B Pericarp **50**
49. Oval shape in transection with size of
    0.58 × 0.43 ± 0.28 mm² ....*Sporobolus diander*
49. Round shape in transection with size of
    0.31 × 0.25 ± 0.16 mm² ....*Sporobolus indicus*
50. Embryo occupied more than 50% (66.42 ± 4.27) .... *Eragrostis viscosa*
50. Embryo occupied less than 50% **51**
51. Circular to oval shaped in transection with size of
    0.18 × 0.18 ± 0.05 mm² ....*Eragrostis ciliaris*
51. Oblong shaped in transection with size of
    0.48 × 0.40 ± 0.27 mm² .... *Eragrostis unioloides*
52. Epiblast is present... **53**
52. Epiblast is absent... **54**
53. Scutellum type V7 .... *Dactyloctenium aegyptium*
53. Scutellum type V4-2.................................... *Tetrapogon villosus*

85. V4-2-shaped scutellum...*Chrysopogon fulvus*

86. V3-6-shaped scutellum.............................. *Heteropogon contortus var. genuinus sub var. typicus*

86. V3-1-shaped scutellum.. **87**

87. Oval shaped in transection with size $0.67 \times 0.76 \pm 0.14$ µm.......................................*Ischaemum molle*

87. Kidney shape in transection with size $1.87 \times 1.69 \pm 0.27$ µm.......................................*Sehima sulcatum*

88. Type D Pericarp.. *Cymbopogon martini*

88. Type C/A Pericarp **89**

89. Type A Pericarp **90**

89. Type C Pericarp **94**

90. V5-shaped scutellum.. *Eriochloa procera*

90. V1/V3-shaped scutellum.. **91**

91. V1-shaped scutellum.. **92**

91. V3-shaped scutellum.. **93**

92. V1-1-shaped scutellum.. *Sorghum halepense*

92. V1-2-shaped scutellum.............................*Paspalidium geminatum*

93. Simple polygonal shaped starch grains, thickness of endosperm: 546.31 ± 7.82 µm.................................. *Themeda triandra*

93. Simple circular shaped starch grains, thickness of endosperm: 877.3 ± 7.63 µm............................ *Themeda qurqdrivalvis*

94. V5-shaped scutellum ....*Eremopogon foveolatus*

94. V3/V4-shaped scutellum **95**

95. V4-2-shaped scutellum..*Digitaria granularis*

95. V3-shaped scutellum.. **96**

96. V3-3-shaped scutellum..*Brachiaria reptans*

96. V3-4/5/6-shaped scutellum **97**

97. V3-4-shaped scutellum..*Andropogon pumilus*

97. V3-5/6-shaped scutellum.. **98**

98. V3-6-shaped scutellum.................................*Brachiaria eruciformis*

98. V3-5-shaped scutellum.. **99**

99. Endosperm thickness: 335.8 ± 11.21 µm ....*Brachiaria distachya*

99. Endosperm is very thick, almost three times more (897.71 ± 23.63 µm) ....*Brachiaria ramosa*

3.4 CLUSTER ANALYSIS

The software used displays a single tree among the possible ones (Fig. 3.57). In the dendrogram based on the anatomical characters of the caryopses, the following features are taken into consideration for the preparation of dendrogram:

I.   Caryopsis type (5 criteria)
      True panicoids (1), Chloridoid–Eragrostoid (2), Bambusoid (3), Oryzoid–Olyroid (4), Arundinoid–Danthonioid (5)

II.   Caryopsis formula (5 criteria)
      P-PP (1), P+PF (2), P+PP (3), F+PP (4), P-PF (5)

III. Angle of vascularization (4 criteria)
      Type-A (1), Type-B (2), Type-C (3), Type-D (4)

IV.   Aleurone layer (4 criteria)
      Vertically elongated (1), Tangentially elongated (2), Square shaped (3), Square shaped and horizontal shaped together (4)

V.   Embryo type (3 criteria)
      Type-A (1), Type-B (2), Type-C (3)

VI.   Pericarp type (4 criteria)
      Type-A (1), Type-B (2), Type-C (3), Type-D (4)

VII. Epiblast type (6 criteria)
      Type-I (1), Type-II (2), Type-III (3), Type-IV (4), Type-V (5), Absent (6)

VIII. Scutellum type (29 criteria)
      Type V1-1 (1), Type V1-2 (2), Type V2 (3), Type V3-1 (4), Type V3-2 (5), Type V3-3 (6), Type V3-4 (7), Type V3-5 (8), Type V3-6 (9), Type V4-1 (10), Type V4-2 (11), Type V5 (12), Type V6 (13), Type V7 (14), Type V8-1 (15), Type V8-2 (16), Type V8-3 (17), Type V9 (18), Type U1-1 (19), Type U1-2 (20), Type U2-1 (21), Type U2-2 (22), Type U2-3 (23), Type Δ1 (24), Type Δ2 (25), Type Δ3 (26), Type Δ4 (27), Type Δ5 (28), Type Δ6 (29)

Two major clusters with the similarity at 70.69% could be obtained. Under one major cluster seven sub-clusters are present, that is, cluster 1, 2, 5, 7, 8, 9, and 10, while under second major cluster, three sub clusters are present, that is, cluster 3, 4, and 6. The observations per clusters are presented in Table 3.2 and dendrogram tree was shown in Figure 3.57.

TABEL 3.2 Distribution of Caryopses into Clusters on the Basis of Their Qualitative Features.

Cluster number	Number of observation (grass species) per cluster	Maximum value from centroid	Avarage distance from centroid
1	1	0.00	0.00
2	7	1.92	1.51
3	20	3.47	2.44
4	19	3.37	2.04
5	28	3.41	2.12
6	10	2.70	1.85
7	4	2.07	1.50
8	2	0.00	0.00
9	7	2.83	1.89
10	2	0.71	0.71

Cluster 1 is a simplicifolius cluster, that is, it has only a single species in cluster. It has *Chionachne koenigiiand* showing similarity acircular 94%. It has maximum value from centroid, which is 0.00.

Clusters 8 and 10 are having two species. Cluster 8 is having 0.00 values from centroid, while cluster 10 has 0.71 distances from the centroid. Cluster 8 has *Aristida adscensionis* and *Aristida funiculate,* and they show almost 100% similarity with other clusters. Cluster 10 has *Melanocenchris jaeque-montii* and *Oropetium villosulum* and shows 99.5% similarity with other clusters.

Clusters 2 and 9 are having seven species. Cluster 2 has maximum distance from centroid is 1.92 and having similarity with other clusters acircular 99%. Cluster 9 has maximum distance from centroid is 2.83 and having similarity 98.5%.

Cluster 5 is a major cluster, highest number of species than the other clusters. It has 28 different species. The basic common feature among all 28 species is that they have U-shaped scutellum. This cluster has maximum value from centroid, 3.41, and shows similarity acircular 98% with other clusters.

Cluster 7 contains four species. Cluster 7 has maximum value from centroid 2.07 and has about 98.5% similarities with other clusters. This cluster contains species like *Imperata cylindrica, Vetivaria zinzanoides, Dactyloctenium scindicum, Ischaemum pilosum* etc. These species show a

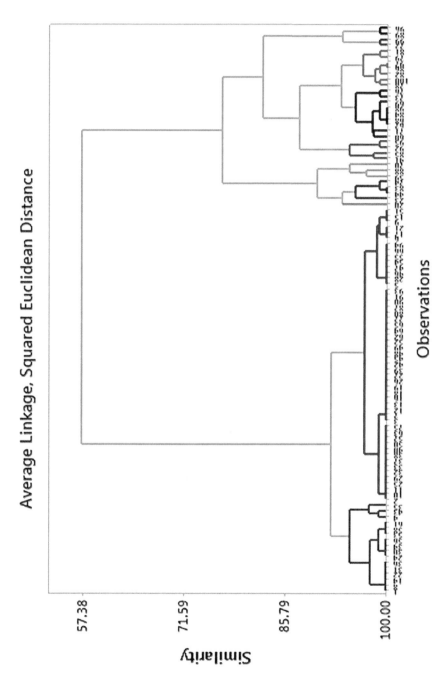

FIGURE 3.57 A dendrogram representing clustering of different grass species on basis of caryopses anatomical features.

few common features like absence of epiblast, belongs to the true panicoids caryopsis type, and also have embryo Type C.

Cluster 3 is a second major cluster among the 10 clusters. It belongs to the other second major cluster. It contains 20 species and has maximum value from centroid that is 3.47. It shows acircular 98% similarity with other clusters. Cluster 3 contains species like *Sorghum halepense, Ophiuros exaltatus, Thelepogn elegans, Cenchrus biflorus, Panicum trypheron, Apluda mutica* etc., and shows common features like V-shaped scutellum.

Cluster 4 contains 19 different species. Species like *Setaria glauca, Alloteropsis cimicina, Echinochloa stagnina, Iseilema laxum, Cenchrus setigerus, Tetrapogon villosus* etc. are from cluster 4 and shown common features like tangentially elongated rectangular shaped aleurone cells, true panicoids group. This cluster has maximum value from centroid, 3.37, and shows acircular 98.7% similarity with other clusters.

Cluster 6 has 10 species. It has maximum value from centroid, 2.70, and has acircular 99% similarity with other clusters. Cluster 10 has species like *Brachiaria eruciformis, Brachiaria distachya, Brachiaria ramosa, Eremopogon foveolatus, Eriochloa procera, Chrysopogon fulvus* etc. They shared one common feature, which is angle of vascularization, and is Type C.

Cluster analysis of anatomical features of caryopsis was done to see the relationship between different groups of species. For cluster analysis, Square Euclidean distance showed good resolution of the genera based on qualitative characters. Dendrogram clearly showed two major groups/clusters. This dendrogram showed similarity with the dichotomous key that was prepared by using anatomical characters. One major cluster contained 51 species and other major cluster contained 49 species. These two major clusters showed 70.69% similarity between them. Clusters 1, 2, 5, 7, 8, 9, and 10 forming one major cluster group and they showed around 92% similarity between them, and the species shared common character like "V"-shaped scutellum, likewise clusters 3, 4, and 6 forming second major cluster and they showed around 87.3% similarity between them and these clusters species shared character like they had "U"- or "Δ"-shaped scutellum.

From the first major cluster, cluster 1 that was simplicifolius cluster and showed around 94% similarity with other clusters of the group. Clusters 8 and 10 were close together and separated from other clusters like 2, 5, 7, and 9. Cluster 5 was the major cluster among the group, containing 28 species and showed around 98% similarity with other clusters. The species of cluster 5 showed common character like "U"-shaped scutellum. Among the species of cluster 5 other few species belonged to the Panicoid type (caryopsis formula: P-PP) and others belonged to the Chloridoid–Eragrostoid type (caryopsis formula: P+PF).

In the second major cluster, cluster 3 was separated with other two clusters at around 87.3% similarity. Cluster 3 species shared common characters like absence of epiblast and had a Panicoid type of caryopsis. Clusters 4 and 6 together showed 97% similarity with other clusters, they showed common characters like "V"-shaped scutellum, tangentially elongated rectangular shaped aleurone cells. Cluster 4 had acircular 98.7% similarity with others and cluster 6 had acircular 99% similarity with other clusters.

Cladogram describes similarities between species. On the basis of this, we can conclude that all the studied species shared around 71% similar characteristic features. Species of clusters 1, 2, 5, 7, 8, 9, and 10 forming one group, showing few similar features while clusters 3, 4, and 6 that were forming other group showing some different characteristic features than the first group. *Chionachne koengii* was showed unique characteristic features and distinctly separating it and this single species form one cluster. *Heteropogon contortus, Chrysopogon fulvus, Thelepogon elegans,* and *Eragrostis* species were present in cluster 5, which were closer to each other, that is, these species shared many similar characteristic features. *Eragrostis* species, *Tragus biflorus, Hackelochloa granularis, Capellipedium hugeliim Sehima nervosum, Triplopogon ramosissimus,* and *Paspalidium flavidum* also shared few similar features, that is, they all were much closer species. *Imperata cylindrica, Vetiveria zinzanoides, Ischaemum pilosum,* and *Saccharum spontanum* belonging to group Panicoideae, shared 90% similar characteristic features forms one cluster.

KEYWORDS

- **caryopsis**
- **anatomy**
- **scutellum**
- **embryo**
- **endosperm**
- **epiblast**
- **vascularization**

REFERENCES

Arber A. *The Gramineae: A Study of Cereal, Bamboo and Grass*; Cambridge University Press: Cambridge, UK, 2010; pp 506.

Artschwager, E.; Brandes, E. W.; Starrett R. C. Development of Flower and Seed of Some Varieties of Sugar Cane. *J. Agr. Res.* **1929,** *39* (1), 130.

Barker, N. P.; Linder, H. P.; Harley, E. H. Sequences of the Grass-Specific Insert in the Chloroplast rpoC2 gene Elucidate Generic Relationships of the Arundinoideae (Poaceae). *Syst. Bot.* **1999,** *23,* 327–350.

Bechtel, D. B.; Pomeranz, Y. Ultrastructure of the Mature Ungerminated Rice (*Oryza sativa*) Caryopsis: The Caryopsis Coat and the Aleurone Cells. *Am. J. Bot.* **1977,** *64* (8), 966–973.

Brown, W. V. The Kranz Syndrome and Its Subtypes in Grass Systematics. *Mem. Torrey Bot. Club.* **1977,** *23* (3), 1–97.

Brown, W. V. Leaf Anatomy in Grass Systematics. *Bot. Gaz.* **1958,** *119* (3), 170–178.

Burton, R. A.; Fincher, G. B. Evolution and Development of Cell Walls in Cereal Grains. *Front. Plant Sc.* **2014,** *5,* 456.

Carlquist, S.; Schneider, E. L. Origins and Nature of Vessels in Monocotyledons. I. *Acorus. Int. J. Plant Sci.* **1997,** *158* (1), 51–56.

Choi, H.; Kim, W.; Shin, M. Properties of Korean Amaranth Starch Compared to Waxy Millet and Waxy Sorghum Starches. *Starch-Stärke* **2004,** *56* (10), 469–477.

Clayton, W. D.; Renvoize, S. A. *Genera Graminum.* Kew Bulletin Additional Series XIII, London. **1986,** pp 389.

Clayton, W. D.; Vorontsova, M. S.; Harman, K. T.; Williamson, H. World Grass Species. *Rodriguésia* **2002,** *63* (1), 089–100.

Dallwitz, M. J.; Paine, T. A.; Zurcher, E. J.. *User's Guide to the DELTA System: A General System for Processing Taxonomic Descriptions*, 4th ed., 1993.

Decker, H. F. Affinities of the Grass Genus *Ampelodesmos. Brittonia* **1964,** *16* (1), 76–79.

Eichemberg, M. T.; Scatena, V. L. Morphology and Anatomy of the Diaspores and Seedling of *Paspalum* (Poaceae, Poales). *Anais da Academia Brasileira de Ciências* **2013,** *85* (4), 1389–1396.

Ellis, R. P. A Review of Comparative Leaf Blade Anatomy in the Systematics of the Poaceae: the Past Twenty-Five Years. In *Grass Systematics and Evolution*: an International Symposium held at the Smithsonian Institution, Washington, DC, Jul 27–31, 1986; Soderstrom, T. R. and others, Eds.; Smithsonian Institution Press: Washington, DC, London, 1987.

Esau, K. *Plant Anatomy*; John Wiley and Sons: New York, 1953; pp 407.

Gopal, B. H.; Ram, H. M. Systematic Significance of Mature Embryo of Bamboos. *Plant Syst. Evol.* **1985,** *148* (3-4), 239–246.

GPWG. Phylogeny and Subfamiliad Classification of the Grasses (Poaceae). *Ann. Missouri Bot. Garden* **2001,** *88* (3), 373–457.

Guerin, M. P. On the Development of Seminal Integuments and Pericarp of Gramineae. *Bull. Bot. Soc. France* **1898,** *45* (5), 405–411.

Hands, P.; Kourmpetli, S.; Sharples, D.; Harris, R. G.; Drea S. Analysis of Grain Characters in Temperate Grasses Reveals Distinctive Patterns of Endosperm Organization Associated with Grain Shape. *J Exp. Bot.* **2012,** *63* (17), 6253–6266.

Harrington, G. T.; Crocker, W.; Plant, P. Structure, Physical Characteristics, and Com-Position of the Pericarp and Integument of Johnson Grass Seed in Relation to its Physiology. *J. Agric. Res.* **1923,** *23* (3), 193–222.

Hattersley, P. W.; Wong, S. C.; Perry, S.; Roksandic, Z. Comparative Ultrastructure and Gas Exchange Characteristics of the C_3–C_4 Intermediate Neurachne Minor ST Blake (Poaceae). *Plant Cell Environ.* **1986**, *9* (3), 217–233.

Hinton, J. J. C. The Distribution of Vitamin B 1 in the Rice Grain. *Brit. J. Nutr.* **1948**, *2* (3), 237–241.

Hinton, J. J.; Shaw, B. The Distribution of Nicotinic Acid in the Rice Grain. *Brit. J. Nutr.* **1954**, *8*, 65–71.

Irving, D. W. Anatomy and Histochemistry of *Echinochloa turnerana* (Channel Millet) Spikelet. *Cereal Chem.* **1983**, *60* (2), 155–160.

Izaguirre de Artucio, P.; Laguardia, A. Un Nuevo Enfoque Hacia La Definicion Del Fruto De Las Gramineas. *Uruguay Fac. Agron. Bot. Invest.* **1987**, 3, 1–16.

Jane, W. N. Ultrastructure of Endosperm Development in *Arundo formosana* Hack. (Poaceae) from Differentiation to Maturity. *Bot. Bull. Acad. Sinica.* **2004**, *45*, 69–85.

Kar, R. K.; Misra, N. M.; Kabi, T. *Textbook on Fundamentals of Botany*, 4th ed.; Kalyani Publishers: India, 2002.

Kiesselbach, T. A. The Structure and Reproduction of Corn. University of Nebraska, College of Agriculture, Agricultural Experiment Station. *Research Bulletin*, 161. University of Nebraska College of Agriculture, Agricultural Experiment Station, Lincoln, 1949, pp 96.

Mann, A.; Harlan, H. V. Morphology of the Barley Grain with Reference to its Enzyme-Secreting Areas. *J. Inst. Brew.* **1916**, *22* (2), 73–108.

Morrison, I. N.; Dushnicky, L. Structure of the Covering Layers of the Wild Oat *Avena fatua*caryopsis. *Weed Sci.* **1982**, *30*, 352–359.

Narayanaswami, S. The Structure and Development of the Caryopsis in Some Indian Millets. 1. Pennisetumtyphoideum Rich. *Phytomorphology* **1953**, *3*, 98–112.

Narayanaswami, S. The Structure and Development of the Caryopsis in Some Indian Millets. III. *Paspalum scrobiculatum* L. *Bull. Torrey Bot. Club.* **1954**, *81*, 288299.

Narayanswami, S. The Structure and Development of the Caryopsis in Some Millets 4. *Echinochloa frumentaacea* Link. *Phytomorphology* **1955a**, *5*, 161–170.

Narayanswami, S. The Structure and Development of the Caryopsis in Some Indian Millets 5. *Eleusine coracana* Gaertn. *Pap. Mich. Acad. Sci.* **1955a**, *11*, 33–46.

Narayanaswami, S. The Structure and Development of the Caryopsis in Some Indian Millets 3. *Panicum millare* Lank. and *P. miliacemum* Linn. *Lloydia* **1955c**, *18*, 61–73.

Narayanaswami, S. Structure and Development of the Caryopsis in Some Indian Millets. VI. *Setaria italica. Bot. Gaz.* **1956**, *118*, 112–122.

Negbi, M. The Structure and Function of the Scutellum of the Gramineae. *Bot. J. Linnean Soc.* **1984**, *88* (3), 205–222.

Nonogaki, H.; Bradford, K. J. Mechanisms and Genes Involved in Germination Sensu Stricto. In *Seed Development, Dormancy and Germination*; Bradford, K.J., Nonogaki, H., Eds.; Blackwell Publishing Ltd.: Oxford, 2007; pp 264–304.

Olsen, O. A. Endosperm Development: Cellularization and Cell Fate Specification. *Ann. Rev. Plant Biol.* **2001**, *52* (1), 233–267.

Olsen, O. A.; Linnestad, C.; Nichols, S. E. Developmental Biology of the Cereal Endosperm. *Trends Plant Sci.* **1999**, *4* (7), 253–257.

Osman, A. K.; Zaki, M. A.; Hamed, S. T.; Hussein, N. R. Comparative Anatomical Study on Fruits of Some Tribes of Family Gramineae from Egypt. *Pak. J. Bot.* **2012**, *44* (2), 599–618.

Reeder, J. R. Affinities of the Grass Genus Beckmannia Host. *Bull. Torrey Bot. Club.* **1953**, *80*, 187–196.

Reeder, J. R. The Embryo in Grass Systematics. *Am. J. Bot.* **1957,** *44* (9), 756–768.

Renvoize, S. Grass Anatomy. *Flora Aus.* **2002,** *43*, 71–132.

Rost, T. L. The Anatomy of the Caryopsis Coat in Mature Caryopses of the Yellow Foxtail Grass (*Setaria lutescens*). *Bot. Gaz.* **1973,** *134* (1), 32–39.

Rost, T. L. Fine Structure of Endosperm Protein Bodies in Setarialutescens (Gramineae). *Protoplasma* **1971,** *73* (3-4), 475–479.

Rost, T. L.; De Artucio, P. I.; Risley, E. B. Transfer Cells in the Placental Pad and Caryopsis Coat of Pappophorum Subbulbosum Arech.(Poaceae). *Am. J. Bot.* **1984,** *71* (7), 948–957.

Rost, T. L.; de Artucio, P. I.; Risley, E. B. Anatomy of the Caryopsis of *Briza maxima* (Poaceae). *Am. J. Bot.* **1990,** *77* (1), 69–76.

Rudall, P. J.; Stuppy, W.; Cunniff, J.; Kellogg, E. A.; Briggs, B. G. Evolution of Reproductive Structures in Grasses (Poaceae) Inferred by Sister-Group Comparison with their Putative Closest Living Relatives, Ecdeiocoleaceae. *Am. J. Bot.* **2005,** *92* (9), 1432–1443.

Sanders, E. H. Developmental Morphology of the Kernal of Grain Sorghum. *Cereal Chem.* **1955,** *32*, 12–25.

Sargant, E.; Robertson A.; Hill Q. The Anatomy of the Scutellum in *Zeamais*. *Ann. Bot.* **1905,** *19* (73), 115–123.

Shah, C. K.; Sreekumari, S. B. Developmental Morphology and Homologies in Grass Embryo. *Curr. Sci.* **1980,** *49* (7), 284–285.

Shapter, F. M.; Henry, R. J.; Lee, L. S. Endosperm and Starch Granule Morphology in Wild Cereal Relatives. *Plant Genet. Resour.* **2008,** *6* (2), 85–97.

Stebbins, G. L. Taxonomy and the Evolution of Genera, with Special Reference to the Family Gramineae. *Evolution* **1956,** *10* (3), 235–245.

Stebbins, G. L.; Crampton, B. A Suggested Revision of the Grass Genera of Temperate North America. *Recent Adv. Bot.* **1961,** *1*, 133–144.

Swift, J. G.; O'brien, T. P. The Fine Structure of the Wheat Scutellum before Germination. *Aus. J. Biol. Sci.* **1972,** *25* (1), 9–22.

Taiz, L.; Zeiger, E. Photosynthesis: Physiological and Ecological Considerations. In *Plant Physiology*; Sinauer Associates: Massachusetts, 2002; pp 171–192.

Thornton, M. L. Seed Dormancy in Tall Wheat Grass (*Agropyronelongatum*). In *Proceedings of the Association of Official Seed Analysts.* 2004, 56, 116–119.

Tillich, H. J. Seedling Diversity and the Homologies of Seedling Organs in the Order Poales (Monocotyledons). *Ann. Bot.* **2007,** *100* (7), 1413–1429.

Weatherwax, P. The Endosperm of *Zea* and *Coix*. *Am. J. Bot.* **1930,** *17*, 371–380.

Wolf, M. J.; Buzan, C. L.; Mac Masters, M. M.; Rist, C. E. Structure of the Mature Corn Kernel. Gross Anatomy and Structural Relationships. *Cereal Chem.* **1952,** *29* (5), 321–333.

Zeleznak, K.; Varriano-Marston, E. Pearl Millet (*Pennisetum americanum* (L.) Leeke) and Grain Sorghum (*Sorghum bicolor* (L.) Moench) Ultrastructure. *Am. J. Bot.* **1982,** *69* (8), 1306–1313.

Index

Printed and bound by CPI Group (UK) Ltd, Croydon, CR0 4YY

23/10/2024

01777705-0019